Application of LC-MS/MS in the Mycotoxins Studies

Application of LC-MS/MS in the Mycotoxins Studies

Special Issue Editors

Laura Gámiz-Gracia
Ana M. García-Campaña
Natalia Arroyo-Manzanares

MDPI • Basel • Beijing • Wuhan • Barcelona • Belgrade • Manchester • Tokyo • Cluj • Tianjin

Special Issue Editors
Laura Gámiz-Gracia
Department of Analytical Chemistry, Faculty of Sciences, University of Granada
Spain

Ana M. García-Campaña
Department of Analytical Chemistry, Faculty of Sciences, University of Granada
Spain

Natalia Arroyo-Manzanares
Department of Analytical Chemistry, Faculty of Chemistry, University of Murcia
Spain

Editorial Office
MDPI
St. Alban-Anlage 66
4052 Basel, Switzerland

This is a reprint of articles from the Special Issue published online in the open access journal *Toxins* (ISSN 2072-6651) (available at: https://www.mdpi.com/journal/toxins/special_issues/lcms_mycotoxins).

For citation purposes, cite each article independently as indicated on the article page online and as indicated below:

LastName, A.A.; LastName, B.B.; LastName, C.C. Article Title. *Journal Name* **Year**, *Article Number*, Page Range.

ISBN 978-3-03936-206-6 (Pbk)
ISBN 978-3-03936-207-3 (PDF)

Contents

About the Special Issue Editors

Laura Gámiz-Gracia is a full-time professor at the Department of Analytical Chemistry at the University of Granada (Spain). She completed her degree in Chemistry at the University of Granada (Spain, 1994) and received her PhD Degree in Analytical Chemistry from the University of Cordoba (Spain, 1999). Her main research interests are the development and application of new sample treatments and analytical strategies based on separation techniques (as capillary electrophoresis, capillary-liquid chromatography and ultra-high performance liquid chromatography), coupled to different detection systems (tandem mass spectrometry, high resolution mass spectrometry and fluorescence), for the determination of mycotoxins and other contaminants in food and feed.

Ana M. García-Campaña is a full-time professor and Head of the Department of Analytical Chemistry at the University of Granada (Spain). She completed her PhD in 1995 at the University of Granada (Spain). Currently, she is the President of the Spanish Society of Chromatography and Related Techniques (SECyTA). Since 2000, she has been the lead of the research group "FQM302: Quality in Food, Environmental and Clinical Analytical Chemistry," and her research fields are related to the use of liquid chromatography and capillary electrophoresis, coupled with UV-Vis, fluorescence and mass spectrometry detection, using innovative sample treatments, in accordance with green chemistry principles, mainly focused on food safety and quality control, drug monitoring in biological fluids and the determination of contaminants and residues (pesticides, veterinary drugs, mycotoxins, cyanotoxins, etc.) in foods, feed, nutraceuticals and environmental samples.

Natalia Arroyo-Manzanares has been an associate professor at the University of Murcia since 2018. She received her PhD in 2013 from the University of Granada (Spain), focusing her research on food safety and the development of analytical methods for the determination of mycotoxins. In 2014, she started a postdoctoral fellowship at Ghent University (Belgium), broadening her knowledge on the use of high-resolution mass spectrometry to explore the metabolome of fungi, and their capability of producing both known and unknown mycotoxins. She was also a postdoctoral researcher at the University of Córdoba (Spain), focusing on the study of Ion Mobility Spectrometry.

toxins

MDPI

Editorial

Application of LC–MS/MS in the Mycotoxins Studies

Laura Gámiz-Gracia [1,*], Ana M. García-Campaña [1] and Natalia Arroyo-Manzanares [2]

1 Department of Analytical Chemistry, Faculty of Sciences, University of Granada, 18071 Granada, Spain; amgarcia@ugr.es
2 Department of Analytical Chemistry, Faculty of Chemistry, University of Murcia, 30100 Murcia, Spain; natalia.arroyo@um.es
* Correspondence: lgamiz@ugr.es; Tel.: +34-958-248-594

Received: 31 March 2020; Accepted: 20 April 2020; Published: 23 April 2020

Mycotoxins are secondary metabolites produced by fungi of different species (mainly *Aspergillus, Fusarium,* and *Penicillium*) with toxic effects for humans and animals that can contaminate food and feed. Notifications on the Rapid Alert System for Food and Feed (RASFF) concerning mycotoxins are becoming frequent, being among the "top 10" hazards reported on food products, mainly cereals and nuts [1]. The European Union (EU), among others, has established maximum permitted or recommended levels for several mycotoxins, including aflatoxins (AF) B1, B2, G1, G2 and M1, ochratoxin A (OTA), patulin (PAT), deoxynivalenol (DON), zearalenone (ZEN), fumonisins (FB1 and FB2), HT-2 and T-2 toxins, or citrinin (CIT) in different food commodities [2]. However, there are other mycotoxins not included in regulations, the so-called "emerging mycotoxins", whose toxicity is still not clear, that could also pose a risk to humans and animals. This group includes *Alternaria* toxins, sterigmatocystin (STC), phomopsins, and *Fusarium* toxins as enniatins (ENN) or beauvericin (BEA), among others [3]. Moreover, the so-called "modified or masked mycotoxins" (produced as a consequence of a detoxification strategy of the host plant of the fungus or during food processing) could be equally or even more toxic than the original mycotoxin, and should also be taken into account [4].

All these facts make it necessary to develop analytical methods for the accurate determination of mycotoxins in different food matrices and feeds. In this sense, liquid chromatography tandem mass spectrometry (LC–MS/MS) is a powerful tool for the unique identification and quantification of analytes. Moreover, the use of high-resolution mass spectrometry (HRMS) allows the identification of novel mycotoxins and targeted/untargeted approaches for their study.

This Special Issue is dedicated to recent applications of LC–MS/MS in mycotoxin studies, including the development and validation of new analytical methods for their identification and quantification in different food matrices and feed, occurrence studies and biomonitoring of mycotoxins, and their metabolites in biological fluids.

Although most of the papers propose methods for the simultaneous determination of several mycotoxins, Bertuzzi et al. focus their work on the improvement of determination of moniliformin (MON) by adding lanthanide ions to the mobile phase during chromatographic separation [5]. MON is weakly retained in reversed-phase chromatography, making the separation difficult. The addition of La^{3+}, Tb^{3+} or Eu^{3+} to the mobile phase is investigated, improving peak shape and MON separation, probably due to the formation of coordination complexes lanthanide–MON. After extraction and purification, the method is applied to the determination of MON in maize and wheat samples.

The potential of LC–MS/MS for multi-mycotoxin determination in different food matrices is shown in different contributions of this Special Issue. For instance, Guo et al. propose a modified QuEChERS (Quick, Easy, Cheap, Efficient, Rugged, and Safe) procedure for the determination of 6 *Alternaria* toxins in 56 grape samples [6]. A total of 40 samples (incidence of 71.4%) were contaminated with *Alternaria* toxins, tentoxin being the most frequently found mycotoxin (37.5%). Cannabidiol food supplements (a scarcely studied matrix) are analyzed by Narváez et al., who develop an LC–HRMS method for the quantification of 16 mycotoxins produced by major *C. sativa* fungi [7]. They report that up to six different *Fusarium* mycotoxins were found in seven of the ten analyzed samples, the most prevalent being ZEN (60%) and ENN B1 (30%). Co-occurrence was observed in four samples, including one with ENN B1, ENN A, and ENN A1. Moreover, 46 pesticides were detected after a retrospective analysis. Considering that cereals are one of the most susceptible matrices for contamination by mycotoxins, the brewing industry is an interesting field to be considered in these studies. In this context, Piacentini et al. characterize the *Fusarium* isolates from brewing barley, and the phylogenetic study shows that the majority of the strains clustered with *F. graminearum*, *F. poae*, and *F. avenaceum* [8]. Moreover, DON and ZEN were detected in 90.6% and 87.5% of the samples, respectively, and 86% of the samples contaminated with ZEN were above the limits set by the EU and Brazilian regulations. The same research group studied the stability of DON and ZEN during all steps of the malting and brewing processes [9]. The results show that during the malting process, DON levels decreased significantly in the steeping, germination, and malting steps (62%, 51.5%, and 68%, respectively). Considering ZEN, when the levels were compared between barley and the last step of the process, a significant decrease was observed. Another paper by Nafusa et al. reports the variation of fungal metabolites in 81 sorghum malts used to prepare Namibian traditional fermented beverages (omalodu and otombo malts) [10]. It must be highlighted that 102 metabolites were quantified in both malts (96% in omalodu malts and 93% in otombo malts). An average of 48 metabolites (including emerging mycotoxins, AF precursors, and ergot alkaloids) was quantified in otombo malts while an average of 67 metabolites was quantified in omalodu malts. Co-occurrence of EU-regulated mycotoxins, such as PAT, AFs (B1, B2, and G2), and fumonisins (FB1, FB2, and FB3) was detected in both malts with a prevalence range of 2–84%. All these studies accentuate the need to control mycotoxins in cereals intended for brewing, their resistance to food processing treatments, and their fate in the beverages.

Considering the high toxicity of mycotoxins, occurrence studies and exposure assessment in different regions of the world (especially those without a food safety regulation system) are relevant aspects, LC–MS/MS being the technique of choice for multi-mycotoxin analysis. In the framework of the first multi-center Sub-Saharan Africa Total Diet Study (SSA-TDS), 194 composite samples of cereals, tubers, legumes, vegetables, nuts and seeds, dairy, oils, beverages and miscellaneous were tested for mycotoxins [11]. The results show that populations from North Cameroon and Benin may suffer chronic and simultaneous exposure to AFB1, FB1, STC, OTA, and CIT, which are prevalent in their diet. In the same line, Ijeoma-Ekwomadu et al. investigate the variation of different *Fusarium* metabolites (including regulated, modified, and emerging mycotoxins) in 123 maize samples from various regions of South Africa [12]. The relationship between the maize producing regions, the maize type, as well as the mycotoxins, was established. The results reveal that all maize types exhibited a mixture of free, modified and emerging mycotoxins contamination across the regions with an average of 5 mycotoxins, and up to 24 out of 42 investigated *Fusarium* mycotoxins, including 1 to 3 modified forms at the same time, fumonisins (FB1, FB2, FB3, FB4, and FA1) being the most prevalent mycotoxins. Significant differences in the contamination pattern were observed between the agricultural regions and maize types. Finally, the presence of 15 mycotoxins in 120 cereal samples (including barley, maize, rice, and wheat) from Algeria is evaluated by Mahdjoubi et al. [13]. Analytical results show that 78 cereal samples (65%) were contaminated with at least one toxin, while 50% were contaminated with three to nine mycotoxins, T-2 toxin, CIT, BEA, and DON being the most frequently found. Moreover, the exposure assessment shows that the high levels of fumonisins (FB1 and FB2) in maize, and DON, ZEN, HT-2 and T-2 toxins in wheat, could represent a health risk for the average adult consumer in Algeria.

Contamination of mycotoxins in feed samples is also a matter of concern, as this is the first link in the food chain, with negative effects not only on the health of animals but also on humans consuming animal-derived products. Arroyo-Manzanares et al. carry out a survey, including 228 pig feed samples from different Spanish farms and suppliers, exploring the occurrence of 19 mycotoxins [14]. Although 97% of samples were in accordance with EU regulations, the study shows the high occurrence of emerging mycotoxins (100% samples were contaminated with ENN B and more than 90% were contaminated with BEA), as well as the high co-occurrence of different mycotoxins in the same sample (40% samples were contaminated with more than five mycotoxins). Moreover, Facorro et al. develop a methodology for the simultaneous determination of 26 mycotoxins in 97 mixed feed rations collected in 20 dairy cows farms [15]. LC–HRMS based on data-independent acquisition was employed. All unified samples analyzed showed co-occurrence of two or more mycotoxins, recurrently ZEN, FB1, and β-zearalenol, with an occurrence frequency ranging from 60% to 90%. Additional screening based on a retrospective approach enabled the identification of non-targeted mycotoxins, most of them originated from *Fusarium* fungi. The results of these studies show that, in order to ensure feed safety, emerging and modified mycotoxins, as well as the co-occurrence of different mycotoxins, should be considered as a potential risk.

In addition to foodstuff and feed, the determination of mycotoxins in biological fluids is also one of the topics addressed in this Special Issue. Debevere et al. develop and fully-validate a UHPLC–MS/MS method for the quantitative determination in the rumen fluid of some of the most relevant mycotoxins found in maize silage [16]. Extraction recovery and matrix effects were determined in the rumen fluid with and without maize silage. Differences in extraction recovery and matrix effect between rumen fluid alone and rumen fluid with maize silage highlight the importance of using matrix-matched calibration curves for the quantification of mycotoxins in this complex sample. This method can be applied to investigate the degradation of the reported mycotoxins by rumen microbiota. Related to this topic, another aspect that is attracting great interest is biomonitoring of biological fluids as an alternative approach to assessing health risk. However, metabolic pathways have not been investigated thoroughly for all mycotoxins, and, consequently, most of the existing analytical LC–MS methods used for the assessment of human exposure focus only on the detection of parent compounds. In this context, Slobodchikova et al. perform human in vitro microsomal incubations of 17 mycotoxins in order to characterize all resulting metabolites using LC–HRMS [17]. The results allowed them to build a comprehensive LC–MS library. Moreover, the method can also screen 188 additional metabolites, including 100 metabolites reported for the first time. This contribution represents one of the most comprehensive LC–HRMS methods for mycotoxin biomonitoring or metabolism/fate studies. In other work covering this topic, Lauwers et al. propose the combination of LC–MS/MS to determine regulated and emerging mycotoxins in biological matrices of pigs and broiler chickens, with LC–HRMS to determine mycotoxin metabolites for which analytical standards are not always commercially available [18]. After in-house validation, the methods were applied to plasma, urine, feces, and/or excreta samples obtained during in vivo toxicokinetic studies with DON and ZEN in pigs, and with DON, AF B1 and OTA in broiler chickens, and during a pilot field screening study to monitor exposure to mycotoxins.

Most of the papers reported so far use solid–liquid or liquid–liquid extractions, or QuEChERS-based extractions as sample treatments. However, novel extraction methods are also proposed in this Special Issue, as in the paper of Arroyo-Manzanares et al. that explores the potential of dispersive magnetic solid-phase extraction (DMSPE) for the determination of emerging mycotoxins (ENN A, ENN A1, ENN B, ENN B1, and BEA) [19]. After its characterization, a magnetic multiwalled carbon nanotube (Fe_3O_4@MWCNT) composite was selected for the extraction and preconcentration of these mycotoxins in human urine samples. Analysis of the extracts was performed by LC–HRMS. Matrix-matched calibration curves were necessary, obtaining limits of quantification between 0.04 and 0.1 μg/L, relative standard deviation lower than 12%, and recoveries between 89.3% and 98.9%.

Method validation is a very important aspect in the development of an analytical methodology. The proposed methods must assure compliance with the requirements and recommendations of mycotoxins determination in order to ensure high-quality results. Most of the published papers have carried out in-house validation. On the other hand, Bessaire et al. present the results of an intercollaborative study (including 23 entities) organized to evaluate the performance characteristics of an LC–MS/MS procedure for the determination of 12 mycotoxins in food, based on QuEChERS and immunoaffinity column cleanup [20]. Each participant analyzed 28 incurred and/or spiked blind samples composed of spices, nuts, milk powder, dried fruits, cereals, and baby food using the protocol given. The results show that the method is applicable regardless of the food, the regulated mycotoxin, and the concentration level (even at low regulated levels for foods intended for infants and young children), therefore being an excellent candidate for future standardization.

Funding: This research received no external funding.

Acknowledgments: The editors are really grateful to all the authors who contributed their high-quality works to this Special Issue. Special thanks go to the expert peer reviewers, whose rigorous evaluations of all of the submitted manuscripts have, no doubt, improved this Special Issue. Last but not least, the editors really appreciate the good organization and constant support of the MDPI editorial team and staff.

Conflicts of Interest: The authors declare no conflict of interest.

References

1. Rapid Alert System for Food and Feed (RASFF). Available online: https://ec.europa.eu/food/safety/rasff_en (accessed on 28 March 2020).
2. European Commission. Mycotoxins. Available online: https://ec.europa.eu/food/safety/chemical_safety/contaminants/catalogue/mycotoxins_en (accessed on 28 March 2020).
3. Gruber-Dorninger, C.; Novak, B.; Nagl, V.; Berthiller, F. Emerging mycotoxins: Beyond traditionally determined food contaminants. *J. Agric. Food Chem.* **2017**, *65*, 7052–7070. [CrossRef] [PubMed]
4. Freire, L.; Sant'Ana, A.S. Modified mycotoxins: An updated review on their formation, detection, occurrence, and toxic effects. *Food Chem. Toxicol.* **2018**, *111*, 189–205. [CrossRef] [PubMed]
5. Bertuzzi, T.; Rastelli, S.; Mulazzi, A.; Pietri, A. LC-MS/MS and LC-UV Determination of moniliformin by adding lanthanide ions to the mobile phase. *Toxins* **2019**, *11*, 570. [CrossRef] [PubMed]
6. Guo, W.; Fan, K.; Nie, D.; Meng, J.; Huang, Q.; Yang, J.; Shen, Y.; Tangni, E.K.; Zhao, Z.; Wu, Y.; et al. Development of a QuEChERS-based UHPLC-MS/MS method for simultaneous determination of six *Alternaria* toxins in grapes. *Toxins* **2019**, *11*, 87. [CrossRef] [PubMed]
7. Narváez, A.; Rodríguez-Carrasco, Y.; Castaldo, L.; Izzo, L.; Ritieni, A. Ultra-high-performance liquid chromatography coupled with quadrupole Orbitrap high-resolution mass spectrometry for multi-residue analysis of mycotoxins and pesticides in botanical nutraceuticals. *Toxins* **2020**, *12*, 114. [CrossRef] [PubMed]
8. Piacentini, K.C.; Rocha, L.O.; Savi, G.D.; Carnielli-Queiroz, L.; De Carvalho Fontes, L.; Correa, B. Assessment of toxigenic *Fusarium* species and their mycotoxins in brewing barley grains. *Toxins* **2019**, *11*, 31. [CrossRef] [PubMed]
9. Piacentini, K.C.; Běláková, S.; Benešová, K.; Pernica, M.; Savi, G.D.; Rocha, L.O.; Hartman, I.; Čáslavský, J.; Corrêa, B. *Fusarium* mycotoxins stability during the malting and brewing processes. *Toxins* **2019**, *11*, 257. [CrossRef] [PubMed]
10. Nafuka, S.N.; Misihairabgwi, J.M.; Bock, R.; Ishola, A.; Sulyok, M.; Krska, R. Variation of fungal metabolites in sorghum malts used to prepare Namibian traditional fermented beverages *Omalodu* and *Otombo*. *Toxins* **2019**, *11*, 165. [CrossRef] [PubMed]
11. Ingenbleek, L.; Sulyok, M.; Adegboye, A.; Hossou, S.E.; Koné, A.Z.; Oyedele, A.D.; Kisito, C.S.K.J.; Dembélé, Y.K.; Eyangoh, S.; Verger, P.; et al. Regional Sub-Saharan Africa total diet study in Benin, Cameroon, Mali and Nigeria reveals the presence of 164 mycotoxins and other secondary metabolites in foods. *Toxins* **2019**, *11*, 54. [CrossRef] [PubMed]
12. Ekwomadu, T.I.; Dada, T.A.; Nleya, N.; Gopane, R.; Sulyok, M.; Mwanza, M. Variation of Fusarium free, masked, and emerging mycotoxin metabolites in maize from agriculture regions of South Africa. *Toxins* **2020**, *12*, 149. [CrossRef] [PubMed]
13. Mahdjoubi, C.K.; Arroyo-Manzanares, N.; Hamini-Kadar, N.; García-Campaña, A.M.; Mebrouk, K.; Gámiz-Gracia, L. Multi-mycotoxin occurrence and exposure assessment approach in foodstuffs from Algeria. *Toxins* **2020**, *12*, 194. [CrossRef] [PubMed]
14. Arroyo-Manzanares, N.; Rodríguez-Estévez, V.; Arenas-Fernández, P.; García-Campaña, A.M.; Gámiz-Gracia, L. Occurrence of mycotoxins in swine feeding from Spain. *Toxins* **2019**, *11*, 342. [CrossRef] [PubMed]
15. Facorro, R.; Llompart, M.; Dagnac, T. Combined (d)SPE-QuEChERS extraction of mycotoxins in mixed feed rations and analysis by high performance liquid chromatography-high-resolution mass spectrometry. *Toxins* **2020**, *12*, 206. [CrossRef] [PubMed]

16. Debevere, S.; De Baere, S.; Haesaert, G.; Rychlik, M.; Fievez, V.; Croubels, S. Development of an UPLC-MS/MS method for the analysis of mycotoxins in rumen fluid with and without maize silage emphasizes the importance of using matrix-matched calibration. *Toxins* **2019**, *11*, 519. [CrossRef] [PubMed]
17. Slobodchikova, I.; Sivakumar, R.; Rahman, M.S.; Vuckovic, D. Characterization of Phase I and glucuronide Phase II metabolites of 17 mycotoxins using liquid chromatography—High-resolution mass spectrometry. *Toxins* **2019**, *11*, 433. [CrossRef] [PubMed]
18. Lauwers, M.; De Baere, S.; Letor, B.; Rychlik, M.; Croubels, S.; Devreese, M. Multi LC-MS/MS and LC-HRMS methods for determination of 24 mycotoxins including major Phase I and II biomarker metabolites in biological matrices from pigs and broiler chickens. *Toxins* **2019**, *11*, 171. [CrossRef] [PubMed]
19. Arroyo-Manzanares, N.; Peñalver-Soler, R.; Campillo, N.; Viñas, P. dispersive solid-phase extraction using magnetic carbon nanotube composite for the determination of emergent mycotoxins in urine samples. *Toxins* **2020**, *12*, 51. [CrossRef] [PubMed]
20. Bessaire, T.; Mujahid, C.; Mottier, P.; Desmarchelier, A. Multiple mycotoxins determination in food by LC-MS/MS: An international collaborative study. *Toxins* **2019**, *11*, 658. [CrossRef] [PubMed]

toxins

MDPI

Article

LC-MS/MS and LC-UV Determination of Moniliformin by Adding Lanthanide Ions to the Mobile Phase

Terenzio Bertuzzi *, Silvia Rastelli, Annalisa Mulazzi and Amedeo Pietri

Department of Animal Science, Food and Nutrition-DIANA, Faculty of Agriculture, Food and Environmental Sciences, Università Cattolica del Sacro Cuore, Via Emilia Parmense, 84-29122 Piacenza, Italy; silvia.rastelli@unicatt.it (S.R.); annalisa.mulazzi@unicatt.it (A.M.); amedeo.pietri@unicatt.it (A.P.)

* Correspondence: terenzio.bertuzzi@unicatt.it; Tel.: +39-0523-599262

Received: 6 September 2019; Accepted: 27 September 2019; Published: 29 September 2019

check for updates

Abstract: An innovative chromatographic analysis was developed for the determination of moniliformin (MON). Because of its ionic nature, MON is weakly retained in reversed-phase chromatography and the separation may be tricky. Nevertheless, this technique is normally used either with the formation of ion pairs or employing specific RP columns for polar compounds, or combining anion exchange and hydrophobic interactions. Hydrophilic interaction chromatography (HILIC) was also used, but a non-negligible peak tailing was observed. Besides its ionic nature, MON is a di-ketone and di-ketones, mainly β-di-ketones, can easily form complexes with lanthanide ions. Then, in this work the addition of lanthanide ions to the mobile phase was investigated, aiming at improving peak shape and MON separation. La^{3+}, Tb^{3+} or Eu^{3+} aqueous solutions were used as mobile phase and MON was chromatographed using a LC-NH_2 column. The probable formation of coordination complexes lanthanide-MON in the HPLC mobile phase allowed to obtain a symmetrical peak shape and a satisfactory chromatographic separation by both mass spectrometry (MS/MS) and UV detection. Finally, a suitable extraction and purification method for MON determination in cereal samples was developed.

Keywords: moniliformin; lanthanide complexes; LC-MS/MS; LC-UV

Key Contribution: Simple and innovative LC-MS/MS and LC-UV methods for the determination of MON were developed.

---

1. Introduction

Moniliformin (MON) is a *Fusarium* mycotoxin often occurring in cereals; it is mainly produced by *F. avenaceum, proliferatum, subglutinans, tricinctum* and *verticilloides* [1]. MON is a small, highly polar, acidic molecule and because of the low pKa value (<1.7) of the free acid, MON occurs as a water-soluble sodium or potassium salt [2]. MON is toxic to experimental animals, causing myocardial degeneration, muscular weakness and respiratory distress [3,4]; the European Food Safety Authority (EFSA) Panel on Contaminants in the Food Chain (CONTAM) indicated haematotoxicity and cardiotoxicity as major adverse health effects of MON [5]. At present, no specific maximum levels for MON in food and feed have been set by EU legislation. MON was detected worldwide in several cereal crops at different concentration levels with values up to 2606 μg kg^{-1} in maize and 326 μg kg^{-1} in wheat produced, respectively, in Italy and in the Netherlands being reported [1,6–11]. However, EFSA recommended the collection of more occurrence data on MON in foods and feeds to enable a comprehensive risk assessment for humans, and for farm and companion animals [5]. The same report recommended the development of validated analytical methods for MON determination. Because of its ionic nature, MON is weakly retained by reversed-phase (RP) chromatography. Nevertheless, this technique is normally used either with the formation of ion pairs [12] or employing specific RP columns for polar compounds, or combining anion exchange and hydrophobic interactions [9,13,14]. Hydrophilic interaction chromatography was also used, but a non-negligible peak tailing was observed [8]. In this work, a new approach was evaluated; besides its ionic nature, MON is a di-ketone (1-hydroxycyclobut-1-ene-3,4-dione. As seen in Figure 1, di-ketones, mainly β-di-ketones, can easily form complexes with lanthanide ions (as Lanthanum La^{3+}, Terbium Tb^{3+} or Europium Eu^{3+}), linking the metallic ion through the oxygen atoms [15]. Generally, three di-ketones are linked to one lanthanide ion (Figure 2).

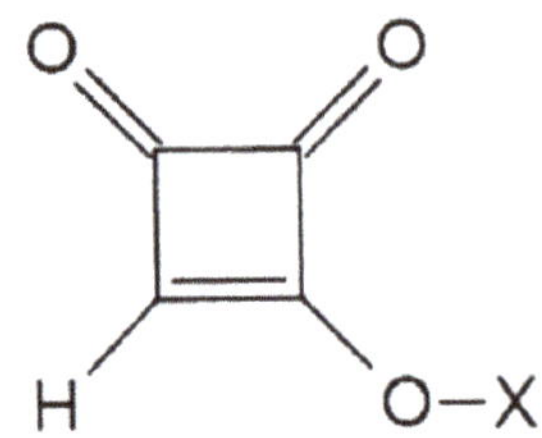

Figure 1. Chemical structure of moniliformin (X = H, Na or K).

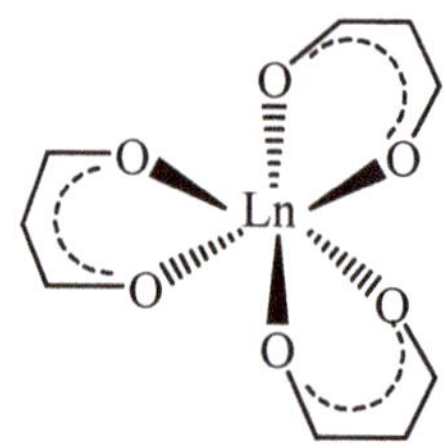

Figure 2. Structure of a [Ln-(β-di-ketone)$_3$] complex (Ln: lanthanide ion).

Then, in this work the addition of lanthanide ions to the mobile phase was investigated, aiming at improving peak shape and MON separation. MON determination was carried out using both UV and mass spectrometry (MS/MS) detection. UV detection was used for its major flexibility during the development of the method and MS/MS detection for its major accuracy and lower detection limits. Finally, a suitable extraction and purification method for MON determination in cereal samples was developed.

2. Results and Discussion

2.1. Development of the Chromatographic Method

It is known that MON is weakly retained in RP chromatography, recently, specific columns for polar compounds were used for its determination. Initially, four columns were tested using an HPLC-UV instrument: an RP-8 (Lichrospher, 5 µm particle size, 125 × 4 mm i.d., Merck, Darmstadt, Germany), a X-Select HSS T3 (RP-18 with low ligand density, 2.5 µm particle size, 100 × 2.1 mm i.d., Waters Corporation, Milford, MA, USA), a XBridge BEH Amide column (2.5 µm particle size, 100 × 2.1 mm i.d., Waters Corporation, Milford, MA, USA) and a Supelcosil LC-NH_2 column (250 × 3 mm, 5 µm, Supelco, Bellefonte, PA, USA); a mixture acetonitrile:water 20 + 80 *v/v* at a flow rate of 1.0 mL min^{-1} was used as mobile phase. A MON standard at 1000 µg L^{-1} was injected; MON was weakly retained by RP-8, X-Select HSS T3 and BEH Amide columns (retention time lower than 2.0 min); on the other hand, MON was strongly retained by the LC-NH_2 column (no peak was observed until 30 min). Successively, the mobile phase was substituted with acetonitrile: 10 mm $LaCl_3 \cdot 7H_2O$ 20 + 80 *v/v*; no relevant difference of the retention time was observed for RP-8, X-Select HSS T3 and BEH Amide column, while MON was eluted by the LC-NH_2 column at 4.9 min (Figure 3).

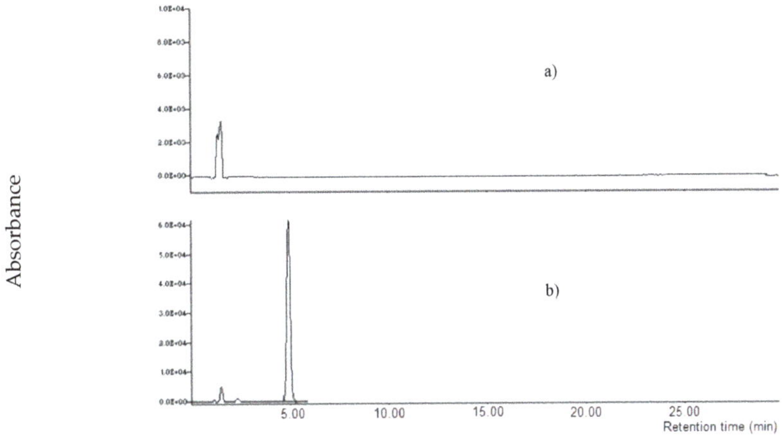

Figure 3. Chromatographic separation of moniliformin (MON) standard (1000 µg l^{-1}) on a LC-NH_2 column using water (**a**) or 10 mM La^{3+} aqueous solution (**b**) in the mobile phase.

Five calibration standards were injected (20, 100, 250, 500 and 1000 µg L^{-1}), showing a satisfactory calibration curve (R^2 = 0.998). Successively, increasing concentrations from 1 to 50 mM of $LaCl_3{\cdot}7H_2O$ in the mobile phase were tested and a decrease of the retention time was observed (Figure 4).

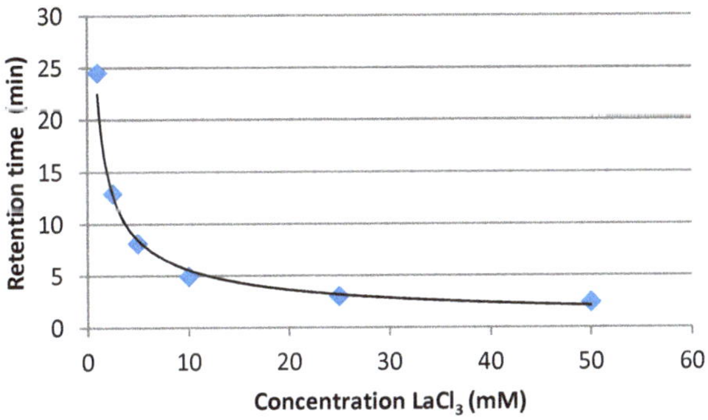

Figure 4. Variation of retention time depending on concentration of $LaCl_3$ (mM) in the mobile phase.

Finally, a satisfactory chromatographic separation was obtained applying a linear gradient acetonitrile:2.5 mM $LaCl_3{\cdot}7H_2O$ (Figure 5); a low $LaCl_3{\cdot}7H_2O$ concentration was preferred in order to delay MON elution and avoid co-elution with other substances.

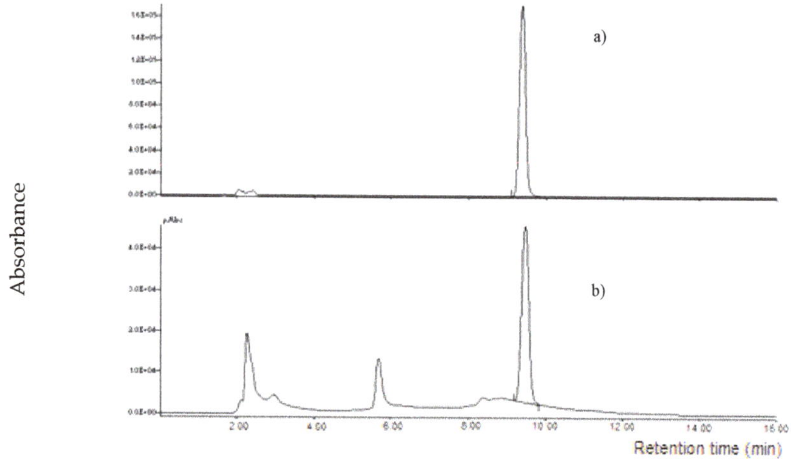

Figure 5. Chromatograms of: (**a**) MON standard (1000 μg L^{-1}); (**b**) maize spiked extract (250 μg L^{-1}). Chromatographic separation was carried out using linear gradient acetonitrile: 2.5 mM $LaCl_3 \cdot 7H_2O$ solution; detection at 260 nm (UV).

Similar results were obtained using Tb^{3+} or Eu^{3+}; MON was detected at a retention time of 7.0 min using acetonitrile:2.5 mM $TbCl_3 \cdot 6H_2O$ 60 + 40 *v/v* (Figure 6). Because of its easy availability and lower cost, it was preferred to use $LaCl_3 \cdot 7H_2O$ for the following quantitative MON analyses.

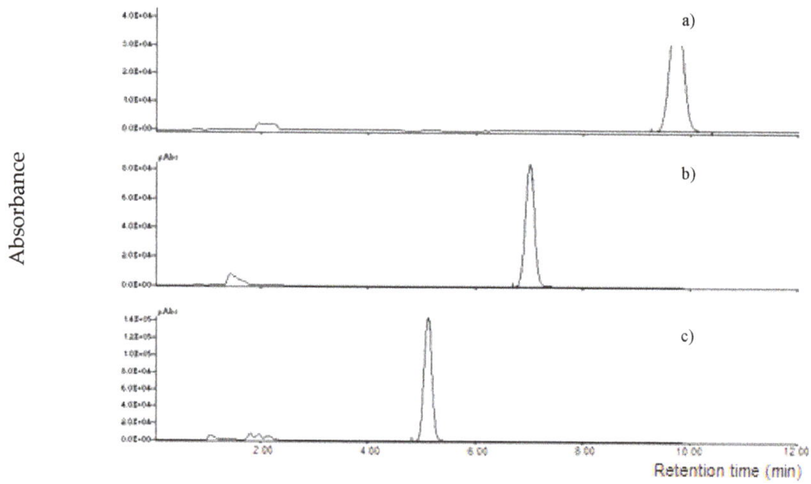

Figure 6. Chromatographic separation of MON standard (1000 μg L^{-1}) on LC-NH_2 column using: (**a**) 2.5 mM Tb^{3+} aqueous solution:acetonitrile 4 + 6 *v/v*; (**b**) 2.5 mM Tb^{3+} aqueous solution:acetonitrile 6 + 4 *v/v*; (**c**) 2.5 mM Tb^{3+} aqueous solution:acetonitrile 8 + 2 *v/v*, as mobile phase.

The chromatograms indicate a coordination of MON with lanthanide metals during the separation, resulting in a stronger affinity of MON for the mobile phase and a consequent faster elution from the LC-NH_2 column. Furthermore, increasing the concentration of La^{3+}, a higher MON affinity for the mobile phase is favoured and therefore shorter retention times are shown. It is known that lanthanide ions can form coordination complexes with di-ketones [15], as well as with other organic compounds.

Recently, a coordination between cyclopiazonic acid, a neurotoxin and lanthanide metals was reported by Maragos [16].

Generally, the luminescence of some lanthanides, such as Terbium and Europium, can be enhanced by interaction with selected molecules. For example, Terbium was used in post-column HPLC separation to increase the luminescence of ochratoxin A [17]. The luminescence is greatly influenced by the solvent and in particular by the presence of water, as well as by the lanthanide concentration and the chelate/lanthanide ratio [16]. Using a mobile phase 2.5 mM Tb^{3+} aqueous solution:acetonitrile 4 + 6 *v/v*, we replaced the UV with a fluorimetric detector (λ_{ex} = 260 nm; λ_{em} = 550 nm) in order to evaluate a possible fluorescence. However, the results showed a negative peak at the retention time of MON detected by the fluorimeter, showing a decrease of fluorescence of the baseline (Figure 7). During the chromatographic separation, water and lanthanide concentrations are very much higher than MON concentration. These conditions do not seem to favour an increase of fluorescence, as already reported [16,18,19].

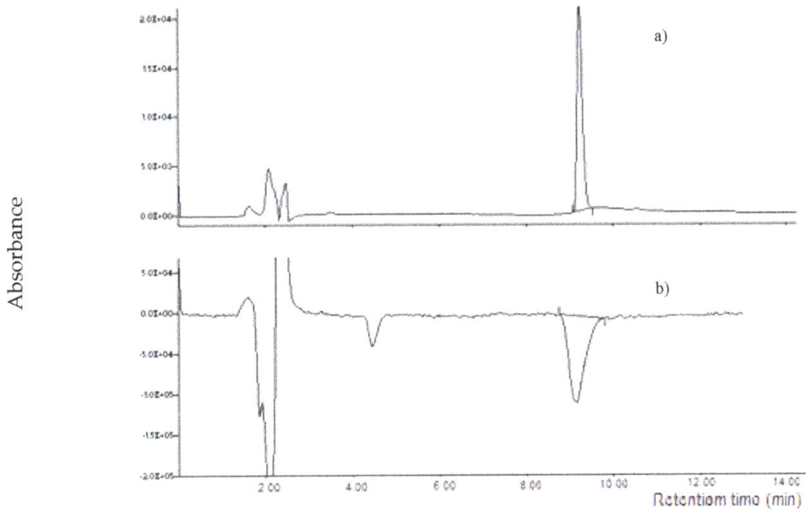

Figure 7. Chromatograms of MON standard (1000 μg L^{-1}) detected by UV at 260 nm (**a**) and fluorimeter detector at λ_{ex} = 260 nm; λ_{em} = 550 nm (**b**).

The process of MON separation adding lanthanide ions in the mobile phase can be speculated considering the theory of ligand exchange chromatography (LEC), a chromatographic process in which complex-forming compounds are separated through the formation and breaking of labile coordinate bonds to a central metal atom, coupled with partition between a mobile and a stationary phase [20]. Based on LEC theory, the coordination of lanthanide ions with water in the mobile phase, can be partially replaced with the coordination to MON, resulting in the formation of mixed coordination complexes (Figure 8). These coordination complexes are kinetically weak, their formation and dissociation is fast and can be described by the following reversible reaction:

$$\text{MON} + [\text{Ln}(\text{H}_2\text{O})_n] \leftrightarrow [\text{Ln}(\text{MON})(\text{H}_2\text{O})_{n-1}] + \text{H}_2\text{O} \quad \text{Ln—lanthanide}$$

This process could explain the higher MON affinity for the mobile phase and the faster elution, when lanthanide concentration is increased. In absence of lanthanide ions in the mobile phase, MON is strongly retained by the stationary phase of the LC-NH_2 column.

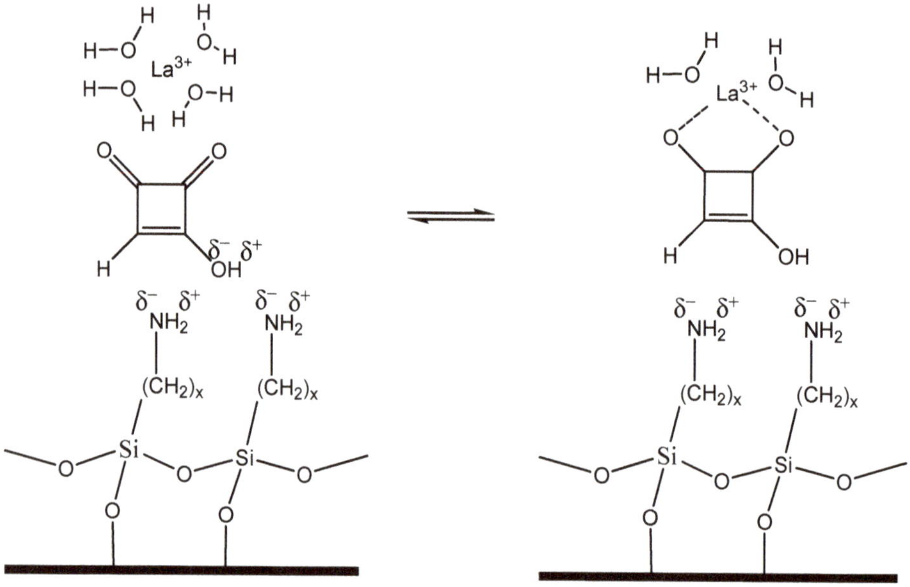

Figure 8. Possible structure of the coordination complex during chromatographic separation in presence of La^{3+} in the mobile phase.

Finally, the chromatographic analysis was carried out using a mass spectrometric detector (MS/MS, triple quadrupole, Thermo Fisher Scientific, San Jose, CA, USA) in order to obtain high accuracy and lower detection limits; a very low $LaCl_3{\cdot}7H_2O$ concentration was used (1.25 mM) to avoid troubles during the ionisation. MON was chromatographed on a 75 × 3 mm, 3 µm LC-NH_2 column (Supelco) and detected at 4.4 min, improving the limit of detection (Figure 9).

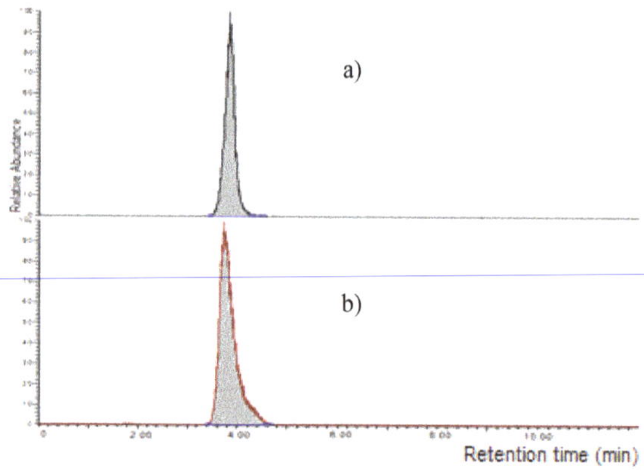

Figure 9. Chromatograms of: (**a**) MON standard (100 µg L^{-1}); (**b**) naturally contaminated maize sample (1637 µg kg^{-1}). Chromatographic separation was carried out using a linear gradient methanol: 25 mM ammonium acetate containing 1.25 mM $LaCl_3{\cdot}7H_2O$; detection by MS/MS (MRM transition *m/z* 97 >41).

2.2. Development of an Extraction and Purification Method for Cereal Samples

Generally, MON was extracted using a mixture acetonitrile:water 84:16 (as for tricothecenes). Considering that MON is highly water-soluble, Barthel et al. [10] and Herrera et al. [9] recently increased the percentage of water in the extraction mixture, using acetonitrile:water 50 + 50 *v/v* or pure water, obtaining higher extraction yields. Since in our tests the extract using water 100% was turbid, probably for the presence of other large polar molecules, the mixture acetonitrile:water 50 + 50 *v/v* was preferred. As regards the purification step, several authors used either a SAX-like SPE [9] or a MycoSep® MON 240 column [8,13]. However, Herrera et al. [10] did not obtain satisfactory recoveries using these columns and consequently no clean-up step was carried out in their method. In this work, we previously tested MycoSep® MON 240 (Romer Labs, Getzersdorf, Austria), MAX, WAX and HLB OASIS columns (Waters Corporation, Milford, MA, USA) and the Quechers procedure; in all tests, unsatisfactory purification or low recoveries were obtained. Moreover, we confirmed that the evaporation under N_2 flow decreased the recovery, as reported by Herrera et al. [10]. Finally, a LC-NH_2 column (Supelco) was tested, in order to retain MON and elute it using a lanthanide ion solution, as for the HPLC separation. This clean-up step was only introduced for HPLC-UV determination, for LC-MS/MS analysis, the presence of high concentration of La^{3+} (12.5 mM) in the purified extract worsened MON detection.

2.3. Performances of the Method

The result of the considered parameters are shown in Table 1. Despite the omission of clean-up step for LC-MS/MS analysis, the matrix effect, calculated at the concentration of 100 μg L^{-1} (800 μg kg^{-1} for a cereal sample), was low, close to 4% and 6% for wheat and maize samples, respectively.

Table 1. Validation parameters of MON analysis using UV and LC-MS/MS detector (three replicates).

	HPLC-UV	LC-MS/MS
Matrix effect (at 800 μg kg^{-1}) *	/	5% (wheat) 8% (maize)
Calibration range *	160–8000 μg kg^{-1}	20–2000 μg kg^{-1}
LOD *	80 μg kg^{-1}	10 μg kg^{-1}
LOQ *	200 μg kg^{-1}	25 μg kg^{-1}
Average recovery		
Wheat	88.2% ± 5.5%	97.1% ± 4.3%
Maize	84.8% ± 5.4%	96.4% ± 5.1%

LOD: limit of detection; LOQ: limit of quantification. * Data referred to cereal sample.

A satisfactory linearity was calculated for both LC-MS/MS (R^2 = 0.997) and HPLC-UV (R^2 = 0.998). The limit of detection (LOD), for a cereal sample, was 10 and 80 μg kg^{-1} using LC-MS/MS and HPLC-UV, respectively. The decision limit (CCα) and the detection capability (CCβ) were 10 and 18 μg kg^{-1} for LC-MS/MS, 80 and 136 μg kg^{-1} for LC-UV.

Skipping the clean-up step (LC-MS/MS analysis), the global average recoveries were 97.1% ± 4.3% and 96.4% ± 5.1% for wheat and maize, respectively. Using purification step (UV analysis), the average recoveries were: 86.4% ± 6.4% and 89.8% ± 4.5% for wheat (at 100 and 500 μg kg^{-1}, respectively), 83.6% ± 5.9% and 86.1% ± 5.1% for maize (at 250 and 1000 μg kg^{-1}, respectively).

Finally, the standard deviation obtained by repeatability tests was less than 9.2%.

2.4. MON Occurrence in Cereal Samples

MON occurred in 100% and 50% of maize and wheat samples, respectively. The levels of contamination, corrected for the recovery percentage, ranged between 38 and 3629 μg kg^{-1} (the last value was obtained after dilution of the sample extract) for maize (in 4 samples MON exceeded 1000 μg kg^{-1}), <10 and 481 μg kg^{-1} for wheat samples (Figure 10).

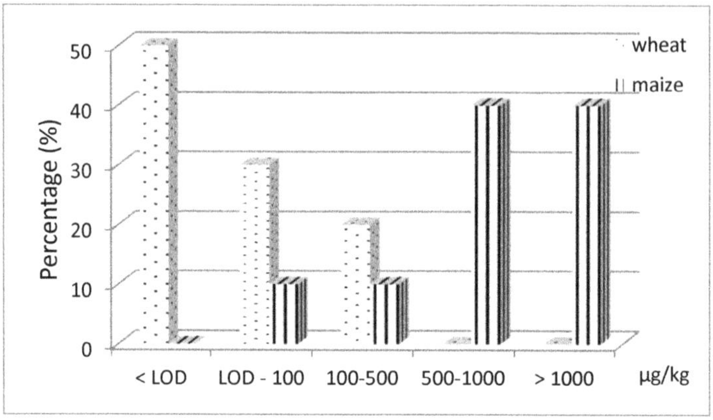

Figure 10. Frequency of occurrence (%) of MON in wheat (n = 10) and maize (n = 10) collected in northern Italy.

3. Materials and Methods

3.1. Reagents and Standards

Chemicals and solvents used for extraction and clean-up were ACS grade or equivalent (Carlo Erba, Milan, Italy); deionised water was purified through a Milli-Q treatment system (Millipore, Bedford, MA, USA). For HPLC and LC-MS/MS analysis, water, methanol and acetonitrile were HPLC grade (Merck, Darmstadt, Germany). MON (as sodium salt), Lanthanum (III) chloride heptahydrate ($LaCl_3 \cdot 7H_2O$), Terbium (III) chloride hexahydrate ($TbCl_3.6H_2O$) and Europium (III) chloride hexahydrate ($EuCl_3 \cdot 6H_2O$) were obtained from Sigma-Aldrich (St. Louis, MO, USA). A MON stock standard solution was prepared in acetonitrile at a concentration of 100 mg L^{-1}; working solutions were obtained by dilution using water:methanol 15 + 85 *v/v*. All the solutions were stored at −20 °C when not in use.

3.2. LC-MS/MS Analysis for Moniliformin Determination

MON was extracted from cereal samples (10 g each) with 40 mL of a mixture acetonitrile: water 50 + 50 *v/v* using a rotary-shaking stirrer for 60 min. After filtration on a folded filter paper, the extract was diluted (1 + 1) with methanol:water 85 + 15 *v/v* and injected into the LC-MS/MS system (20 μL). The HPLC-MS/MS system consisted of a LC 1.4 Surveyor pump (Thermo Fisher Scientific, San Jose, CA, USA), a PAL 1.3.1 sampling system (CTC Analytics AG, Zwingen, Switzerland) and a Quantum Discovery Max triple quadrupole mass spectrometer; the system was controlled by an Excalibur 1.4 software (Thermo-Fisher). MON was chromatographed on a Supelcosil LC-NH_2 column (75 × 3 mm, 3 μm, Supelco, Bellefonte, PA, USA) and separated using a gradient elution with 25 mM ammonium acetate containing 1.25 mM $LaCl_3 \cdot 7H_2O$, and methanol as mobile phase A and B, respectively. The gradient program was linear gradient from 15% to 35% of solvent A in 3 min, then isocratic for 1 min; column conditioning lasted 7 min. The flow rate was 0.3 mL min^{-1}. The ionisation was carried out with an ESI interface (Thermo-Fisher) in negative mode as follows: spray capillary voltage was 3.5 kV, sheath and auxiliary gas 40 and 15 psi, respectively; skimmer 9 V, temperature of the heated capillary 350 °C. The mass spectrometric analysis was performed in selected reaction monitoring (SRM). For fragmentation of the $[M–H]^-$ ion (97 *m/z*), the argon collision pressure was set to 1.2 mTorr and the collision energy to 21 V. The detected and quantified fragment ion was 41 *m/z*. Quantitative determination was performed by an LC-Quan 2.0 software (Thermo-Fisher, Waltham, MA, USA).

3.3. HPLC-UV Analysis for Moniliformin Determination

After extraction as for LC-MS/MS analysis, a purification step was introduced before HPLC-UV analysis; a 2 mL aliquot of the extract was purified through a LC-NH_2 column (500 mg, 3 mL, Supelco, Bellefonte, PA, USA), previously conditioned with 2 mL acetonitrile:water 50 + 50 *v/v*. The column was washed with acetonitrile (2 mL), deionized water (2 mL) and 1 mL of a 12.5 mM $LaCl_3{\cdot}7H_2O$ aqueous solution; then, MON was eluted in a graduated glass vial with additional 2 mL of the 12.5 mM $LaCl_3{\cdot}7H_2O$ aqueous solution. The purified extract was diluted (1 + 1) using acetonitrile and injected into a HPLC-UV system (30 µL). The HPLC system consisted of a Jasco PU 1580 pump (Jasco Corp., Tokyo, Japan) equipped with an AS 2055 sampling system and a UV 1575 detector set at 219 and 260 nm. The system was governed by a Borwin 1.5 software (Jasco). MON was chromatographed on a Supelcosil LC-NH_2 column (250 × 3 mm, 5 µm, Supelco) and separated using a gradient elution with 2.5 mM $LaCl_3{\cdot}7H_2O$ (or $TbCl_3{\cdot}6H_2O$) and acetonitrile as mobile phase A and B, respectively. The gradient program was 15% solvent A for 1 min, linear gradient to 35% solvent A in 3 min, then isocratic for 5 min; column conditioning lasted 7 min. The flow rate was 0.7 mL min^{-1}; the column was thermostated at 25 °C.

3.4. Method Validation

For method validation, different parameters were considered. As regards LC-MS/MS, the matrix effect was examined; this effect is due to the presence of compounds that can co-elute, affecting the ionisation of the analyte. It was defined as the difference between the mass spectrometric response for the analyte in standard solution and the response for the same analyte at the same concentration in matrix extract. A comparison between the mass spectrometric response of MON in standard solution and in matrix extract at 100 µg l^{-1} was conducted; the matrix extract was evaluated by spiking an uncontaminated wheat or maize extract (950 µL) with MON standard (50 µL). Linearity of both HPLC-UV and LC-MS/MS measurement was established through five calibration standards in solvent, at concentrations between 2.5 and 250 µg L^{-1} for LC-MS/MS (20 and 2000 µg kg^{-1} for a cereal sample) and between 20 and 1000 µg L^{-1} (160 and 8000 µg kg^{-1} for a cereal sample) for HPLC-UV. The limit of detection (LOD) was defined as the level corresponding to a signal-to-noise ratio (S/N) of three, while the limit of quantification (LOQ) as the lowest level for which the repeatability of the analysis was below 10%. The decision limit (CCα) and the detection capability (CCβ) was calculated as reported by Commission Decision 2002/657/EC [21]. The accuracy of the proposed method was evaluated by determination of the recovery. Recovery experiments were performed by spiking wheat or maize flour in triplicate at two levels, 100 and 500 µg kg^{-1} for wheat and 250 and 1000 µg kg^{-1} for maize. The matrices were also analysed without spiking, as reagent blank. The spiked samples were allowed to stand for two hours at ambient temperature under a fume hood to allow any residual solvent to evaporate. Finally, the method repeatability was evaluated. Four samples (2 wheat and 2 maize flours) were extracted and analysed three times in different days.

3.5. Real Sample Collection and Analysis

A total of ten samples of maize and ten of wheat (2 durum wheat) were collected in northern Italy; samples were dried at 65 °C, milled using a cyclone hammer mill (1 mm sieve, Pulverisette, Fritsch GmbH, Idar-Oberstein, Germany) and homogenised. Then, an aliquot (1 kg) of the sample was taken and stored at −20 °C until the time of analysis. Quantification was carried out by LC-MS/MS, to detect lower contamination values.

4. Conclusions

Simple and suitable LC-MS/MS and LC-UV methods for MON determination were developed, fulfilling the wishes of the EFSA report. The simple addition of lanthanide ions to the mobile phase allowed an easy determination of MON using either mass spectrometric or UV detection. Data on

MON occurrence in food and feed can be easily obtained using this method with a satisfactory accuracy. Further studies are needed to confirm the formation of mixed coordination complexes during the chromatographic separation. Finally, this chromatographic technique could be applied to the analysis of other mycotoxins, and also to the development of chiral separations.

Author Contributions: The manuscript was written through contributions of all authors. All authors have given approval to the final version of the manuscript. T.B., A.M. and S.R. designed the research. A.M. and S.R. performed the experiment. T.B. analyzed the data and wrote the manuscript. T.B. and A.P. supervised the research and edited and approved the final manuscript.

Funding: This research received no external funding.

Conflicts of Interest: The authors declare no conflicts of interest.

References

1. Uhlig, S.; Torp, M.; Jarp, J.; Parich, A.; Gutleb, A.C.; Krska, R. Moniliformin in Norwegian grain. *Food Addit. Contam.* **2004**, *21*, 598–606. [CrossRef] [PubMed]
2. Steyn, M.; Thiel, P.G.; Van Schalkwyk, G.C. Isolation and purification of moniliformin. *J. Assoc. Off. Anal. Chem.* **1978**, *61*, 578–580. [PubMed]
3. Kriek, N.; Marasas, W.; Steyn, P.; Van Rensburg, S.; Steyn, M. Toxicity of a moniliformin-producing strain of fusarium moniliforme var. subglutinans isolated from maize. *Food Cosmet. Toxicol.* **1977**, *15*, 579–587. [CrossRef]
4. Jestoi, M. Emerging Fusarium-Mycotoxins Fusaproliferin, Beauvericin, Enniatins, and Moniliformin—A Review. *Crit. Rev. Food Sci. Nutr.* **2008**, *48*, 21–49. [CrossRef] [PubMed]
5. EFSA Panel Contam. Risks to human and animal health related to the presence of moniliformin in food and feed. *EFSA J.* **2018**, *16*, 5082.
6. Rokka, M.; Yli-Mattila, T.; Parikka, P.; Rizzo, A.; Peltonen, K.; Jestoi, M. Presence and concentrations of theFusarium-related mycotoxins beauvericin, enniatins and moniliformin in Finnish grain samples. *Food Addit. Contam.* **2004**, *21*, 794–802.
7. Van Asselt, E.; Azambuja, W.; Moretti, A.; Kastelein, P.; De Rijk, T.; Stratakou, I.; Van Der Fels-Klerx, H. A Dutch field survey on fungal infection and mycotoxin concentrations in maize. *Food Addit. Contam. Part A* **2012**, *29*, 1556–1565. [CrossRef] [PubMed]
8. Scarpino, V.; Blandino, M.; Negre, M.; Reyneri, A.; Vanara, F. Moniliformin analysis in maize samples from North-West Italy using multifunctional clean-up columns and the LC-MS/MS detection method. *Food Addit. Contam.* **2013**, *30*, 876–884. [CrossRef] [PubMed]
9. Von Bargen, K.W.; Lohrey, L.; Cramer, B.; Humpf, H.-U. Analysis of the Fusarium mycotoxin moniliformin in cereals samples using 13C2-moniliformin and High- Resolution Mass Spectrometry. *J. Agric. Food Chem.* **2012**, *60*, 3586–3591. [CrossRef] [PubMed]
10. Herrera, M.; Van Dam, R.; Spanjer, M.; De Stoppelaar, J.; Mol, H.; De Nijs, M.; López, P. Survey of moniliformin in wheat- and corn-based products using a straightforward analytical method. *Mycotoxin Res.* **2017**, *33*, 333–341. [CrossRef]
11. Barthel, J.; Rapp, M.; Holtmannspötter, H.; Gottschalk, C. A rapid LC-MS/MS method for the determination of moniliformin and occurrence of this mycotoxin in maize products from the Bavarian market. *Mycotoxin Res.* **2018**, *349*, 9–12. [CrossRef] [PubMed]
12. Shepherd, M.J.; Gilbert, J. Method for the analysis in maize of the fusarium mycotoxin moniliformin employing ion-pairing extraction and high-performance liquid chromatography. *J. Chromatogr. A* **1986**, *358*, 415–422. [CrossRef]
13. Lim, C.W.; Lai, K.Y.; Yeo, J.F.; Tai, S.H.; Chan, S.H. Quantitative assessment of moniliformin in cereals via alternative precipitation pathways, aided by LC-LIT-MS and LC-Q-TOF-MS. *Food Chem.* **2015**, *174*, 372–379. [CrossRef] [PubMed]
14. Hallas-Møller, M.; Frisvad, J.C.; Nielsen, K.F. Production of the Fusarium Mycotoxin Moniliformin by Penicillium melanoconidium. *J. Agric. Food Chem.* **2016**, *64*, 4505–4510. [CrossRef] [PubMed]

15. Binnemans, K. Rare-earth beta-diketonates. In *Handbook of the Physics and Chemistry of Rare Earths*; Gschneidner, K.A., Jr., Bünzli, J.-C.G., Pecharsky, V.K., Eds.; North Holland: Amsterdam, Holland, 2005; Volume 35, pp. 107–272.
16. Maragos, C.M. Complexation of the Mycotoxin Cyclopiazonic Acid with Lanthanides Yields Luminescent Products. *Toxins* **2018**, *10*, 285. [CrossRef] [PubMed]
17. Vazquez, B.; Fente, C.; Franco, C.; Cepeda, A.; Prognon, P.; Mahuzier, G. Simultaneous high-performance liquid chromatographic determination of ochratoxin A and citrinin in cheese by time-resolved luminescence using terbium. *J. Chromatogr. A* **1996**, *727*, 185–193. [CrossRef]
18. Armelao, L.; Quici, S.; Barigelletti, F.; Accorsi, G.; Bottaro, G.; Cavazzini, M.; Tondello, E. Design of luminescent lanthanide complexes: From molecules to highly efficient photo-emitting materials. *Coord. Chem. Rev.* **2010**, *254*, 487–505. [CrossRef]
19. Rocha, J.; Carlos, L.D.; Paz, F.A.; Ananias, D. Luminescent multifunctional lanthanides-based metal-organic frameworks. *Chem. Soc. Rev.* **2011**, *40*, 926–940. [CrossRef] [PubMed]
20. Davankov, V.A.; Navratil, J.D.; Walton, H.F. *Ligand Exchange Chromatography*; CRC Press: Boca Raton, FL, USA, 1988.
21. Commission Decision of 12 August 2002 implementing Council Directive 96/23/EC concerning the performance of analytical methods and the interpretation of results 2002/657/EC. *Off. J. Eur. Communities* **2002**, *L221*, 8–36.

toxins

MDPI

Article

Development of a QuEChERS-Based UHPLC-MS/MS Method for Simultaneous Determination of Six *Alternaria* Toxins in Grapes

Wenbo Guo [1,†], Kai Fan [1,†], Dongxia Nie [1], Jiajia Meng [1], Qingwen Huang [1,2], Junhua Yang [1], Yuanyuan Shen [1], Emmanuel K. Tangni [3], Zhihui Zhao [1], Yongjiang Wu [2,*] and Zheng Han [1,*]

1 Institute for Agro-food Standards and Testing Technology, Shanghai Key Laboratory of Protected Horticultural Technology, Shanghai Academy of Agricultural Sciences, Shanghai 201403, China; guowenbo@saas.sh.cn (W.G.); fankai1983@gmail.com (K.F.); niedongxia@saas.sh.cn (D.N.); mengjiajia@saas.sh.cn (J.M.); huangqingwen@zju.edu.cn (Q.H.); yangjunhua@saas.sh.cn (J.Y.); shenyuanyuan@saas.sh.cn (Y.S.); zhao9912@hotmail.com (Z.Z.);

2 College of Pharmaceutical Sciences, Zhejiang University, Hangzhou 310058, China;

3 Organic Contaminants and Additives, Sciensano, Tervuren 3080, Belgium; emmanuel.tangni@sciensano.be

* Correspondence: yjwu@zju.edu.cn (Y.W.); hanzheng@saas.sh.cn (Z.H.); Tel.: +86-571-88208455 (Y.W.); +86-21-62203612 (Z.H.);

† These authors contributed equally to this work.

Received: 28 December 2018; Accepted: 28 January 2019; Published: 1 February 2019

check for updates

Abstract: A simple and reliable analytical method for the simultaneous determination of alternariol (AOH), altenuene (ALT), tentoxin (TEN), altenusin (ALS), tenuazonic acid (TeA), and alternariol monomethyl ether (AME) in grapes was developed by ultra-high-performance liquid chromatography–tandem mass spectrometry (UHPLC-MS/MS). A modified QuEChERS (quick, easy, cheap, effective, rugged, and safe) procedure with the extraction by acetonitrile and purification by sodium chloride (0.5 g) and anhydrous magnesium sulfate (0.5 g) was established to recover the six *Alternaria* toxins. After validation by determining the linearity ($R^2 > 0.99$), recovery (77.8–101.6%), sensitivity (limit of detection in the range of 0.03–0.21 μg kg^{-1}, and limit of quantification in the range of 0.09–0.48 μg kg^{-1}), and precision (relative standard deviation (RSD) ≤ 12.9%), the analytical method was successfully applied to reveal the contamination state of *Alternaria* toxins in grapes. Among 56 grape samples, 40 (incidence of 71.4%) were contaminated with *Alternaria* toxins. TEN was the most frequently found mycotoxin (37.5%), with a concentration range of 0.10–1.64 μg kg^{-1}, followed by TeA (28.6%) and AOH (26.8%). ALT (10.7%), AME (3.6%), and ALS (5.4%) were also detected in some samples. To the best of our knowledge, this is the first report about the *Alternaria* toxins contamination in grapes in China.

Keywords: *Alternaria* toxins; grape; modified QuEChERS; UHPLC-MS/MS

Key Contribution: An accurate and reliable UHPLC-MS/MS method was developed for simultaneous determination of six *Alternaria* toxins in grapes for the first time.

1. Introduction

Alternaria toxins, secondary metabolites produced by *Alternaria spp*, e.g., *A. alternata*, *A. tenuissima*, and *A. infectoria*—are frequently found in cereals, fruits, and vegetables [1,2]. The most important members include alternariol (AOH), alternariol monomethyl ether (AME), altenuene (ALT), altenusin (ALS), tentoxin (TEN), and tenuazonic acid (TeA) [3]. Acute and chronic ingestion of *Alternaria* toxins can be associated with carcinogenesis [4], teratogenesis [5,6], and cytotoxicity [7], as well as reproductive and developmental toxicities [8,9], and therefore pose high potential risks to human health. The European Food Safety Authority (EFSA) has established the threshold of toxicological concern (TTC) values as 2.5 ng kg^{-1} body weight per day for AOH and AME, and 1500 ng kg^{-1} for TeA [10]. Considering its widespread occurrence and intense toxicity, TeA, the most prevalent *Alternaria* toxin, has been registered as a toxic chemical by the Occupational Safety and Health Act (OSHA), and a maximal limit of 500 μg kg^{-1} was proposed by German federal state Bavaria in sorghum/millet-based infant food [11].

As a worldwide grown fruit, global grape production is over 3 million tons annually, of which 71% are used for wine-making, 27% for fresh consumption, and 2% for raisin producing [12]. Grapes and its derived products are susceptible to the infection of *Alternaria* spp during maturation, as well as post-harvest and during processing, when they are improperly stored. AOH and AME have been found in red and white wine with concentrations in the range of 0.03–19.4 ng mL^{-1} and 0.01–0.23 ng mL^{-1}, respectively, and also in red and white grape juice samples with concentrations of 0.03–0.46 ng mL^{-1} and 0.01–39.5 ng mL^{-1}, respectively [13]. TeA, AOH, and AME have also been found in raisins with the concentrations in the range of 6.9–594.4 μg kg^{-1}, 0.3–13.5 μg kg^{-1} and 3.5–15.6 μg kg^{-1}, respectively [14]. However, until now, the literature has been sparse on the presence of *Alternaria* toxins contaminating grapes, and it is ambiguous whether the risks come from the production process or from the original fruits. Hence, it is important to develop a reliable and accurate method for simultaneous determination of multiple *Alternaria* toxins, and to investigate their real contamination levels in grapes.

A variety of analytical methods, i.e., thin-layer chromatography (TLC) [15], gas chromatography [GC] [16], and high-performance liquid chromatography (HPLC) with different detectors [17,18], have been developed for determination of *Alternaria* toxins. The most frequently used technique for toxin separation is HPLC, as it combines high resolution with increasingly sophisticated automation. The availability of different ionization sources, i.e., electrospray (ESI) and atmospheric pressure chemical ionization (APCI), has drastically improved the possibilities of employing HPLC-tandem mass spectrometry (HPLC-MS/MS) in *Alternaria* toxin analysis, resulting in enhanced performance, providing additional selectivity, and generating information with a high degree of structural specificity. Hitherto, most of the previous reports only focused on TeA, AOH, and AME [13,17,19], and other important *Alternaria* toxins (e.g., TEN, ALS, and ALT) were not investigated. Recently, an UHPLC-MS/MS method was established for the detection of various *Alternaria* toxins in wine, vegetable juices, and fruit juices [20]. The low recovery of ALS and tedious sample pretreatment procedures made this method unsuitable for analysis of the targeted *Alternaria* toxins in grapes.

The major objective of this study is to develop a rapid and reliable ultra-high performance liquid chromatography tandem mass spectrometry (UHPLC-MS/MS) method for simultaneous determination of AOH, AME, ALT, ALS, TEN, and TeA based on a simple sample preparation of modified QuEChERS (quick, easy, cheap, effective, rugged and safe) approach, and to explore the actual contamination situations of *Alternaria* toxins in grapes for the first time in the world.

2. Results and Discussion

2.1. Optimization of the Ultra-High-Performance Liquid Chromatography–Tandem Mass Spectrometry Conditions

MS/MS parameters were optimized by flow injection analysis of an individual *Alternaria* toxin standard at a concentration of 50–200 ng mL^{-1}. The precursor ions and cone voltage were optimized by MS scan acquisition, both in positive and negative ionization modes. Then, MS/MS scan acquisitions were applied to find the optimum product ions and collision energies (CE), cone voltages, and dwell time (Figure S1). As shown in Table 1, the majority of *Alternaria* toxins displayed better specificity and selectivity in ESI^+, except for ALS, which showed stronger signals and lower background interference in ESI^-.

To obtain good chromatographic separation with symmetry and a sharp peak shape for targeted analytes, the compositions of the mobile phase and the chromatographic columns were optimized.

Different additives—e.g., formic acid, ammonium formate, and ammonium acetate—were primarily evaluated. The highest ionization efficiency and sensitivity were obtained for all *Alternaria* toxins when methanol (A) and water containing 5 mmol L^{-1} aqueous ammonium acetate methanol (B) were used as the mobile phase. Subsequently, various chromatographic columns, i.e., the Waters ACOUITY UPLC BEH C_{18} column (100 mm × 2.1 mm, 1.7 μm), Agilent Proshell EC_{18} column (50 mm × 2.1 mm, 2.7 μm), and Waters ACQUITY UPLC® HSS T3 (50 mm × 2.1 mm, 2.7 μm), which represented different stationary phases, were compared. Considering the efficient separation and response values, shown in Figure 1, the Proshell EC_{18} column (50 mm × 2.1 mm, 2.7 μm) was finally selected.

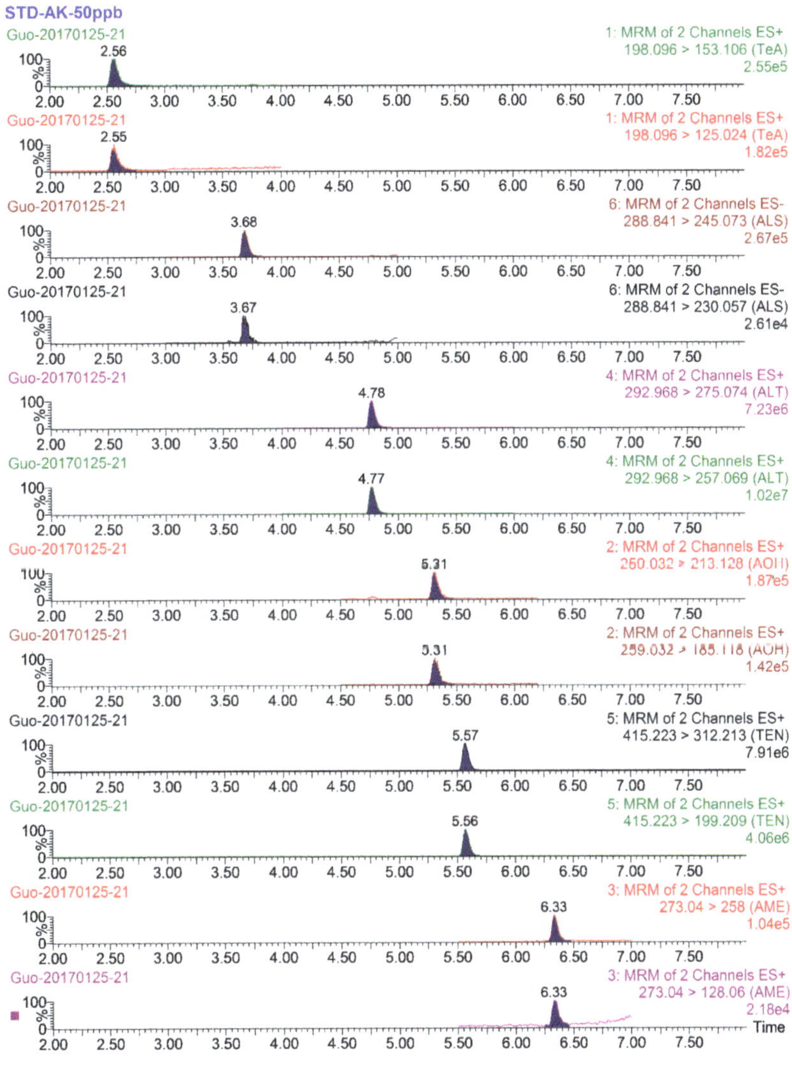

Figure 1. Multiple reaction monitoring (MRM) chromatograms of the six *Alternaria* toxins in the mixed standard solution. The concentration is 50 ng mL^{-1}.

Table 1. Tandem mass spectrometry (MS/MS) parameters for the determination of six *Alternaria* toxins.

Alternaria Toxins	Precursor Ions (*m*/*z*)	Product Ions (*m*/*z*)	Dwell Time (s)	Cone Voltage (V)	Collision Energy (eV)
AOH	259.0 $[M+H]^+$	185.1 *	0.025	64	28
		213.1	0.025	64	24
AME	273.0 $[M+H]^+$	128.1 *	0.030	54	26
		258.0	0.030	54	25
ALT	293.0 $[M+H]^+$	257.1 *	0.025	4	14
		275.1	0.025	4	8
TEN	415.2 $[M+H]^+$	199.2 *	0.029	32	12
		312.2	0.029	32	18
TeA	198.1 $[M+H]^+$	125.0 *	0.025	42	16
		153.1	0.025	42	12
ALS	288.8 $[M-H]^-$	235.1 *	0.029	−8	−20
		245.1	0.029	−8	−16

* Primary product ion.

2.2. Optimization of the Sample Pretreatment Method

Establishment of an efficient sample pretreatment method is always the bottleneck for the development of an accurate and sensitive analytical method, due to the diversity of the physical and chemical properties of various *Alternaria* toxins, along with the complex sample matrices. In this study, five different frequently used extraction solvents, including methanol, methanol/water (80/20, *v*/*v*), acetonitrile, acetonitrile/water (84/16, *v*/*v*), and acetonitrile/acetic acid (99/1, *v*/*v*) were evaluated by using the blank grape samples spiked with 50 μg kg^{-1} of the targeted *Alternaria* toxins. As shown in Figure 2A, when methanol or methanol/water (80/20, *v*/*v*) were used, unsatisfactory recoveries (48.0–76.0%) were obtained for AOH, TeA, and AME. In addition, the extraction was emulsified with high contents of pigments and sugar. When acetonitrile was used as the extraction solvent, the recoveries were significantly increased (92.7–102.2%), especially for AOH and AME, against the recoveries of 70.4% and 66.2% for acetonitrile/water (84/16, *v*/*v*), and 69.8% and 70.9% for acetonitrile/acetic acid (99/1, *v*/*v*). As a consequence, acetonitrile was selected as the optimal extraction solvent.

For sample clean-up, a modified QuEChERS method was developed to enrich the targeted analytes, and to remove the co-extractives as completely as possible. Different materials, including graphitized carbon black (GCB) (0.5 g), primary secondary amine (PSA) (0.5 g), C_{18} (0.5 g), $MgSO_4$ (0.5 g), and NaCl (0.5 g) were tested for their purification efficiency. Unsatisfactory recoveries of 18.2–77.5% and 3.4–51.2% were obtained (Figure 2B) by using GCB and PSA, respectively, which had been used frequently to remove chlorophylls from fruits and vegetables in the previous studies [21]. The poor purification effects in the current work might be because of the π–π interactions through the sp^2 hybrid orbitals of GCB and the planar aromatic compounds (i.e., AOH, AME, and ALT), as well as the ionic affinity between the amines in PSA and the carboxyl group in ALS [22–24]. With regard to C_{18}, this material showed poor recoveries in the range of 15.0–61.5%. Finally, the salting-out step with anhydrous $MgSO_4$ and NaCl was employed with the highest recoveries. in the range of 77.9–98.5% (Figure 2B), and lowest matrix effects, in the range of 82.8–102.3% (Figure 3).

Different membrane filters for filtering the re-dissolved solutions before injection into UHPLC-MS/MS, including nylon, poly tetra fluoroethylene (PTFE), mixed cellulose membrane (MCM), and polyvinylidene difluoride (PVDF), were compared. As shown in Figure S2, all membrane filters could be used for filtering ALT, TEN, ALS, and TeA. However, when nylon, MCM, and PVDF were applied, the recoveries were unsatisfactory for AOH and AME (3.6–19.0%). Satisfactory recoveries in the range of 86.1–100.5% for all *Alternaria* toxins were achieved when a PTFE membrane filter was selected.

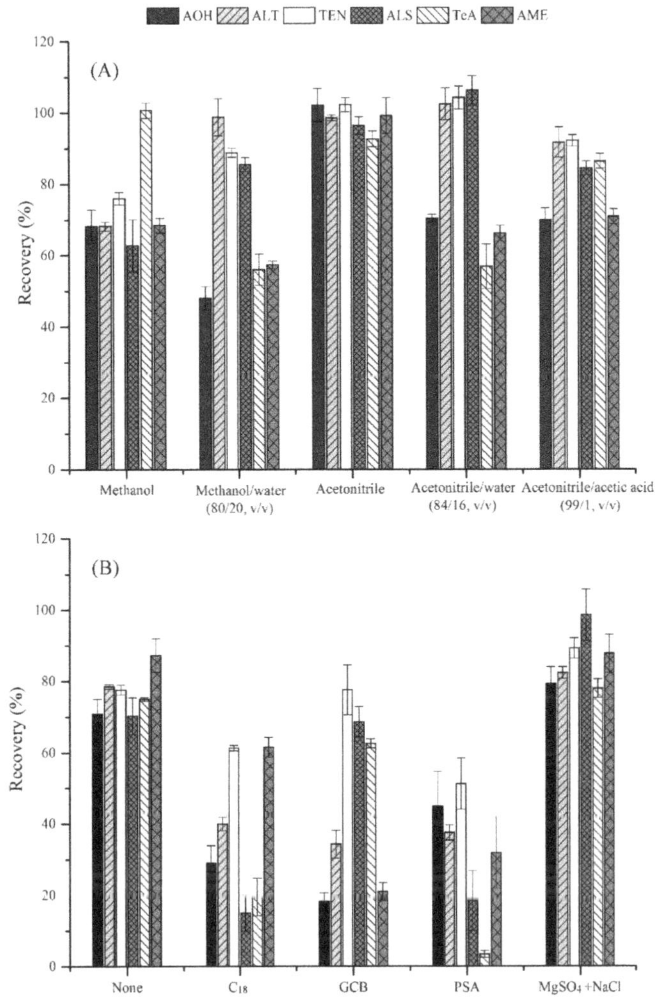

Figure 2. Comparison of the extraction efficiencies of different solvents (**A**) and purification efficiencies of different materials (**B**), using spiked grape samples. The concentration is 50 μg kg^{-1} (n = 6).

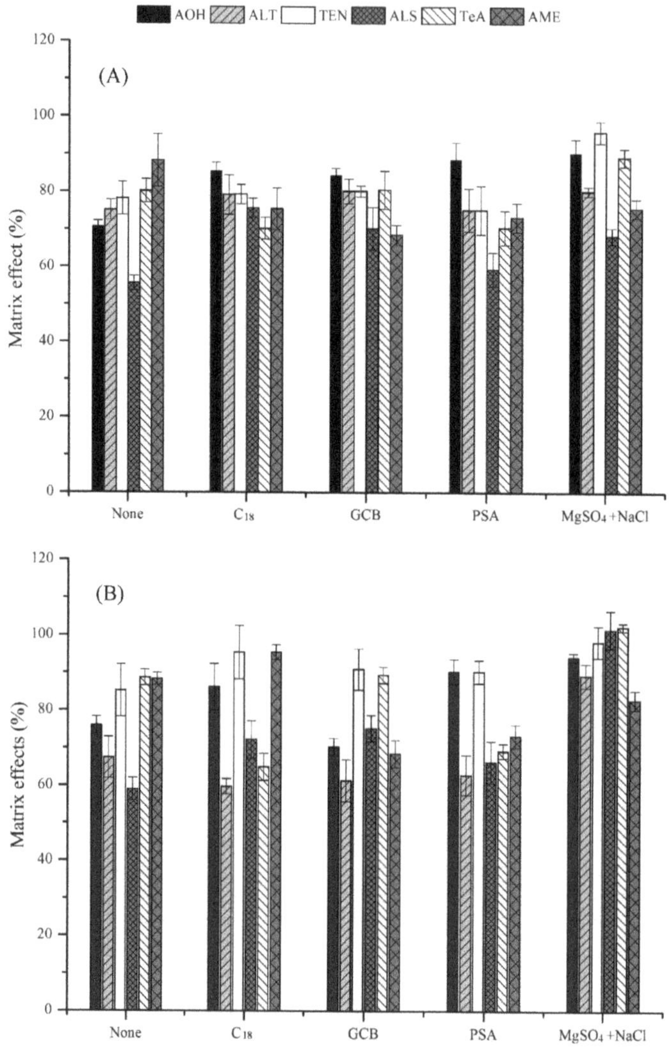

Figure 3. Matrix effects of the six *Alternaria* toxins in green grapes (**A**) and red grapes (**B**), purified by different materials (n = 6).

2.3. Method Validation

The linearity of the six analytes in neat solvent and in matrix is shown in Table 2. Good linear relationships with correlation coefficients $R^2 > 0.99$ were obtained. The limit of detection (LOD) and limit of quantification (LOQ) values were in the range of 0.03–0.21 μg kg^{-1} and 0.09–0.48 μg kg^{-1}, respectively. The recoveries and precisions for the six *Alternaria* toxins at the three fortified levels are listed in Table 3. The mean recovery values ranged from 78.4% to 101.6% for green grapes, and 77.8% to 100.1% for red grapes. Intra- and inter-day precision was in the range of 2.5–12.2% and 3.7–12.9% for green grapes, and 1.9–11.4% and 2.9–10.8% for red grapes. Overall, the validation data indicated that the accuracy, repeatability, and sensitivity of the proposed method were acceptable and in agreement with the requirements of European Commission Decision 2002/657/EC (EC 2002). The current method could be used for the accurate detection of six *Alternaria* toxins in grapes.

Table 2. Linearity, limit of detection (LOD) and limit of quantification (LOQ) of different *Alternaria* toxins.

Alternaria Toxins	Neat Solvent				Green Grape				Red Grape			
	Linear Range (ng mL^{-1})	*Correlation Coefficient* (R^2)	LOD (μg kg^{-1})	LOQ (μg kg^{-1})	Linear Range (ng mL^{-1})	*Correlation Coefficient* (R^2)	LOD (μg kg^{-1})	LOQ (μg kg^{-1})	Linear Range (ng mL^{-1})	*Correlation Coefficient* (R^2)	LOD (μg kg^{-1})	LOQ (μg kg^{-1})
AOH	0.1–200	0.996	-	-	0.2–200	0.997	0.08	0.18	0.3–200	0.994	0.12	0.29
AME	0.1–200	0.994	-	-	0.3–200	0.996	0.11	0.28	0.3–200	0.999	0.13	0.30
ALT	0.1–200	0.999	-	-	0.1–200	0.995	0.05	0.10	0.2–200	0.999	0.08	0.18
TEN	0.1–200	0.997	-	-	0.1–200	0.999	0.03	0.09	0.1–200	0.994	0.04	0.10
TeA	0.1–200	0.999	-	-	0.2–200	0.999	0.09	0.19	0.3–200	0.999	0.11	0.25
ALS	0.1–200	0.993	-	-	0.3–200	0.992	0.11	0.30	0.5–200	0.996	0.21	0.48

Table 3. Recoveries and precisions of the six *Alternaria* toxins in grapes ($n = 6$).

Alternaria Toxins	Spiked Levels (μg kg^{-1})	Green Grape			Red Grape		
		Recovery (Mean ± SD, %)	Intra-Day Precision (RSD, %)	Inter-Day Precision (RSD, %)	Recovery (Mean ± SD, %)	Intra-Day Precision (RSD, %)	Inter-Day Precision (RSD, %)
AOH	10	79.8 ± 8.1	2.5	5.1	81.4 ± 4.6	5.1	5.7
	50	100.4 ± 2.9	3.8	12.9	95.6 ± 5.7	2.8	3.4
	100	88.3 ± 4.4	11.9	3.7	88.2 ± 2.1	9.6	6.6
AME	10	92.1 ± 3.1	5.4	12.0	94.1 ± 1.9	4.9	7.4
	50	89.7 ± 4.9	2.7	5.1	94.6 ± 5.8	6.8	8.1
	100	101.6 ± 1.8	3.0	3.8	100.1 ± 3.3	5.1	3.8
ALT	10	78.4 ± 6.9	5.9	8.7	86.7 ± 5.7	3.8	10.4
	50	80.8 ± 5.4	12.2	5.4	93.4 ± 6.1	10.1	6.7
	100	82.9 ± 1.2	4.8	7.8	90.1 ± 2.1	2.8	5.5
TEN	10	92.1 ± 2.8	5.4	5.3	98.6 ± 0.9	9.9	6.9
	50	89.4 ± 4.5	4.7	6.5	94.5 ± 1.4	6.8	10.8
	100	93.0 ± 9.1	9.2	3.7	96.7 ± 2.7	2.7	5.6
TeA	10	79.5 ± 5.6	4.1	6.9	87.2 ± 7.5	1.9	3.9
	50	81.7 ± 4.9	2.9	4.1	77.8 ± 3.4	5.4	8.9
	100	100.8 ± 2.7	5.8	5.8	95.7 ± 5.5	3.8	4.1
ALS	10	79.2 ± 5.9	9.8	6.9	86.3 ± 1.1	11.4	2.9
	50	80.0 ± 5.1	6.4	7.1	79.8 ± 2.3	5.6	4.1
	100	81.5 ± 2.8	3.7	4.0	90.2 ± 1.9	8.7	8.1

2.4. Method Application

The validated method was applied to detect six *Alternaria* toxins in 56 grape samples randomly collected from the different markets and vineyards in Shanghai. MRM chromatograms of a typically contaminated grape sample are shown in Figure 4. The occurrences and concentration levels of the six *Alternaria* toxins are summarized in Table 4.

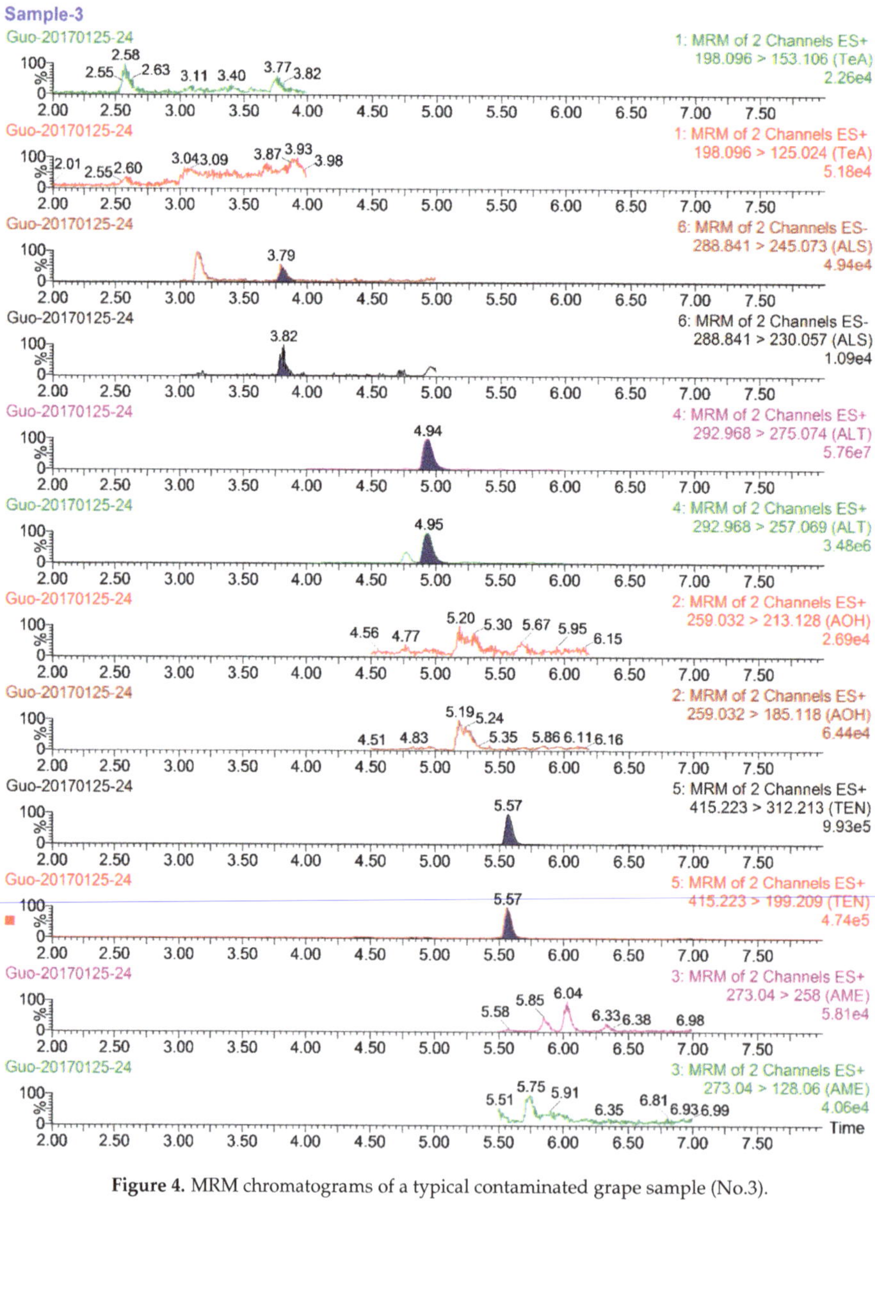

Figure 4. MRM chromatograms of a typical contaminated grape sample (No.3).

Table 4. Occurrence of the six *Alternaria* toxins in various grape samples.

Grape Variety	Total Amount	AOH		AME		ALT		TEN		TeA		ALS	
		Positive	Range ($\mu g\ kg^{-1}$)	Positive	Range ($\mu g\ kg^{-1}$)	Positive	Range ($\mu g\ kg^{-1}$)	Positive	Range ($\mu g\ kg^{-1}$)	Positive	Range ($\mu g\ kg^{-1}$)	Positive	Range ($\mu g\ kg^{-1}$)
Kyoho	6	0	nd [a]	0	nd	0	nd	0	nd	1	0.76	0	nd
Summer Black	16	2	0.12~7.15	0	nd	4	0.10~0.32	10	0.10~1.64	7	0.32~46.97	1	0.22
Shenhua	5	2	0.09~0.12	0	nd	0	nd	3	0.29~0.59	0	nd	0	nd
Hupei No.1	3	1	0.80	0	nd	0	nd	1	0.15	1	1.15	0	nd
Shenfeng	2	1	0.13	0	nd	0	nd	2	0.44~1.39	0	nd	0	nd
Muscat Hamburg	4	2	0.13~0.37	0	nd	0	nd	0	nd	1	0.60	0	nd
Shenyu	5	1	0.09	0	nd	0	nd	1	0.29	1	0.35	1	0.21
Zuijinxiang	10	4	0.11~0.28	2	0.11~0.15	1	0.53	2	0.28~0.55	3	0.25~4.39	1	0.42
Gold Finger	5	2	0.23~0.44	0	nd	1	0.18	2	0.17~0.31	2	0.38~4.61	0	nd

[a] nd = not detected.

Among the 56 samples, 40 (incidence of 71.4%) were contaminated *Alternaria* toxins. TEN was the most frequently found mycotoxin (incidence of 37.5%), with concentrations in the range of 0.10–1.64 μg kg^{-1}, followed by TeA (28.6%) and AOH (26.8%), with concentrations of 0.25–46.97 μg kg^{-1} and 0.09–7.15 μg kg^{-1}, respectively. ALT (10.7%), AME (3.6%), and ALS (5.4%) were also detected in some samples. The contamination situations of *Alternaria* toxins revealed in the current work were in great agreement with the microbial status reported in the previous studies, in which, the toxigenic fungi, especially *Alternaria* spp., that could produce ALT, AOH, AME and TeA, were recovered from grapes [25,26]. It is not surprising to find so many *Alternaria* toxins in grapes, since large amounts of these toxins have been detected in grape juice and red/white wine with the incidences of almost 100% [13,17,19,20,27]. Compared to the contaminations in grape derivatives, the obviously lower incidences of AOH, TeA, and TEN in the original fruit were possibly because the production of *Alternaria* toxins occurred during the processing and storage processes. It is noteworthy that the grapes are frequently contaminated with multiple *Alternaria* toxins, and there is a need to improve prevention and control strategies during pre- and post-harvest procedures.

3. Conclusions

An accurate and reliable UHPLC-MS/MS method based on a modified QuEChERS technique was developed for the simultaneous determination of six *Alternaria* toxins in grapes for the first time. The method was proven to be simple, efficient, and accurate after validation by the determination of linearity, accuracy, and precision, and is feasible in practical grape samples. The survey results strongly suggested that the grape is a favorable matrix for *Alternaria spp* producing *Alternaria* toxins, and emphasizes the necessity of the current established method, which could be used for continuous monitoring of *Alternaria* toxins and reducing the health risk to consumers in China.

4. Materials and Methods

4.1. Chemicals and Reagents

The analytical standards (stock solutions) of AOH (100.0 μg mL^{-1}), AME (100.3 μg mL^{-1}), TEN (100.4 μg mL^{-1}), and TeA (101.1 μg mL^{-1}), dissolved in acetonitrile, were purchased from Romer labs (Union, MO, USA). Solid portions of ALT (99.4%) and ALS (98.0%) standards were purchased from AdipoGen (Liestal, Basel, Switzerland). The chemical structures of the six *Alternaria* toxins are shown in Figure S3.

Acetonitrile and methanol (HPLC grade) from Merck (Darmstadt, Germany) were used. Anhydrous magnesium sulfate ($MgSO_4$, analytical grade), sodium chloride (NaCl, analytical grade), and ammonium acetate (HPLC grade) were supplied by ANPEL (Shanghai, China). Water used throughout the whole experiment was prepared by a Milli-Q system (Millipore, Billerica, MA, USA).

4.2. Preparation of Standard Solution

Solid portions of the ALT and ALS standards were dissolved in acetonitrile to prepare 100.0 μg mL^{-1} of stock solutions. A mixed standard solution of AOH, AME, ALT, ALS, TEN, and TeA with a concentration of 1 μg mL^{-1} was prepared in acetonitrile by diluting and mixing appropriate amounts of stock solutions of *Alternaria* toxins, and stored at −20 °C until use.

4.3. Samples Collection

A total of 56 grape samples, including 9 varieties (Kyoho, Summer Black, Shenhua, Hupei No.1, Shenfeng, Muscat Hamburg, Shenyu, Zuijinxiang, and Gold Finger) were randomly collected from different markets and vineyards in Shanghai. Approximately 0.5 kg of each sample was collected and mashed by a food processer (Midea, Guangdong, China). All samples were stored in a freezer at −20 °C until analysis.

4.4. Sample Preparation

The homogenized grape samples (2.0 g) were weighted into a 50 mL centrifuge tube and 10 mL of acetonitrile was added. The mixture was shaken at 200 rpm for 30 min. Subsequently, 0.5 g anhydrous magnesium sulfate and 0.5 g sodium chloride were added to the slurry and vigorously shaken for 30 s immediately. After centrifugation at 4500 rpm for 10 min, 5 mL of the supernatant was collected and evaporated under a soft stream of nitrogen gas at 40 °C. The residue was re-dissolved with 1 mL acetonitrile/water containing 5 mmol L^{-1} ammonium acetate (20/80 *v*/*v*), and filtered through a 0.22 μm PTFE membrane filter to be ready for analysis.

4.5. UHPLC–MS/MS Analysis

UHPLC analysis was performed on a Waters ACQUITY Ultra High-Performance LC system (Waters, Milford, MA, USA). Chromatographic separation was achieved on a Proshell EC_{18} column (50 mm × 2.1 mm, 2.7 μm). The mobile phase was consisted of methanol (A) and water containing 5 mmol L^{-1} ammonium acetate (B). A linear gradient elution program was set as follows: initial 10% A; 1 min, 10% A; 5 min, 90% A; 6 min, 90% A; 6.5 min, 10% A; 8 min, 10% A. The flow rate was 0.4 mL min^{-1}. The injection volume was 3 μL, and the column temperature was 35 °C.

For MS/MS detection, a Waters T-QS mass spectrometer system (Waters, Milford, MA, USA) was used both in positive electrospray ionization mode (ESI^+) and in negative electrospray ionization mode (ESI^-) with the following parameters: interface voltages of capillary, 2.5 kV(ESI^+) and 1.5 kV(ESI^-); desolvation temperature, 500 °C; and source temperature, 150 °C. The gas flow rates were 7.0 bar for nebulizing gas and 1000 L h^{-1} for desolvation gas, respectively. Multiple reaction monitoring (MRM) mode was used for the quantification and confirmation of the *Alternaria* toxins with the parameters shown in Table 1.

4.6. Method Validation

The proposed method was validated by determination of the linearity, sensitivity, recovery, precision, and matrix effect according to the recommendations of European Commission Decision 2002/657/EC [28]. Different concentrations (0.1, 0.2, 0.3, 0.5, 1, 5, 10, 50, 100, and 200 ng mL^{-1}) of *Alternaria* toxins were freshly prepared by diluting the working solution step by step with acetonitrile and a blank matrix, respectively. The calibration curves were constructed by plotting the responses versus analyte concentrations. The sensitivity was evaluated by determining the limit of detection (LOD) and limit of quantification (LOQ), which were designed as the concentrations of the toxins that produced signal-to-noise ratios (S/N) of 3 and 10 in matrix, respectively. The recoveries were tested using non-contaminated grape samples spiked with low, intermediate, and high concentration levels (10, 50, and 100 μg kg^{-1}) of *Alternaria* toxins. The intra- and inter-day precisions were evaluated through the relative standard deviations (RSDs), using the non-contaminated samples spiked with different concentrations (10, 50, and 100 μg kg^{-1}) of *Alternaria* toxins in the same day and in five successive days, respectively. All experiments were performed in sextuplicate.

Signal suppression and enhancement (SSE) was used to evaluate the matrix effect, which was calculated according to Equation (1) [21]:

$$\text{SSE }(\%) = 100 \times \text{slope}_{\text{matrix}} / \text{slope}_{\text{solvent}} \quad (1)$$

where slope $_{\text{matrix}}$ is the slope of matrix-matched calibration curve, and slope $_{\text{solvent}}$ is the slope of standard calibration curve.

Supplementary Materials: The following are available online at http://www.mdpi.com/2072-6651/11/2/87/s1, Figure S1: MS/MS spectra for AOH with the collision energy of 28 eV (A), for AME with the collision energy of 26 eV (B), for ALT with the collision energy of 14 eV (C), for ALS with the collision energy of −20 eV (D), for TEN with the collision energy of 12 eV (E) and for TeA with the collision energy of 16 eV (F). The concentration for all *Alternaria* toxins was 200 ng mL^{-1}. Figure S2: Recoveries of the six *Alternaria* toxins in a standard solution filtered by different membrane filters. The concentration is 50 μg kg^{-1} (n = 6). Figure S3: Chemical structures of the six *Alternaria* toxins.

Author Contributions: Z.H. and W.G. conceived and designed the experiments; W.G. and K.F. performed the experiments; D.N., E.K.T. and J.Y. interpreted the results; J.M., Y.S., and Q.H. contributed to the sample preparation; Z.H., Y.W., Z.Z., and W.G. wrote the manuscript.

Funding: This work was financially supported by the National Natural Science Foundation of China [No. 31671950], the Shanghai Municipal Commission of Science and Technology (18395810100), the Youth Talent Development Plan of Shanghai Agriculture Committee of China [Grant No. 2017(1-30)] and [Grant No. 2016(1-7)], and the Shanghai Rising-Star Program (17QB1403100).

Conflicts of Interest: The authors declare no conflict of interest.

References

1. Huang, F.; Fu, Y.; Nie, D.; Stewart, J.E.; Peever, T.L.; Li, H. Identification of a novel phylogenetic lineage of *Alternaria alternata* causing citrus brown spot in China. *Fungal Biol.* **2014**, *119*, 320–330. [CrossRef] [PubMed]
2. Melki, B.F.S.; Chebil, S.; Lebrihi, A.; Lasram, S.; Ghorbel, A.; Mliki, A. Occurrence of pathogenic fungal species in Tunisian vineyards. *Int. J. Food Microbiol.* **2007**, *113*, 245–250. [CrossRef] [PubMed]
3. Fraeyman, S.; Croubels, S.; Devreese, M.; Antonissen, G. Emerging fusarium and *Alternaria* mycotoxins: Occurrence, toxicity and toxicokinetics. *Toxins* **2017**, *9*, 228. [CrossRef]
4. Ostry, V. *Alternaria* mycotoxins: an overview of chemical characterization, producers, toxicity, analysis and occurrence in foodstuffs. *World Mycotoxin J.* **2008**, *1*, 175–188. [CrossRef]
5. Puppel, N.; Tjaden, Z.; Fueller, F.; Marko, D. DNA strand breaking capacity of acrylamide and glycidamide in mammalian cells. *Mutat. Res.* **2005**, *580*, 71–80. [CrossRef] [PubMed]
6. Solhaug, A.; Torgersen, M.L.; Holme, J.A.; Lagadic-Gossmann, D.; Eriksen, G.S. Autophagy and senescence, stress responses induced by the DNA-damaging mycotoxin alternariol. *Toxicology* **2014**, *326*, 119–129. [CrossRef] [PubMed]
7. Oviedo, M.S.; Ramirez, M.L.; Barros, G.G.; Chulze, S.N. Effect of environmental factors on tenuazonic acid production by alternaria alternata on soybean-based media. *J. Appl. Microbiol.* **2010**, *107*, 1186–1192. [CrossRef]
8. Solhaug, A.; Eriksen, G.S.; Holme, J.A. Mechanisms of action and toxicity of the mycotoxin Alternariol: A review. *Basic Clin. Pharmacol.* **2016**, *119*, 533–539. [CrossRef]
9. Lehmann, L.; Wagner, J.; Metzler, M. Estrogenic and clastogenic potential of the mycotoxin alternariol in cultured mammalian cells. *Food Chem. Toxicol.* **2006**, *44*, 398–408. [CrossRef]
10. European Food Safety Authority. Scientific Opinion on risks for animal and public health related to the presence of nivalenol in food and feed. *Efsa J.* **2013**, *11*, 3262. [CrossRef]
11. Rychlik, M.; Lepper, H.; Weidner, C.; Asam, S. Risk evaluation of the *Alternaria* mycotoxin tenuazonic acid in foods for adults and infants and subsequent risk management. *Food Control* **2016**, *68*, 181–185. [CrossRef]
12. Somma, S.; Perrone, G.; Logrieco, A.F. Diversity of black *Aspergilli* and mycotoxin risks in grape, wine and dried vine fruits. *Phytopathol. Mediterr.* **2012**, *51*, 131–147.
13. Scott, P.M.; Lawrence, G.A.; Lau, B.P. Analysis of wines, grape juices and cranberry juices for *Alternaria* toxins. *Mycotoxin Res.* **2006**, *22*, 142–147. [CrossRef] [PubMed]
14. Wei, D.; Wang, Y.; Jiang, D.; Feng, X.; Li, J.; Wang, M. Survey of *Alternaria* toxins and other mycotoxins in dried fruits in China. *Toxins* **2017**, *9*, 200. [CrossRef] [PubMed]
15. Matysik, G.; Giryn, H. Gradient thin-layer chromatography and densitometry determination of *Alternaria* mycotoxins. *Chromatographia* **1996**, *42*, 555–558. [CrossRef]
16. Harvan, D.J.; Pero, R.W. Gas chromatographic analysis of the *Alternaria* metabolite, tenuazonic acid. *J. Chromatogr. A* **1974**, *101*, 222–224. [CrossRef]

17. Fan, C.; Cao, X.; Liu, M.; Wang, W. Determination of *Alternaria* mycotoxins in wine and juice using ionic liquid modified countercurrent chromatography as a pretreatment method followed by high-performance liquid chromatography. *J. Chromatogr. A* **2016**, *1436*, 133–140.
18. Müller, M.E.H.; Korn, U. *Alternaria* mycotoxins in wheat—A 10 years survey in the Northeast of Germany. *Food Control* **2013**, *34*, 191–197. [CrossRef]
19. Lau, B.P.Y.; Scott, P.M.; Lewis, D.A.; Kanhere, S.R.; Chantal, C.; Roscoe, V.A. Liquid chromatography-mass spectrometry and liquid chromatography-tandem mass spectrometry of the altemaria mycotoxins alternariol and alternariol monomethyl, ether in fruit juices and beverages. *J. Chromatogy A* **2003**, *998*, 119–131. [CrossRef]
20. Zwickel, T.; Klaffke, H.; Richards, K.; Rychlik, M. Development of a high performance liquid chromatography tandem mass spectrometry based analysis for the simultaneous quantification of various *Alternaria* toxins in wine, vegetable juices and fruit juices. *J. Chromatogr. A* **2016**, *1455*, 74–85.
21. Azaiez, I.; Giusti, F.; Sagratini, G.; Mañes, J.; Fernández-Franzón, M. Multi-mycotoxins analysis in dried fruit by LC/MS/MS and a modified QuEChERS procedure. *Food Anal. Methods* **2014**, *7*, 935–945. [CrossRef]
22. Pizzutti, I.R.; Kok, A.D.; Scholten, J.; Righi, L.W.; Cardoso, C.D.; Rohers, G.N.; Silva, R.C.D. Development, optimization and validation of a multimethod for the determination of 36 mycotoxins in wines by liquid chromatography–tandem mass spectrometry. *Talanta* **2014**, *129*, 352–363. [CrossRef] [PubMed]
23. Tamura, M.; Uyama, A.; Mochizuki, N. Development of a multi-mycotoxin analysis in beer-based drinks by a modified QuEChERS method and ultra-high-performance liquid chromatography coupled with tandem mass spectrometry. *Anal. Sci.* **2011**, *27*, 629–635. [CrossRef] [PubMed]
24. Dong, H.; Xian, Y.; Xiao, K.; Wu, Y.; Zhu, L.; He, J. Development and comparison of single-step solid phase extraction and QuEChERS clean-up for the analysis of 7 mycotoxins in fruits and vegetables during storage by UHPLC-MS/MS. *Food Chem.* **2019**, *274*, 471–479. [CrossRef] [PubMed]
25. Tančinová, D.; Mašková, Z.; Rybárik, L'.; Felšöciová, S.; Císarová, M. Colonization of grapes berries by alternaria sp. and their ability to produce mycotoxins. *Potravinarstvo* **2016**, *10*, 7–13. [CrossRef]
26. Prendes, L.P.; Merín, M.G.; Andreoni, M.A.; Ramirez, M.L.; de Ambrosini, V.I.M. Mycobiota and toxicogenic alternaria spp. strains in malbec wine grapes from doc san rafael, mendoza, argentina. *Food Control* **2015**, *57*, 122–128. [CrossRef]
27. Myresiotis, C.K.; Testempasis, S.; Vryzas, Z.; Karaoglanidis, G.S.; Papadopoulou-Mourkidou, E. Determination of mycotoxins in pomegranate fruits and juices using a QuEChERS-based method. *Food Chem.* **2015**, *182*, 81–88. [CrossRef]
28. Muscarella, M.; Lo, M.S.; Palermo, C.; Centonze, D. Validation according to European Commission Decision 2002/657/EC of a confirmatory method for aflatoxin M1 in milk based on immunoaffinity columns and high performance liquid chromatography with fluorescence detection. *Anal. Chim. Acta* **2007**, *594*, 257–264. [CrossRef]

toxins

MDPI

Article

Ultra-High-Performance Liquid Chromatography Coupled with Quadrupole Orbitrap High-Resolution Mass Spectrometry for Multi-Residue Analysis of Mycotoxins and Pesticides in Botanical Nutraceuticals

Alfonso Narváez [1], Yelko Rodríguez-Carrasco [2,*], Luigi Castaldo [1,3], Luana Izzo [1] and Alberto Ritieni [1]

1 Department of Pharmacy, University of Naples "Federico II", Via Domenico Montesano 49, 80131 Naples, Italy; alfonso.narvaezsimon@unina.it (A.N.); luigi.castaldo2@unina.it (L.C.); luana.izzo@unina.it (L.I.); ritialb@unina.it (A.R.)

2 Laboratory of Food Chemistry and Toxicology, Faculty of Pharmacy, University of Valencia, Av. Vicent Andrés Estellés s/n, Burjassot, 46100 València, Spain

3 Department of Clinical Medicine and Surgery, University of Naples "Federico II", Via S. Pansini 5, 80131 Naples, Italy

* Correspondence: yelko.rodriguez@uv.es

Received: 24 January 2020; Accepted: 10 February 2020; Published: 12 February 2020

check for updates

Abstract: Cannabidiol (CBD) food supplements made of *Cannabis sativa* L. extracts have quickly become popular products due to their health-promoting effects. However, potential contaminants, such as mycotoxins and pesticides, can be coextracted during the manufacturing process and placed into the final product. Accordingly, a novel methodology using ultra-high-performance liquid chromatography coupled with quadrupole Orbitrap high-resolution mass spectrometry (UHPLC-Q-Orbitrap HRMS) was developed to quantify 16 mycotoxins produced by major *C. sativa* fungi, followed by a post-target screening of 283 pesticides based on a comprehensive spectral library. The validated procedure was applied to ten CBD-based products. Up to six different *Fusarium* mycotoxins were found in seven samples, the most prevalent being zearalenone (60%) and enniatin B1 (30%), both found at a maximum level of 11.6 ng/g. Co-occurrence was observed in four samples, including one with enniatin B1, enniatin A and enniatin A1. On the other hand, 46 different pesticides were detected after retrospective analysis. Ethoxyquin (50%), piperonyl butoxide (40%), simazine (30%) and cyanazine (30%) were the major residues found. These results highlight the necessity of monitoring contaminants in food supplements in order to ensure a safe consumption, even more considering the increase trend in their use. Furthermore, the developed procedure is proposed as a powerful analytical tool to evaluate the potential mycotoxin profile of these particular products.

Keywords: mycotoxins; pesticides; Q-Exactive Orbitrap; CBD capsule; nutraceutical

Key Contribution: The first multi-class analysis of CBD-based supplements regarding mycotoxins and pesticide residues using high-resolution mass spectrometry techniques.

1. Introduction

Nutrition is known to be an essential component of the health state, so having an unbalanced diet can lead to several disorders and diseases [1]. Due to current lifestyles, new and fast ways to maintain proper dietary habits are required. Nutraceuticals have emerged as an alternative to increase the input of nutrients, contributing to an improvement in health. These products are bioactive compounds naturally occurring in food or produced de novo in human metabolism, biologicals or botanicals, each intended to impart a physiological or medicinal effect after ingestion [2]. They can be delivered either in foods and beverages or in other non-conventional forms, such as capsules, tablets, powders or liquid extracts. In terms of marketing, nutraceuticals include a large number of different products packaged for specific groups by age, gender, physical conditions and activity level. The global market was valued at US$109 billion in 2015 and is projected to reach US$180 billion by 2020 [3].

Inside the variety of products classified as nutraceuticals, food supplements based on botanical ingredients represent the second largest segment, behind vitamins and minerals. Most recently, cannabidiol (CBD) dietary supplements made of *Cannabis sativa* L. extracts have quickly become popular products. CBD is a phytocannabinoid present in the resin secreted from trichomes in female *C. sativa* plants, and is mainly found in inflorescences. The bioactivity of this compound has been related to an enhancement of its antioxidant and neurological activity, among others, by the promotion of several metabolic pathways [4–6]. However, the European Union (EU) does not consider CBD supplements as a novel food [7] and lets member states set their own rules over its marketing, leading to a convoluted situation in terms of regulation. Despite several ambiguities in its legislation, the European market for CBD-based supplements was valued at US$318 million in 2018 and with a strong growth projection [8].

Due to the complex nature of *C. sativa* and other botanicals, potential contaminants can be coextracted during the different stages of the manufacturing process and placed into the final product. Among all the potential non-desirable compounds in herbal-based supplements, mycotoxins and pesticides are the most commonly reported [9,10]. Mycotoxins are secondary metabolites mainly produced by the fungi genera *Fusarium*, *Aspergillus*, *Penicillium*, *Claviceps* and *Alternaria*. These compounds can be present in food and feed commodities and display immunosuppressive, nephrotoxic or carcinogenic effects, among others [11]. According to their carcinogenic potential, some mycotoxins, like aflatoxins, have been included in the classification list of human carcinogens provided by the International Agency for Research on Cancer (IARC) [12]. These mycotoxins are produced by the genera *Aspergillus*, which has been categorized as a major fungus occurring in *C. sativa* inflorescences alongside other mycotoxin producing fungi, like *Fusarium* spp., so different mycotoxins could be also expected [13,14]. On the other hand, pesticides include a broad range of compounds routinely applied to protect crops from different pests. However, residues coming from these products can accumulate in plants intended for human consumption, leading to several health issues related to neurotoxicity, carcinogenicity and pulmonotoxicity, as well as developmental and reproductive disorders [15–18].

In terms of regulation, maximum residue limits (MRL) for different types of contaminants have been set by the EU. Regulation (EC) No. 396/2005 [19] establishes limits for pesticides, whereas Regulation (EC) No. 1881/2006 [20] covers mycotoxins, attaching maximum limits in food and feeds. Nevertheless, nutraceutical products are not considered by the legislation yet but, due to a potential carryover during the manufacturing process, contamination could be expected not only in raw material, but also in other by-products. Moreover, several studies have reported the sole presence of pesticides [21,22], mycotoxins [23,24] and both types of contaminants [25–27] in diverse food supplements, remarking the necessity to evaluate the contamination profile of these products considering their rising consumption and popularity.

To overcome this point, the development of analytical procedures is needed. Concerning the extraction of contaminants, QuEChERS (quick, easy, cheap, effective, rugged and safe) [21,23,24] and "dilute and shoot" procedures have been recently applied to food supplements delivered as gelatin capsules, traditional capsules, tablets, powder extracts or liquid presentations [25–27]. Analytical methods used in the detection and quantification of contamination include ELISA detection [28], gas chromatography (GC) coupled with mass spectrometry (MS) [22] and ultra-high-performance liquid chromatography (UHPLC) coupled with tandem mass spectrometry (MS/MS) [23,24] and high-resolution Orbitrap mass spectrometry (Q-Orbitrap HRMS) [25–27]. Due to its high resolving power, sensitivity and accurate mass measurement, high-resolution mass spectrometry stands as a suitable alternative for evaluating a large number of contaminants present in complex matrices at low concentrations. Therefore, the aim of the present study was to provide an analysis of pesticide residues and mycotoxins produced by major *C. sativa* fungi occurring in CBD-based food supplements, using ultra-high-performance liquid chromatography coupled with high-resolution Orbitrap mass spectrometry. To achieve this, a novel methodology was developed in order to identify and quantify 16 mycotoxins after evaluating different extraction procedures, followed by a post-target screening of 283 pesticides based on a comprehensive spectral library. To the best of the authors' knowledge, this is the first multi-class analysis of CBD-based supplements through the use of high-resolution mass spectrometry techniques.

2. Results

2.1. Optimization of Extraction Procedure

The molecular complexity of this matrix demands an effective extraction in order to detect and quantify several mycotoxins in a reliable way. A QuEChERS methodology previously developed on this typology of sample [24] was selected as the starting point, whereas different volumes of extraction solvent and the type of sorbent for clean-up was tested.

2.1.1. Evaluation of the Volume of Extraction Solvent

The extraction procedure was first evaluated in triplicate by spiking the sample at 10 ng/g using the following volumes of extraction solvent per gram of sample: 2.5, 5, 7.5 and 10 mL.

The extraction performed with 2.5 mL showed recovery values below the minimum limit (70%) for the vast majority of the studied analytes as a consequence of solvent saturation (Figure 1a). Satisfactory recoveries (70–120%) were obtained after performing the extraction with 5 mL of solvent for the majority of compounds, with the exception of β-ZEL (155%) and ZAN (150%), which were significantly more efficient than the other volumes tested ($p < 0.05$). On the other hand, the extractions performed with 7.5 and 10 mL showed a gradual decrease in recoveries due to the larger dilution of the analytes. Therefore, 5 mL of AcN was selected as the optimal volume of extraction solvent for this type of CBD capsule.

2.1.2. Evaluation of the Type of Sorbent for Clean-Up

The molecular composition of the soft gel capsules mainly consists on fatty acids and proteins. Because of the complex nature of this matrix, an efficient clean-up is required in order to avoid interference with the analytes. To achieve this, clean-up with different sorbents (100 mg), including C18, as previously suggested [24], GCB, Z-Sep+ and PSA was performed.

PSA exhibited a good performance for the vast majority of analytes (Figure 1b) but was unable to recover other important mycotoxins, such as AFB1 and AFG1. The moderate affinity of PSA with polar compounds may explain low recoveries for aflatoxins, being consistent with other works based on oily matrices [29,30]. Similarly, extraction with C18 was efficient for most compounds and only some low-polarity mycotoxins showed recoveries out of the range set, like ZAN (150%) and β-ZEL (155%). Clean-up using GCB showed poor results, allowing us to detect only NEO (85%), HT-2 (89%) and T-2 (89%). This sorbent is able to retain planar molecules and mycotoxin adsorption has been previously reported [31], which might be the reason for the low recoveries obtained here. Finally, extraction performed with Z-Sep+ showed satisfactory recoveries (70–120%) for all the mycotoxins studied.

On the other hand, the influence of the matrix was minimal (80% ≤ signal suppression/enhancement (SSE) ≤ 120%) for all targeted analytes when using Z-Sep+ and PSA. Clean-up based on Z-Sep+ has been successfully applied to the extraction of analytes from lipid matrices [32,33]. Furthermore, Z-Sep+ is also able to form irreversible links with carboxylic groups present in proteins [34], standing as the most suitable sorbent for the here-analyzed matrix. Similarly, the use of PSA has been suggested to remove coextracted fatty acids and other ionic lipids [35].

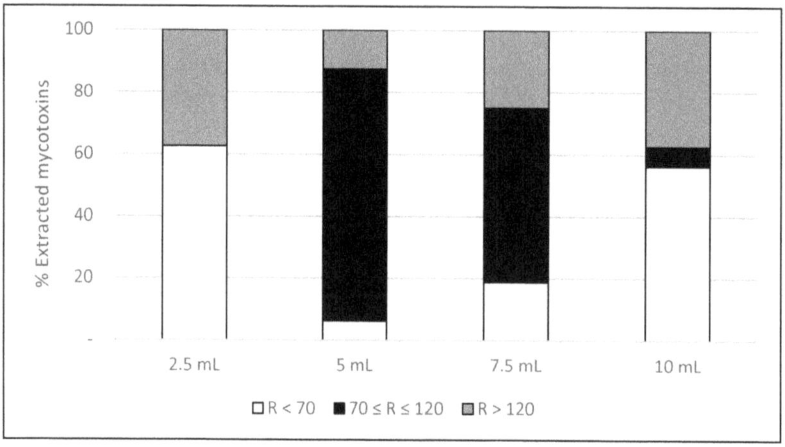

(**a**)

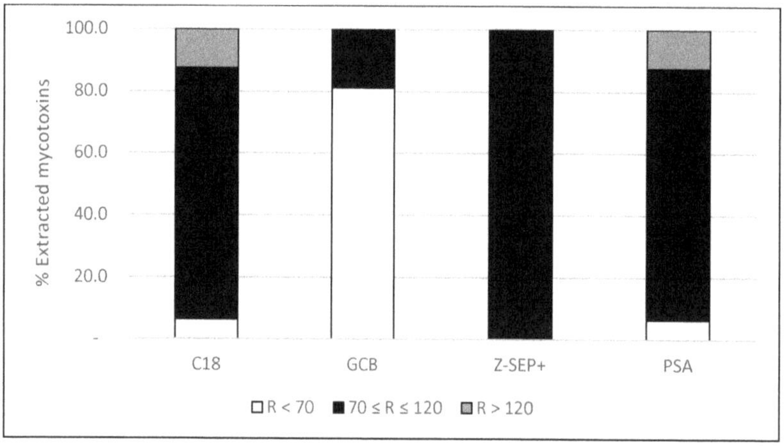

(**b**)

Figure 1. Percentage of mycotoxins extracted with a recovery value (R) below 70% (white), between 70% and 120% (black) and above 120% (grey), corresponding to extractions performed with: (**a**) different volumes of solvent at a spiking level of 10 ng/g; (**b**) different sorbents for clean-up at a spiking level of 10 ng/g.

On the contrary, a strong matrix effect was evidenced for half the analytes when using C18 and GCB. Signal suppression was detected after using C18, obtaining SSE ranging from 40% to 69%, whereas signal enhancement occurred after GCB clean-up, with SSE increasing from 128% to 167%. Since both sorbents have a preferential affinity for non-polar compounds, matrix interferents were not fully removed but coextracted. The presence of these coextracted species can change the ionization efficiency, leading to improper SSE and preventing a reliable quantification. Although no significant differences were observed between the use of Z-Sep+ and PSA ($p > 0.05$), Z-Sep+ was chosen because of its better performance minimizing matrix interference.

2.2. Analytical Method Validation

The optimized method was validated for the simultaneous extraction of 16 mycotoxins in CBD-based products. Results are shown in Table 1. Good linearity was observed for all analytes in the range assessed (0.20–100 ng/g), with regression coefficients (r^2) above 0.990 and a deviation ≤20% for each level of the calibration curve. Comparison between calibration curves built in a blank matrix and in neat solvent showed a minimal interference in the matrix (±20%) for the studied analytes. Hence, external calibration curves were used for quantification purposes. Limits of quality (LOQs) obtained for all studied analytes were between 0.20 and 6.25 ng/g. Regarding trueness, recovery values corresponding to a fortification level of 20 ng/g ranged between 63 and 103% and between 63 and 113% for the lowest fortification level (10 ng/g). Referring to the additional spiking level (2 ng/g) for aflatoxins, recoveries ranged between 63% and 86%. Precision study revealed both RSD_r and RSD_R values below 20% for all the mycotoxins analyzed. These results confirmed that the optimized procedure is suitable for a reliable quantification of the mycotoxins analyzed, fulfilling the criteria set by Commission Decision 2002/657/EC [36]. Table 2 reviews the available literature regarding mycotoxins in herbal-based supplements. As shown, the here-obtained LOQs were lower than the ones reported in previous studies using UHPLC-Q-Orbitrap HRMS. As established by Regulation (EC) No. 1881/2006 [20], maximum limits for aflatoxins in many food matrices must not reach levels which are below those LOQs (5 ng/g), whereas LOQs obtained in this study were between 5 and 25 times lower. Other analytical methods based on low resolution mass spectrometry [37] required longer and more complicated extraction procedures than the QuEChERS developed here. Even ELISA detection has been used for quantification of mycotoxins in medicinal herbs [28], but a very specific extraction had to be performed for different groups of analytes using several multi-functional columns. The QuEChERS procedure developed in this study, in combination with UHPLC-Q-Orbitrap mass spectrometry, was extremely simple and reliable, allowing for the quantification of all mycotoxins with high sensitivity.

Table 1. Method performance: linearity, matrix effect (SSE %), recovery and LOQ.

Analyte	Linearity (r^2)	SSE (%)	Recovery (%)				Precision (%) [RSD_r, (RSD_R)]				LOQ (ng/g)
			2 ng/g [1]	10 ng/g	20 ng/g	50 ng/g	2 ng/g [1]	10 ng/g	20 ng/g	50 ng/g	
AFG2	0.9975	111	78	77	81	98	16 (19)	5 (6)	6 (6)	5 (4)	0.78
AFG1	0.9982	106	81	86	86	105	12 (9)	16 (19)	7 (6)	11 (10)	1.56
AFB1	0.9984	115	71	91	98	107	14 (13)	10 (8)	4 (4)	4 (3)	0.20
AFB2	0.9998	111	86	88	91	103	18 (15)	10 (8)	7 (5)	5 (4)	0.20
NEO	0.9988	112		88	93	104		18 (14)	16 (18)	17 (18)	0.78
HT-2	0.9984	108		113	101	92		12 (14)	16 (11)	12 (15)	6.25
T-2	0.9990	83		89	98	110		19 (13)	9 (7)	7 (10)	0.78
α-ZEL	0.9943	81		81	94	100		11 (11)	10 (14)	5 (16)	6.25
β-ZEL	0.9985	84		106	103	89		8 (18)	15 (16)	9 (11)	3.13
ZAN	0.9992	108		111	100	105		15 (13)	18 (11)	5 (13)	1.56
ZEN	0.9991	109		104	103	93		5 (16)	15 (14)	10 (19)	3.13
ENN B	0.9998	102		63	63	65		18 (19)	18 (18)	6 (7)	6.25
ENN B1	0.9982	99		83	89	85		12 (11)	8 (6)	8 (8)	1.56
ENN A	0.9942	84		96	91	80		11 (9)	14 (17)	11 (12)	3.13
ENN A1	0.9972	87		92	101	90		12 (14)	9 (6)	7 (14)	1.56
BEA	0.9971	119		80	71	63		18 (17)	10 (18)	10 (19)	6.25

[1] Additional fortification level only for AFs.

2.3. Application to Commercial CBD-Based Products

The validated UHPLC-Q-Orbitrap HRMS procedure was applied to ten commercially available samples in order to evaluate the occurrence of mycotoxins. Results are shown in Table 3. A considerable occurrence of mycotoxins was observed, since contamination with at least one analyte was found in 70% of the samples. Up to six different mycotoxins (T-2, ZAN, ZEN, ENNB1, ENNA, ENNA1) were quantified at a range from below LOQ to 11.6 ng/g, all produced by *Fusarium* genera, reported as a major *C. sativa* pathogen fungus [14]. Previous studies regarding mycotoxins in different herbal-based extracts have revealed the occurrence of similar mycotoxins independently of the matrix and the dosage form (Table 2). Despite the fact that the percentage of positive samples varied among the different studies (19–99%), when the sensitivity of the analytical method increased, reaching lower LOQs, the number of positive samples dramatically increased. This indicated that mycotoxin contamination in herbal-based products at low levels is frequent.

In the here-analyzed samples, ZEN appeared to be the most common mycotoxin, with an incidence of 60% and concentration levels ranging from 4.2 to 11.6 ng/g (mean level = 6.9 ng/g). A high incidence of ZEN has also been previously reported in supplements made of different herbals from Czech and US retail markets (84%, n = 69) at a wide range of concentrations (5–824 ng/g, mean value = 75.7 ng/g) [24]. Moreover, ZEN was previously found in 96% of medicinal herbals from Spain (n = 84) as well, but in a tighter range (1–44.1 ng/g, mean value = 8.9 ng/g) [28].

Referring to T-2, results reported contamination in one sample at 2.0 ng/g, in contrast with the prevalent presence of T-2 in 78% (n = 69) of the same Czech and US samples, at concentrations rising from 69 to 1,870 ng/g (mean value = 162 ng/g) [24]. High levels of T-2 were also observed in milk thistle samples from Spain (363–453.9, mean value = 408.9 ng/g) in only two out of seven samples [38]. In the other hand, T-2 was quantified in 98% (n = 84) of the Spanish medicinal herbals, but in much lower concentrations (0.6–256 ng/g, mean value = 22.645 ng/g) [28].

Similarly, ZAN was quantified in one sample at 1.9 ng/g. This mycotoxin has been scarcely targeted in dietary supplement studies, but has been previously quantified at similar concentrations as those here-reported in two samples of Chinese medicinal herbals (n = 33) [39].

Results also showed ENN contamination. ENNB1, ENNA and ENNA1 were found in the same sample at 11.6, 4.2 and 5.8 ng/g, respectively, whereas ENNB1 was detected in two other samples below the LOQ (1.56 ng/g). These emerging *Fusarium* mycotoxins have been previously found in herbal products (84–91%, n = 69) widely ranging from 5 ng/g up to 10,900 ng/g (mean value = 354 ng/g) [24]. Similarly, ENNB1 was the most common toxin out of these emerging *Fusarium* mycotoxins, being consistent with the results here obtained.

All the mycotoxins found in the present study correspond to low- to non-polar compounds, which should be prevalently expected due to the nature of the matrix.

Co-occurrence of at least two mycotoxins was also observed in four out of ten samples. Results showed the presence of ZEN in combination with ENNs B1, A and A1, ZAN or T-2, which are common associations found by previous studies in herbal-based supplements [24,28]. It must be highlighted that synergic or additive effects have been observed as a consequence of these combinations in in vitro assays [40]. Based on what has been discussed and considering the uprising trend of *C. sativa*-based products, alongside the use of environment-friendly raw materials cultivated without pesticides, quality controls regarding mycotoxins should be set for these products in order to ensure safe consumption.

Table 2. Available methods for measurement of mycotoxins in herbal-based supplements [1].

Samples Procedence (no.)	Positives Samples (%)	Major Analytes Detected	Concentration Reported (ng/g)	Determination		
				Sensitivity (LOQ, ng/g)	Detection Method	Reference
Medicinal or aromatic herbs (84)	99	ZEN T-2 DON[3] CIT[3]	1.0–44.1 0.6–256.9 20.5–343.5 14.9–354.8	0.14 0.28 14.8 16.5	ELISA detection (EIA reader, SIRIO S)	[28]
Traditional Chinese herbs (60)	83	ZEN AFs[3] MPA[3]	2.1–15.5 0.2–19.5 0.2–22.7	0.4 0.1 0.02	QQQ (Applied Biosystems) ESI+ MRM mode	[37]
Milk thistle (83)	19	AFB1	0.04–1.9	0.03	LC-FLD (Waters)	[41]
Green coffee bean (50)	36	OTA[3] OTB[3] FB1[3] MPA	1–136.9 1–20.2 50–415 5–395	2.5 2.5 100 10	QQQ (AB SCIEX) ESI$^+$ and ESI$^-$ MRM mode	[23]
Milk thistle (7)	29	T-2 HT-2	363–453.9 826.9–943.7	30.5 43.8	QQQ (AB SCIEX) ESI+ MRM mode	[38]
Herbals (69)	96	ZEN T-2 HT-2 ENNB ENNB1 ENNA ENNA1	5–824 69–1870 59–1530 5–9260 5–10,900 5–8340 5–2340	10 10 50 5 5 5 5	QQQ (AB SCIEX) ESI$^+$ and ESI$^-$ MRM mode	[24]
Gingko biloba (8)	50	AFB1 AFB2 T-2	5.0–54 4–300 18–20	5 10 30.5	Q-Orbitrap (Exactive, Thermo FisherScientific) ESI+ and ESI- HRMS	[25]
Green tea (10) Royal jelly (8)	10 0	AFB1	5.4	5	Q-Orbitrap (Exactive, Thermo FisherScientific) ESI+ and ESI- HRMS	[26]
Soy (11)	27	AFB1	8.2–17.1	5	Q-Orbitrap (Exactive, Thermo FisherScientific) ESI+ and ESI- HRMS	[27]
		AFG2	6.4	5		
Cannabis sativa (10)	70	ZEN ENNB1	4.2–11.6 <LOQ–11.6	3.13 1.56	Q-Orbitrap (Exactive, Thermo FisherScientific) ESI+ and ESI- HRMS	Current study

[1] ESI+ = positive ion mode; ESI− = negative ion mode; HRMS = high-resolution MS; LOQ = limit of quantification; MRM = multiple reaction monitoring; QQQ = triple quadrupole. [2] Range of LOQs referring to the analyzed mycotoxins. [3] AFs = aflatoxins; DON = deoxynivalenol; CIT = citrinin; FB1 = fumonisin B1; MPA = mycophenolic acid; OTA = ochratoxin A; OTB = ochratoxin B.

Table 3. Occurrence of studied mycotoxins in the analyzed samples.

Sample	Mycotoxin (ng/g)					
	T-2	ZAN	ZEN	ENN B1	ENN A	ENN A1
1			11.6	11.6	4.2	5.8
4			6.5			
5				<LOQ		
7			8.1			
8		1.9	4.7			
9			4.2	<LOQ		
10	2.0		6.3			

2.4. *Identification of Non-Target Compounds through Retrospective Analysis in Studied Samples*

The post-target screening approach allowed us to detect pesticide residues in the analyzed samples using a spectral library. Results are shown in Figure 2. Up to 46 different pesticides were tentatively identified based on the pesticides mass spectral library. Ethoxyquin was putatively found in five samples, being the most prevalent pesticide. The main function of ethoxyquin is to avoid fungal contamination during the postharvest stage of the plant through its scald-preventive properties [42]. Surprisingly, the use of this pesticide is forbidden by the European Commission (EC) Decision 2011/143/EU. Piperonyl butoxide was found in four samples. This compound is not a pesticide by itself but can inhibit the resistance mechanisms of insects, being widely used in combination with other different pesticides [42]. The tentative presence of cyanazine and simazine, both found in three different samples, must also be noted. The use of these pesticides was prohibited by EC Regulation No. 1107/2009 [43] and Commission Decision 2004/247/EC [44], respectively. Therefore, the occurrence of forbidden pesticides found in the here-analyzed samples highlights the necessity of monitoring potential contaminants in *C. sativa*-derived products acquired from online shops.

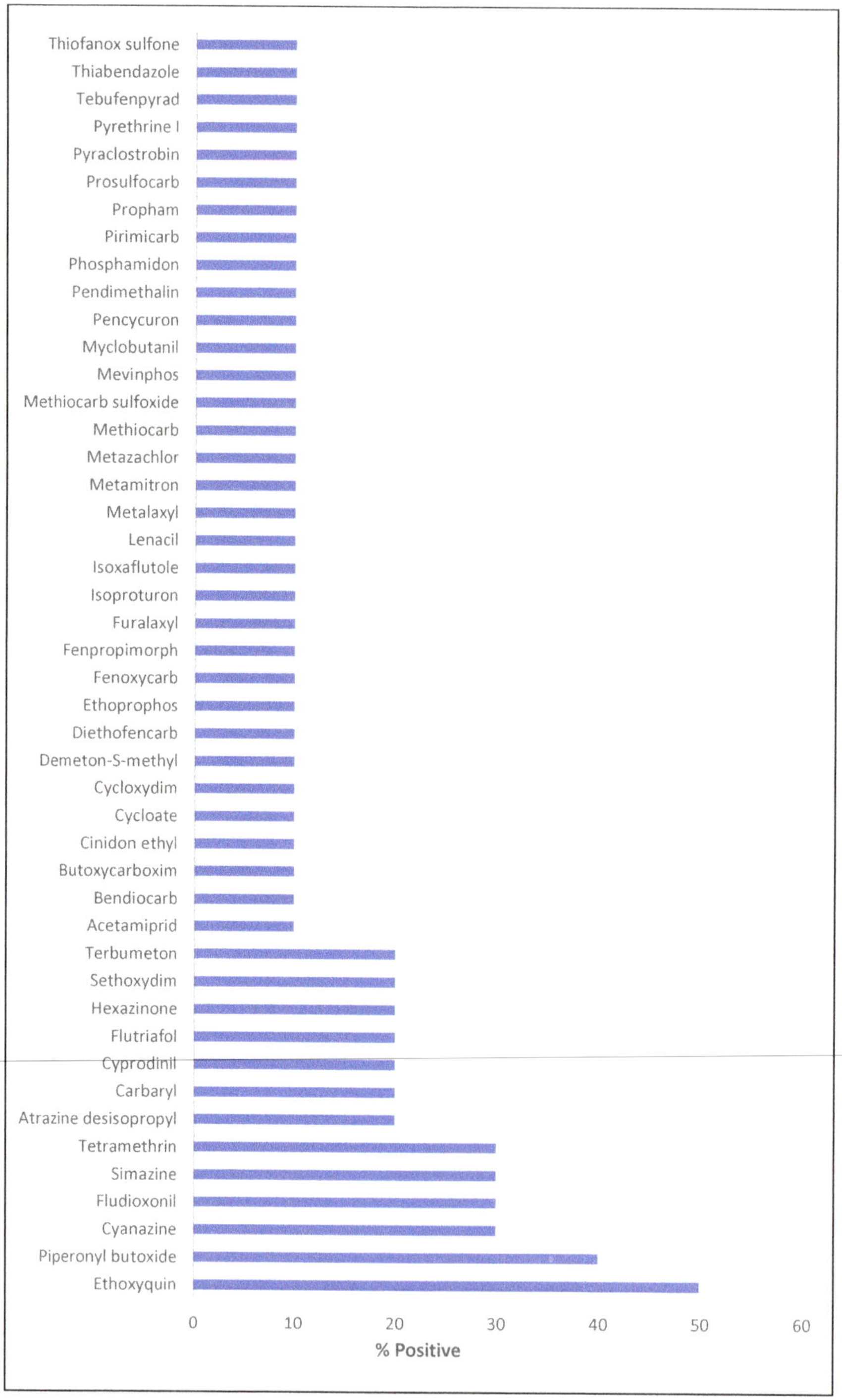

Figure 2. Occurrence of non-target pesticides in analyzed samples after post-run retrospective screening.

3. Conclusions

A sample preparation procedure based on a QuEChERS followed by UHPLC coupled with high-resolution Q-Orbitrap mass spectrometry was optimized in order to determine and quantify 16 mycotoxins in *C. sativa*-based supplements. The proposed methodology was validated following the EU criteria, ensuring a proper specificity, selectivity, linearity, trueness and precision with a fast chromatography run performance (8 min). The validated procedure was applied to ten CBD-based supplements that are commercially available online, allowing us to quantify up to six different *Fusarium* mycotoxins in 70% of samples. ZEN was the most prevalent mycotoxin (60%) found at a maximum level of 11.6 ng/g (mean value = 6.9 ng/g). Co-occurrence was observed in four out of ten samples, including one sample with ENNB1, ENNA and ENNA1. Additionally, a retrospective analysis of pesticide residues was performed. Up to 46 different pesticides were tentatively detected, including some forbidden in *C. sativa* cultivation. Considering the uprising trend of CBD-based products, quality controls regarding contaminants should be set for these products in order to ensure a safe consumption. Furthermore, the developed procedure is proposed as a powerful analytical tool to evaluate the potential mycotoxin profile of these particular products.

4. Materials and Methods

4.1. Chemicals and Reagents

Acetonitrile (AcN), methanol (MeOH), and water for LC mobile phase (HPLC grade) were acquired from Merck (Darmstadt, Germany). Formic acid and ammonium formate were obtained from Fluka (Milan, Italy). Sodium chloride (NaCl), magnesium sulfate ($MgSO_4$), octadecyl carbon chain-bonded silica (C18), graphitized carbon black (GCB), primary-secondary amine (PSA) and zirconium oxide (Z-Sep$^+$) were obtained from Sigma Aldrich (Milan, Italy).

Mycotoxin standards and metabolites, namely aflatoxins (AFB1, AFB2, AFG1, and AFG2), HT-2 toxin (HT-2), T-2 toxin (T-2), neosolaniol (NEO), zearalenone (ZEN), α-zearalenol (α-ZEL), β-zearalenol (β-ZEL), zearalanone (ZAN), beauvericin (BEA) and enniatins (ENNA, ENNA1, ENNB, and ENNB1) were purchased from Sigma Aldrich (Milan, Italy). Individual stock solutions of all analytes were prepared by diluting 1 mg of each mycotoxin in 1 mL of methanol. The working standard solution including all the mycotoxins was made by adequate diluting in MeOH:H_2O (70:30 *v/v*) 0.1% formic acid to reach the required concentrations for performing the spike experiments: 20, 10 and 2 μg/mL. All solutions were kept in safe conditions at −20 °C.

4.2. Sampling

For the analysis of real samples, ten different CBD gelatin capsules were obtained from online shops based in different European countries. The capsules are made of gel mass, which contains gelatin, water, glycerin and other minor additives whereas the fill formulation consists of olive oil mixed with hemp oil containing CBD at certain concentrations. The weight of each capsule depended on the manufacturer; there were 0.25, 0.5 and 1 g capsules. Only soft gel capsules were studied since it was the prevalent presentation available for CBD supplements. On the other hand, one sample of CBD supplements delivered as soft gel capsules was acquired from a local store (Naples, Italy). After confirming the absence of contaminants, they were used for preparing fortified samples for recovery assays and matrix-matched standards for calibration purposes. All the samples were conserved in dark and cool conditions, as recommended by the manufacturer, until further analysis.

4.3. Sample Preparation

The sample preparation procedure developed by Veprikova et al. [24] was selected as a starting point and then slightly modified, as follows: 1 g of sample was weighed into a 50 mL polytetrafluorethylene (PTFE) tube and mixed with 5 mL of 1% aqueous formic acid. The mixture was placed in an SKO-D XL orbital shaker (Argo Lab, Italy) for 30 min at 294 × *g*. Then, 5 mL of

AcN were added and the mixture was shaken for an additional 30 min at 294 × *g*. After that, 0.5 g of sodium chloride and 2 g of magnesium sulfate were added and the tube was shaken for 1 min by hand, followed by centrifugation at 4907 × *g* for 15 min in an SL 16R centrifuge (Thermo Fisher Scientific LED GmbH, Germany). A 2 mL aliquot of the upper acetonitrile layer was taken for dispersive solid phase extraction (d-SPE) cleanup in a 15 mL PTFE tube containing 100 mg of Z-Sep+ sorbent and 300 mg of magnesium sulfate. The tube was vortexed for 1 min and then centrifuged at 4907 × *g* for 15 min. An aliquot of the supernatant (1 mL) was collected and filtered through a 0.2 µm PTFE filter (Phenomenex, Italy) into a vial prior to UHPLC-Q-Orbitrap HRMS analysis.

4.4. UHPLC-Q-Orbitrap HRMS Analysis

The qualitative and quantitative profiles of the mycotoxins were obtained using an ultra-high-pressure liquid chromatograph (UHPLC, Thermo Fisher Scientific, Waltham, MA, USA) equipped with a degassing system, a Dionex Ultimate 3000, a Quaternary UHPLC pump working at 1250 bar, an auto sampler device and a thermostated (T = 30 °C) Luna Omega 1.6 µm (50 × 2.1 µm) column.

The eluent consisted of two different phases: A (H_2O containing 0.1% formic acid and 5 mM ammonium formate) and B (MeOH containing 0.1% formic acid and 5 mM ammonium formate). The gradient elution for LC-Orbitrap HRMS analyses was applied as follows: an initial 0% of phase B was held for 1 min, which linearly went up to 95% B over 1 min and held for 0.5 min. Next, the gradient decreased to 75% B over 2.5 min and then decreased again to 60% B over 1 min. Finally, the gradient turned to 0% B over 0.5 min and then the column was equilibrated for 1.5 min at 0% B. The total run time was 8 min, at a flow rate of 0.4 mL/min. A total of 5 µL of the sample was injected. Detection was performed using a Q-Exactive mass spectrometer. The mass spectrometer was operated in both positive and negative ion mode using fast polarity switching by setting two scan events (full ion MS and all ion fragmentation (AIF)). Full scan data were acquired at a resolving power of 35,000 FWHM at *m/z* 200.

The ion source parameters were: spray voltage 4 kV (-4 kV in ESI− mode); capillary temperature 290 °C; S-lens RF level 50; sheath gas pressure ($N_2 > 95\%$) 35, auxiliary gas ($N_2 > 95\%$) 10, and auxiliary gas heater temperature 305 °C. The value for automatic gain control (AGC) target was set at 1×10^6, a scan range of *m/z* 100 to 1000 was selected and the injection time was set to 200 ms. The scan rate was set at 2 scans/s. For the scan event of AIF, the parameters in the positive and negative ion mode were: mass resolving power = 17,500 FWHM; maximum injection time = 200 ms; scan time = 0.10 s; ACG target = 1×10^5; scan range = 100–1000 *m/z*, isolation window to 5.0 *m/z*, and retention time window to 30 s. The Orbitrap-MS parameters were optimized in a previous work [45]. The exact mass for the studied compounds, including elemental composition, retention time (RT), theoretical masses and accurate mass errors for the detected ions are shown in Table 4. A mass error below 5 ppm, referring to the molecular ions, was set for identification. Retrospective screening was carried out on spectral data collected using a pesticide spectral library (Pesticide Spectral Library Version 1.1 for LibraryView™ Software, AB SCIEX, Framingham, USA). For accurate mass measurement, identification and confirmation were performed at a mass tolerance of 5 ppm for the molecular ion and for both fragments at the intensity threshold of 1000. Data analysis and processing were performed using the Xcalibur software, v. 3.1.66.10.

Table 4. Retention times, accurate mass and mass accuracy of mycotoxins evaluated.

Analyte	Retention Time (min)	Elemental Composition	Adduct Ion	Theoretical Mass (m/z)	Measured Mass (m/z)	Accuracy (Δ ppm)
NEO	4.25	$C_{19}H_{26}O_8$	$[M+NH_4]^+$	400.1966	400.1963	−0.67
AFG2	4.50	$C_{17}H_{14}O_7$	$[M+H]^+$	331.0812	331.0808	−1.36
AFG1	4.52	$C_{17}H_{12}O_7$	$[M+H]^+$	329.0656	329.0655	−0.27
AFB2	4.58	$C_{17}H_{14}O_6$	$[M+H]^+$	315.0863	315.0862	−0.51
AFB1	4.62	$C_{17}H_{12}O_6$	$[M+H]^+$	313.0707	313.0705	−0.42
HT-2	4.74	$C_{22}H_{32}O_8$	$[M+NH_4]^+$	442.2435	442.2432	−0.7
α-ZEL	4.83	$C_{18}H_{24}O_5$	$[M-H]^-$	319.1551	319.1550	−0.31
T-2	4.85	$C_{24}H_{34}O_9$	$[M+NH_4]^+$	484.2541	484.2543	0.39
β-ZEL	4.97	$C_{18}H_{24}O_5$	$[M-H]^-$	319.1551	319.1550	−0.31
ZAN	4.98	$C_{18}H_{24}O_5$	$[M-H]^-$	319.1551	319.1549	−0.6
ZEN	5.01	$C_{18}H_{22}O_5$	$[M+H]^+$	317.1395	317.1393	−0.54
ENN B	5.56	$C_{33}H_{57}N_3O_9$	$[M+NH_4]^+$	657.4433	657.4435	0.26
ENN B1	5.68	$C_{34}H_{59}N_3O_9$	$[M+NH_4]^+$	671.4599	671.4594	−0.76
BEA	5.73	$C_{45}H_{57}N_3O_9$	$[M+NH_4]^+$	801.4433	801.4432	−0.16
ENN A1	5.82	$C_{35}H_{61}N_3O_9$	$[M+NH_4]^+$	685.4746	685.4745	−0.18
ENN A	5.99	$C_{36}H_{63}N_3O_9$	$[M+NH_4]^+$	699.4903	699.4899	−0.56

4.5. Validation Parameters

An in-house validation study was conducted following the EU Commission Decision 2002/657/EC [36]. The parameters evaluated were selectivity, specificity, linearity, trueness, repeatability (intra-day precision), within-reproducibility (inter-day precision), limit of quantification (LOQ) and limit of detection (LOD). The selectivity and specificity of the method were evaluated by analyzing both standard solutions and samples, comparing the retention time of the peaks corresponding to the analytes of interest alongside the determination of its precursor and product ion, with a mass error below 5 ppm. For linearity, standard solutions built in neat solvent and matrix-matched calibration were analyzed by spiking blank samples at eight concentration levels from 0.2 to 100 ng/g. The slopes of each linear calibration function were compared in order to detect a signal suppression/enhancement (SSE) effect due to the matrix interference. This effect was quantified following the equation: SSE (%) = matrix-matched calibration slope/solvent calibration slope x 100. An SSE value of 100% was interpreted as no matrix interference in the concentration range evaluated. An SSE value above 100% revealed signal enhancement whereas a value below 100% indicated signal suppression. For trueness, recovery studies were evaluated by spiking three blank samples at three different levels. Additionally, a lower spike level was used only for aflatoxins. Intra-day precision (RSD_r) was expressed as the relative standard deviation after three determinations in a single day (n = 3). Inter-day precision was calculated by repeating the measurements in triplicate on three non-consecutive days (n = 9) and expressed as relative standard deviation (RSD_R). The LOD was defined as the minimum concentration where the molecular ion can be identified by the instrument (mass error value below 5 ppm) and the LOQ as the minimum concentration where a linear response (mass error value below 5 ppm) can be observed with an accuracy and precision of ≤ 20%.

4.6. Statistical Analysis

Validation experiments were performed in triplicate and the results expressed as the average values alongside relative standard deviation (RSD, %). The Saphiro–Wilk test was applied to evaluate normality and multivariant analysis was performed using a non-parametric Kruskal–Wallis test, considering p values < 0.05 as significant. Analysis of data was carried out using IBM SPSS version 25 statistical software package (SPSS, Chicago, IL, USA).

Author Contributions: Conceptualization, Y.R.-C. and A.N.; methodology, L.I. and A.N.; validation, L.C. and A.N.; formal analysis, A.N.; investigation, L.C., L.I. and A.N.; resources, A.R.; writing—original draft preparation, A.N.; writing—review and editing, Y.R.-C.; supervision, Y.R.-C. and A.R.; project administration, Y.R.-C. and A.R.; funding acquisition, A.R. All authors have read and agreed to the published version of the manuscript.

Funding: This research received no external funding.

Acknowledgments: The authors want to acknowledge Alice Valli for donating sample material.

Conflicts of Interest: The authors declare no conflict of interest.

References

1. Afshin, A.; Sur, P.J.; Fay, K.A.; Cornaby, L.; Ferrara, G.; Salama, J.S.; Mullany, E.C.; Abate, K.H.; Abbafati, C.; Abebe, Z.; et al. Health effects of dietary risks in 195 countries, 1990–2017: A systematic analysis for the Global Burden of Disease Study 2017. *Lancet* **2019**, *393*, 1958–1972. [CrossRef]
2. Almada, A.L. Nutraceuticals and functional foods: Innovation, insulation, evangelism, and evidence. In *Nutraceutical and Functional Food Regulations in the United States and around the World*, 3rd ed.; Academic Press: Dana Point, CA, USA, 2019; pp. 3–11.
3. Binns, C.W.; Lee, M.K.; Lee, A.H. Problems and Prospects: Public Health Regulation of Dietary Supplements. *Annu. Rev. Public Health* **2018**, *39*, 403–420. [CrossRef] [PubMed]
4. Zuardi, A. Cannabidiol: From an inactive cannabinoid to a drug with wide spectrum of action. *Rev. Bras. Psiquiatr.* **2008**, *30*, 271–280. [CrossRef] [PubMed]
5. Casares, L.; García, V.; Garrido-Rodríguez, M.; Millán, E.; Collado, J.A.; García-Martín, A.; Peñarando, J.; Calzado, M.A.; de la Vega, L.; Muñoz, E. Cannabidiol induces antioxidant pathways in keratinocytes by targeting BACH1. *Redox Biol.* **2019**, *28*, 101321. [CrossRef]
6. Maroon, J.; Bost, J. Review of the neurological benefits of phytocannabinoids. *J. Surg. Neurol. Int.* **2018**, *9*, 91. [CrossRef]
7. *Regulation (EU) 2015/2283 of the European Parliament and of the Council of 25 November 2015 on Novel Foods, Amending Regulation (EU) No 1169/2011 of the European Parliament and of the Council and Repealing Regulation (EC) No 258/97 of the European Parliament and of the Council and Commission Regulation (EC) No 1852/2001*; European Parliament and the Council: Strasbourg, France, 2015.
8. *International CBD and Cannabis Market Landscape*; BrightField Group Database; Brightfield Group: Chicago, IL, USA, 2019.
9. Santini, A.; Cammarata, S.M.; Capone, G.; Ianaro, A.; Tenore, G.C.; Pani, L.; Novellino, E. Nutraceuticals: Opening the debate for a regulatory framework. *Br. J. Clin. Pharmacol.* **2018**, *84*, 659–672. [CrossRef]
10. Gulati, O.P.; Ottaway, P.B.; Jennings, S.; Coppens, P.; Gulati, N. Botanical nutraceuticals (food supplements and fortified and functional foods) and novel foods in the EU, with a main focus on legislative controls on safety aspects. In *Nutraceutical and Functional Food Regulations in the United States and around the World*, 3rd ed.; Academic Press: Brussels, Belgium, 2019; pp. 277–321.
11. Rodríguez-Carrasco, Y.; Fattore, M.; Albrizio, S.; Berrada, H.; Mañes, J. Occurrence of Fusarium mycotoxins and their dietary intake through beer consumption by the European population. *Food Chem.* **2015**, *178*, 149–155. [CrossRef]
12. Ostry, V.; Malir, F.; Toman, J.; Grosse, Y. Mycotoxins as human carcinogens—The IARC Monographs classification. *J. Mycotoxin Res.* **2017**, *33*, 65–73. [CrossRef]
13. McKernan, K.; Spangler, J.; Zhang, L.; Tadigotla, V.; Helbert, Y.; Foss, T.; Smith, D. Cannabis microbiome sequencing reveals several mycotoxic fungi native to dispensary grade Cannabis flowers. *F1000Res* **2015**, *4*, 1422. [CrossRef]
14. McHardy, I.; Romanelli, A.; Harris, L.J.; Opp, G.; Gaudino, R.; Torres, A.; Polage, C.R.; Tuscano, J.M.; Thompson, G.R. Infectious risks associated with medicinal Cannabis: Potential implications for immunocompromised patients? *J. Infect.* **2018**, *76*, 500–501. [CrossRef]
15. Atapattu, S.N.; Johnson, K.R.D. Pesticide analysis in cannabis products. *J. Chromatogr. A* **2019**, *1612*, 460656. [CrossRef] [PubMed]
16. Choudri, B.S.; Charabi, Y. Pesticides and herbicides. *Water Environ. Res.* **2019**, *91*, 1342–1349. [CrossRef] [PubMed]

17. Mostafalou, S.; Abdollahi, M. Pesticides: An update of human exposure and toxicity. *Arch. Toxicol.* **2017**, *91*, 549–599. [CrossRef] [PubMed]
18. Ye, M.; Beach, J.; Martin, J.W.; Senthilselvan, A. Pesticide exposures and respiratory health in general populations. *J. Environ. Sci.* **2017**, *51*, 361–370. [CrossRef]
19. *Regulation (EC) No. 396/2005 of the European Parliament and of the Council of 23 February 2005 on Maximum Residue Levels of Pesticides in or on Food and Feed of Plant and Animal Origin and Amending Council Directive 91/414/EEC*; European Parliament and the Council: Strasbourg, France, 2005.
20. *Commission Regulation (EC) No. 1881/2006 of the European Parliament and the Council of 19 December 2006 Setting Maximum Levels for Certain Contaminants in Foodstuffs*; European Parliament and the Council: Brussels, Belgium, 2006.
21. Jeong, M.L.; Zahn, M.; Trinh, T.; Brooke, F.A.; Ma, W. Pesticide residue analysis of a dietary ingredient by gas chromatography/selected-ion monitoring mass spectrometry using neutral alumina solid-phase extraction cleanup. *J. AOAC Int.* **2008**, *91*, 630–636. [CrossRef]
22. González-Martín, M.I., Revilla, I.; Betances-Salcedo, E.V.; Vivar-Quintana, A.M. Pesticide residues and heavy metals in commercially processed propolis. *Microchem. J.* **2018**, *143*, 423–429. [CrossRef]
23. Vaclavik, L.; Vaclavikova, M.; Begley, T.; Krynitsky, A.; Rader, J. Determination of Multiple Mycotoxins in Dietary Supplements Containing Green Coffee Bean Extracts Using Ultra-High Performance Liquid Chromatography-Tandem Mass Spectrometry (UHPLC-MS/MS). *J. Agric. Food Chem.* **2013**, *61*, 4822–4830. [CrossRef]
24. Veprikova, Z.; Zachariasova, M.; Dzuman, Z.; Zachariasova, A.; Fenclova, M.; Slavikova, P.; Vaclavikova, M.; Mastovska, K.; Hengst, D.; Hajslova, J. Mycotoxins in Plant-Based Dietary Supplements: Hidden Health Risk for Consumers. *J. Agric. Food Chem.* **2015**, *63*, 6633–6643. [CrossRef]
25. Martínez-Domínguez, G.; Romero-González, R.; Garrido Frenich, A. Determination of toxic substances, pesticides and mycotoxins, in ginkgo biloba nutraceutical products by liquid chromatography Orbitrap-mass spectrometry. *Microchem. J.* **2015**, *118*, 124–130. [CrossRef]
26. Martínez-Domínguez, G.; Romero-González, R.; Garrido Frenich, A. Multi-class methodology to determine pesticides and mycotoxins in green tea and royal jelly supplements by liquid chromatography coupled with Orbitrap high resolution mass spectrometry. *Food Chem.* **2016**, *197*, 907–915. [CrossRef]
27. Martínez-Domínguez, G.; Romero-González, R.; Arrebola, F.J.; Garrido Frenich, A. Multi-class determination of pesticides and mycotoxins in isoflavones supplements obtained from soy by liquid chromatography coupled with Orbitrap high resolution mass spectrometry. *Food Control* **2016**, *59*, 218–224. [CrossRef]
28. Santos, L.; Marín, S.; Sanchis, V.; Ramos, A.J. Screening of mycotoxin multicontamination in medicinal and aromatic herbs sampled in Spain. *J. Sci. Food Agric.* **2009**, *89*, 1802–1807. [CrossRef]
29. Zhao, H.; Chen, X.; Shen, C.; Qu, B. Determination of 16 mycotoxins in vegetable oils using a QuEChERS method combined with high-performance liquid chromatography-tandem mass spectrometry. *Food Addit. Contam.* **2017**, *34*, 255–264. [CrossRef] [PubMed]
30. Hidalgo-Ruiz, J.L.; Romero-González, R.; Martínez Vidal, J.L.; Garrido Frenich, A. A rapid method for the determination of mycotoxins in edible vegetable oils by ultra-high performance liquid chromatography-tandem mass spectrometry. *Food Chem.* **2019**, *288*, 22–28. [CrossRef] [PubMed]
31. Myresiotis, C.K.; Testempasis, S.; Vryzas, Z.; Karaoglanidis, G.S.; Papadopoulou-Mourkidou, E. Determination of mycotoxins in pomegranate fruits and juices using a QuEChERS-based method. *Food Chem.* **2015**, *182*, 81–88. [CrossRef] [PubMed]
32. Han, L.; Matarrita, J.; Sapozhnikova, Y.; Lehotay, S.J. Evaluation of a recent product to remove lipids and other matrix co-extractives in the analysis of pesticide residues and environmental contaminants in foods. *J. Chromatogr. A* **2016**, *1449*, 17–29. [CrossRef]
33. Rajski, Ł.; Lozano, A.; Uclés, A.; Ferrer, C.; Fernández-Alba, A.R. Determination of pesticide residues in high oil vegetal commodities by using various multi-residue methods and clean-ups followed by liquid chromatography tandem mass spectrometry. *J. Chromatogr. A* **2013**, *1304*, 109–120. [CrossRef]
34. Lozano, A.; Rajski, Ł.; Uclés, S.; Belmonte-Valles, N.; Mezcua, M.; Fernández-Alba, A.R. Evaluation of zirconium dioxide-based sorbents to decrease the matrix effect in avocado and almond multiresidue pesticide analysis followed by gas chromatography tandem mass spectrometry. *Talanta* **2014**, *118*, 68–83. [CrossRef]

35. Tuzimski, T.; Szubartowski, S. Method Development for Selected Bisphenols Analysis in Sweetened Condensed Milk from a Can and Breast Milk Samples by HPLC-DAD and HPLC-QqQ-MS: Comparison of Sorbents (Z-SEP, Z-SEP Plus, PSA, C18, Chitin and EMR-Lipid) for Clean-Up of QuEChERS Extract. *Molecules* **2019**, *24*, 2093. [CrossRef]
36. *Commission Decision 2002/657/EC of 12 August 2002 implementing Council Directive 96/23/EC Concerning the Performance of Analytical Methods and the Interpretation of Results (Text with EEA Relevance)*; Brussels, Belgium, 2002. Available online: https://op.europa.eu/en/publication-detail/-/publication/ed928116-a955-4a84-b10a-cf7a82bad858/language-en (accessed on 12 February 2020).
37. Han, Z.; Ren, Y.; Zhu, J.; Cai, Z.; Chen, Y.; Luan, L.; Wu, Y. Multianalysis of 35 Mycotoxins in Traditional Chinese Medicines by Ultra-High-Performance Liquid Chromatography–Tandem Mass Spectrometry Coupled with Accelerated Solvent Extraction. *J. Agric. Food Chem.* **2012**, *60*, 8233–8247. [CrossRef]
38. Arroyo-Manzanares, N.; García-Campaña, A.M.; Gámiz-Gracia, L. Multiclass mycotoxin analysis in Silybum marianum by ultra high performance liquid chromatography–tandem mass spectrometry using a procedure based on QuEChERS and dispersive liquid–liquid microextraction. *J. Chromatogr. A* **2013**, *1282*, 11–19. [CrossRef] [PubMed]
39. Han, Z.; Ren, Y.; Zhou, H.; Luan, L.; Cai, Z.; Wu, Y. A rapid method for simultaneous determination of zearalenone, α-zearalenol, β-zearalenol, zearalanone, α-zearalanol and β-zearalanol in traditional Chinese medicines by ultra-high-performance liquid chromatography–tandem mass spectrometry. *J. Chromatogr. B* **2011**, *879*, 411–420. [CrossRef] [PubMed]
40. Smith, M.-C.; Madec, S.; Coton, E.; Hymery, N. Natural Co-Occurrence of Mycotoxins in Foods and Feeds and Their in vitro Combined Toxicological Effects. *Toxins* **2016**, *8*, 94. [CrossRef] [PubMed]
41. Tournas, V.H.; Sapp, C.; Trucksess, M.W. Occurrence of aflatoxins in milk thistle herbal supplements. *Food Addit. Contam.* **2012**, *29*, 994–999. [CrossRef] [PubMed]
42. Pesticides Properties DataBase (PPDB). University of Hertfordshire, United Kingdom. Available online: http://sitem.herts.ac.uk/aeru/ppdb/en/ (accessed on 22 December 2019).
43. *Regulation (EC) No. 1107/2009 of the European Parliament and of the Council of 21 October 2009 Concerning the Placing of Plant Protection Products on the Market and Repealing Council Directives 79/117/EEC and 91/414/EEC*; European Parliament and the Council: Strasbourg, France, 2009.
44. *Commission Decision 2004/247/EC of 10 March 2004 Concerning the Non-Inclusion of Simazine in Annex I to Council Directive 91/414/EEC and the Withdrawal of Authorisations for Plant Protection Products Containing this Active Substance*; Brussels, Belgium, 2004. Available online: https://www.legislation.gov.uk/eur/2009/1107/contents (accessed on 12 February 2020).
45. Castaldo, L.; Graziani, G.; Gaspari, A.; Izzo, L.; Tolosa, J.; Rodríguez-Carrasco, Y.; Ritieni, A. Target Analysis and Retrospective Screening of Multiple Mycotoxins in Pet Food Using UHPLC-Q-Orbitrap HRMS. *Toxins* **2019**, *11*, 434. [CrossRef]

toxins

MDPI

Article

Assessment of Toxigenic Fusarium Species and Their Mycotoxins in Brewing Barley Grains

Karim C. Piacentini [1,*], Liliana O. Rocha [2], Geovana D. Savi [3], Lorena Carnielli-Queiroz [4], Livia De Carvalho Fontes [4] and Benedito Correa [1,4]

1 Biotechnology Department, University of Sao Paulo, Av. Professor Lineu Prestes, Sao Paulo 2415, Brazil; correabe@usp.br
2 Department of Food Science, Food Engineering Faculty, University of Campinas, Av. Monteiro Lobato, 80, Campinas 13083862, Brazil; l.rocha@unicamp.br
3 Department of Materials Sciences and Engineering, University of the Extreme Southern Santa Catarina, Av. Universitaria, 1105 Criciuma, Santa Catarina 88807-400, Brazil; geovanasavi@gmail.com
4 Microbiology Department, University of Sao Paulo, Av. Professor Lineu Prestes, Sao Paulo 1374, Brazil; carnielli@usp.br (L.C.-Q.); livia.fontes@usp.br (L.D.C.F.)
* Correspondence: karim.piacentini@hotmail.com

Received: 14 November 2018; Accepted: 29 December 2018; Published: 10 January 2019

check for updates

Abstract: *Fusarium* species threaten yield and quality of cereals worldwide due to their ability to produce mycotoxins and cause plant diseases. Trichothecenes and zearalenone are the most economically significant mycotoxins and are of particular concern in barley, maize and wheat. For this reason, the aim of this study was to characterize the *Fusarium* isolates from brewing barley and to assess deoxynivalenol and zearalenone contamination in grains. Characterization of the *Fusarium* strains was carried out by the phylogeny based on two loci (EF-1α and RPB2). Mycotoxin detection and quantification were performed by LC-MS. The results show that *Fusarium* was the predominant genus. Phylogenetic study demonstrated that the majority of the strains clustered within the *Fusarium sambucinum* species complex followed by the *Fusarium tricinctum* species complex. The results revealed high incidence of deoxynivalenol (DON) and zearalenone (ZEA) contamination (90.6% and 87.5%, respectively). It was observed that 86% of the samples contaminated with ZEA were above the limits set by the EU and Brazilian regulations. These results may highlight the importance of controlling *Fusarium* toxins in barley, mainly because of its use in the brewing industry and the resistance of various mycotoxins to food processing treatments.

Keywords: cereals; mycotoxigenic fungi; phylogeny; deoxynivalenol; zearalenone

Key Contribution: High contamination of deoxynivalenol and zearalenone was found in barley analyzed. Most of the samples contaminated were above to the limits set by the regulations.

1. Introduction

The *Fusarium* genus includes plant pathogens which are of great concern to agricultural production and food/feed safety worldwide [1], threatening yield and quality of cereals and producing mycotoxins, secondary metabolites that are toxic to humans and other animals [2]. *Fusarium* genus is capable of producing several mycotoxins, including fumonisins, trichothecenes and zearalenone. These are the most economically significant *Fusarium* mycotoxins and are of particular concern in barley, maize and wheat [3].

Trichothecenes have been classified into four groups: types A–D, according to their chemical structure [4], the most important in cereals are types A and B [1]. The B-trichothecenes include the mycotoxins deoxynivalenol (DON), its acetylated derivatives, 3-acetyldeoxynivalenol (3ADON)

and 15-acetyldeoxynivalenol (15ADON) and nivalenol (NIV). DON is the most frequent type-B trichothecene and can be found worldwide. Furthermore, DON inhibits protein synthesis and it has been associated with intoxication of animals through consumption of contaminated feed [5].

Zearalenone (ZEA) is a mycotoxin with estrogenic effects produced by several *Fusarium* species and is usually found in cereal grains. Swine are especially sensitive to the estrogenic effects of ZEA. This toxin has been shown to bind to the estrogenic receptors and to inhibit ovulation. It is, therefore, often involved in hormonal disorders of farm animals and it is also related to hypoestrogenic syndromes in humans [6]. ZEA has been classified into group 3 (non-classifiable due to its carcinogenicity to humans) by IARC (International Agency for Research on Cancer) [7].

During recent years, mycotoxins have attracted international attention not only for their perceived impact on human health but also because of the economic losses accruing from contaminated foods. Considering barley matrix, several international studies have reported on *Fusarium* and its mycotoxins contamination due to the beer gaining increased popularity [8]. The main problem is the characteristics that these compounds have. Some mycotoxins such as trichothecenes, zearalenone and fumonisins are considered stable during the brewing process [9] and can persist to the final product, the beer. Furthermore, *Fusarium* infection causes a negative impact on barley germination rates which results in malting quality and yield reduction. Additionally, this produces gushing and changes in color and flavor of the beer [10].

Barley is considered to have good characteristics for fungi contamination under favorable conditions. In addition, environmental factors associated with barley varieties and agronomic practices influence the *Fusarium* infection and the mycotoxin production. The climate conditions during critical phases of barley plant growth could lead to noticeable difference between the mycotoxin diversity. Tropical conditions, such as those found in Brazil contribute to fungi dissemination and consequently mycotoxin production in barley. For example, 2015 demonstrated a high rainfall average and high levels of humidity, which prompted worse contamination levels found until now [11]. It is necessary to mention that only two *Fusarium* mycotoxins were analyzed. Other fungi mycotoxins should be evaluated along with their masked toxins.

All of the barley harvested in Brazil is destined for the brewing industry and the production continues to increase. Southern Brazil has the largest number of barley-producing regions, therefore, the quality of the grains needs to be monitored and studies about the fungi profile should be taken into account. The knowledge of the current contamination of barley in the region, as well as the constant monitoring, is necessary in order to evaluate if agronomic practices are being duly effective to control the contamination in grain production. The irrigation management, resistant cultivars, harvesting strategies, chemical and biological control and disease forecasting could minimize the loss of grain quality and avoid the disease caused by mycotoxigenic fungi [12].

The current occurrence of mycotoxins in barley could lead to the necessity of developing new strategies or improving those currently in place for more effective management of mycotoxins in the future. This fact is even more relevant when taking into account that the Brazilian regulation for DON in barley will be updated by 2019 [13] and therefore, the information on the occurrence of this toxin is still discussed in the country often.

The regulations have set maximum levels for mycotoxin contamination in grains, in order to avoid further accumulation of mycotoxins in processed food and to control mycotoxin of major concern in unprocessed cereals, such as brewing barley, with 1.250 μg/kg being the maximum permitted for DON and 100 μg/kg for ZEA. From January 2019, DON limits for brewing barley will be set at 1000 μg/kg [13]. Similarly, the limits for DON and ZEA fixed by the European Commission [14] are equal to 1.250 μg/kg and 100 μg/kg for unprocessed cereals.

For the reasons stated above, the aim of the present research was to characterize the *Fusarium* isolates and to assess DON and ZEA contamination in brewing barley grains. These findings provide new insights into the diversity of *Fusarium* species isolated from Brazilian barley and add information to the mycotoxin profile in a source (raw material) destined for the food industry.

2. Results

2.1. Water Activity and Identification of Isolated Fungi

Water activity levels ranged from 0.579 to 0.667 (mean: 0.622 ± 0.02) for all of the 64 brewing barley samples analyzed. Filamentous fungi were isolated from 96.8% of the samples, highlighting that *Fusarium* was the predominant genus (46%), followed by *Alternaria* (28.8%), *Phoma* (15%), *Epicocum* (6.2%), *Penicillium* (2%), *Aspergillus* (1.1%) and *Rhyzopus* (0.9%). Species belonging to *Fusarium* genus were initially identified through the sequencing of the *EF-1α* locus. Sequences were submitted to Blast (basic local alignment) tool on NCBI database (https://blast.ncbi.nlm.nih.gov/Blast.cgi?PAGE_TYPE=BlastSearch).

The identification analysis was carried out with 48 *Fusarium* strains and sequencing analysis determined 56.26% of the *Fusarium* species isolated in this study, were within the *Fusarium sambucinum* species complex (FSAMSC), 31.25% within the *Fusarium tricinctum* species complex (FTSC), 8.33% within the *Fusarium fujikuroi* species complex (FFSC) and 2% within both, *Fusarium incarnatum-equiseti* (FIESC) and *Fusarium oxysporum* species complexes (FOSC) (Table 1).

Table 1. Frequency of the members of each *Fusarium* species complex isolated from brewing barley grains.

Species Complex *	Frequency %	*Fusarium* Species	% Samples Contaminated
FSAMSC	56.25	*F. graminearum*	23.4
		F. poae	17.2
		F. meridionale	1.6
FTSC	31.25	*F. avenaceum*	23.4
FFSC	8.33	*F. proliferatum*	3.1
		F. verticillioides	3.1
FIESC	2	*F. incarnatum-equiseti*	1.6
FOSC	2	*F. oxysporum*	1.6

* FSAMSC: *F. sambucinum* species complex, FTSC: *F. tricinctum* species complex, FFSC: *F. fujikuroi* species complex, FIESC: *F. incarnatum-equiseti* species complex, FOSC: *F. oxysporum* species complex.

2.2. Phylogenetic Study

The phylogenetic study was conducted for the species complexes that have the potential to produce trichothecenes and for the closely related species complex *F. tricinctum*. The concatenated loci *EF-1α* and *RPB2* were used to infer the phylogeny of the species isolated from barley. The data set consisted of 65 taxa, 1292 nucleotides with 452 parsimony-informative characters (PICs) (Figure 1). The analysis resulted in a one most parsimonious tree (CI = 0.69 RI = 0.94). The majority of the isolates clustered within *F. graminearum*, *F. poae* and *F. avenaceum* species complexes with both posterior probability and bootstrap supports (Figure 1).

The method validation and recovery experiments are summarized in Table 2 and Table S2. The reported LODs were set at 5 and 10 µg/kg for DON and ZEA, respectively and LOQs 25 µg/kg for both toxins. The coefficients of correlation (R^2) of the calibration curve were 0.997 and 0.999 for DON and ZEA, respectively. Spiking was performed in triplicates at three levels in the barley matrix. In addition, the spiking experiments for the *F. graminearum* strains grown in culture media were carried out at two levels.

The method is suitable for the determination of mycotoxins in barley as the given LOQ's are lower than the maximum limit set by the Brazilian and EU regulations for the content of DON and ZEA. The parameter linearity, reproducibility, repeatability and recovery obtained were also shown to be adequate.

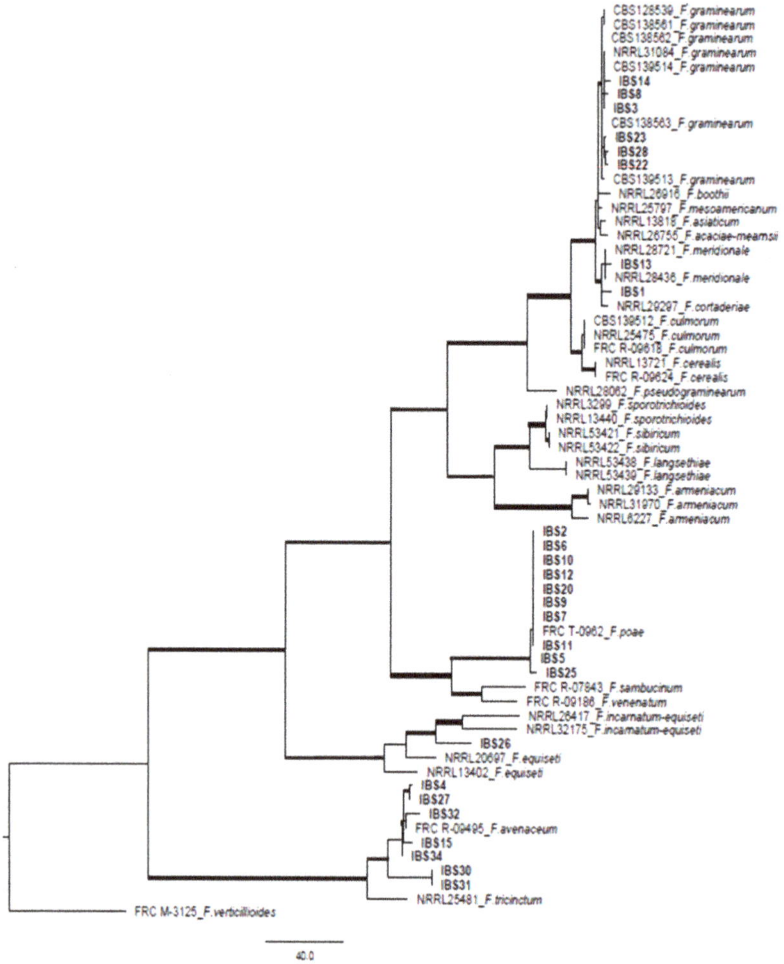

Figure 1. LC-MS/MS method performance.

2.3. Mycotoxins Analysis

The analysis carried out in the present research revealed the occurrence of the two mycotoxins most commonly found in barley. Grain samples that presented levels above the LOQ were considered positive. Mean calculations were performed using Microsoft® Excel 2007 only including positive samples. Deoxynivalenol showed the highest incidence (90.6%) with levels ranging from 45.95 to 1155.21 µg/kg. ZEA also had a high occurrence (87.5%) with values ranging from 82.41 to 423.71 µg/kg. Regarding the Brazilian regulation for DON, only one sample was above the established maximum levels (1000 µg/kg). Nevertheless, for ZEA, 55 samples (86%) were above the regulation (100 µg/kg) (Table 2).

Additionally, the mycotoxin production from toxigenic potential *F. graminearum* strains was observed. The frequency of the DON strain producers was 80% (12 strains) and only 20% (3 strains) for ZEA. The levels ranged from 123.03 to 592.61 µg/kg and 33.64 to 140.58 µg/kg, respectively.

Table 2. DON and ZEA contamination in barley grains.

	Number Samples	Deoxynivalenol				Zearalenone			
		Positive Samples/% *	Range of Positive Samples (μg/kg)	Mean ± SD (μg/kg)	Median (μg/kg)	Positive Samples/% *	Range of Positive Samples (μg/kg)	Mean ± SD (μg/kg)	Median (μg/kg)
Barley grains	64	58/90.6	45.95–1155.21	147.65 ± 167.16	98.68	56/87.5	82.41–423.71	123.24 ± 45.29	119.26

* > LOQ of 25 μg/kg.

3. Discussion

This study has shown that the majority of the *Fusarium* species isolated from brewing barley grains belonged to the FSAMSC, whereas the main trichothecene producing species are clustered. Mycotoxins analysis demonstrated that most of the samples were contaminated with DON (90.6%) and ZEA (87.5%), highlighting the importance of this investigation.

Fusarium species are found in cereal grains, such as barley, wheat, maize and rice worldwide, where mycotoxins can be found in high concentrations [15–18]. This fact, may be worsened by weather conditions, such as high humidity and temperatures that tend to increase *Fusarium* infection in plants [19,20]. Humidity is an important environmental factor and it influences the water activity of the grains. This intrinsic factor is important for fungal growth and has considerable association with mycotoxin production [21].

Generally, *Fusarium* development as well as DON and ZEA production can be seen with higher levels of water activity (0.90) [22]. Nevertheless, the current research showed a low variation of the water activity that was observed among the 64 brewing barley samples, with mean value of 0.622 ± 0.02. In this case, both the germination of fungal spores and the growth of storage fungi are inhibited. However, the grain analyses were carried out after the cleaning and drying stages, explaining the levels found. For safe storage of grain, the grain moisture content must be compatible with the period of time the grain will be stored in order to avoid the fungi growth and therefore, the water activity should be less. To associate *Fusarium* mycotoxins found in this study it is necessary to assume that the DON and ZEA production could be correlated to the fungi presence and high water activity in the growing plant stages. This fact could be explained in our recent study with rice grains, where the levels of ZEA found are associated with the presence of *Fusarium* during pre-harvest, in grains freshly harvested with high levels of moisture content and water activity. After the food processing steps were completed in the industry, there was not any *Fusarium* growth in the grains, however, the ZEA levels remained in the parboiled rice (water activity: 0.64 ± 0.02), resisting the degradation [23].

The phylogenetic study showed that the majority of the strains clustered with *F. graminearum*, *F. poae* and *F. avenaceum*. *Fusarium poae* can produce high levels of nivalenol; therefore, further analyses should be done to investigate the degree of nivalenol contamination in Brazilian barley. Previous studies carried out by [24] in Russia and [25] in Italy have shown a high incidence of *F. avenaceum* in barley as well as other *Fusarium* species. [26] observed a high incidence of *F. avenaceum*, *F. graminearum* and *F. culmorum* in in Finnish barley grains between the years of 2005–14, with high levels of trichothecene contamination. The presence of *F. avenaceum* in barley may indicate the presence of enniatins, moniliformin and beauvericin in the samples. In Brazil, another study in barley has demonstrated that the majority of the isolates belonged to the *F. graminearum* lineage; however, mycotoxin analysis was not performed in that study [27]. To our knowledge, this is the first report of phylogenetic identification of the *Fusarium* species in Brazilian barley and the first correlation with DON and ZEA contamination.

With respect to DON contamination, the current study showed similarities with those found in a survey carried out in Spain with a mean level of 119.9 μg/kg [28]. In addition, the highest results were obtained by [25] reporting incidence and a maximum contamination level of 108.7 μg/kg. Also, [29] in Italy, a study reported lower incidence and concentrations of DON with a maximum level of 35.5 μg/kg; and [30] in Tunisia evaluated its presence with a maximum level of 6.1 μg/kg. In the current study, only one sample presented a high level of DON (1155.21 μg/kg) and was above the established maximum level set by the Brazilian regulation. The samples showed mean values of 147.65 μg/kg and median values of 98.68 μg/kg that also demonstrated low levels.

In contrast, the ZEA contamination found in the present study were of significance, considering the maximum levels established by the Brazilian and international regulation, where 86% of the samples were above [13,31]. The samples showed mean values of 123.24 μg/kg and median values of 119.26 μg/kg. In a study performed by [32] in the Czech Republic, a few samples were contaminated with ZEA with values ranging from 181.2 to 204.4 μg/kg, which is quite similar to the results of our study. Furthermore, high levels (max 985.9 μg/kg) of ZEA were found in another survey carried out in the Czech Republic with samples from the 2011 crop [33].

Two other studies were conducted in Brazil by the current author and can be compared to this study. The first one was from the 2014 crop and low levels of DON were found ranging from 200 μg/kg to 15.000 μg/kg [34]. The other one was from the 2015 crop and higher levels were found for both toxins, DON and ZEA. The last study was considered an issue for the industry due to the levels exceeding the regulation levels established. The mean levels ranged from 1700 to 7500 μg/kg and from 300 to 630 μg/kg for DON and ZEA, respectively.

It is necessary to mention that not only the large-scale brewing industry is increasing but that the craft breweries in Brazil are expanding. These small brewing groups always look for the best sources and have the characteristic of "German beer purity law," meaning that just barley is used for beer production. For these reasons, barley needs to be of higher quality. On the other hand, the large-scale industry in Brazil uses other grains such as corn, rice and sorghum for beer production which are considered low quality grains [35]. Some studies were carried out and they showed contamination with fumonisin B_1 [36] that is commonly found in corn and its derivatives [37].

In the last years the stability of these metabolites has been studied. Deoxynivalenol showed to be a mycotoxin that persists through the process and demonstrated stability in some industry processes, such as cleaning, milling, brewing and extrusion [9]. Zearalenone has had some studies published about it which showed lower stability. However, its levels are high when they are found. A research conducted by [38] showed a considerable reduction of ZEA levels in the presence of *Saccharomyces cerevisiae* yeast. The main point is even if there is a significant reduction of these metabolites, sometimes they still offer a risk, especially when in beer, which is considered one of the most consumed beverages in the world.

Another aspect that should be taken into account is the masked toxins. ZEN-14-sulfate and DON-3-glucoside are most commonly observed in grains. They could be present in the matrix, however masked toxins are either bound to carbohydrates or proteins and, therefore, are not extractable with existing protocols aimed at the extraction of the toxin, or they are not detectable using established chromatography routines; hence their name "masked" mycotoxins [39]. Further studies are being planned to gain more knowledge on these metabolites in barley and also in beer.

The *Fusarium* mycotoxins found in barley grains and the toxigenic potential analysis of the *F. graminearum* strains isolated in this study reinforce the importance of these genera in this relevant commodity. Furthermore, the identification of *F. poae*, *F. avenaceum* and the genus *Alternaria* in barley samples highlights the importance of further research on other mycotoxins in barley and its by-products. *Alternaria* species were recovered from 28% of the barley samples. This may have important implications on other mycotoxins that may be found in high concentrations, such as tenuazonic acid. Further studies should be conducted in order to evaluate the co-occurrence of the mycotoxins produced by this genus [25,40].

In this study, the toxigenic potential of the *F. graminearum* strains isolated in the brewing barley was evaluated and both toxins detected in the samples were produced by them. In total, 80% and 20% of the strains produced DON (mean: 297.02 μg/kg and median: 268.74 μg/kg) and ZEA (mean: 79.7 μg/kg and median: 64.86 μg/kg), respectively. The toxins levels found were lower when compared to that of a study performed by Wu et al. (2017), in which the strains exhibited a production of 1405.05 μg/kg for DON and 4118.31 μg/kg for ZEA. The parameter temperature is crucial for DON production and some studies showed a variation of the optimal value, that varies between 20 and 28 °C [19,20,41]. On the other hand, temperature negatively affected ZEA production. The optimized condition for ZEA production was cultivation at 15 °C [41]. The toxin combination can be related to several mycotoxigenic fungi that contaminate barley in the field. However, the mycotoxin occurrence in Brazilian barley suggests high prevalence of toxigenic *F. graminearum* and related species, which could explain the DON and ZEA levels in almost 85% of the samples.

4. Conclusions

The DON and ZEA contamination in brewing barley grains were detected in 90.6% and 87.5% of the samples. The phylogenetic study showed that the majority of the strains clustered with *F. graminearum*, *F. poae* and *F. avenaceum*. Toxigenic species of *F. graminareum* isolates presented a higher percentage in the samples (31.25%) and can explain the DON and ZEA contamination found in the barley samples. Taking into account the Brazilian regulation of ZEA levels in barley, 86% of the samples were significantly higher than the current maximum limit, while for DON, only one sample was above the established maximum levels. The Brazilian regulation for mycotoxins will be updated by 2019 and therefore, the new maximum limits are still under discussion, based on the analysis of the largest amount of data available of the occurrence of mycotoxins in the grains produced in Brazil. This monitoring data of the toxigenic *Fusarium* and its mycotoxins could lead to greater knowledge of the current situation of the barley contamination in the industry, which can assist Brazilian regulation and the programing of management strategies in order to avoid the toxic effects on human and animal health.

Fungal infection and the presence of mycotoxins in cereals is natural and the prevention of these occurrences is difficult even if good agricultural practices are maintained. The data provided in this study was important for the knowledge on *Fusarium* diversity and toxin contamination. Furthermore, the results highlight the importance of monitoring DON and ZEA contamination in barley grains during pre, post-harvest and in processed food, such as beer, mainly for the development of management strategies. Consequently, serious economic losses and health problems could potentially be avoided.

5. Materials and Methods

5.1. Barley Samples

A total of 64 brewing barley (BRS Brau variety) samples were obtained from the 2016 harvest, from the States of Paraná and Rio Grande do Sul, the largest barley-producing regions in Brazil. Samples were collected from bulk batches, after dirt removal and drying (up to 60 °C) in the storage units. Sampling was performed using a grain auger from different points of the bulk batches, with a minimum final weight of 5 kg. Each sample was homogenized, reduced into portions of 1.0 kg to be representative of the overall sample and further was milled for each analysis. Samples were packed in polyethylene bags and stored at 4 °C and different amounts were used for mycobiota and mycotoxin analyses.

5.2. Water Activity

To perform water activity (aw) analysis, 2 g of each barley sample were submitted to Aqua-Lab 4TE equipment Aqua-Lab 4TE, Decagon Devices (Sao Jose dos Campos, SP, Brazil). Samples were analyzed in triplicate according to the Association of Official Analytical Chemists—[42].

5.3. Mycobiota and Identification of Fungi

The dilution technique was used for fungal isolation, as described by [43]. To summarize briefly, twenty-five grams of each sample were added to 225 mL of 0.1% peptone dissolved in water in sterile conditions. The mixture was stirred on a rotary shaker for 2 min., dilutions of 10^{-1}, 10^{-2}, 10^{-3} and 10^{-4} were obtained, 0.1 mL aliquots of each dilution were spread on the PDA medium [9,10] containing chloramphenicol (100 mg/L) (in duplicate). These were incubated for 5 days, at 25 °C in the dark. The results were expressed into colony forming units per gram (CFU/g) in the dilution 10^{-1}, as the colonies were easily distinguished in this dilution factor. The isolates were identified morphologically according to Pitt and Hocking, (2009).

5.4. Identification of the Fusarium Species

The strains were grown in yeast extract sucrose (YES) agar [44] for 3 days at 25 °C. The DNA was extracted using DNeasy Plant Mini Kit (Qiagen, Hilden, Germany) according to the manufacturer's instructions.

The partial sequences of elongation factor (*EF-1α*) and the second fragment of *RPB2* (*7CF/11AR*) were selected in order to identify the *Fusarium* isolates. The amplification of the *EF-1α* and *RPB2* loci were performed according to [45,46]. Amplicons were purified with ExoSAP-IT (Affymetrix, Santa Clara, CA, USA) and sent to the Centre of Human Genome Studies, University of Sao Paulo, Brazil for sequencing in ABI PRISM 3130 DNA Analyzer (Applied Biosystems, Foster City, CA, USA).

Sequences were aligned using the multiple alignment software ClustalX v. 1.83 plug-in in the software Geneious v. 5.3.6 (Biomatters, Auckland, New Zealand). The alignments were edited using the sequence alignment-editing program Geneious v. 1.83 and each polymorphism was re-examined by checking the chromatograms. The sequences generated in this study were deposited in the GenBank (Table 1 and Table S1).

5.5. Phylogenetic Analysis

Phylogenetic analysis was performed based on the *EF-1α* and *RPB2* combined datasets using the PAUP 4.0b10 (Sinauer Associates, Sunderland, MA, USA) [47]. Phylogenies were obtained by using Unweighted Parsimony analysis and heuristic search option with 1000 random addition sequences and tree bisection reconnection branch swapping in PAUP 4.0b10 [48]. Gaps were treated as missing data. The Consistency Index (CI) and the Retention Index (RI) were calculated to indicate the amount of homoplasy present. Clade stability was assessed via bootstrap analysis in PAUP 4.0b10, using 1000 heuristic search replications with random sequence addition. The data sets were rooted with *Fusarium* sp. as it is considered a suitable out-group [49]. The reference sequences for the *Fusarium* species used in this study were obtained from NCBI (Table S1).

5.6. Mycotoxin Analysis

5.6.1. Chemicals and Reagents

Both standards (DON and ZEA) were purchased from Sigma Aldrich Chemicals (St. Louis, MO, USA). Stock solution standards were prepared in methanol at concentrations of 1 mg/mL for DON and ZEA. From the individual stock standard solutions, a standard mixture was prepared at the following concentrations: 0.025, 0.0375, 0.0625, 0.125, 0.375, 0.500 μg/mL. The standard mixture was prepared in methanol and stored at −18 °C. Methanol and acetonitrile (LC-MS/MS grade) were supplied by J.T Baker (Sao Paulo, SP, Brazil). Acetic acid was obtained from Biotec (Pinhais, PR, Brazil). High-purity Milli-Q water (18.2 MΩ/cm) was obtained from the Millipore Synergy system (Billerica, MA, USA).

5.6.2. DON and ZEA Extraction

Mycotoxin extraction was carried out according to [50], with some minor modifications. Briefly, 2 g of brewing barley were ground and homogenized in 8 mL of acetonitrile:water (80:20 *v*/*v*) and shaken for 60 min. The mixture was then centrifuged for 10 min at 3500 rpm. The supernatant was transferred to an amber vessel and dried using a heating block and a nitrogen stream. The dried extract was resuspended in 500 μL of a mobile phase consisted of 70% of water:methanol:acetic acid (94:5:1, *v*/*v*/*v*) and 30% of water:methanol:acetic acid (2:97:1, *v*/*v*/*v*). Finally, 5 μL was injected in the LC-MS/MS system for analysis.

5.7. Production of DON and ZEA by the Strains

The isolates of *F. graminearum* were grown onto PDA (three agar plugs, 6 mm in diameter) and tested for DON and ZEA production. The culture media was incubated at 24 °C and 15 °C with a moisture content of 90% and 80% for 20 days for DON and ZEA, respectively [51,52]. The mycelium was transferred into an Erlenmeyer flask containing 30 mL of chloroform and shaken for 60 min for mycotoxin extraction, followed by filtration through anhydrous sodium sulfate (Na_2SO_4). The extract was dried and re-suspended with 500 μL of mobile phase. Finally, the extract was filtered with a syringe filter (nylon membrane 0.22 μM). The sample was quantified by liquid chromatography/mass spectrometry (LC-MS/MS) for DON/ZEA.

5.8. Chromatography Conditions

Detection and quantification were carried out using an LC-MS system from Thermo Scientific® (Bremen, Germany) composed of an ACCELA 600 quaternary pump, an ACCELAAS auto-sampler and a triple quadrupole mass spectrometer TSQ Quantum Max.

The chromatographic conditions were performed according to [53]. In short, the following instrumental settings were applied: the triple quadrupole mass spectrometer TSQ Quantum Max was operated at positive polarity and the ionization conditions were 208 °C for capillary temperature, 338 °C for vaporizer temperature, 4500 V for spray voltage and 60 arbitrary units for sheath gas pressure. For selectivity, the mass spectrometer was operated at MRM mode monitoring, three transitions per analyte, using a collision gas pressure of 1.7 mTorr and collision energy (CE) ranging from 11 to 40 eV.

The mass spectrometric conditions were optimized by re-tuning different analytes by direct infusion of each analyte individually. The tube lens potential, collision energies and product ions were optimized and carefully chosen. The most abundant mass-to-charge ratio (m/z) was selected for each compound of interest. The mycotoxins exhibited precursor ions and product ions with reasonably high signal intensities in positive ESI mode (ESI+) and protonated molecules [M + H] were found. Table S2 shows the retention times (tR), MRM transitions as well as the tube lens potential and collision energies optimized for each compound.

Separation was performed on a C8 Luna column, with a particle size of 3 μm, 150 × 2.0 mm, length and diameter, respectively, Phenomenex (Torrance, CA, USA). In the mobile phase, solvent A (water:methanol:acetic acid, 94:5:1, $v/v/v$) and solvent B (water:methanol:acetic acid, 2:97:1, $v/v/v$) were used. The gradient program was applied at a flow rate of 0.2 mL/min under the following conditions: 0–1 min 55% B; 1–3 min 55–100% B; 3.01–7 min 100% B and 7.01–12 min 55% B. The total analytical run time was 7.5 min for the 2 toxins (Table 3).

Table 3. Retention time and mass spectrometric parameters used in the analysis of the mycotoxins.

Mycotoxin	Retention Time (min)	Precursor ion (m/z)	Product ion (m/z) *	CE (V)	TubeLens
DON	2.19	297 [M + H]	203Q	17	71
			175C	18	71
			91C	39	71
ZEA	6.55	319 [M + H]	283Q	11	79
			187C	25	79
			185C	20	79

* Q, Quantification transition C, Confirmation transition.

Method Validation

The methods for extraction of mycotoxins in brewing barley and in culture media with *Fusarium* growth were validated according to the Commission Regulation [54] guideline. To determine the limit of detection (LOD), limit of quantification (LOQ), recovery, repeatability and selectivity/specificity, samples with non-detectable levels of mycotoxins were submitted to spiking experiments.

Considering linearity, a six-point calibration curve was constructed with the following concentrations of the mycotoxin standard mixture (DON and ZEN): 0.025, 0.0375, 0.0625, 0.125, 0.375, 0.500 μg/mL. The LOD and LOQ methods were determined by fortifying blank samples with different concentration levels and the experiments were repeated on three different days. The LOD was defined as the minimum concentration of an analyte in the spiked sample with a signal noise ratio equal to 3 and LOQ with a signal noise ratio equal to 10.

5.9. Data Analysis

Results regarding DON and ZEA in brewing barley samples and *F. graminearum* strains were reported as the mean ± standard deviation and median, using Microsoft office Excel 2007.

Supplementary Materials: The following are available online at http://www.mdpi.com/2072-6651/11/1/31/s1, Table S1: Fusarium species references used in this study for phylogenetic analysis, Table S2: Characteristics of the method performance for extraction of deoxynivalenol and zearalenone in barley grains and culture media.

Author Contributions: The authors' responsibilities were as follows—K.C.P. and L.O.R. designed the research; K.C.P., L.D.C.F. and L.C.-Q.: conducted the research; G.D.S.: guided and performed the statistical analysis; B.C.: Oversight and leadership responsibility for the research activity planning and execution, K.C.P. wrote the manuscript and hold primary responsibility for the final content; and all authors: read and approved the final manuscript.

Funding: The authors thank to the Research support foundation of the state of Sao Paulo (FAPESP/Project 2016/01798-4) for financial support.

Conflicts of Interest: The authors declare no conflict of interest.

References

1. Desjardins, A.E. Fusarium Mycotoxins: Chemistry, Genetics and Biology. *Plant Pathol.* **2007**, *56*, 337. [CrossRef]
2. Ma, L.-J.; Geiser, D.M.; Proctor, R.H.; Rooney, A.P.; O'Donnell, K.; Trail, F.; Gardiner, D.M.; Manners, J.M.; Kazan, K. Fusarium pathogenomics. *Annu. Rev. Microbiol.* **2013**, *67*, 399–416. [CrossRef]
3. FAO. *FAO Statistical Yearbook 2013: World Food and Agriculture*; FAO: Rome, Italy, 2013; ISBN 9789251073964.
4. WHO. *International Programme on Chemical Safety (IPCS): Environmental Health Criteria 105 Selected Mycotoxins. Ochratoxins, Trichothecenes, Ergot. Trichothecenes*; World Health Organization: Geneva, Switzerland, 1990; pp. 71–164.
5. Pestka, J.J. Deoxynivalenol: Mechanisms of action, human exposure, and toxicological relevance. *Arch. Toxicol.* **2010**, *84*, 663–679. [CrossRef] [PubMed]
6. Zinedine, A.; Soriano, J.M.; Moltó, J.C.; Mañes, J. Review on the toxicity, occurrence, metabolism, detoxification, regulations and intake of zearalenone: An oestrogenic mycotoxin. *Food Chem. Toxicol.* **2007**, *45*, 1–18. [CrossRef] [PubMed]
7. International Agency for Research on Cancer. *IARC Monographs on the Evaluation of Carcinogenic Risks to Humans*; International Agency for Research on Cancer: Lyon CEDEX, France, 2002; Volume 96. [CrossRef]
8. Pascari, X.; Ramos, A.J.; Marín, S.; Sanchís, V. Mycotoxins and beer. Impact of beer production process on mycotoxin contamination. A review. *Food Res. Int.* **2018**, *103*, 121–129. [CrossRef]
9. Bullerman, L.B.; Bianchini, A. Stability of mycotoxins during food processing. *Int. J. Food Microbiol.* **2007**, *119*, 140–146. [CrossRef] [PubMed]
10. Piacentini, K.C.; Savi, G.D.; Olivo, G.; Scussel, V.M. Quality and occurrence of deoxynivalenol and fumonisins in craft beer. *Food Control* **2015**, *50*, 925–929. [CrossRef]
11. Piacentini, K.C.; Rocha, L.O.; Savi, G.D.; Carnielli-Queiroz, L.; Almeida, F.G.; Minella, E.; Corrêa, B. Occurrence of deoxynivalenol and zearalenone in brewing barley grains from Brazil. *Mycotoxin Res.* **2018**. [CrossRef]
12. Wegulo, S.N.; Baenziger, P.S.; Hernandez Nopsa, J.; Bockus, W.W.; Hallen-Adams, H. Management of Fusarium head blight of wheat and barley. *Crop Prot.* **2015**, *73*, 100–107. [CrossRef]
13. *Brasil Limites Máximos Tolerados (LMT) Para Micotoxinas*; Agência Nacional Vigilância Sanitária: ANVISA, Brazil, 2017.

14. European Commission. Commission Regulation (EC) No 1881/2006 of 19 December 2006 setting maximum levels for certain contaminants in foodstuffs. *Off. J. Eur. Union* **2006**, *L364*, 5–24.
15. Karlsson, I.; Friberg, H.; Kolseth, A.K.; Steinberg, C.; Persson, P. Agricultural factors affecting Fusarium communities in wheat kernels. *Int. J. Food Microbiol.* **2017**, *252*, 53–60. [CrossRef] [PubMed]
16. Hofer, K.; Geißinger, C.; König, C.; Gastl, M.; Hückelhoven, R.; Heß, M.; Coleman, A.D. Influence of Fusarium isolates on the expression of barley genes related to plant defense and malting quality. *J. Cereal Sci.* **2016**, *69*, 17–24. [CrossRef]
17. Silva, J.J.; Viaro, H.P.; Ferranti, L.S.; Oliveira, A.L.M.; Ferreira, J.M.; Ruas, C.F.; Ono, E.Y.S.; Fungaro, M.H.P. Genetic structure of Fusarium verticillioides populations and occurrence of fumonisins in maize grown in Southern Brazil. *Crop Prot.* **2017**, *99*, 160–167. [CrossRef]
18. Choi, S.; Jun, H.; Bang, J.; Chung, S.H.; Kim, Y.; Kim, B.S.; Kim, H.; Beuchat, L.R.; Ryu, J.H. Behaviour of Aspergillus flavus and Fusarium graminearum on rice as affected by degree of milling, temperature, and relative humidity during storage. *Food Microbiol.* **2015**, *46*, 307–313. [CrossRef] [PubMed]
19. Garcia, D.; Barros, G.; Chulze, S.; Ramos, A.J.; Sanchis, V.; Marín, S. Impact of cycling temperatures on Fusarium verticillioides and Fusarium graminearum growth and mycotoxins production in soybean. *J. Sci. Food Agric.* **2012**, *92*, 2952–2959. [CrossRef] [PubMed]
20. Llorens, A.; Mateo, R.; Hinojo, M.J.; Valle-Algarra, F.M.; Jiménez, M. Influence of environmental factors on the biosynthesis of type B trichothecenes by isolates of *Fusarium* spp. from Spanish crops. *Int. J. Food Microbiol.* **2004**, *94*, 43–54. [CrossRef] [PubMed]
21. Fleurat-Lessard, F. Integrated management of the risks of stored grain spoilage by seedborne fungi and contamination by storage mould mycotoxins—An update. *J. Stored Prod. Res.* **2017**, *71*, 22–40. [CrossRef]
22. Pitt, J.I.; Hocking, A.D. *Fungi and Food Spoilage*; Springer: Berlin/Heidelberg, Germany, 2009; ISBN 9780387922065.
23. Savi, G.D.; Piacentini, K.C.; Rocha, L.O.; Carnielli-Queiroz, L.; Furtado, B.G.; Scussel, R.; Zanoni, E.T.; Machado-de-Ávila, R.A.; Corrêa, B.; Angioletto, E. Incidence of toxigenic fungi and zearalenone in rice grains from Brazil. *Int. J. Food Microbiol.* **2018**, *270*. [CrossRef]
24. Stakheev, A.A.; Khairulina, D.R.; Zavriev, S.K. Four-locus phylogeny of Fusarium avenaceum and related species and their species-specific identification based on partial phosphate permease gene sequences. *Int. J. Food Microbiol.* **2016**, *225*, 27–37. [CrossRef]
25. Beccari, G.; Caproni, L.; Tini, F.; Uhlig, S.; Covarelli, L. Presence of Fusarium species and other toxigenic fungi in malting barley and multi-mycotoxin analysis by liquid chromatography-high-resolution mass spectrometry. *J. Agric. Food Chem.* **2016**, *64*, 4390–4399. [CrossRef]
26. Hietaniemi, V.; Rämö, S.; Yli-Mattila, T.; Jestoi, M.; Peltonen, S.; Kartio, M.; Sieviläinen, E.; Koivisto, T.; Parikka, P. Updated survey of Fusarium species and toxins in Finnish cereal grains. *Food Addit. Contam. Part A Chem. Anal. Control. Expo. Risk Assess.* **2016**, *33*, 831–848. [CrossRef] [PubMed]
27. Astolfi, P.; dos Santos, J.; Schneider, L.; Gomes, L.B.; Silva, C.N.; Tessmann, D.J.; Del Ponte, E.M. Molecular survey of trichothecene genotypes of Fusarium graminearum species complex from barley in Southern Brazil. *Int. J. Food Microbiol.* **2011**, *148*, 197–201. [CrossRef] [PubMed]
28. Ibáñez-Vea, M.; González-Peñas, E.; Lizarraga, E.; López de Cerain, A. Co-occurrence of mycotoxins in Spanish barley: A statistical overview. *Food Control* **2012**, *28*, 295–298. [CrossRef]
29. Juan, C.; Ritieni, A.; Mañes, J. Occurrence of Fusarium mycotoxins in Italian cereal and cereal products from organic farming. *Food Chem.* **2013**, *141*, 1747–1755. [CrossRef] [PubMed]
30. Juan, C.; Berrada, H.; Maes, J.; Oueslati, S. Multi-mycotoxin determination in barley and derived products from Tunisia and estimation of their dietary intake. *Food Chem. Toxicol.* **2017**, *103*, 148–156. [CrossRef] [PubMed]
31. 401/2006/EC Commision Regulation (EC) No 401/2006 of 23 February 2006 laying down the methods of sampling and analysis for the official control of the levels of mycotoxins in foodstuffs. *Off. J. Eur. Union* **2006**, *L70*, 12–34.
32. Bolechová, M.; Benešová, K.; Běláková, S.; Čáslavský, J.; Pospíchalová, M.; Mikulíková, R. Determination of seventeen mycotoxins in barley and malt in the Czech Republic. *Food Control* **2015**, *47*, 108–113. [CrossRef]
33. Běláková, S.; Benešová, K.; Čáslavský, J.; Svoboda, Z.; Mikulíková, R. The occurrence of the selected fusarium mycotoxins in czech malting barley. *Food Control* **2014**, *37*, 93–98. [CrossRef]
34. Piacentini, K.C.; Savi, G.D.; Pereira, M.E.V.; Scussel, V.M. Fungi and the natural occurrence of deoxynivalenol and fumonisins in malting barley (Hordeum vulgare L.). *Food Chem.* **2015**, *187*. [CrossRef]

35. Poreda, A.; Czarnik, A.; Zdaniewicz, M.; Jakubowski, M.; Antkiewicz, P. Corn grist adjunct—Application and influence on the brewing process and beer quality. *J. Inst. Brew.* **2014**, *120*, 77–81. [CrossRef]
36. Piacentini, K.C.; Rocha, L.O.; Fontes, L.C.; Carnielli, L.; Reis, T.A.; Corrêa, B. Mycotoxin analysis of industrial beers from Brazil: The influence of fumonisin B1and deoxynivalenol in beer quality. *Food Chem.* **2017**, *218*. [CrossRef] [PubMed]
37. Oliveira, M.S.; Rocha, A.; Sulyok, M.; Krska, R.; Mallmann, C.A. Natural mycotoxin contamination of maize (Zea mays L.) in the South region of Brazil. *Food Control* **2016**, 1–6. [CrossRef]
38. Zhang, H.; Dong, M.; Yang, Q.; Apaliya, M.T.; Li, J.; Zhang, X. Biodegradation of zearalenone by Saccharomyces cerevisiae: Possible involvement of ZEN responsive proteins of the yeast. *J. Proteom.* **2016**, *143*, 416–423. [CrossRef] [PubMed]
39. Gareis, M.; Bauer, J.; Thiem, J.; Plank, G.; Grabley, S.; Gedek, B. Cleavage of Zearalenone-Glycoside, a "Masked" Mycotoxin, during Digestion in Swine. *J. Vet. Med. Ser. B* **1990**, *37*, 236–240. [CrossRef]
40. Efsa Scientific Opinion on the risks for animal and public health related to the presence of Alternaria toxins in feed and food. *EFSA J.* **2011**, *9*, 97. [CrossRef]
41. Wu, L.; Qiu, L.; Zhang, H.; Sun, J.; Hu, X.; Wang, B. Optimization for the production of deoxynivalenol and zearalenone by Fusarium graminearum using response surface methodology. *Toxins (Basel)* **2017**, *9*. [CrossRef] [PubMed]
42. AOAC. *Official Methods of Analysis of AOAC International*; AOAC: Rockville, MD, USA, 2005; ISBN 0935584544.
43. Silva, N.D.; Junqueira, V.C.; Silveira, N.F.; Taniwaki, M.; Santos, R.F.; Gomes, R.A. *Manual de Métodos de Análise Microbiológica de Alimentos e Água*, 4th ed.; Português: Sao Paulo, Barzil, 2010.
44. Degola, F.; Berni, E.; Dall'Asta, C.; Spotti, E.; Marchelli, R.; Ferrero, I.; Restivo, F.M. A multiplex RT-PCR approach to detect aflatoxigenic strains of Aspergillus flavus. *J. Appl. Microbiol.* **2007**, *203*, 409–427. [CrossRef] [PubMed]
45. Geiser, D.M.; Jiménez-Gasco, M.D.M.; Kang, S.; Makalowska, I.; Veeraraghavan, N.; Ward, T.J.; Zhang, N.; Kuldau, G.A.; O'Donnell, K. FUSARIUM-ID v. 1.0: A DNA sequence database for identifying *Fusarium*. *Eur. J. Plant Pathol.* **2004**, *110*, 473–479. [CrossRef]
46. O'Donnell, K.; Ward, T.J.; Geiser, D.M.; Kistler, H.C.; Aoki, T. Genealogical concordance between the mating type locus and seven other nuclear genes supports formal recognition of nine phylogenetically distinct species within the Fusarium graminearum clade. *Fungal Genet. Biol.* **2004**, *41*, 600–623. [CrossRef]
47. Swofford, D.L. PAUP* phylogenetic analysis using parsimony (*and other methods). Version 4.0b10. *Sinauer Assoc.* **2002**. [CrossRef]
48. Swofford, D.L. Phylogenetic Analysis Using Parsimony. *Options* **2002**, *42*, 294–307. [CrossRef]
49. O'Donnell, K.; Rooney, A.P.; Proctor, R.H.; Brown, D.W.; McCormick, S.P.; Ward, T.J.; Frandsen, R.J.N.; Lysøe, E.; Rehner, S.A.; Aoki, T.; et al. Phylogenetic analyses of RPB1 and RPB2 support a middle Cretaceous origin for a clade comprising all agriculturally and medically important fusaria. *Fungal Genet. Biol.* **2013**, *52*, 20–31. [CrossRef] [PubMed]
50. Al-Taher, F.; Cappozzo, J.; Zweigenbaum, J.; Lee, H.J.; Jackson, L.; Ryu, D. Detection and quantitation of mycotoxins in infant cereals in the U.S. market by LC-MS/MS using a stable isotope dilution assay. *Food Control* **2017**, *72*, 27–35. [CrossRef]
51. Savi, G.D.; Vitorino, V.; Bortoluzzi, A.J.; Scussel, V.M. Effect of zinc compounds on Fusarium verticillioides growth, hyphae alterations, conidia, and fumonisin production. *J. Sci. Food Agric.* **2013**, *93*, 3395–3402. [CrossRef]
52. Savi, G.D.; Bortoluzzi, A.J.; Scussel, V.M. Antifungal properties of Zinc-compounds against toxigenic fungi and mycotoxin. *Int. J. Food Sci. Technol.* **2013**, *48*, 1834–1840. [CrossRef]
53. Njumbe Ediage, E.; Van Poucke, C.; De Saeger, S. A multi-analyte LC-MS/MS method for the analysis of 23 mycotoxins in different sorghum varieties: The forgotten sample matrix. *Food Chem.* **2015**, *177*, 397–404. [CrossRef] [PubMed]
54. EC Commission regulation (EC) 401/2006 of 23 february 2006 laying down the methods of sampling and analysis for the official control of the levels of mycotoxins in foodstuffs. *Off. J. Eur. Communities* **2000**, *L269*, 1–15.

toxins

MDPI

Article

Fusarium Mycotoxins Stability during the Malting and Brewing Processes

Karim C. Piacentini [1,*], Sylvie Běláková [2], Karolína Benešová [2], Marek Pernica [2], Geovana D. Savi [3], Liliana O. Rocha [4], Ivo Hartman [2], Josef Čáslavský [5] and Benedito Corrêa [1]

1 Biotechnology Department, University of Sao Paulo, Av. Professor Lineu Prestes, Sao Paulo 2415, Brazil; correabe@usp.br

2 Research Institute of Brewing and Malting, Malting Institute Brno, Mostecká 7, 614 00 Brno, Czech Republic; belakova@beerresearch.cz (S.B.); benesova@beerresearch.cz (K.B.); pernica@beerresearch.cz (M.P.); hartman@beerresearch.cz (I.H.)

3 University of the Extreme Southern Santa Catarina, Av. Universitaria, Criciuma 1105, Santa Catarina, Brazil; geovanasavi@gmail.com

4 Department of Food Science, Food Engineering Faculty, University of Campinas, Av. Monteiro Lobato, 80., Campinas 13083-862, Brazil; lrocha@unicamp.br

5 Institute of Food Chemistry and Biotechnology, Faculty of Chemistry, Brno University of Technology, Purkyňova 118, 612 00 Brno, Czech Republic; caslavsky@fch.vut.cz

* Correspondence: karim.piacentini@hotmail.com; Tel.: + 55-11-3091-7728

Received: 15 April 2019; Accepted: 4 May 2019; Published: 7 May 2019

Abstract: Mycotoxins are widely studied by many research groups in all aspects, but the stability of these compounds needs further research for clarification. The objective of this study is to evaluate deoxynivalenol and zearalenone stability during all steps of the malting and brewing processes. The levels of these compounds decreased significantly during the production process (barley to beer). During the malting process, the DON levels decreased significantly in the steeping, germination, and malting steps (62%, 51.5%, and 68%, respectively). Considering ZEN, when the levels were compared between barley and the last step of the process, a significant decrease was observed. Most of the mycotoxins produced were transferred to the rootlets and spent grains, which is advantageous considering the final product. Furthermore, the mycotoxin dietary intake estimation was included in this study. The results proved that if the concentrations of target mycotoxins in raw material are under the limits established by the regulations, the levels decrease during the malting and brewing processes and make the beer secure for consumers. The quality of the five commodities involved in the beer process plays a decisive role in the creation of a safe final product.

Keywords: mycotoxins; stability; malting; brewing; beer

Key Contribution: Deoxynivalenol and Zearalenone shown to decrease in levels over the entire process with significant difference. This study showed that if the raw material is under the limits established by the regulations, the levels can decrease during the process and make the beer safe for consumers.

1. Introduction

Fusarium contamination in cereals has increased in recent years, mainly in barley, wheat, corn, and oats [1–5]. This can be related to changes in climate condition, but is also due to problems during cultivation and harvest. Along with fungi contamination, temperature, and changes in forecast, the production of secondary metabolites occurs.

Mycotoxins are being widely studied by many research groups in all aspects, such as the occurrence of these compounds in food, effects on animals and also on humans, prevention, and mitigation [2,6,7]. The stability of mycotoxins is also studied, but all aspects need further research for clarification and perhaps help in the control of contamination.

In this context, the beer industry has several concerns about fungi and mycotoxin contamination due to the loss of raw material, economic impact, and, one of the most significant issues, loss of quality.

Two *Fusarium* toxins that are considered hazardous to the entire industry are deoxynivalenol (DON) and zearalenone (ZEN). These toxins are very common and can be found throughout the entire malting and brewing processes [8].

DON, also named vomitoxin, is a trichothecene B-type that elicits a complex spectrum of toxic effects. Chronic exposure to low doses of DON can lead to anorexia, impaired weight gain, and immunotoxicity. Acute exposure to high doses can cause diarrhea, vomiting, leukocytosis, circulatory shock, and, ultimately, death [9].

Trichothecenes are products of the sesquiterpenoid metabolism of some genera in the order Hypocreales [10] and, among them, deoxynivalenol (DON) is notorious for its worldwide occurrence. This mycotoxin is mainly produced by phylogenetic species within the *Fusarium graminearum* species complex, which includes the *F. graminearum sensu stricto* [11]. This species is encountered on barley, maize, and wheat causing Fusarium head blight, a disease that leads to yield losses in small grain cereals.

ZEN is another mycotoxin that should be taken into account due to the high contamination found in recent monitoring studies in grains [2,3,12]. It is treated as a non-steroidal estrogenic compound because its structure is similar to the hormone estrogen and it competes with 17β-estradiol for binding of the estrogen receptor, resulting in infertility and reproductive problems [13].

Regulations have been established worldwide to fix the maximum levels allowed of mycotoxins in grains. In Europe, the levels for DON and ZEN in barley are set at 1250 μg/kg and 100 μg/kg, respectively [14]. In Brazil, the regulations are being updated and in 2019 the DON levels for barley will be set at 1000 μg/kg for barley and 750 μg/kg for malt. Considering ZEN, the levels will be maintained at 100 μg/kg [15].

Some studies have shown a high contamination of these two mycotoxins in beer, but no studies have been developed to control or decrease the levels. The beer industry requires more information about the stability of these compounds during the entire process.

For the reasons stated above, the objective of the current study is to evaluate the stability of DON and ZEN during all steps of the malting and brewing processes, considering that barley grains constantly pass through different conditions, which can increase or decrease the levels of these metabolites. To our knowledge, this will be the first study carried out considering mycotoxins in the entire beer chain.

2. Results

2.1. Method Validation

The methods for extraction of mycotoxin grains and liquid matrices during the malting and brewing process were validated according to the Commission Regulation [16] guidelines. To determine the recovery, repeatability, and selectivity/specificity, solid and liquid samples with non-detectable levels of mycotoxins were submitted to spiking experiments.

The blank solid malt control sample was used (produced from the barley variety Kangoo) with no targeted mycotoxins detected. The spiking levels were chosen considering the contamination levels of examined samples. The blank liquid sample used was wort (from the above-mentioned malt, according to Section 4.3). The spiking levels of the liquid matrix were chosen considering the predicted contamination levels of examined samples of produced wort and beer. The standards were applied to the matrix in methanol:water (1:1 v:v). Average recovery was calculated using triplicate analyses conducted for each level. The experiments were repeated on three different days.

The limit of detection (LOD) was defined as the minimum concentration of an analyte in the spiked sample with a signal–noise ratio equal to 3 and limit of quantification (LOQ) with a signal–noise ratio equal to 10. Considering linearity, a seven-point calibration dependence was constructed with the following concentrations of the mycotoxin standard mixture (DON and ZEN): 5, 10, 20, 50, 100, 200, and 1000 ng/mL. The coefficients of determination (r^2) for DON and ZEN were 0.9994 and 0.9992, respectively. The parameters achieved are shown in the Table 1.

Table 1. Validation parameters of the method.

Analytes	Matrix	Spiking Level (ng/mL)	Recovery (%)	RSD [c] (%)	LOD *[d]	LOQ *[e]
DON [a]	Grains (malt)	100	92.1	9.2	3.5	11.6
		500	92.9	2.9		
		1000	87.6	4.4		
	Liquid (wort)	50	90.6	1.4	0.5	1.6
		100	91.5	2.0		
		250	94.6	1.4		
ZEN [b]	Grains (malt)	50	89.6	5.3	2.8	9.2
		250	99.8	2.9		
		500	105.7	0.4		
	Liquid (wort)	10	85.8	1.1	2.0	6.7
		50	89.2	0.9		
		100	97.9	0.8		

* grain: µg/kg; liquid: µg/L; [a] DON: deoxynivalenol; [b] ZEN: zearalenone; [c] RSD: relative standard deviation; [d] LOD: limit of detection; [e] LOQ: limit of quantification.

2.2. DON and ZEN Stability during the Malting and Brewing Processes

The stability of the mycotoxins was evaluated in the steps considered as the most important ($N = 13$) of the malting and brewing processes (Figure 1). First of all, as aforementioned, the barley used for the analysis was naturally contaminated and the levels among ten samples had an average of 3835 µg/kg and 1069 µg/kg for DON and ZEN, respectively (Table 2).

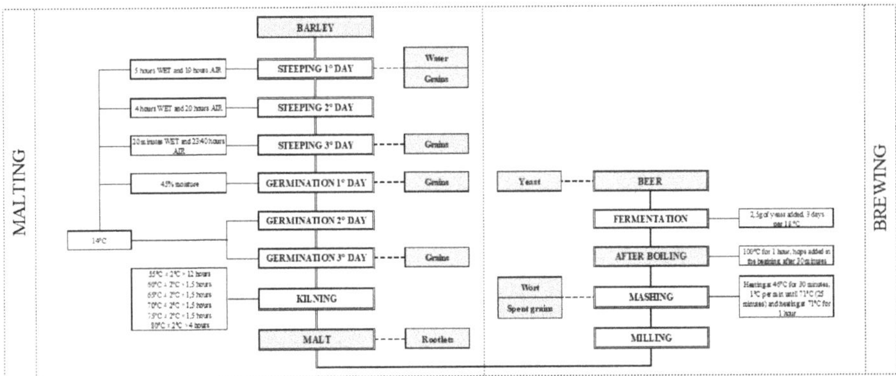

Figure 1. Map of malting and brewing process sampling (blocks in gray) carried out in the study.

Table 2. Average, minimum, and maximum values for all of the steps of the malting and brewing processes.

Process	Steps	DON (μg/kg)			ZEN (μg/kg)		
		Average	Maximum	Minimum	Average	Maximum	Minimum
Malting	Barley	3835	5204	2687	1070	3596	175
	Water 1st day	599	851	470	12	21	3
	Steeping 1st day	1319	2906	438	222	1045	53
	Steeping 3rd day	1442	2103	754	799	2602	15
	Germination 1st day	1308	2641	617	618	2385	57
	Germination 3rd day	1858	2957	822	1172	2854	140
	Malt	1211	1780	728	392	1386	95
	Rootlets	1797	2181	1431	1122	1735	206
Brewing	Wort	1105	1505	808	25	64	9
	Spent grains	1068	1742	590	1429	3188	200
	After boiling	1132	1483	840	<LOQ	<LOQ	<LOQ
	After fermentation/beer	1089	1414	863	<LOQ	<LOQ	<LOQ
	Yeast	166	241	89	<LOQ	<LOQ	<LOQ

Considering DON in the malting process, water was analyzed first and an average of 599 μg/kg was found. In this experiment, the mass balance of mycotoxins DON and ZEN was always related to 1 kg of the original barley and the reported values in malt, wort, beer after boiling, and beer after fermentation were recalculated according to the weight balance, where barley is 100%.

Detected levels in byproducts (yeasts, spent grains, and rootlets) were reported in μg/kg and were not included in the mass balance.

Comparing DON levels found in barley with levels found on the first and third day of steeping and the third day of germination and malt, DON had a decrease of 66%, 62%, 51.5%, and 68%, respectively, and were significantly different (*** $p < 0.001$) (Figure 2). On the other hand, DON decreased significantly (38%) between the third day of germination and malt (* $p < 0.05$).

The ZEN levels during the malting process decreased significantly (79%) when barley was compared to the first day of steeping (* $p < 0.05$). Nevertheless, an increase could be seen between the first day of steeping and the third day of germination (173%), with a significant difference (** $p < 0.01$).

Further analyses were completed in brewing. Considering the entire process (barley to beer), DON showed a significant (*** $p < 0.001$) decrease of 71.6%. In the same context, when ZEN levels were compared, a significant decrease was also observed (** $p < 0.01$).

It is necessary to highlight that all of the results for the liquid matrix were recalculated to 1 kg.

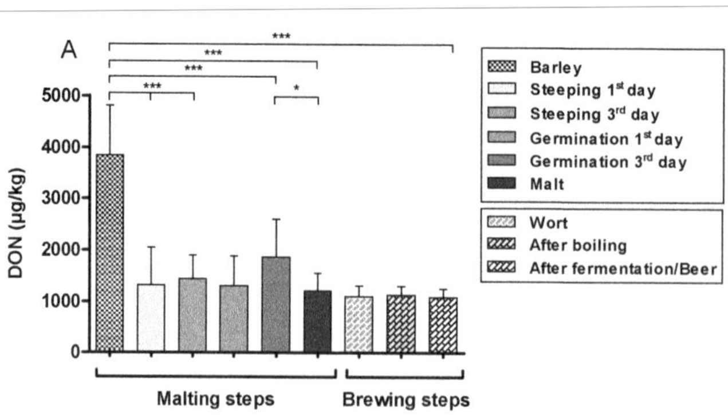

Figure 2. *Cont.*

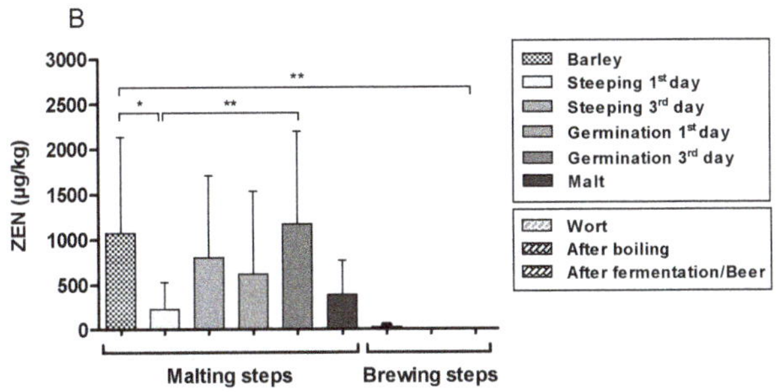

Figure 2. Steps of the malting and brewing processes showing (**A**) DON and (**B**) ZEN levels expressed as mean ± SD and different significance between the steps, according to ANOVA followed by Tukey test (* $p < 0.05$; ** $p < 0.01$, *** $p < 0.001$).

2.3. Mycotoxin Dietary Intake Estimation from Beer Consumption

Using an average adult body weight of 60 kg and the mean mycotoxin data from this study, an estimation of dietary exposure to DON from the last step of the process while following the levels established by the regulation was carried out (Table 3).

Table 3. Mycotoxin dietary intake estimation from beer consumption.

	DON	
	Level Study	Level Regulation
Mean (μg/L)	272.17	88.75
Daily average exposure (μg/kg/bw)	0.77	0.25
Tolerable daily intake (μg/kg/bw)	1	1
% of tolerable daily intake	77	25

It is important to mention that the level of the mycotoxin in the last step was 1089 μg, recalculated to 1 kg. Based on the literature [17], 1 kg of barley is able to produce 4 L of beer. 272 μg/L of DON in beer is equivalent to 1088 μg/kg DON in barley. This corresponds to a 71.6% reduction of DON content in the final product, whereas the average DON concentration in the original barley was 3835 μg/kg. Applying the same model to barley contaminated with 1250 μg/kg DON (the maximum allowable limit), the resulting DON concentration in beer will be 89 μg/L.

3. Discussion

As the literature shows, mycotoxins are highly stable compounds (resistant to high temperatures and extreme pH levels) [18]. Although the malting and brewing processes have maximum operation temperatures below the ones able to destroy the mycotoxins, they may influence mycotoxin concentration due to physical, chemical, and biochemical changes that take place [17,19]

In the current study, mycotoxin levels were determined in seven steps of the malting process. First, barley and water were analyzed and the results showed levels for DON in both matrices (barley average 3835 μg/kg, water average 599 μg/kg). The amount of DON found in the water can be explained by the soluble characteristic of this mycotoxin; for this reason, it was eluted from the matrix [20]. On the other hand, low levels for ZEN in water were found.

Furthermore, when DON levels of barley were compared to the first and third day of steeping, a decrease in the levels was observed. These results could also be related to the loss of mycotoxins in the water.

The next step involved in malting is germination. In this study, the samples were taken from the first and third day of this process. The levels of DON decreased significantly between barley and the third day of germination. A possible reason for this reduction (DON glycosylation) is due to the increase in glucose content, which might activate the barley enzyme responsible for the respective reaction, transforming DON into DON-3-glucoside. Some studies have shown that 50% of DON is converted after a few days of germination [21]. To our knowledge, there is no study carried out considering ZEN glycosylation during germination in barley. However, the results of the current study could suggest this reaction (between barley and the first day of germination), transforming ZEN in α- and β-zearalenol and other masked toxins, even though the decrease in the levels is considered without significant difference [22].

Additionally, during germination, fungal biomass and mycotoxin production may grow, most likely as a result of cross-contamination from the residual steeping water or because of a latent barley grain infection, which may be activated with the increase in humidity during this process [23]. This can be seen in the current study when ZEN levels from the first day of steeping and the third day of germination are compared.

The final procedure in malting is kilning, where the germination is interrupted and the green malt is prepared for storage and transportation. This step is characterized by several temperature scales (55–100 °C) and it is considered crucial to malt flavor and color formation [19]. In the current study, DON levels were shown to decrease significantly between the third day of germination and malt. This could be related to the high temperatures, but also due to the separation of the rootlets from the grain.

Malt and rootlets were analyzed for both toxins and it is necessary to highlight the eminent amount of both toxins found in rootlets. For the beer industry, it was considered a good result, as most of the mycotoxins are released in this part of the grain. On the other hand, it should be a concern for agriculture, as this matrix is frequently used to feed animals [24–26].

After barley is malted, the grains are milled, and water is added to start the brewing process. In the mashing phase, the water is heated to about 62–71 °C and enzymes, such as α- and β-amylases, are activated to allow the conversion of starches into fermentable sugars [17]. Concerning the influence on mycotoxin levels at this step, there is a possible release of DON conjugated to protein structures and, as a result, there is an increase in total DON concentration [27]. This fact could be seen in the current study for both mycotoxins, not only for DON. Being more specific, DON and ZEN had an average malt level of 1211 μg/kg and 391 μg/kg, respectively. In mashing, if the results of wort and spent grains are added, the value is almost twice as high for DON and almost four times as high for ZEN (DON: 2173 μg/kg; ZEN: 1460 μg/kg). Also, the milling process could interfere in the levels due to the mycotoxin homogeneous spread into the entire malt batch and its later solubilization into mashing water [8].

Boiling is the next stage and it is characterized by the enzyme inactivation, hops addition, isomerization of hop alfa acids, evaporation of water and volatile compounds (dimethyl sulphides, undesirable in beer), protein precipitation, sterilization, Maillard reactions, and flavour modulation [17]. Regarding the impact of wort boiling on mycotoxin content, the temperature in this process, (above 100 °C) and the average time of boiling (1 h) could cause a decrease in mycotoxin concentration [8,28]. This experience can be seen in the results of the present study, even in low concentrations.

On the other hand, a few authors have shown that DON and ZEN are very stable compounds and the melting points in food processing are around 153 °C and 165 °C, respectively [18,29]. Another thing that should be mentioned in this stage is the addition of adjuncts, such as corn, wheat, and sorghum, which might represent an additional source of mycotoxins [1,30].

After the boiling process, the wort is cooled to 21 °C and the fermentation of wort is initiated by the yeast (*Saccharomyces* genus). In general, the fermentation temperature ranges from 2 °C to 30 °C, over 7–9 days [31,32]. DON was shown to be very stable in this process and the results did not show significant differences (after boiling: 1132 μg/kg; after fermentation: 1089 μg/kg = 3.8% of apparent reduction). In research performed by [33], similar results were found, where there was a reduction of only 11.6% of DON.

The adsorption of mycotoxins by yeasts have been studied and this phenomenon occurs due to the binding of the toxins to ß-glucans from yeast to the cell wall [34]. In the current study, the yeasts were also analyzed after fermentation and DON was shown to be adsorbed. For ZEN, any amount of this mycotoxin found after boiling made this determination impossible in the present experiments. Nevertheless, in a study conducted by [33], ZEN was shown to have a very high ratio of adsorption (75.1%) on the beer fermentation residues.

Finally, considering the results of tolerable daily intake, the consumption of 62 L/year per capita [35] (0.17 L/day of beer) resulted in an exposure of 0.77 μg/kg/bw/day from the beer in this study and 0.25 μg/kg/body weight (bw)/day from beer using the levels established by the regulation for a 60 kg adult. Both these estimated daily intake (DI) results are still lower than the provisional maximum tolerable daily intake (PMTDI) of 1 μg/kg bw/day for DON stipulated by the Joint Expert Committee on Food Additives (JECFA) [36].

4. Material and Methods

4.1. Samples

A total of 10 barley samples (*Hordeum vulgare* L.) with a weight of 500 g each were used from a 2015 harvest and were supplied by the Brazilian agricultural research corporation (Embrapa). It is important to mention that the samples studied had natural mycotoxins levels acquired in the field without artificial interference.

The samples used in the current study were sown in May 2015 and harvested in December 2015. During this period, above-average rainfall with a mean precipitation of 95 mm was recorded [37]. Furthermore, the humidity was also considered significant, with an average value of 80% [38]. Another parameter included was the high temperature (average of 29 °C). These barley samples contained a concentration average of 3835 μg/kg for DON and 1070 μg/kg for ZEN, according to our previous study [2]; the environmental conditions presumably affected the mycotoxin production, thereby justifying the levels found.

The samples were used in the entire processes of malting and brewing (microscale) and aliquots were taken in the selected steps for mycotoxin analysis.

4.2. Chemicals and Reagents

Deoxynivalenol and zearalenone standards were purchased from Sigma Aldrich (Vienna, Austria). The stock solutions standards were prepared in methanol:water (1:1) at a concentration of 1 μg/mL.

All reagents used in the following analyses were analytical and LC/MS-MS grade. For quality assurance in the mycotoxin analyses, wheat flour certified reference materials (CRM) (Trilogy, Washington) were used. The concentrations for DON and ZEN were 700 ± 100 μg/kg and 454.2 ± 37.6 μg/kg, respectively.

4.3. Malting and Brewing Process

All of the malting and brewing trials were carried out in the Research Institute of Malting and Brewing, Brno, Czech Republic. The procedures used for malting were according to the [39] methodology and amendments were approved by the barley and malt European Brewery Convention [32].

The first step of malting was steeping, carried out as follows: On the 1st day of steeping, water was added to the grains for 5 h, followed by 19 h of air rest. On the 2nd day, grains were submitted to 4 h in the water, followed by 20 h of air rest. Finally, on the 3rd day of steeping, the grains were in the water for 20 min and then submitted to air rest for 23 h and 40 min. The germination step was performed over 3 days (72 h), with a subsequent kilning step. The grains in the steeping and germination processes were always maintained at a controlled temperature (14 °C) and the moisture content of the grains was measured in each step, being controlled to 45%. The total kilning time was 22 h, with a pre-kilning temperature of 55 °C for 12 h and a kilning temperature of 80 °C for 4 h.

Furthermore, the methodology used for wort production was according to [39] and the [32], with slight modifications. Fifty grams of milled malt was weighed into the mashing beaker and 200 mL of water (45 °C) was added. The beaker was placed into a programmed mashing bath with automatic stirrers. The temperature was maintained for 30 min under constant stirring (100 rpm). Following this period, the temperature of the mashing bath was increased 1 °C per minute for 25 min. Once the temperature reached 70 °C, 100 mL of water at 70 °C was added and maintained at 70 °C for 1 h. The beaker was then removed and cooled to room temperature. The contents of the beaker were adjusted to 450 g with water, stirred, and the contents filtered through filter paper. Wort and spent grains were used for further analyses.

The methods carried out for the final steps of the brewing process, such as boiling and fermentation, were developed by the Research Institute of Brewing and Malting and also according to available literature [31,32], in an attempt to simulate the real brewing process. In short, each sample with 200 mL of the wort was boiled with 8 g of traditional hops for one hour. It is necessary to mention that the hops were added at two different times, in the beginning (4 g) and after 30 min of boiling (4 g). The boiled wort was cooled to 21 °C for the yeast (W 34/70Starobrno Lager Yeast, *Saccharomyces pastorianus*) addition. The fermentation process was carried out over the course of 3 days at a controlled temperature (18 °C).

The sampling for mycotoxin analysis was performed in 8 steps of the malting process and 5 steps in the brewing process, as shown in Figure 1.

4.4. Extraction and Mycotoxin Cleanup

To accomplish the procedure for extraction and cleanup of DON and ZEN analysis, immunoaffinity columns (DZT MS-Prep—R. Biopharm, Glasgow, Scotland) for both mycotoxins were used. This step was performed according to the DZT MS-Prep protocol with some modifications. In short, for grains, 10 g of milled barley, malt, rootlets, and spent grains were mixed with methanol 70% for 50 min, followed by 15 min of centrifugation at 4500 rpm. Then, 2 mL of the extract were added to 48 mL of PBS buffer (pH 7.4, adjusted with 2M NaOH) [40] and shacked manually. Twenty mL of the diluted extract were passed through the column at one drop per second. In addition, to wash the column, 20 mL of distilled water was added. The elution was performed with 2 mL of 100% methanol. For injection, the extract was dried and then dissolved with 1 mL of 50% methanol in water. For the yeast extract, the procedure described above was applied to 2 g per sample.

The DZT MS-Prep columns were also used for mycotoxin extraction from a liquid matrix. In summary, 2 mL of wort or beer was added to 18 mL of PBS buffer and then loaded to the column. For washing, 20 mL distilled water was used, followed by elution with 2 mL of 100% methanol. Finally, the extract was dried and dissolved in 1 mLof 50% methanol for injection.

4.5. LC/MS-MS Method for Analysis

To accomplish the identification and quantification of mycotoxins (DON and ZEN), LC/MS-MS system consisting of Finnigan Surveyor HPLC coupled to the ion trap LCQ Advantage mass spectrometer (Thermo-Fisher, USA) with atmospheric pressure chemical ionization (APCI) was used.

The chromatographic conditions were adopted according to the procedure developed by [41]. Chromatographic separation was performed with Synergi Hydro RP 80A column (3.0 × 150 mm, 4.0 µm particle size) equipped with a Security Guard™ cartridge C18 (4.0 × 3.0 mm, 4.0 µm) at 30 °C using gradient elution. The mobile phase was comprised of solvent A (water containing 10 mM ammonium acetate) and solvent B (methanol). The gradient program was applied at a flow rate of 0.5 mL/min under the following conditions: 0.1 min 90% A; 2 min 50% A; 10 min 20% A; 15 min 20% A; 16 min 90%; 25 min 90% A. The total analytical run time was 25 min for the two toxins.

The APCI interface was operated at negative polarity and the following ionization conditions were used: Capillary temperature, 160 °C; source heater temperature of 450 °C; nitrogen sheath gas flow of 35 L/min; nitrogen auxiliary gas flow of 10 L/min; source voltage of 6.0 kV; collision gas was helium. For selectivity, the mass spectrometer was operated in MRM mode and two transitions per analyte were monitored (Table 4).

Table 4. Retention time and mass spectrometric parameters used in the analysis of the mycotoxins.

Mycotoxin	Retention Time (min.)	Precursor ion (m/z)	Product ion (m/z) *	Normalized CE (%)	Tube Lens
DON	6.3	355	295^C	34	15
			265^Q		
ZEN	16.2	317	273^C	64	15
			299^Q		

* C: Confirmation transition; Q: Quantification transition.

The mass spectrometry conditions were optimized by re-tuning different analytes via direct infusion of each analyte individually. The cone voltages, collision energies, and product ions were optimized and carefully chosen.

4.6. Estimation of the Average Tolerable Daily Intake (TDI) and Maximum Tolerable Daily Intake

The mycotoxin dietary intake estimation was calculated using the mean level of DON found in the end of the process (after fermentation/beer) which was divided by four, considering that 1 kg of barley is able to produce 4 L of beer [17]. The mean level obtained was multiplied by the daily consumption of beer according to [42] and divided by 60 kg (body weight). The TDI calculation was based on the tolerable intake 1 µg/kg body weight/day for DON [36].

4.7. Statistical Analysis

Results regarding DON and ZEN during the malting and brewing processes were reported as mean, maximum, and minimum using Microsoft Office Excel 2007. Also, analysis of variance (repeated measures ANOVA) using the Tukey test was conducted on the obtained data. The results are presented as mean ± standard deviation and values of $p < 0.05$ were considered statistically significant.

5. Conclusions

The malting and brewing processes may impact the stability of DON and ZEN. Both mycotoxins were shown to decrease in levels over the entire process (barley to beer), with significant difference. It is necessary to highlight that rootlets and spent grains were shown to discard most of the mycotoxins produced in the previous processes. However, it should be a concern for the agriculture, due to this matrix being destined for animal feeding.

The quality of the five commodities involved (barley, hops, water, yeast, and adjuncts) in the beer process plays a decisive role in the creation of a safe final product. Finally, according to mycotoxin dietary intake estimation, this study showed that if the raw material is under the limits established by the regulations, the levels can decrease during the process and make the beer safe for consumers.

Author Contributions: The authors' responsibilities were as follows—K.C.P. and S.B. designed the research; K.C.P., S.B., K.B., M.P., and I.H.: conducted the research; G.D.S., and L.O.R.: guided and performed the statistical analysis; J.C., S.B., and B.C.: Oversight and leadership responsibility for the research activity planning and execution, K.C.P. wrote the manuscript and hold primary responsibility for the final content; and all authors: read and approved the final manuscript.

Funding: The authors would like to thank FAPESP (Sao Paulo Research Foundation) grant numbers 2016/01798-4, 2018/01795-0 and also TACR TE020001777 "Centre for Innovative Use and Strengthening of Competitiveness of Czech Brewery Raw Materials and Products" for the financial support.

Acknowledgments: Furthermore, the authors would like to thank the technician involved, Pavel Mezulanik, in the analysis of this study.

Conflicts of Interest: The authors declare no conflict of interest.

References

1. Oliveira, M.S.; Rocha, A.; Sulyok, M.; Krska, R.; Mallmann, C.A. Natural mycotoxin contamination of maize (Zea mays L.) in the South region of Brazil. *Food Control* **2017**, *73*, 127–132. [CrossRef]
2. Piacentini, K.C.; Rocha, L.O.; Savi, G.D.; Carnielli-Queiroz, L.; Almeida, F.G.; Minella, E.; Corrêa, B. Occurrence of deoxynivalenol and zearalenone in brewing barley grains from Brazil. *Mycotoxin Res.* **2018**, *34*, 173–178. [CrossRef] [PubMed]
3. Tralamazza, S.M.; Bemvenuti, R.H.; Zorzete, P.; De Souza Garcia, F.; Corrêa, B. Fungal diversity and natural occurrence of deoxynivalenol and zearalenone in freshly harvested wheat grains from Brazil. *Food Chem.* **2016**, *196*, 445–456. [CrossRef] [PubMed]
4. Bolechová, M.; Benešová, K.; Běláková, S.; Čáslavský, J.; Pospíchalová, M.; Mikulíková, R. Determination of seventeen mycotoxins in barley and malt in the Czech Republic. *Food Control* **2015**, *47*, 108–113. [CrossRef]
5. Xu, W.; Han, X.; Li, F. Co-occurrence of multi-mycotoxins in wheat grains harvested in Anhui province, China. *Food Control* **2019**, *96*, 180–185. [CrossRef]
6. Savi, G.D.; Piacentini, K.C.; Bittencourt, K.O.; Scussel, V.M. Ozone treatment efficiency on Fusarium graminearum and deoxynivalenol degradation and its effects on whole wheat grains (*Triticum aestivum* L.) quality and germination. *J. Stored Prod. Res.* **2014**, *59*, 245–253. [CrossRef]
7. da Rocha, M.E.B.; Freire, F.D.C.O.; Maia, F.E.F.; Guedes, M.I.F.; Rondina, D. Mycotoxins and their effects on human and animal health. *Food Control* **2014**, *36*, 159–165. [CrossRef]
8. Pascari, X.; Ramos, A.J.; Marín, S.; Sanchís, V. Mycotoxins and beer. Impact of beer production process on mycotoxin contamination. A review. *Food Res. Int.* **2018**, *103*, 121–129. [CrossRef]
9. Pestka, J.J.; Smolinski, A.T. Deoxynivalenol: Toxicology and Potential Effects on Humans. *J. Toxicol. Environ. Heal. Part B* **2005**, *8*, 39–69. [CrossRef]
10. Rocha, L.O.; Laurence, M.H.; Proctor, R.H.; McCormick, S.P.; Summerell, B.A.; Liew, E.C.Y. Variation in type A trichothecene production and trichothecene biosynthetic genes in Fusarium goolgardi from Natural Ecosystems of Australia. *Toxins (Basel)* **2015**, *7*, 4577–4594. [CrossRef]
11. van der Lee, T.; Zhang, H.; van Diepeningen, A.; Waalwijk, C. Biogeography of Fusarium graminearum species complex and chemotypes: A review. *Food Addit. Contam. Part A* **2015**, *32*, 453–460.
12. Savi, G.D.; Piacentini, K.C.; Rocha, L.O.; Carnielli-Queiroz, L.; Furtado, B.G.; Scussel, R.; Zanoni, E.T.; Machado-de-Ávila, R.A.; Corrêa, B.; Angioletto, E. Incidence of toxigenic fungi and zearalenone in rice grains from Brazil. *Int. J. Food Microbiol.* **2018**, *270*, 5–13. [CrossRef]
13. Takemura, H.; Shim, J.Y.; Sayama, K.; Tsubura, A.; Zhu, B.T.; Shimoi, K. Characterization of the estrogenic activities of zearalenone and zeranol in vivo and in vitro. *J. Steroid Biochem. Mol. Biol.* **2007**, *103*, 170–177. [CrossRef]
14. Commission Regulation (EC) No 1881/2006 Maximum levels for certain contaminants in food. *Off. J. Eur. Union* **2006**, *2006*, 5–24.
15. BRASIL Limites máximos tolerados (LMT) para micotoxinas. *Agência Nac. Vigilância Sanitária ANVISA* **2017**, *2017*, 3.
16. 401/2006/EC Commision Regulation (EC) No 401/2006 of 23 February 2006 laying down the methods of sampling and analysis for the official control of the levels of mycotoxins in foodstuffs. *Off. J. Eur. Union* **2006**, *L70*, 12–34.
17. Briggs, D.E.; Boulton, C.A.; Brookes, P.A.; Stevens, R. *Brewing: Science and Practice*, 1st ed.; Woodhead Publishing: Cambridge, UK, 2004; ISBN 9781855734906.
18. Kabak, B. The fate of mycotoxins during thermal food processing. *J. Sci. Food Agric.* 2009. [CrossRef]
19. Briggs, D.E. *Malts and Malting*, 1st ed.; Blackie Academic & Professional: London, UK, 1998; ISBN 0412298007.
20. Schwarz, P.B.; Casper, H.H.; Beattie, S. Fate and Development of Naturally Occurring FusariumMycotoxins During Malting and Brewing. *J. Am. Soc. Brew. Chem.* **1995**, *53*, 101–127.
21. Maul, R.; Müller, C.; Rieß, S.; Koch, M.; Methner, F.J.; Irene, N. Germination induces the glucosylation of the Fusarium mycotoxin deoxynivalenol in various grains. *Food Chem.* **2012**, *131*, 274–279. [CrossRef]
22. Schneweis, I.; Meyer, K.; Engelhardt, G.; Bauer, J. Occurrence of Zearalenone-4-b-D-glucopyranoside in Wheat. *J. Agric. Food Chem.* **2002**, *50*, 1736–1768. [CrossRef] [PubMed]
23. Vegi, A.; Schwarz, P.; Wolf-Hall, C.E. Quantification of Tri5 gene, expression, and deoxynivalenol production during the malting of barley. *Int. J. Food Microbiol.* **2011**, *150*, 150–156. [CrossRef]

24. Cavaglieri, L.R.; Keller, K.M.; Pereyra, C.M.; Gonz??lez Pereyra, M.L.; Alonso, V.A.; Rojo, F.G.; Dalcero, A.M.; Rosa, C.A.R. Fungi and natural incidence of selected mycotoxins in barley rootlets. *J. Stored Prod. Res.* **2009**, *45*, 147–150. [CrossRef]
25. Zachariasova, M.; Dzuman, Z.; Veprikova, Z.; Hajkova, K.; Jiru, M.; Vaclavikova, M.; Zachariasova, A.; Pospichalova, M.; Florian, M.; Hajslova, J. Occurrence of multiple mycotoxins in european feedingstuffs, assessment of dietary intake by farm animals. *Anim. Feed Sci. Technol.* **2014**, *193*, 124–140. [CrossRef]
26. Mastanjević, K.; Šarkanj, B.; Warth, B.; Krska, R.; Sulyok, M.; Mastanjević, K.; Šantek, B.; Krstanović, V. Fusarium culmorum multi-toxin screening in malting and brewing by-products. *LWT* **2018**, *98*, 642–645. [CrossRef]
27. Wolf-Hall, C.E. Mold and mycotoxin problems encountered during malting and brewing. *Int. J. Food Microbiol.* **2007**, *119*, 89–94. [CrossRef]
28. Habler, K.; Geissinger, C.; Hofer, K.; Schüler, J.; Moghari, S.; Hess, M.; Gastl, M.; Rychlik, M. Fate of fusarium toxins during brewing. *J. Agric. Food Chem.* **2017**, *65*, 190–198. [CrossRef] [PubMed]
29. Milani, J.; Maleki, G. Effects of processing on mycotoxin stability in cereals. *J. Sci. Food Agric.* **2014**, *94*, 2372–2375. [CrossRef]
30. Piacentini, K.C.; Rocha, L.O.; Fontes, L.C.; Carnielli, L.; Reis, T.A.; Corrêa, B. Mycotoxin analysis of industrial beers from Brazil: The influence of fumonisin B1 and deoxynivalenol in beer quality. *Food Chem.* **2017**, *218*, 64–69. [CrossRef] [PubMed]
31. Basarova, G.; Savel, J.; Petr, B.; Lejsek, T. *The Comprehensive Guide to Brewing: From Raw Materials to Packaging*; Fachverlag Hans Carl: Nuremberg, Germany, 2017.
32. *EBC European Brewery Convention - Analytica EBC*; Fachverlag Hans Carl: Nuremberg, Germany, 2009; ISBN 3-418-00759-7.
33. Campagnollo, F.B.; Franco, L.T.; Rottinghaus, G.E.; Kobashigawa, E.; Ledoux, D.R.; Daković, A.; Oliveira, C.A.F. In vitro evaluation of the ability of beer fermentation residue containing Saccharomyces cerevisiae to bind mycotoxins. *Food Res. Int.* **2015**, *77*, 643–648. [CrossRef]
34. Jouany, J.; Yiannikouris, A.; Bertin, G. The chemical bonds between mycotoxins and cell wall components of Saccharomyces cerevisiae have been identified. *Arch. Zootech.* **2005**, 26–50.
35. Renato Matos Lopes, P.; Marin Morales, E.; Nallin Montagnolli, R. Cerveja brasileira: do campo ao copo. *Rev. Agron. Bras.* **2018**, *1*. [CrossRef]
36. EFSA Deoxynivalenol in food and feed. occurrence and exposure. *EFSA J.* **2013**.
37. CPTEC Centro de Previsão de Tempo e Estudos Climáticos—CPTEC/INPE. Available online: http://www.cptec.inpe.br/ (accessed on 5 September 2017).
38. INMET Instituto Nacional de Meteorologia. Available online: http://www.inmet.gov.br/portal/ (accessed on 4 December 2017).
39. MEBAK. *Raw Materials—Barley, Adjuncts, Malt, Hops and Hop Products*; MEBAK: Freising-Weihenstephan, Germany, 2011.
40. Běláková, S.; Benešová, K.; Mikulíková, R.; Svoboda, Z. Determination of ochratoxin A in brewing materials and beer by ultra performance liquid chromatography with fluorescence detection. *Food Chem.* **2011**, *126*, 321–325. [CrossRef]
41. Běláková, S.; Benešová, K.; Čáslavský, J.; Svoboda, Z.; Mikulíková, R. The occurrence of the selected fusarium mycotoxins in czech malting barley. *Food Control* **2014**, *37*, 93–98. [CrossRef]
42. Lopes, P.R.M.; Morales, E.M.; Montagnolli, R.N. Cerveja brasileira: do campo ao copo. *Rev. Agron. Bras.* **2018**, *1*, 18–21.

toxins

MDPI

Article

Variation of Fungal Metabolites in Sorghum Malts Used to Prepare Namibian Traditional Fermented Beverages *Omalodu* and *Otombo*

Sylvia N. Nafuka [1,*], Jane M. Misihairabgwi [2], Ronnie Bock [1], Anthony Ishola [3], Michael Sulyok [4] and Rudolf Krska [4,5]

1 Department of Biological Sciences, Faculty of Science, University of Namibia, Windhoek 10005, Namibia; rbock@unam.na

2 Department of Biochemistry and Microbiology, School of Medicine, University of Namibia, Windhoek 10005, Namibia; jmisihairabgwi@unam.na

3 Department of Pharmaceutical Chemistry and Phytochemistry, School of Pharmacy, University of Namibia, Windhoek 10005, Namibia; aishola@unam.na

4 Center for Analytical Chemistry, Department of Agro Biotechnology (IFA-Tulln), University of Natural Resources and Life Sciences Vienna (BOKU), Konrad Lorenz Str. 20, 3430 Tulln, Austria; michael.sulyok@boku.ac.at (M.S.); rudolf.krska@boku.ac.at (R.K.)

5 Institute for Global Food Security, School of Biological Sciences, Queen's University Belfast, University Road, Belfast BT7 1NN, Northern Ireland, UK

* Correspondence: snafuka@unam.na

Received: 23 December 2018; Accepted: 12 March 2019; Published: 16 March 2019

check for updates

Abstract: Sorghum malts, which are important ingredients in traditional fermented beverages, are commonly infected by mycotoxigenic fungi and mycotoxins may transfer into the beverages, risking consumers' health. Liquid chromatography–tandem mass spectrometry was used to determine variation of fungal metabolites in 81 sorghum malts processed for brewing of Namibian beverages, *otombo* (n = 45) and *omalodu* (n = 36). Co-occurrence of European Union (EU)-regulated mycotoxins, such as patulin, aflatoxins (B_1, B_2, and G_2), and fumonisins (B_1, B_2, and B_3) was detected in both malts with a prevalence range of 2–84%. Aflatoxin B_1 was quantified in *omalodu* (44%) and *otombo* malts (14%), with 20% of *omalodu* malts and 40% of *otombo* malts having levels above the EU allowable limit. Fumonisin B_1 was quantified in both *omalodu* (84%) and *otombo* (42%) malts. Emerging mycotoxins, aflatoxin precursors, and ergot alkaloids were quantified in both malts. Notably, 102 metabolites were quantified in both malts, with 96% in *omalodu* malts and 93% in *otombo* malts. An average of 48 metabolites were quantified in *otombo* malts while an average of 67 metabolites were quantified in *omalodu* malts. The study accentuates the need to monitor mycotoxins in sorghum malts intended for brewing and to determine their fate in the beverages.

Keywords: traditional sorghum malts; mycotoxins; aflatoxins; *Aspergillus*; LC/MS/MS

Key Contribution: This paper reports the first results on LC/MS/MS based fungal metabolite quantification in sorghum malt varieties used to prepare traditional fermented beverages, *otombo* and *omalodu*, in Namibia. The quantification of EU-regulated mycotoxins, emerging mycotoxins, aflatoxin precursors, and ergot alkaloids in both malt varieties necessitates adequate malt processing under hygienic conditions to reduce fungal contamination and possible transfer into popularly consumed beverages.

1. Introduction

Sorghum is a genus of cereals in the family Poaceae of approximately 30 species. One species, *Sorghum bicolor*, is native to Africa and is the world's fifth most important cultivated cereal crop [1], with many significant uses, such as being a staple food in some sub-Saharan countries, the main ingredient in the production of beverages, and animal feed [2]. In Namibia, sorghum is mainly cultivated by rural subsistence farmers of the northern regions and is mainly used for the brewing of traditional beverages [3]. Specifically, unground malted sorghum grains are used as the main ingredients in the brewing of the traditional alcoholic drink locally known as *otombo*, while malted sorghum flour is used for the brewing of the non-alcoholic beverages *omalodu* and oshikundu [3]. *Otombo*, which is mainly brewed nationwide for income generation, is sold at shebeens in rural areas and at open markets in urban areas, and, due to its alcoholic content, is generally consumed by elders. *Omalodu* is also a popular ceremonial traditional beverage in Namibian Oshiwambo and Rukwangali communities. In both communities, *omalodu* is primarily consumed at sociocultural ceremonies.

Sorghum malts used in this study vary depending on the malting process and the milling stages. According to [4], malting is defined as the germination of grains to promote the development of hydrolytic enzymes which were inactive in the raw grain. Generally, the malting process involves three main processes: Steeping, germination, and drying [5]. In Namibia, particularly among the Oshiwambo and Rukwangali communities, the malting process is carried out at the household level and is similar, with minor differences due to cultural specifications and weather conditions. Basically, the process of sorghum malting involves steeping the cleaned grains in water for 24 to 48 h, draining, and germination in sealed plastics, jute sacks, or metal trays for 1 to 2 weeks. Some Oshiwambo community members may add sandy soil to facilitate the germination process. The germinated grains are then air dried at ambient temperature, then the dry malted grains, including the root fragments, are used for *otombo* brewing. Milling of the dry malts used for *omalodu* brewing is usually carried out in a hut or an open area by pounding with strong wooden sticks in a wooden traditional mill. The pounding is continued until all grains are pulverized with intermittent sifting using a circular basket made from palm leaves. The initial round of sifted coarse sorghum flour with grains and root fragments is reserved for *omalodu* brewing. *Omalodu* malts are mainly prepared for brewing at the household level and for family use, while some may also be transported and sold at open markets in urban areas. *Otombo* malts are prepared for brewing at shebeens and for selling at open markets.

Due to the warm, moist, and likely unhygienic conditions during the traditional malting and milling processes, the growth of mycotoxigenic fungi is stimulated [6]. In addition, mycotoxigenic fungi can infiltrate deep into sorghum matrices and produce mycotoxins during the pre-harvest, storage, transportation, processing, and marketing stages [7]. Mycotoxins are fungal secondary metabolites representing natural contaminants in raw materials, foods, and feeds [8]. The most dangerous mycotoxins are aflatoxins, ochratoxins, fumonisins, patulin, and ergot alkaloids, produced by fungi belonging to *Aspergillus*, *Penicillium*, *Claviceps*, and *Fusarium* genera [9]. The toxins are known to have carcinogenic, mutagenic, teratogenic, cytotoxic, neurotoxic, nephrotoxic, estrogenic, dermotoxic, and immunotoxic effects in humans [10,11]. Many parts of the world are regulating mycotoxins by creating maximum allowable limits in different foods and feed. According to a global food prevalence mycotoxin survey by [12], 72% of the food samples, including cereals, contained detectable amounts of mycotoxins addressed by regulatory limits in the European Union (EU). However, other fungal metabolites, such as beauvericin (BEA), moniliformin (MON), sterigmatocystin (STE), emodin (EMO), alternariol (AOH), tenuazonic acid (TeA), and 3-Nitropropionic acid, (3-NPA), are now frequently quantified in a variety of foods and feed in different parts of the world [13]. There are also indications that the incidence of these so-called emerging mycotoxins, which are neither routinely determined nor legislatively regulated, is rapidly increasing [14–16].

The quality of raw materials used to prepare traditional beverages influences the final product safety. Many reports on the occurrence and quantities of fungal metabolites in sorghum malts and related products from countries neighboring Namibia have been documented in South Africa [17],

Zimbabwe [18], and Botswana [19]. A previous study in Namibia, which determined the diversity of fungal metabolites in sorghum malt samples used for oshikundu beverage production and their transfer rates into the beverage, reported that, although EU-legislated mycotoxins were not quantifiable in the beverage, transfer rates into the beverage were above 50% for most of the other fungal metabolites [20]. Due to the diversity of malting and processing methods for different beverages, the types and quantities of fungal metabolites in the specific malts may vary. Consequently, consumers' exposure to the metabolites will also vary. There is, therefore, a need to investigate the quality of raw materials used to prepare other traditional Namibian beverages, such as *otombo* and *omalodu*, which are consumed daily and by many people. This study, therefore, aimed at determining the occurrence and variation of fungal metabolites in sorghum malts intended for the brewing of *omalodu* and *otombo* beverages.

2. Results and Discussion

2.1. Occurrence of Fungal Metabolites in Sorghum Malts for Omalodu and Otombo Brewing

Only metabolite concentration levels that were above the limits of quantification (LOQ) were quantified for both malts. Hence, a total of 102 fungal and bacterial metabolites, including seven regulated mycotoxins, were quantified in sorghum malts for both *omalodu* and *otombo* beverages (Tables 1–3). Only 4% of the 102 metabolites were of non-fungal origin. The occurrence of fungal metabolites, including mycotoxins in sorghum malts intended for the beverage oshikundu, quantified using the same analytical technique and method, was previously reported in Namibia [20], with a total of 98 metabolites quantified. Other reports on the occurrence and quantities of fungal metabolites in sorghum malts and grains from Southern Africa have been documented in South Africa [17], Malawi [21], Zimbabwe [18], and Botswana [19].

The total number of metabolites was higher in *omalodu* malts (n = 101) than in *otombo* malts (n = 96). An average of 48 metabolites was quantified in *otombo* malts, while an average of 67 metabolites was quantified in *omalodu* malts, respectively. Although some metabolites were quantified with low prevalence rates and quantities, the risks of exposure to these complex mixtures of metabolites by consumption of brews should be studied in more detail in order to minimize the possible synergistic and/or additive effects during brewing.

The metabolites detected were representatives of the following mycotoxigenic fungal genera: *Aspergillus* 34%, (Tables 1 and 2) *Penicillium* 16%, (Table 1), *Fusarium* 15%, (Tables 1 and 2), *Alternaria* 7% (Tables 1 and 2), and *Claviceps* 6% (Table 1), while 22% were non-fungal metabolites or produced by unspecified and uncommon fungal genera (Table 3). According to [22], these genera are known to be associated with sorghum malts and grains in Nigeria, Botswana [19], and Ethiopia [23] and also known as the main mycotoxigenic fungal genera [24,25]. Only 2% of the metabolites were 100% prevalent in *otombo* malts, whereas 15% of the metabolites were 100% prevalent in *omalodu* malts. Of the quantified fungal metabolites, 64%, 69%, 65%, and 85% of the *Aspergillus*, *Fusarium*, *Penicillium*, and *Alternaria* metabolites had higher average concentrations in *omalodu* malts than in *otombo* malts, respectively.

2.2. Variation of Regulated Mycotoxins and Aflatoxins Precursors

Seven mycotoxins addressed by regulatory limits in the EU (i.e., aflatoxin B_1, B_2, and G_1, fumonisins B_1, B_2, and B_3, and patulin) were quantified in both malts (Table 1). Due to the absence of such limits in Namibia, the limits fixed by the EU [26] were used as the basis for discussion in the present study. The same mycotoxins have been recently reported in sorghum malts from Namibia using the same analytic method by [20], with the exception of patulin found in the present study and fumonisin B_4 quantified only in the previous study. Overall, 98% and 76% of the *omalodu* and *otombo* samples were contaminated with at least one of the EU regulated toxins, respectively. Comparing the two groups of malt samples, 71% of the of the EU-regulated mycotoxins had higher incidences in *omalodu* malts than in *otombo* malts, while 47% of these mycotoxins had higher average concentrations in *otombo* malts than in *omalodu* malts.

Table 1. Regulated mycotoxins, aflatoxin precursors, ergot alkaloids, and emerging mycotoxins quantified in sorghum malts for the production of *omalodu* and *otombo* beverages.

Compounds	Types	Origin	*Omalodu* Malts $n = 45$			*Otombo* Malts $n = 36$		
			Prevalence (%)	Range (µg/kg)	Average (µg/kg)	Prevalence (%)	Range (µg/kg)	Average (µg/kg)
Aflatoxin B_1	Regulated	*Aspergillus*	44	0.61–28.3	2.87 ± 2.93	14	0.56–54.2	15.1 ± 22.9
Aflatoxin B_2			9	0.14–2.35	0.15 ± 0.44	5	0.5–4.48	2.49 ± 2.8
Aflatoxin G_1			17	0.39–6.95	1.19 ± 1.10	3	0.4	0.4
Patulin			2	57.7	57.7	6	81.8–284.3	183.1 ± 143.2
Fumonisin B_1		*Fusarium*	84	12–500.2	61.4 ± 70	42	8.17–88.3	29.12 ± 25.7
Fumonisin B_2			66	7.55–79.46	17.56 ± 12.1	22	5.92–46.8	16.4 ± 13
Fumonisin B_3			7	21.6–136.6	60.14 ± 66.3	3	22	22
Averufanin	Aflatoxin precursors	*Aspergillus*	29	13.5–384	37.8 ± 47.3	N/D	N/D	N/D
Averufin			N/D	N/D	N/D	83	0.09–103	6.73 ± 20.2
Versicolorin C			13	89.8–200	1534 ± 33.5	24	29.8–2815	444 ± 846
Sterigmatocystin	Aflatoxin precursor and emerging mycotoxin		89	377–1690	4.30 ± 7.98	17	29.8–2810	6.24 ± 4.53
3-Nitropropionic acid	Emerging mycotoxins		100	83.7–10,200	3290 ± 5000	94	7.61–14,900	2530 ± 2860
Alternariol		*Alternaria*	91	1.24–318	14 ± 30	72	0.45–71.42	9.99 ± 18.09
Alternariolmethylether			84	1.27–564	45.7 ± 90.4	42	1.61–80.2	23.6 ± 26.4
Tenuazonic acid			73	132.4–13,400	1925.6 ± 3406.4	81	4.84–11,400	999 ± 2170
Beauvericin		*Fusarium*	97	0.23–30.4	5.08 ± 7.01	39	0.24–5.65	1.60 ± 1.72
Emodin		Plants and *Fusarium*	84	2.16–79.2	23.7 ± 17.7	97	0.35–93.4	19.6 ± 31.9
Moniliformin		*Fusarium*	100	11.3–1550	348 ± 511	94	4.58–728.2	75.6 ± 135.9
Agroclavine	Ergot alkaloids	*Claviceps*	96	18.7–20,500	733 ± 2760	50	6.7–95.4	43.8 ± 31.3
Chanoclavin			98	0.37–188	46.3 ± 40.7	72	0.39–49.7	10.8 ± 14.9
Elymoclavine			38	0.89–153	10.4 ± 27.4	3	1.48	1.48
Festuclavine			100	25.7–11,400	1690 ± 1750	83	1.23–5660	570 ± 1120
Fumigaclavine A			96	0.004–613	89.4 ± 118	89	0.55–118	20.7 ± 29.7
Fumigaclavine C			100	6.49–6040	1060 ± 1260	86	3.26–1159.2	228.4 ± 332.5

N/D = Not detected.

Table 2. Unregulated metabolites quantified in sorghum malt samples for the production of *omalodu* and *otombo* beverages.

Compounds	*Omalodu* Malts $n = 45$			*Otombo* Malts $n = 36$		
	Prevalence (%)	Range [(μg/kg) or Peak Area^2]	Average [(μg/kg) or Peak Area^2]	Prevalence (%)	Range [(μg/kg) or Peak Area^2]	Average [(μg/kg) or Peak Area^2]
Aspergillus						
Asperfuran	96	1980–669,000	46,000 ± 110,000	89	14.8–428,000	60,000 ± 88,000
Asterric acid	22	1.09–170	43.6 ± 104	3	62.8	62.8 ± 0.00
Bis (methylthio) gliotoxin	87	4.77–699	103 ± 128	61	4.07–229.9	34 ± 49.9
Bisdethio (methylthio) gliotoxin	64	6.04–263	77.9 ± 62.9	67	1.14–285	35.4 ± 57.6
Gliotoxin	78	3.55–193.7	54.3 ± 55.7	44	3.12–44.9	13.9 ± 11.4
Cyclopiazonic acid	69	55.17–2070	456.15 ± 652.18	39	60.4–486	122 ± 134
Cytochalasin E	84	1.66–96.7	47.5 ± 96.7	42	2.24–521	74.5 ± 135.6
Deoxynortryptoquivalin	91	2.59–727	67.70 ± 122.94	47	2.15–467	57.6 ± 114
Deoxytryptoquivaline A	87	2.32–894	46.57 ± 116.81	44	1.15–142	30.2 ± 47.9
Dihydrocitrinone	41	2.63–184	21.17 ± 49.4	17	2.98–274	50.7 ± 109.6
Flavoglaucin	89	0.51–949	79 ± 222	69	0.16–2810	306 ± 654
Fumagillin	91	6.76–2220	478.85 ± 814.68	44	36–1910	321.3 ± 465.7
Fumiquinazolin A	89	11–979	267 ± 256	61	3.37–224	85.4 ± 85.4
Fumiquinazolin D	100	5.95–3140	826 ± 745	89	1.06–837	175 ± 237
Fumitremorgin C	89	1.70–1140	142 ± 286	56	0.71–411	51.8 ± 90.7
Trypacidin	11	0.41–20.3	3.74 ± 4.58	ND	ND	ND
Tryprostatin B *	88	259,000–130,000,000	12,100,000 ± 24,000,000	6	17,200–22,400	19,800 ± 3680
Tryptoquivaline A	77	1.62–1040	54.7 ± 178	39	1.49–442	75 ± 125.4
Tryptoquivaline F *	88	356,000–14,300,000	4,250,000 ± 3,220,000	20	80,500–4,710,000	2,240,000 ± 1,440,000
Helvolic acid	87	21.7–2860	696 ± 680	47	16.9–2350	329.6 ± 556.2
Kojic acid	100	631–182,000	41,000 ± 48,000	56	1594–52,296	17,712.5 ± 14,125.8
Nigragillin *	100	244,000–113,000,000	27,800,000 ± 31,200,000	94	112,000–92,900,000	7,310,000 ± 16,500,000
Phenopyrrozin	100	9.43–35.8	805 ± 891	64	10.6–7.14	3.03 ± 1.64
Pseurotin A	91	8.54–4400	805 ± 902	64	10.58–764.3	198.9 ± 216.1
Pseurotin D *	82	24,900–996,000	243,000 ± 199,000	N/D	N/D	N/D
Iso-Rhodoptilometrin	91	0.13–6.91	1.70 ± 1.91	78	0.11–4.28	0.71 ± 0.87
Pyrophen	29	1.30–6.45	3.09 ± 1.59	14	1.04–95.1	23.7 ± 36.2

Table 2. *Cont.*

Compounds	*Omalodu* Malts $n = 45$			*Otombo* Malts $n = 36$		
	Prevalence (%)	Range [(μg/kg) or Peak Area^2]	Average [(μg/kg) or Peak Area^2]	Prevalence (%)	Range [(μg/kg) or Peak Area^2]	Average [(μg/kg) or Peak Area^2]
Penicillium						
Aurantine	9	1.25–16.4	5.41 ± 7.33	3	1.38	1.38
Barceloneic acid	87	7.75–2630	316 ± 497	56	13.3–2220	1060 ± 2230
Citreorosein	87	2.77–104.6	27.04 ± 25.7	59	1.91–79.544	17.61 ± 23.2
Brefeldin A	11	41–1150	786 ± 528	3	289	289 ± 0.00
Citreohybridinol	69	1.19–22,600	35.9 ± 36.6	17	9.38–114	1630 ± 5050
Curvularin	100	9.77–5780	403.7 ± 755	100	4.48–3080	754.7 ± 1080
Dechlorogriseofulvin	16	2.90–53.7	11.26 ± 16.52	22	1.6–18.6	6.60 ± 5.92
Dehydrocurvularin	18	104–758	588 ± 459	14	138–1340	247 ± 157
Dichlordiaportin	93	5.63–482	70 ± 104	72	4.20–435	147 ± 170
Griseofulvin	62	0.58–14.1	4.50 ± 6.59	31	0.55–13.3	4.92 ± 4.65
Herquline A	29	0.52–1.88	1.97 ± 1.41	N/D	N/D	N/D
Hydroxycurvularin	64	41.1–697	43.3 ± 37.4	25	11.8–137	132 ± 133
Pinselin	49	1.28–25.5	6.79 ± 7.46	36	0.81–26.5	6.29 ± 6.12
Quinolactacin A	22	0.87–84.1	18.9 ± 28.8	11	0.53–68.6	9.20 ± 18.3
Thielavin B	40	1–3.8	3.86 ± 3.52	19	0.40–7.70	0.87 ± 0.86
Fusarium						
Aminodimethyloctadecanol	7	1570–2420	1877 ± 387	N/D	N/D	N/D
Antibiotic Y	11	34.2–103.4	64.8 ± 29.2	17	41.46–616	287.8 ± 236.3
Aurofusarin *	98	10.7–4230	672 ± 791	86	10.5–89	669 ± 1644
Bikaverin	100	18.7–2390	618 ± 645	50	2.43–3920	360 ± 874
Epiequisetin	N/D	N/D	N/D	11	0.84–30.4	12.9 ± 13.6
Equisetin	31	0.23–5.40	1.41 ± 1.33	67	0.79–103	22.3 ± 24.4
Fuscofusarin *	47	4.03–2720	141,000 ± 94,800	53	19,000–1,400,000	190,000 ± 322,000
Sambucinol	N/D	N/D	N/D	3	27.86	27.86
Siccanol *	80	55,500–1,510,000	714,000 ± 747,000	69	52,000–14,300,000	1,210,000 ± 2,780,000
Alternaria						
Altersetin	87	3.23–381	54.09 ± 100.87	61	5.09–618	55.6 ± 132.18
Altersolanol	42	428.6–21,300	4710.70 ± 6568.10	25	670.5–14100	3725.3 ± 4333.3
Macrosporin	87	0.79–154	24.58 ± 30.28	78	1.58–84.1	25.6 ± 21.2
Pyrenophorol	51	3.57–30.6	11.8 ± 8.6	25	3.50–31.3	11.4 ± 9.04

For metabolites indicated by * no quantitative standards were available, therefore numbers denote LC-MS/MS peak area in order to enable relative comparison. N/D = Not detected.

Table 3. Metabolites produced by unspecified, uncommon fungal genera and other organisms quantified in sorghum malts.

Compounds	Origin	*Omalodu* Malts $n = 45$			*Otombo* Malts $n = 36$		
		Prevalence (%)	Range [(μg/kg) or Peak Area^2]	Average [(μg/kg) or Peak Area^2]	Prevalence (%)	Range [(μg/kg) or Peak Area^2]	Average [(μg/kg) or Peak Area^2]
Abscisic acid	*Botrytis* and plants	2	2380	2380	N/D	N/D	N/D
Antibiotic PF 1052	*Phoma*	4	27.7–274	90 ± 57.8	5	9.63–46.8	28.2 ± 18.6
Asperglaucide	Unspecific	24	0.13–17.4	2.67 ± 4.37	19	0.12–1.37	0.64 ± 0.45
Bassianolide	*Cladosporium*	69	0.10–5.29	0.67 ± 0.89	17	0.10–0.33	0.21 ± 0.09
Brevianamide F	Fungi and bacterial	100	37.6–427	144 ± 92.6	94	23.09–312	90.1 ± 56.1
Calphostin	*Metarhizium*	2	14.23	14.23	17	11.10–50.1	20.5 ± 14.9
Chloramphenicol	Bacterial	84	11.4–3173.3	484.9 ± 745.5	28	0.14–0.90	0.44 ± 0.26
Cyclo (L-Pro-L-Tyr)	Unspecific	100	48.2–48,200	121 ± 33.8	94	32.5–165	75.6 ± 31.5
Cyclo (L-Pro-L-Val)	Unspecific	100	42.7–345	456 ± 158	100	2.13–420	121 ± 86.4
Destruxin A	*Metarhizium*	29	0.25–2.06	0.73 ± 0.30	22	0.29–12.7	2.39 ± 3.90
Destruxin-Ed Derivative	*Metarhizium*	7	0.83–7.1	3.32 ± 2.72	8	0.81–13.4	5.59 ± 5.56
Dihydroxymellein	Unspecific	N/D	N/D	N/D	8	31.2–107	68.2 ± 31.3
Fellutanine A	Unspecific	82	7.14–25.5	16.1 ± 5.79	75	3.39–22.6	11.2 ± 4.46
Heptelidic acid	*Phoma*	7	33.2–87.09	53.9 ± 28.9	6	37.7–60.7	49.9 ± 16.2
Monactin	Bacterial	40	0.27–5.16	1.50 ± 1.26	19	0.45–1.83	0.89 ± 0.54
Monocerin	Unspecific	100	293–1120	139 ± 226	89	1.28–560	47.5 ± 100
Orsellinic acid	Unspecific	98	115–21,000	4080 ± 4748	47	1090–17,000	3780 ± 3690
Phomalactone	*Trichoderma*	18	1.12–7.83	1.65 ± 1.21	8	2.05–5.49	3.52 ± 1.77
Rugulusovin	Unspecific	100	27.7–157	108 ± 47.6	94	5.40–254.3	47.8 ± 51.4
Tryptophol	Unspecific	N/D	N/D	N/D	81	16.7–352	110 ± 80.5
Skyrin	Unspecific	91	0.50–23.9	0.69 ± 0.42	47	0.38–2.08	0.73 ± 0.42
Siccanin	*Helmintosporum*	98	3.19–9.95	7.01 ± 1.45	14	3.73–10.6	6.11 ± 2.67

N/D = Not detected.

AFB_1 was quantified in *omalodu* (prevalence = 44%, average = 2.87 ± 2.93 μg/kg) and *otombo* malts (prevalence = 14%, average = 15.1 ± 22.9 μg/kg) with 20% of *omalodu* malts and 40% of *otombo* malts having levels above the EU regulatory limit of 5 μg/kg. An independent-samples *t*-test conducted to compare average AFB_1 levels in *omalodu* and *otombo* malts showed that the difference was not statistically significant ($p > 0.05$). Differences in averages of AFB_2, AFG_1, FB_1, FB_2, and FB_3 levels between *omalodu* and *otombo* malts were also not statistically significant ($p > 0.05$). These results suggest that the different malt preparation methods may not have a significant effect on levels of aflatoxins. In the previous study [20], AFB_1 was quantified with higher prevalence rate (50%) and level (average: 4.5 ± 5.5 μg/kg) in the sorghum malt flour samples compared with *omalodu* malts (44% prevalence; average: 2.87 ± 2.93 μg/kg).

Other reports regarding aflatoxin occurrence in sorghum malt grain samples intended for beverage production in Africa analyzed using other techniques were conducted: In Malawi [21], total aflatoxin contents were identified via immunoaffinity column and were reportedly higher (408 ± 68 μg/kg) in sorghum malts prepared for beer brewing than in the current study (5.47 ± 13.8 μg/kg). In Burkina Faso [27], aflatoxin B_1 and ochratoxin A were purified with immunoaffinity columns and analyzed using high-performance liquid chromatography (HPLC), revealing a higher AFB_1 level (97.6 ± 88.2 μg/kg) for malt samples than the present study (8.49 ± 16.9 μg/kg). In South Africa [17], mycotoxins were identified using a multi-mycotoxin thin-layer chromatography method and quantified via HPLC to screen sorghum malt grains intended for traditional beers (Utshwala). The former study revealed the presence of zearalenone and absence of AFB_1 in sorghum malt grains, while the current study revealed contrasting results. FB_1 is one of the common toxicologically important mycotoxins and was quantified in *omalodu* malts with 84% prevalence (average concentration: 61.4 ± 70 μg/kg) and in *otombo* malts with a prevalence of 42% (average concentration: 29.1 ± 25.7 μg/kg) (Table 1).

FB_1 was quantified with a higher prevalence rate (75%) in sorghum flour malts from Namibia prepared for oshikundu beverage [20] compared to the current study. In Botswana, FB_1 was detected at a lower prevalence rate of 6% in sorghum malt samples, with concentrations ranging from 47 to 1316 μg/kg [19]. All sorghum malt samples had FB_1 concentrations below the EU regulatory level of 2000 μg/kg.

Patulin, which is normally found in fruits and vegetables, particularly apple and its products [28–30], was quantified in a single sample of *omalodu* (average = 57.7 μg/kg) and only two samples of *otombo* malts (average = 183.1 ± 143.2 μg/kg). However, the average concentrations for *omalodu* and *otombo* malts were higher than those fixed by the EU at 50 μg/kg for patulin in apple juice. Patulin is of concern because it is produced by many fungal genera and is suspected of being clastogenic, mutagenic, teratogenic, genotoxic, and cytotoxic [31]. The co-occurrence of different regulated toxins suggests synergistic toxic effects that raise concerns on the health hazards associated with these malts.

Several metabolites from the biosynthetic pathway of aflatoxins, namely averufanin, averufin, STE, and versicolorin C (Table 1), were quantified in the sorghum malts. The same aflatoxin precursors were reported by [20] in sorghum malts prepared for oshikundu, with the exception of versicolorin C, found only in the present study. Averufanin was quantified in 27% of the *omalodu* malts and not in the *otombo* malts, while averufin was quantified in 84% of the *otombo* malts but not in *omalodu* malts. A high incidence rate of 91% (average = 95.49 μg/kg) was recorded for STE from *omalodu* malts. STE is classified as a possible human carcinogen by the International Agency of Research in Cancer [32]. In addition, in vitro genotoxic and cytotoxic studies of STE revealed that it is genotoxic to liver hepatocellular cells [33] and cytotoxic to immortalized ovarian hamster cells [34] and liver hepatocellular cells [35].

2.3. Emerging Mycotoxins and Ergot Alkaloids Quantified in Sorghum Malts

As shown in Table 1, some emerging mycotoxins, especially 3-NPA, EMO, AOH, AME, TeA, MON, STE, and BEA, occurred with prevalence ranges of 84–100% in *omalodu* malts and 17–100% in *otombo* malts. A magnetic resonance imaging study by [36] reported that 3-NPA is a potent mitochondrial toxin and neurotoxin. 3-NPA was observed in all sorghum flour malt samples, similar to previously reported results in Namibia [20]. The average concentration of 3-NPA was lower (2530 ± 2860 μg/kg) in malts for oshikundu [20] than that observed in malts for *otombo* (3290 ± 5000 μg/kg). MON was not quantified in sorghum flour malts in the previous study [20], but it was observed in all samples of *omalodu* malts and quantified with a concentration of 348 ± 511 μg/kg and with concentration of 75.6 ± 135.9 μg/kg in *otombo* malts. In vivo toxicity investigations suggest that MON can induce cardiotoxicity [37] and cause immunosuppression, muscular weakness, and intestinal problems in poultry [38]. EMO is both a fungal and plant metabolite and it was quantified with average concentrations of 23.7 ± 17.7 μg/kg and 19.6 ± 31.9 μg/kg in *omalodu* and *otombo* malts, respectively. Several studies demonstrated that EMO has anticancer [39,40], antiviral [41,42], and antibacterial activity [43]. BEA was quantified at concentrations of 5.08 ± 7.01 μg/kg and 1.60 ± 1.72 μg/kg in *omalodu* and *otombo* malts, respectively. BEA has shown cytotoxic effects on human cell lines [44]. Among the *Alternaria* toxins, TeA was quantified with high amounts in both *omalodu* (1925.6 ± 3406.4 μg/kg) and *otombo* (999 ± 2170 μg/kg) malts, compared to lower amounts of AOH (14 ± 30 μg/kg) and (9.99 ± 18.09 μg/kg) in *omalodu* and *otombo* malts, respectively. Some in vivo studies of TeA revealed that it is toxic to animals, such as mice and rats [45], beagle dogs, monkeys [46], and chickens [47]. According to an in vitro study [48], AOH and AME are mutagenic to hamster lung fibroblast cells lines.

Six clavine ergot alkaloids, synthesized mainly by fungal species of *Claviceps* genera were quantified in both malt samples. Elymoclavine was observed in 3% and 38% of *otombo* and *omalodu* malt samples, respectively. Other alkaloids were observed with high prevalence (50–100%) in both malt samples (Table 1). Ergot alkaloids are typically important because chronic poisoning by these toxins through consumption of contaminated grain products causes ergotism. Sorghum crops are also vulnerable to ergot disease during cultivation. In Africa, the pathogen is recognized as a distinct species, *Claviceps africana* [49]. Damages caused by *C. africana* have been recognized as a major cause for decreased quality and nutritive value of sorghum grains [50].

2.4. Other Fungal Metabolites Quantified in Sorghum Malts

The prevalence and concentrations of other fungal metabolites quantified in both malt samples are indicated in Table 2. Metabolites produced solely by *Aspergillus* genus were mostly quantified at a prevalence of 49%; 19 of these metabolites were quantified in 80% to 100% of *omalodu* samples analyzed, as opposed to three metabolites of the same prevalence rate quantified in *otombo* malts. Secondary metabolites of *Aspergillus* are representatives from the following groups: Gliotoxins, fumitremorgins, fumagillins and fumiquinazolines, helvolic acids, tryptoquivalines, and pseurotins. Although there are no regulations in force for these metabolites, some of them have their in vitro toxicities reported. An example is gliotoxin, an epipolythiodioxopiperazine produced by *Aspergillus fumigatus* and quantified in both *omalodu* (78%) and *otombo* (44%) malts. Its disulfide bridge may cause immunosuppressive properties and apoptosis in macrophages and monocytes [51]. Bis (methylthio) gliotoxin is an inactive derivative of gliotoxin, proposed as a stable biomarker for invasive aspergillosis [52]. Another toxic metabolite from *Aspergillus flavus* quantified in both *omalodu* (69%) and *otombo* (39%) malts is cyclopiazonic acid, an indole tetramic. Cyclopiazonic acid causes degenerative changes and necrosis in the liver, spleen, pancreas, kidney, salivary glands, myocardium, and skeletal muscles, based on toxic effects observed in male and female rats [53]. Higher prevalence rates of *Aspergillus* metabolites are an indication of the higher contamination by storage mycotoxigenic fungi such as *Aspergillus fumigatus*, *Aspergillus clavatus* and *Aspergillus niger.* The latter findings are expected because both malts are traditionally processed under likely unhygienic conditions. The poor storage conditions for prolonged times at homes and markets makes the malts susceptible to fungal contamination. Other metabolites synthesized by *Fusarium*, *Penicillium*, and *Alternaria* species were quantified as well.

2.5. Other Metabolites Quantified in Sorghum Malts

Twenty-two metabolites synthesized by non-fungal organisms and uncommon fungal species were quantified in both *omalodu* and *otombo* malts (Table 3). Dihydroxymellein and tryptophol were absent in *omalodu* malts, while abscisic acid was absent in *otombo* malts. Four diketopiperazines synthesized by fusion of 2 different amino acids, namely Cyclo (L-Pro-L-Tyr), or maculosin, cyclo (L-Pro-L-Val), brevianamide F, or cyclo (L-Trp-L-Pro), Fellutanine A or cyclo (L-Trp-L-Trp), were quantified in both *omalodu* and *otombo* malts. Cyclo (L-Pro-L-Tyr) is formed by the fusion of tyrosine and proline and has been reported as a secondary metabolite of various fungi [54] and bacteria [55]. Additionally, it is identified as a host-specific phytotoxin produced by *Alternaria alternata* [54]. It was quantified at a prevalence of 100% in *omalodu* malts and at 94% in *otombo* and at concentration ranges of 48.2–48,200 μg/kg in *omalodu* and 32.5–165 μg/kg in *otombo* malts, respectively. Cyclo (L-Pro-L-Val) is formed by the fusion of valine and proline and synthesized by marine *Penicillium* species [56]. It was quantified at higher average concentrations (456 ± 158 μg/kg) in *omalodu* malts than in *otombo* malts (121 ± 86.4 μg/kg) and had a maximum prevalence of 100% in both malts. Brevianamide F is the simplest member and the biosynthetic precursor of prenylated tryptophan-proline 2.5-diketopiperazines that are produced mainly by *Aspergillus fumigatus* and other *Aspergillus* species [57]. In addition, Brevianamide F is produced by many *Penicillium* species and intermediaries of many fungal species. The brevianamide F average concentration of these malts was highest in *omalodu* malts (144 ± 92.6 μg/kg) the lowest in *otombo* malts (90.1 ± 56.1 μg/kg). Fellutanine A is bio-active, naturally occurring, 2.5 diketopiperazine alkaloid synthesized by *Penicillium fellutanum* and *Penicillium simplicissimum*, [58]. It is also understood to be a non-annulated analogue of "cis" cyclic dipeptide, cyclo (L-Trp-L-Trp). The concentration range of this metabolite in the samples is indicated in Table 3. Tryptophol is an aromatic alcohol that induces sleep in humans and is produced by the trypanosomal parasite in wine as a secondary product of alcoholic fermentation [59]. Tryptophol is also formed from tryptophan during fermentation as well. *Otombo* malts had the highest concentration of 110 ± 80.5 μg/kg, but it was not quantified in *omalodu* malts.

2.6. Method Performance Characteristics

The values reported in Table 4 are for the LC/MS/MS method validation characteristics, such as limits of detection (LOD) and LOQ, apparent recoveries (i.e., spiked samples vs. solvent standards), and relative standard deviations (RSD). LOD were observed from 0.02 to 124 ng/g while LOQ were observed from 0.03 to 421 ng/g. Deviations from the target range of 50–120% of apparent recoveries, set by the Commission Regulation (EC) No 401/2006, are mainly caused by matrix effects, whereas the recovery of the extraction step has been determined to be in this target range for the majority of all investigated compounds in other matrices (manuscript in preparation). In addition, the determination of the apparent recoveries was hampered by the fact that 15% of the metabolites (e.g., 3-nitropropionic acid, kojic acid) none of the samples were true blanks. This resulted in apparent recoveries significantly larger than 120%, despite a correction for the concentration in the blank samples being performed. The same holds true for large values for the respective RSD not complying with the <20% criterion that are set for replicate analysis, whereas in this study, three different individual samples were spiked, which potentially resulted in higher values for the repeatability and the combined method uncertainty [60].

Table 4. Performance characteristics of the method for some metabolites quantified in sorghum malts.

Compounds	*Omalodu* and *Otombo* Malts		*Omalodu* Malts		*Otombo* Malts	
	LOD (ng/g)	LOQ (ng/g)	Apparent Recovery (%)	RSD (%) (n = 3)	Apparent Recovery (%)	RSD (%) (n = 3)
3-Nitropropionic acid	0.71	2.4	147	47.13	219	36.75
Abscisic acid	15	50	192	31.79	N/D	N/D
Aflatoxin B1	0.17	0.57	40	1.52	40	3.11
Aflatoxin B2	0.04	0.13	40	1.7	50	2.88
Aflatoxin G1	0.1	0.35	46	1.3	45	2.63
Agroclavine	0.1	0.32	81	3.85	92	8.14
Alternariol	0.1	0.32	45	9.17	40	6.97
Alternariolmethylether	0.11	0.38	75	10.37	64	2.16
Altersetin	0.89	3	163	17.35	158	16.75
Altersolanol	126	421	248	113.25	125	73.16
Antibiotic PF 1052	2.5	8.2	120	1.88	176	30.81
Antibiotic Y	6.9	23	154	10.67	154	30.57
Asperfuran	3.7	12	94	8.91	106	0
Asperglaucide	0.03	0.12	66	6.52	78	2.31
Asterric acid	0.15	0.5	186	12.25	208	12.17
Aurantine	0.26	0.85	62	5.95	58	1.85
Averufin	0.02	0.07	N/D	N/D	51	3.9
Barceloneic acid	0.55	1.8	439	252.56	381	180.88
Bassianolide	0.02	0.08	75	16.97	82	12.94
Beauvericin	0.06	0.2	82	28.57	81	0
Bikaverin	0.46	1.5	82	0	129	0
Bis (methylthio)gliotoxin	0.86	2.9	60	2.36	65	1.9
Brefeldin A	20	66	129	35.89	87	10.58
Brevianamid F	0.35	1.2	110	5.75	111	0
Calphostin	3.4	11	115	15.42	127	11.35
Chanoclavin	0.02	0.07	82	6.38	78	6.98
Chloramphenicol	0.03	0.09	84	3.53	58	18.4
Citreohybridinol	0.09	0.31	77	8.18	91	21.22
Citreorosein	0.74	2.5	66	5.85	64	15.5
Curvularin	0.36	1.2	111	0	110	0
Cyclo (L-Pro-L-Tyr)	8.5	28	74	5.16	99	28.37
Cyclo (L-Pro-L-Val)	1.2	3.9	268	16.23	255	0
Cyclopiazonic acid	15	50	137	13.65	128	10.12
Cytochalasin E	0.43	1.4	99	7.21	113	0
Dechlorogriseofulvin	0.43	1.4	111	11.61	93	0.36

Table 4. *Cont.*

Compounds	*Omalodu* and *Otombo* Malts		*Omalodu* Malts		*Otombo* Malts	
	LOD (ng/g)	LOQ (ng/g)	Apparent Recovery (%)	RSD (%) (n = 3)	Apparent Recovery (%)	RSD (%) (n = 3)
Dehydrocurvularin	0.96	3.2	87	7.54	60	0
Demethylsulochrin	0.58	1.9	118	11.63	120	0
Deoxynortryptoquivalin	0.62	2.1	106	30.01	77	0
Deoxytryptoquivaline A	0.27	0.89	87	11.75	76	0
Destruxin A	0.07	0.24	68	2.19	70	0.82
Destruxin-ed derivative	0.66	2.2	74	2	78	0.99
Dichlordiaportin	0.64	2.1	110	15.2	130	0
Dihydrocitrinone	0.77	2.6	121	6.31	143	0
Dihydroxymellein	0.51	1.7	N/D	N/D	136	16.66
Elymoclavine	0.18	0.59	65	3.54	76	8.55
Emodin	0.06	0.2	104	11.21	104	3.59
Epiequisetin	0.1	0.32	N/D	N/D	132	14.5
Equisetin	0.1	0.33	138	16.46	148	10.02
Fellutanine A	0.48	1.6	102	10.17	113	3.86
Festuclavine	0.02	0.07	76	3.92	86	17
Flavoglaucin	0.03	0.1	149	14.27	127	0
Fumagillin	9	30	52	7.91	59	16.34
Fumigaclavine C	0.83	2.8	87	3.06	95	4.61
Fumiquinazolin A	0.18	0.59	84	5.57	93	11.4
Fumiquinazolin D	0.27	0.9	74	1.22	97	19.86
Fumitremorgin C	0.19	0.62	42	4.36	42	8.33
Fumonisin B1	2.4	8	70	11.34	77	6.52
Fumonisin B2	1.7	5.6	74	10	83	4.89
Fumonisin B3	5.8	19	74	10.56	80	6.64
Gliotoxin	0.91	3	42	8.79	47	11.15
Griseofulvin	0.14	0.46	71	1.57	70	0.33
Helvolic acid	2.1	6.9	119	2.97	138	15.56
Heptelidic acid	8.7	29	135	23.55	112	1.27
Herquline A	0.06	0.22	66	5.15	53	8.88
Hydroxycurvularin	0.94	3.1	108	20.09	90	13.67
Iso-rhodoptilometrin	0.03	0.09	55	6.08	60	5.26
Kojic acid	20	68	288	152.08	1013	426.7
Macrosporin	0.13	0.44	55	1.04	57	11.02
Moniliformin	1	3.4	106	13	139	14.96
Monocerin	0.06	0.19	85	0	89	2.58
Patulin	11	36	71	13.25	55	0
Phenopyrrozin	0.31	1	96	15.14	164	28.84

Table 4. *Cont.*

Compounds	*Omalodu* and *Otombo* Malts		*Omalodu* Malts		*Otombo* Malts	
	LOD (ng/g)	LOQ (ng/g)	Apparent Recovery (%)	RSD (%) (n = 3)	Apparent Recovery (%)	RSD (%) (n = 3)
Phomalactone	0.61	2	64	5.16	60	5.98
Pinselin	0.6	2	65	2.01	70	3.84
Pseurotin A	1.5	5	83	2.38	92	4.95
Pyrenophorol	0.96	3.2	73	4.2	77	8.29
Quinolactacin A	0.01	0.03	62	5.24	72	8.77
Rugulusovin	0.45	1.5	81	11.08	136	34.61
Sambucinol	4.5	15	87	7.33	92	17.49
Siccanin	0.93	3.1	69	1.87	76	6.35
Skyrin	0.08	0.26	94	13.05	105	5.66
Sterigmatocystin	0.06	0.19	50	2.01	57	3.15
Tenuazonic acid	30	100	321	0	275	0
Thielavin B	0.29	0.98	44	6.65	58	3.08
Trypacidin	0.09	0.29	53	17.64	N/D	N/D
Tryprostatin B	1.5	5.1	43	2.3	38	3.85
Tryptophol	3.5	12	N/D	N/D	60	5.95
Tryptoquivaline A	0.48	1.6	74	1.91	84	0
Tryptoquivaline F	0.67	2.2	79	10.81	104	23
Versicolorin C	0.13	0.45	94	26.14	53	15.77

N/D = Not detected.

3. Conclusions

The present study reports data on the variation of fungal metabolites in two different sorghum malts as raw materials for the brewing of two indigenous and popular traditional Namibian beverages, *otombo* and *omalodu*. Both malts were substantially contaminated with fungal metabolites produced by major mycotoxigenic fungal genera. The study found little contamination variation between the two malts. Regulated mycotoxins, emerging mycotoxins, aflatoxin precursors, and ergot alkaloids were quantified in both malts. Generally, the study findings were that *omalodu* malts were mostly contaminated with fungal metabolites and health risk mycotoxin groups than *otombo* malts. Based on the high incidence of mycotoxins and other metabolites in both malt samples, adequate milling and processing conditions (low moisture) must be ensured to reduce the prevalence of these toxins. The present study on two sorghum malts provides three major findings: First is the co-occurrence of seven EU-regulated mycotoxins in both malts, particularly toxic AFB_1, which was quantified in 20% of *omalodu* malts and 40% of *otombo* malts at levels above the EU allowable limit of 5 μg/kg. Second is the high occurrence of several fungal metabolites in both malts and the existing knowledge gap on the effects of such intricate metabolite mixtures in humans. The third is the high incidence of emerging mycotoxins such as 3-NPA, MON, STE, and TeA and pending risk assessment studies for these toxins in humans. Since traditional malting and processing are likely carried out by mycotoxin-unaware traditional processors, it is also advised to educate the public on the health risks of mycotoxins and possible methods to alleviate fungal contamination and on hygienic conditions during malting and storage. Data from the present study serves as a foundation for more detailed mycotoxin-related studies, such as further investigation on the fate of mycotoxins during the brewing processes of these beverages, considering the possible formation of masked/bound mycotoxins which may not have been quantifiable in the present study. Investigations of the occurrence of fungal metabolites in other indigenous food commodities from Namibia are necessary, as well as the determination of exposure to mycotoxins and their health effects in the Namibian population.

4. Materials and Methods

4.1. Sorghum Malts Collection

A total of 81 sorghum malt samples, purchased in November 2017 at open-markets in Oshana region, Namibia, were collected for this study. The sorghum malt samples were purchased based on availability at the open markets, hence, 45 sorghum flour malt samples intended for *omalodu* brewing and 36 un-milled sorghum malts grain samples intended for *otombo* brewing. Approximately, 500 g of the samples were collected following the sampling procedure described by [61]. Samples were placed in paper bags, transported to the Centre for Analytical Chemistry, Department of Agrobiotechnology, (IFA-Tulln), University of Natural Resources and Life Sciences, Austria, and stored at −20 °C until analysis.

4.2. Metabolites Extraction and Analysis by LC/MS/MS

Sorghum malt samples were extracted for the presence of targeted multi-metabolites, including regulated, conjugated, and emerging mycotoxins. The extraction was done according to the methods described by [26]. Briefly, 5 g of each milled sample and 20 mL of acetonitrile/water/acetic acid (79:20:1, *v*/*v*/*v*) was agitated in a 50 mL polypropylene tube for 90 min at 180 rpm using a rotary shaker (GFL 3017, Burgwedel, Germany). The mixture was then settled and supernatants/extracts were stored at −20 °C until further analysis.

The occurrences of fungal metabolites were detected and quantified using the procedure described by [62]. Briefly, 500 μL of each extract was diluted with an equal volume of acetonitrile/water/acetic acid (79:20:1, *v*/*v*/*v*) and 5 μL was directly injected into the LC/MS/MS system consisting of an Agilent (Waldbronn, Germany) 1290 HPLC and an AB Sciex 5500 QTrap MS/MS with an electrospray ionization (ESI) triple quadrupole. Chromatographic separation was performed on a Phenomenex

Gemini C18 column (150 × 4.6 mm, 5 μm) equipped with a C18 (4 × 3 mm) i.d. security guard cartridge, eluted with a gradient of methanol/water containing ammonium acetate and acetic acid.

Data acquisition was achieved in the time-scheduled multiple reactions monitoring (MRM) mode both in positive and negative polarities in two separate chromatographic runs per sample. The expected retention time of the MRM detection window of each metabolite was set at about 27 s and about 48 s for both positive and negative modes, respectively. Data were analyzed using MultiQuant™ 3.0.3 software (AB Sciex, Foster City, CA, USA). Quantification of metabolites was performed using external calibration based on serial dilution of a multi-metabolites stock solution. Results were corrected for apparent recoveries based on relative responses of the two matrices by spiking three different approximately blank samples at three concentration levels. Limits of detection and limits of quantification were determined following the Eurachem guide described by [63]. The accuracy of the method is verified on a routine basis by participation in interlaboratory testing schemes including a broad variation of matrices of grains, nuts, dried fruits, spices, baby food, and animal feed. Satisfactory z-scores between −2 and 2 have been obtained for >94% of the >1000 results submitted so far and for 11 of the 12 results submitted for sorghum, respectively.

Confirmation of positive metabolite identification was attained by the acquisition of two MRMs per metabolite (apart from moniliformin and 3-nitropropionic acid, which displayed only one fragment ion). This generated 4.0 identification points according to Ref. [63]. In addition, the LC retention time and the intensity ratio of the two MRM transitions agreed with the related values of a true standard within 0.03 min and 30% relatively and singly.

4.3. Data Analysis

Data evaluation, averages, and range calculations were performed in Microsoft® Excel 2010. An independent-samples *t*-test was done using the Statistical Package for the Social Sciences (SPSS) software, version 21.0 (SPSS, Inc., Chicago, IL, USA).

Author Contributions: S.N.N., J.M.M., R.B., A.I., M.S., and R.K. conceived and designed the experiments. S.N.N., J.M.M., A.I., and M.S. performed various stages of the experiments. S.N.N., J.M.M., and M.S. analyzed the data. M.S. and R.K. contributed reagents/materials/analysis tools. S.N.N., J.M.M., R.B., A.I., M.S., and R.K. contributed to the writing of the paper, reviewing, and fine tuning.

Funding: This research was funded by the Austrian Academic Exchange Department through an OeAD Ernst Mach grant award (ICM-2017-06547).

Conflicts of Interest: The authors declare no conflicts of interest.

References

1. Roger, D.D. Deoxynivalenol (DON) and fumonisins B_1 (FB_1) in artisanal Sorghum opaque beer brewed in north Cameroon. *Afr. J. Microbiol. Res.* **2011**, *5*, 1565–1567. [CrossRef]
2. Dicko, M.H.; Gruppen, H.; Traoré, A.S.; Voragen, A.G.; Van Berkel, W.J. Sorghum grain as human food in Africa: Relevance of content of starch and amylase activities. *Afr. J. Biotechnol.* **2006**, *5*, 384–395.
3. Misihairabgwi, J.; Cheikhyoussef, A. Traditional fermented foods and beverages of Namibia. *J. Ethn. Foods* **2017**, *4*, 145–153. [CrossRef]
4. Adebiyi, J.A.; Obadina, A.O.; Adebo, O.A.; Kayitesi, E. Fermented and malted millet products in Africa: Expedition from traditional/ethnic foods to industrial value-added products. *Crit. Rev. Food Sci. Nutr.* **2018**, *58*, 463–474. [CrossRef] [PubMed]
5. Onesmo, N.M. Effects of Malting and Fermentation on the Composition and Functionality of Sorghum Flour. Master's Thesis, University of Nebraska-Lincoln, Lincoln, NE, USA, April 2011.
6. Tangni, E.K.; Larondelle, Y. Malts, moulds and mycotoxins. Bacteria, Yeasts and Moulds in Malting and Brewing. In Proceedings of the Xth Symposium "Chair J. de Clerck", Leuven, Belgium, 15–18 September 2002.
7. Wagacha, J.M.; Muthomi, J.W. Mycotoxin problem in Africa: Current status, implications to food safety and health and possible management strategies. *Int. J. Food Microbiol.* **2008**, *124*, 1–12. [CrossRef] [PubMed]

8. Hajslova, J.; Zachariasova, M.; Cajka, T. Analysis of multiple mycotoxins in food. *Methods Mol. Biol.* **2011**, *747*, 233–258. [CrossRef] [PubMed]
9. Reddy, K.R.N.; Salleh, B.; Saad, B.; Abbas, H.K.; Abel, C.A.; Shier, W.T. An overview of mycotoxin contamination in foods and its implications for human health. *Toxin Rev.* **2010**, *29*, 3–26. [CrossRef]
10. Bennett, J.; Klich, M. Mycotoxins. *Clin. Microbiol. Rev.* **2003**, *16*, 497–516. [CrossRef] [PubMed]
11. Richard, J.L. Some major mycotoxins and their mycotoxicoses—An overview. *Int. J. Food Microbial.* **2007**, *119*, 3–10. [CrossRef] [PubMed]
12. Schatzmayr, G.; Streit, E. Global occurrence of mycotoxins in the food and feed chain: Facts and figures. *World Mycotoxin J.* **2013**, *6*, 213–222. [CrossRef]
13. Gruber-Dorninger, C.; Novak, B.; Nagl, V.; Berthiller, F. Emerging mycotoxins: Beyond traditionally determined food contaminants. *J. Agric. Food Chem.* **2016**, *65*, 7052–7070. [CrossRef] [PubMed]
14. Vaclavikova, M.; Malachova, A.; Veprikova, Z.; Dzuman, Z.; Zachariasova, M.; Hajslova, J. 'Emerging'mycotoxins in cereals processing chains: Changes of enniatins during beer and bread making. *Food Chem.* **2013**, *136*, 750–757. [CrossRef]
15. Kovalsky, P.; Kos, G.; Nährer, K.; Schwab, C.; Jenkins, T.; Schatzmayr, G.; Sulyok, M.; Krska, R. Co-occurrence of regulated; masked and emerging mycotoxins and secondary metabolites in finished feed and maize—An extensive survey. *Toxins* **2016**, *8*, 363. [CrossRef] [PubMed]
16. Lee, H.J.; Ryu, D. Worldwide occurrence of mycotoxins in cereals and cereal-derived food products: Public health perspectives of their co-occurrence. *J. Agric. Food Chem.* **2017**, *65*, 7034–7051. [CrossRef] [PubMed]
17. Odhav, B.; Naicker, V. Mycotoxins in South African traditionally brewed beers. *Food Addit. Contam.* **2002**, *19*, 55–61. [CrossRef] [PubMed]
18. Gamanya, R.; Sibanda, L. Survey of *Fusarium moniliforme* (*F. verticillioides*) and production of fumonisin B_1 in cereal grains and oilseeds in Zimbabwe. *Int. J. Food Microbiol.* **2001**, *71*, 145–149. [CrossRef]
19. Nkwe, D.O.; Taylor, J.E.; Siame, B.A. Fungi, Aflatoxins, Fumonisin B_1 and Zearalenone Contaminating Sorghum-based Traditional Malt, Wort and Beer in Botswana. *Mycopathologia* **2005**, *160*, 177–186. [CrossRef]
20. Misihairabgwi, J.M.; Ishola, A.; Quaye, I.; Sulyok, M.; Krska, R. Diversity and fate of fungal metabolites during the preparation of *oshikundu*, a Namibian traditional fermented beverage. *World Mycotoxin J.* **2018**, *11*, 471–481. [CrossRef]
21. Matumba, L.; Monjerezi, M.; Khonga, E.B.; Lakudzala, D.D. Aflatoxins in sorghum; sorghum malt and traditional opaque beer in southern Malawi. *Food Control* **2011**, *22*, 266–268. [CrossRef]
22. Garba, M.H.; Makun, H.A.; Jigam, A.A.; Muhammad, H.L.; Patrick, B.N. Incidence and toxigenicity of fungi contaminating sorghum from Nigeria. *World J. Microbiol.* **2017**, *4*, 105–114.
23. Chala, A.; Taye, W.; Ayalew, A.; Krska, R.; Sulyok, M.; Logrieco, A. Multi-mycotoxin analysis of sorghum (*Sorghum bicolor* L. Moench) and finger millet (*Eleusine coracana* L. Garten) from Ethiopia. *Food Control* **2014**, *45*, 29–35. [CrossRef]
24. Alshannaq, A.; Yu, J.H. Occurrence, toxicity, and analysis of major mycotoxins in food. *Int. J. Environ. Res.* **2017**, *14*, 632. [CrossRef] [PubMed]
25. Marin, S.; Ramos, A.J.; Cano-Sancho, G.; Sanchis, V. Mycotoxins: Occurrence; toxicology; and exposure assessment. *Food Chem. Toxicol.* **2013**, *60*, 218–237. [CrossRef]
26. European Commission (EC). Commission Regulation (EC) No 1881/2006 of 19 December 2006 setting maximum levels for certain contaminants in foodstuffs. *Off. J. Eur. Union* **2006**, *L364*, 5–24.
27. Bationo, J.F.; Nikiéma, P.A.; Koudougou, K.; Ouédraogo, M.; Bazié, S.R.; Sanou, E.; Barro, N. Assessment of aflatoxin B_1 and ochratoxin A levels in sorghum malts and beer in Ouagadougou. *Afr. J. Food Sci.* **2015**, *9*, 417–420. [CrossRef]
28. Zbyňovská, K.; Petruška, P.; Kalafová, A.; Capcarová, M. Patulin-a Contaminant of Food and Feed: A review. *Acta Fytotech. Zootechn.* **2016**, *19*, 64–67. [CrossRef]
29. Assunção, R.; Martins, C.; Dupont, D.; Alvito, P. Patulin and ochratoxin A co-occurrence and their bioaccessibility in processed cereal-based foods: A contribution for Portuguese children risk assessment. *Food Chem. Toxicol.* **2016**, *96*, 205–214. [CrossRef] [PubMed]
30. Ioi, J.D.; Zhou, T.; Tsao, R.F.; Marcone, M. Mitigation of patulin in fresh and processed foods and beverages. *Toxins* **2017**, *9*, 157. [CrossRef]
31. Glaser, N.; Stopper, H. Patulin: Mechanism of genotoxicity. *Food Chem. Toxicol.* **2012**, *50*, 1796–1801. [CrossRef]

32. International Agency for Research on Cancer. *IARC Monographs on the Evaluation of Carcinogenic Risks to Human: Some Naturally Occurring Substances*; IARC Press: Lyon, France, 1976; Volume 10, pp. 245–251.
33. Gao, W.; Jiang, L.; Ge, L.; Chen, M.; Geng, C.; Yang, G.; Li, Q.; Ji, F.; Yan, Q.; Zou, Y.; et al. Sterigmatocystin-induced oxidative DNA damage in human liver-derived cell line through lysosomal damage. *Toxicol. In Vitro* **2015**, *29*, 1–7. [CrossRef]
34. Zouaoui, N.; Mallebrera, B.; Berrada, H.; Abid-Essefi, S.; Bacha, H.; Ruiz, M.J. Cytotoxic effects induced by patulin., sterigmatocystin and beauvericin on CHO–K1 cells. *Food Chem. Toxicol.* **2016**, *89*, 92–103. [CrossRef] [PubMed]
35. Liu, Y.; Du, M.; Zhang, G. Proapoptotic activity of aflatoxin B_1 and sterigmatocystin in HepG2 cells. *Toxicol. Rep.* **2014**, *1*, 1076–1086. [CrossRef]
36. Lee, W.T.; Chang, C. Magnetic resonance imaging and spectroscopy in assessing 3-nitropropionic acid-induced brain lesions: An animal model of Huntington's disease. *Prog. Neurobiol.* **2004**, *72*, 87–110. [CrossRef] [PubMed]
37. Carvajal-Moreno, M. Mycotoxins that affect the human cardiovascular system. *Pharm. Anal. Acta* **2015**, *6*, 2. [CrossRef]
38. Li, Y.C.; Ledoux, D.R.; Bermudez, A.J.; Fritsche, K.L.; Rottinghaus, G.E. The individual and combined effects of fumonisin B_1 and moniliformin on performance and selected immune parameters in turkey poults. *Poult. Sci.* **2000**, *79*, 871–878. [CrossRef] [PubMed]
39. Huang, P.H.; Huang, C.Y.; Chen, M.C.; Lee, Y.T.; Yue, C.H.; Wang, H.Y.; Lin, H. Emodin and aloe-emodin suppress breast cancer cell proliferation through ERα inhibition. *Evid. Based Complement. Altern. Med.* **2013**, *2013*. [CrossRef]
40. Wang, Q.J.; Cai, X.B.; Liu, M.H.; Hu, H.; Tan, X.J.; Jing, X.B. Apoptosis induced by emodin is associated with alterations of intracellular acidification and reactive oxygen species in EC-109 cells. *Int. J. Biochem. Cell Biol.* **2010**, *88*, 767–774. [CrossRef]
41. Liu, Z.; Wei, F.; Chen, L.J.; Xiong, H.R.; Liu, Y.Y.; Luo, F.; Hou, W.; Xiao, H.; Yang, Z.Q. In Vitro and in Vivo Studies of the Inhibitory Effects of Emodin Isolated from Polygonum cuspidatum on Coxsackievirus B4. *Molecules* **2013**, *18*, 11842–11858. [CrossRef]
42. Hsiang, C.Y.; Ho, T.Y. Emodin is a novel alkaline nuclease inhibitor that suppresses herpes simplex virus type 1 yields in cell cultures. *Br. J. Pharmacol.* **2008**, *155*, 227–235. [CrossRef]
43. Chukwujekwu, J.C.; Coombes, P.H.; Mulholland, D.A.; Van Staden, J. Emodin, an antibacterial anthraquinone from the roots of *Cassia occidentalis*. *S. Afr. J. Bot.* **2006**, *72*, 295–297. [CrossRef]
44. Jow, G.M.; Chou, C.J.; Chen, B.F.; Tsai, J.H. Beauvericin induces cytotoxic effects in human acute lymphoblastic leukaemia cells through cytochrome c release, caspase 3 activation: The causative role of calcium. *Cancer Lett.* **2004**, *216*, 165–173. [CrossRef] [PubMed]
45. Smith, E.R.; Fredrickson, T.N.; Hadidian, Z. Toxic effects of the sodium and the N, N′-dibenzylethylenediamine salts of tenuazonic acid (NSC-525816 and NSC-82260). *Cancer Chemother. Rep.* **1968**, *52*, 579. [PubMed]
46. Patriarca, A.; Azcarate, M.P.; Terminiello, L.; Pinto, V.F. Mycotoxin production by *Alternaria* strains isolated from Argentinean wheat. *Int. J. Food Microbiol.* **2007**, *119*, 219–222. [CrossRef] [PubMed]
47. Giambrone, J.J.; Davis, N.D.; Diener, U.L. Effect of tenuazonic acid on young chickens. *Poult. Sci.* **1978**, *57*, 1554–1558. [CrossRef] [PubMed]
48. Fleck, S.C.; Burkhardt, B.; Pfeiffer, E.; Metzler, M. *Alternaria* toxins: Altertoxin II is a much stronger mutagen and DNA strand breaking mycotoxin than alternariol and its methyl ether in cultured mammalian cells. *Toxicol. Lett.* **2012**, *214*, 27–32. [CrossRef] [PubMed]
49. Bandyopadhyay, R.; Butler, D.R.; Chandrashekar, A.; Reddy, R.K.; Navi, S.S. Biology, epidemiology, and management of sorghum grain mould. In *Technical and Institutional Options for Sorghum Grain Mould Management: Proceedings of an International Consultation*; Andhra Pradesh: Patancheru, India, 2000; pp. 18–19.
50. Bandyopadhyay, R.; Frederickson, D.E.; McLaren, N.W.; Odvody, G.N.; Riley, M.J. Ergot: A new disease threat to sorghum in the Americas and Australia. *Plant Dis.* **1998**, *82*, 356–367. [CrossRef] [PubMed]
51. Kwon-Chung, K.J.; Sugui, J.A. What do we know about the role of gliotoxin in the pathobiology of *Aspergillus fumigatus*? *Med. Mycol.* **2009**, *47* (Suppl. 1), S97–S103. [CrossRef]

52. Domingo, M.P.; Colmenarejo, C.; Martínez-Lostao, L.; Müllbacher, A.; Jarne, C.; Revillo, M.J.; Delgado, P.; Roc, L.; Meis, J.F.; Rezusta, A.; et al. Bis (methyl) gliotoxin proves to be a more stable and reliable marker for invasive aspergillosis than gliotoxin and suitable for use in diagnosis. *Diagn. Microbiol. Infect. Dis.* **2012**, *73*, 57–64. [CrossRef] [PubMed]
53. Ostry, V.; Toman, J.; Grosse, Y.; Malir, F. Cyclopiazonic acid: 50th anniversary of its discovery. *World Mycotoxin J.* **2018**, *11*, 135–148. [CrossRef]
54. Stierle, A.C.; Cardellina, J.H.; Strobel, G.A. Maculosin. A host-specific phytotoxin for spotted knapweed from *Alternaria alternata*. *Proc. Natl. Acad. Sci. USA* **1988**, *85*, 8008–8011. [CrossRef] [PubMed]
55. Puopolo, G.; Cimmino, A.; Palmieri, M.C.; Giovannini, O.; Evidente, A.; Pertot, I. *Lysobacter capsici* AZ78 produces cyclo (L-Pro-L-Tyr), a 2, 5-diketopiperazine with toxic activity against sporangia of *P. phytophthora infestans* and *P. lasmopara viticola*. *J. Appl. Microbiol.* **2014**, *117*, 1168–1180. [CrossRef]
56. Capon, R.J.; Stewart, M.; Ratnayake, R.; Lacey, E.; Gill, J.H. Citromycetins and bilains A-C: New aromatic polyketides and diketopiperazines from Australian marine-derived and terrestrial *Penicillium* spp. *J. Nat. Prod.* **2007**, *70*, 1746–1752. [CrossRef]
57. Borthwick, A.D. 2, 5-Diketopiperazines: Synthesis, reactions, medicinal chemistry and bioactive natural products. *Chem. Rev.* **2012**, *112*, 3641–3716. [CrossRef]
58. Kozlovskii, A.G.; Vinokurova, N.G.; Adanin, V.M. Diketopiperazine alkaloids from the fungus *Penicillium piscarium* Westling. *Prikl. Biokhim. Mikrobiol.* **2000**, *36*, 317–321.
59. Cornford, E.M.; Bocash, W.D.; Braun, L.D.; Crane, P.D.; Oldendorf, W.H.; MacInnis, A.J. Rapid distribution of tryptophol (3-indole ethanol) to the brain and other tissues. *J. Clin. Investig.* **1979**, *63*, 1241–1248. [CrossRef] [PubMed]
60. Stadler, D.; Sulyok, M.; Schuhmacher, R.; Berthiller, F.; Krska, R. The contribution of lot-to-lot variation to the measurement uncertainty of an LC-MS-based multi-mycotoxin assay. *Anal. Bioanal. Chem.* **2018**, *410*, 4409–4418. [CrossRef] [PubMed]
61. Malachová, A.; Sulyok, M.; Beltrán, E.; Berthiller, F.; Krska, R. Optimization and validation of a quantitative liquid chromatography-tandem mass spectrometric method covering 295 bacterial and fungal metabolites including all regulated mycotoxins in four model food matrices. *J. Chromatogr. A* **2014**, *1362*, 145–156. [CrossRef] [PubMed]
62. European Union Commission. Decision 2002/657/EC of 12 August 2002 implementing Council Directive 96/23/EC concerning the performance of analytical methods and the interpretation of results. *Off. J. Eur. Union* **2002**, *1*, 8–36.
63. Magnusson, B.; Örnemark, U. *Eurachem Guide: The Fitness for Purpose of Analytical Methods—A Laboratory Guide to Method Validation and Related Topics*, 2nd ed.; 2014; ISBN 978-91-87461-59-0. Available online: www.eurachem.org/images/stories/Guides/pdf/MV_guide_2nd_ed_EN.pdf (accessed on 16 October 2014).

toxins

MDPI

Article

Regional Sub-Saharan Africa Total Diet Study in Benin, Cameroon, Mali and Nigeria Reveals the Presence of 164 Mycotoxins and Other Secondary Metabolites in Foods

Luc Ingenbleek [1,2], Michael Sulyok [3], Abimbola Adegboye [4], Sètondji Epiphane Hossou [5], Abdoulaye Zié Koné [6], Awoyinka Dada Oyedele [4], Chabi Sika K. J. Kisito [7], Yara Koreissi Dembélé [8], Sara Eyangoh [1], Philippe Verger [9], Jean-Charles Leblanc [10,*], Bruno Le Bizec [2] and Rudolf Krska [3,11]

1. Centre Pasteur du Cameroun (CPC), Yaoundé BP1274, Cameroon; luc.ingenbleek@gmail.com (L.I.); eyangoh@pasteur-yaounde.org (S.E.)
2. LABERCA, Oniris, INRA, 44307 Nantes, France; Bruno.LeBizec@oniris-nantes.fr
3. Department IFA-Tulln, University of Natural Resources and Life Sciences, Vienna (BOKU), 3430 Tulln, Austria; michael.sulyok@boku.ac.at (M.S.); Rudolf.Krska@boku.ac.at (R.K.)
4. National Agency for Food and Drug Administration and Control (NAFDAC), Abuja 900288, Nigeria; adegboye.a@nafdac.gov.ng (A.A.); oyedele.dada@nafdac.gov.ng (A.D.O.)
5. Agence Béninoise de Sécurité Sanitaire des Aliments (ABSSA), Cotonou BP 362, Benin; hossepfr@yahoo.fr
6. Agence Nationale de la Sécurité Sanitaire des Aliments (ANSSA), Bamako BP 2362, Mali; ngolona@gmail.com
7. Laboratoire Central de Sécurité Sanitaire des Aliments (LCSSA), Cotonou BP 6874, Benin; kinnousika@yahoo.fr
8. Laboratoire de Technologie Alimentaire (LTA), Bamako BP 258, Mali; ykoreissidemb@gmail.com
9. World Health Organization (WHO), 1211 Geneva, Switzerland; vergerp@who.int
10. Food and Agriculture Organization of the United Nations (FAO), 00153 Rome, Italy
11. Institute for Global Food Security, School of Biological Sciences, Queens University Belfast, Belfast BT7 1NN, Northern Ireland, UK

* Correspondence: JeanCharles.Leblanc@fao.org

Received: 22 December 2018; Accepted: 15 January 2019; Published: 17 January 2019

check for updates

Abstract: In the framework of the first multi-centre Sub-Saharan Africa Total Diet Study (SSA-TDS), 2328 commonly consumed foods were purchased, prepared as consumed and pooled into 194 composite samples of cereals, tubers, legumes, vegetables, nuts and seeds, dairy, oils, beverages and miscellaneous. Those core foods were tested for mycotoxins and other fungal, bacterial and plant secondary metabolites by liquid chromatography, coupled with tandem mass spectrometry. The highest aflatoxin concentrations were quantified in peanuts, peanut oil and maize. The mean concentration of the sum of aflatoxins AFB1, AFB2, AFG1 and AFG2 (AF_{tot}) in peanut samples (56.4 μg/kg) exceeded EU (4 μg/kg) and Codex (15 μg/kg) standards. The AF_{tot} concentration (max: 246.0 μg/kg) was associated with seasonal and geographic patterns and comprised, on average, 80% AFB1, the most potent aflatoxin. Although ochratoxin A concentrations rarely exceeded existing Codex standards, it was detected in unregulated foods. One palm oil composite sample contained 98 different metabolites, including 35.4 μg/kg of ochratoxin A. In total, 164 different metabolites were detected, with unspecific metabolites like asperglaucide, cyclo(L-pro-L-val), cyclo (L-pro-L-tyr), flavoglaucin, emodin and tryptophol occurring in more than 50% of composite samples. Aflatoxin B1 (AFB1), fumonisin B1 (FB1), sterigmatocystin (STC), ochratoxin A (OTA), citrinin (CIT) and many other secondary fungal metabolites are frequent co-contaminants in staple foods, such as maize and sorghum. Populations from North Cameroon and from Benin may, therefore, suffer chronic and simultaneous exposure to AFB1, FB1, STC, OTA and CIT, which are prevalent in their diet.

Keywords: Sub-Saharan Africa; aflatoxins; mycotoxins; total diet study; food contaminants; LC-MS/MS

1. Introduction

Mycotoxins are secondary metabolites produced by filamentous fungi in food commodities due to inadequate pre- or post-harvest conditions and practices. These fungal toxins are, therefore, naturally-occurring chemical hazards. Since they are structurally stable, mycotoxins are likely to persist in foods, even if toxin-producing moulds are eliminated during the food preparation process. Consumption of mycotoxin-contaminated food may result in acute or chronic affections, including non-communicable diseases. A particularly severe record of acute toxicity was reported after a major outbreak struck Kenya in 2004, resulting in 317 aflatoxicosis cases including 125 deaths [1]. This episode was the consequence of high exposure to aflatoxins due to the consumption of extensively-contaminated maize [2]. Long-term exposure to aflatoxin B1 or its precursors has been associated with genotoxicity and hepatocellular carcinoma [3,4]. Fumonisin B1 was associated with oesophageal cancer incidence in South Africa and some areas of China [5,6]. Growth impairment, the main indicator for child chronic malnutrition, is also associated with mycotoxin exposure [7–10]. Of the world's 161 million stunted children in 2013, about half live in Asia and over one-third live in Africa [11]. Although often overlooked as a possible cause of retarded growth, mycotoxins may contribute a significant public health burden in less developed countries [12].

An additive or synergistic effect of fumonisin and aflatoxin co-exposure in the development of preneoplastic lesions or hepatocellular carcinoma was suggested in laboratory animals [13–15].

Mycotoxins form the group of food chemicals which triggered the most cases of border rejection (489) recorded in the EU Rapid Alert System on Food and Feed [16]. According to the European Commission Regulation 1881/2006 [17], the maximum level for aflatoxins for peanuts and cereals intended for direct human consumption was set to 2 μg/kg of aflatoxin B1 (AFB1) and 4 μg/kg of the sum of AFB1, AFB2, AFG1 and AFG2. The maximum limit from the international standard [18] is 15 μg/kg of AFB1 or AF_{tot}, which only applies to a variety of nuts (including peanuts) intended for further processing (and 10 μg/kg for ready to eat dried figs, almonds, hazelnuts and pistachios).

In order to assess if the chronic intake of substances is likely to harm consumer health, it is pertinent to assess food safety risks by combining available toxicological studies, as well as food contamination and food consumption data.

One way of assessing the dietary exposure of populations to food chemicals, such as mycotoxins is the Total Diet Study (TDS) approach [19–24]. Two specific aspects characterize a TDS—(1) the representativeness of the sampling, and (2) the preparation of the samples "as consumed"—so that it represents a pertinent public health risk assessment tool as far as food safety and nutrition are concerned.

The World Health Organization (WHO) and the Food and Agriculture Organization of the United Nations (FAO) endorse the TDS methodology, which is both cost-effective and more accurately characterizes human exposure to food chemicals than mere occurrence studies [23].

Following a first experience in Sub-Saharan Africa [25,26], a regional TDS was implemented by FAO in Benin, Cameroon, Mali and Nigeria (2014 and 2018) by four national food safety authorities, in close collaboration with WHO and Centre Pasteur of Cameroon [27]. The purpose of this project is to assess the typical contamination levels of eight African population groups. The dietary exposure of those population groups will then be compared with existing health-based guidance values or end points.

The study methodology was described elsewhere [28].

In this paper, we are presenting the occurrence of mycotoxins and selected fungal, bacterial and plant toxins quantified in composite samples of foods prepared as consumed. The 194 composites result from the pooling of 12 subsamples, representative of the food consumption habits of three study centres located in coastal areas (Duala, the Littoral of Benin and Lagos) and five study centres in non-coastal areas (Bamako, the Borgou region of Benin, Kano, North Cameroon and Sikasso).

2. Results

Since we are dealing with pooled samples (12 sub-samples per composite) of foods prepared as consumed in this study, we will not always be able to conclude with regard to the conformity of food commodities to selected standards [17,18], which, in most cases, apply to raw food commodities. This comparison is nonetheless useful, particularly when the mean concentration (quantified in a composite sample) exceeds or is close to the maximum legal limit of the substances of interest, because this means that at least one subsample out of 12 may have exceeded this limit.

Additionally, since these data will be used for a dietary exposure assessment, they are presented with (1) lower bound (i.e., LB: concentration of non-detected analytes set to zero and to the LOD for detected but non-quantified analytes) and (2) upper bound (i.e., UB: concentration of non-detected analytes set to LOD for non-detected analytes and to the limit of quantification (LOQ) for detected but non-quantified analytes) scenarios. This means that the uncertainty due to censored data will be taken into consideration. When LB–UB is not specified, it is meant that the difference between LB and UB concentrations in not perceptible or less than 0.1 μg/kg. Maximum concentration values are systematically UB concentrations.

Mycotoxins of public health and economic interest (including aflatoxins, fumonisins, ochratoxin A, zearalenone, deoxynivalenol and citrinin) represented 9% of the detected metabolites.

2.1. Aflatoxins

2.1.1. Aflatoxins in Maize

Composite samples were prepared with maize from each study centre (8) purchased during the rainy season (October 2017) and again during the dry season or harmattan (February 2018).

The AF_{tot} concentration in maize was significantly higher ($p < 0.05$) during the wet season (detected: 7/8; mean LB–UB: 22.2–22.5 μg/kg; max: 76.6 μg/kg) than during the dry season (detected: 4/8; mean LB–UB: 0.4–0.8 μg/kg; max: 2.7 μg/kg). Overall, we detected that AF_{tot} > LOD in 11 of 16 composites (69%) with a mean LB–UB concentration of 11.3–11.7 μg/kg (Table 1) in ready-to-eat maize samples, which exceeds the EU standard for both processed (4 μg/kg) and unprocessed maize to be subjected to sorting or physical treatment before human consumption or use as an ingredient (10 μg/kg). However, the fact that all maize composites collected during the dry season contained AF_{tot} concentrations that were below 4 μg/kg and, therefore, complied with EU standard needs to be emphasized.

Table 1. Occurrence and concentration of total aflatoxins (μg/kg wet weight) by core food and by study centre.

CORE FOOD		N	n > LOD	% > LOD	n > 4 μg/kg	% > 4 μg/kg	n > 15 μg/kg	% > 15 μg/kg	Mean Conc. *		Max Conc.	
									LB	UB	Season **	UB
Maize		16	11	69	5	31	3	19	11.3	11.7	Rainy	76.6
Peanut		10	8	80	5	50	5	50	56.4	56.7	Rainy	246.0
Peanut oil		2	2	100	2	100	2	100	60.2	60.4	Rainy	105.1
Beans		16	3	19	1	6	1	6	1.2	1.6	Dry	15.8
Sorghum		10	6	60	1	10	0	0	0.9	1.3	Rainy	4.9
Smoked fish		6	1	17	1	17	0	0	0.8	1.1	Rainy	4.9
Other core foods		134	11	8	0	0	0	0	0.1	0.5	Rainy	2.4
Total		194	42	22	15	8	11	6	4.7	5.1	Rainy	246.0
CENTRE		**N**	**n > LD**	**% > LD**	**n > 4 μg/kg**	**% > 4 μg/kg**	**n > 15 μg/kg**	**%> 15 μg/kg**	**Mean Conc. ***		**Max Conc.**	
									LB	**UB**	**Core food**	**UB**
BENIN	Littoral	26	5	19	3	12	3	12	7.6	8.0	Peanut oil	105.1
	Borgou	22	7	32	1	5	1	5	1.2	1.6	Maize	19.7
CAMEROON	Duala	29	3	10	0	0	0	0	0.2	0.6	Beans	3.0
	North	17	8	47	4	24	3	18	14.3	14.6	Peanuts	92.5
MALI	Bamako	27	4	15	2	7	1	4	9.4	9.8	Peanuts	246.0
	Sikasso	21	6	29	1	5	1	5	2.2	2.6	Peanuts	42.7
NIGERIA	Lagos	29	3	10	1	3	0	0	0.2	0.6	Maize	5.4
	Kano	23	6	26	3	13	3	13	5.6	6.0	Peanuts	96.6

* LB: lower-bound scenario where the concentration of non-detected analyte is zero and the concentration of detected but non-quantified analyte is the limit of detection. UB: upper-bound scenario where the concentration of non-detected analyte is the limit of detection and the concentration of detected but non-quantified analyte is the limit of quantification; ** Samples of the rainy season were collected in October 2017 and samples of the dry season were collected in February 2018.

There is currently no Codex standard applicable to aflatoxins in maize. AFB1, the most potent aflatoxin, represented 87.6% of the sum of AFB1, AFB2, AFG1 and AFG2 detected in maize samples (Table 2).

2.1.2. Aflatoxins in Peanut

As displayed in Table 1, the highest AF_{tot} concentration in this study was quantified in one peanut composite sample from Bamako (Mali): 246.0 μg/kg (mean LB–UB: 56.4–56.7 μg/kg). Aflatoxins were detected in 80% of peanut composites (rainy season: 100%, dry season: 60% detection exceeding LOD = 0.1 μg/kg). The mean AF_{tot} concentration in peanuts was 93.7–93.9 μg/kg (rainy season) and 19.1–19.4 μg/kg (dry season). A high variance of AF_{tot} levels in peanut was observed (CV > 100%). It was noted that while 50% of samples contained AF_{tot} concentrations below the EU standard (4 μg/kg) and 50% were above the Codex standard (15 μg/kg), the mean AF_{tot} concentrations exceeded both EU and Codex standards, regardless of the season. The proportion of AFB1 in peanut was 75.8% of the sum of AFB1, AFB2, AFG1 and AFG2 (Table 2).

Table 2. Proportions of aflatoxin B1, B2, G1 and G2 by core food and by weight.

CORE FOOD	AFB1 (%)	AFB2 (%)	AFG1 (%)	AFG2 (%)	Sum (%)
Maize	87.6	6.8	5.6	0.0	100
Peanut	75.8	14.3	9.4	0.5	100
Peanut oil	86.6	13.1	0.3	0.0	100
Other core foods	87.0	4.0	9.1	0.0	100
Total	80.1	12.1	7.5	0.3	100

2.1.3. Aflatoxins in Peanut Oil

Two composite samples of peanut oil were tested (Table 1) and both contained significant amounts of total aflatoxins: 15.8 μg/kg (Kano) and 105.1 μg/kg (Cotonou). There is currently no standard for aflatoxins in oil, and these concentrations exceed Codex standards available for processed and unprocessed peanuts. The proportion of AFB1 in peanut oil was 86.6% of the sum of AFB1, AFB2, AFG1 and AFG2 (Table 2).

2.1.4. Aflatoxins in Other Foods

Aflatoxins were detected in 60% of sorghum and 19% of bean composites. In Table 1, we reported that one bean sample contained 15.8 μg/kg AF_{tot}. One smoked fish composite contained 4.9 μg/kg AF_{tot}. The observed mean concentration of all tested core foods was below 1 μg/kg in Duala and Lagos (detection rate of 10%) but those recorded in North Cameroon exceeded 10 μg/kg (detection rate of 47%).

2.2. *Fumonisins*

FUM_{tot} (sum of fumonisins FB1, FB2, FB3 and FB4) were most concentrated in maize samples in all eight centres (Table 3). Although fumonisins were detected in 94% of ready-to-eat maize composites, all FUM_{tot} concentrations (mean LB–UB: 285.2–288.2 μg/kg; max: 855.9 μg/kg) remained below the Codex standard of 2 mg/kg applying to fumonisins in maize. Although there is no Codex standard for fumonisins in other foods than maize, other core food samples contained FUM_{tot} of up to 159.4 μg/kg. Apart from maize, composites containing fumonisins are sorghum (including a traditional fermented drink from North Cameroon processed from sorghum called bili-bili) and millet and tubers having undergone a drying process prior to being prepared as consumed (cassava and yam), as reported in Table 3. In food composites from Mali (Bamako and Sikasso), the mean FUM_{tot} concentration was three to ten-fold lower than in samples collected in the other study centres, with UB and LB scenario respectively. FUM_{tot} in our samples comprised 67.2% FB1, 18.9% FB2, 8.0% FB3 and 6.0% FB4. This is close to the proportions determined in maize samples (Table 4).

Table 3. Occurrence and concentration of total fumonisins (µg/kg wet weight) by core food and by study centre.

CORE FOOD	N	n > LOD	% > LOD	n > 10 µg/kg	% > 10 µg/kg	n > 400 µg/kg	% > 400 µg/kg	Mean Conc. *		Max Conc.	
								LB	UB	Season **	UB
Maize	16	15	94	15	94	4	25	285.2	288.2	Dry	855.9
Sorghum	10	5	50	5	50	0	0	20.0	36.1	Dry	159.4
Millet	8	1	13	1	13	0	0	5.0	13.6	Rainy	44.8
Traditional fermented drink	4	1	25	1	25	0	0	5.7	14.1	Rainy	29.3
Cassava dry	12	3	25	3	25	0	0	14.8	22.9	Dry	134.6
Yam dry	2	1	50	1	50	0	0	7.4	17.8	Rainy	21.7
Other core foods	142	2	1	0	0	0	0	0.04	9.2	Both	14.6
Total	194	28	14	26	13	4	2	26.4	34.8	Dry	855.9

CENTRE		N	n > LOD	% > LOD	n > 10 µg/kg	% > 10 µg/kg	n > 400 µg/kg	% > 400 µg/kg	Mean Conc. *		Max Conc.	
									LB	UB	Core food	UB
BENIN	Littoral	26	2	8	2	8	0	0	26.8	35.2	Maize	391.3
	Borgou	22	5	23	5	23	0	0	26.3	34.4	Maize	376.5
CAMEROON	Duala	29	5	17	4	14	0	0	19.0	27.1	Maize	241.7
	North	17	12	71	5	29	1	6	64.4	71.6	Maize	670.3
MALI	Bamako	27	3	11	2	7	0	0	2.0	11.0	Maize	40.6
	Sikasso	21	2	10	2	10	0	0	4.1	12.9	Maize	79.0
NIGERIA	Lagos	29	4	14	3	10	1	3	34.9	43.5	Maize	855.9
	Kano	23	3	13	3	13	2	9	45.9	54.1	Maize	589.9

* LB: lower-bound scenario where the concentration of non-detected analyte is zero and the concentration of detected but non-quantified analyte is the limit of detection. UB: upper-bound scenario where the concentration of non-detected analyte is the limit of detection and the concentration of detected but non-quantified analyte is the limit of quantification; ** Samples of the rainy season were collected in October 2017 and samples of the dry season were collected in February 2018.

Table 4. Proportions of fumonisins B1, B2, B3 and B4 by core food and by weight.

CORE FOOD	FB1 (%)	FB2 (%)	FB3 (%)	FB4 (%)	Sum (%)
Maize	65.9	19.3	8.4	6.4	100
Sorghum	76.7	15.8	4.6	2.8	100
Cassava dry	75.4	14.1	6.2	4.2	100
Other core foods	88.2	11.8	0.0	0.0	100
Total	67.2	18.9	8.0	6.0	100

The co-occurrence of FB1 and AFB1 was observed in 11 of 16 maize composites (69%) and four of 10 sorghum composites (40%), as well as one of eight millet composites (13%) and in one of 12 cassava dry samples (8%).

2.3. *Sterigmatocystin (STC)*

STC, which is a known aflatoxin precursor [29] was mostly prevalent in cooking oils (Table 5). STC was quantified in 50% of peanut composites (mean: 0.6 µg/kg; max: 2.9 µg/kg) and in all peanut oil samples (mean: 8.5 µg/kg; max: 8.7 µg/kg), which also contained aflatoxins (Table 1). Interestingly, STC was quantified in 100% of "other vegetable oil" samples (cottonseed oil in most cases), whereas aflatoxins were not detected in those composites (tested with the same limit of detection, LD = 0.1 µg/kg).

Contrarily, STC detection rate in maize was only 13%, whereas aflatoxins were detected in 69% of composite samples.

There is currently no Codex or EU standard for STC in any food commodity.

The co-occurrence of STC, AFB1 and FB1 was observed in four composites samples, all collected during the rainy season:

1. Maize (North Cameroon): 56.6 µg/kg AFB1; 458.5 µg/kg FB1; 1.0 µg/kg STC
2. Maize (Benin Littoral): 71.8 µg/kg AFB1; 179.0 µg/kg FB1; 0.075 (LB = limit of detection)–0.25 µg/kg (UB = limit of quantification) of STC, which was detected below the limit of quantification.
3. Sorghum (Borgou): 1.7 µg/kg AFB1; 33.5 µg/kg FB1; 0.5 µg/kg STC
4. Sorghum (Sikasso): 0.8 µg/kg AFB1; 12.5 µg/kg FB1; 2.4 µg/kg STC

Table 5. Occurrence and concentration of sterigmatocystin (µg/kg wet weight) by core food and study centre.

CORE FOOD		N	n > LOD	% > LOD	n > 1 µg/kg	% > 1 µg/kg	n > 4 µg/kg	% > 4 µg/kg	Mean Conc. *		Max Conc.	
									LB	UB	Season **	UB
Peanut oil		2	2	100	2	100	2	100	8.5	8.5	Rainy	8.7
Peanut		10	5	50	2	20	0	0	0.6	0.6	Rainy	2.9
Palm oil		4	3	75	3	75	1	25	2.0	2.0	Rainy	5.3
Other vegetable oil		4	4	100	3	75	1	25	3.9	3.9	Rainy	9.2
Sorghum		10	3	30	2	20	0	0	0.4	0.5	Rainy	2.4
Millet		8	2	25	1	13	1	13	0.6	0.7	Rainy	4.8
Other core foods		156	10	6	0	0	0	0	0.02	0.1	Rainy	1.0
Total		194	29	15	13	7	5	3	0.3	0.4	Rainy	9.2
CENTRE		N	n > LOD	% > LOD	n > 1 µg/kg	% > 1 µg/kg	n > 4 µg/kg	% > 4 µg/kg	Mean Conc. *		Max Conc.	
									LB	UB	Core food	UB
BENIN	**Littoral**	26	4	15	2	8	1	4	0.4	0.5	Peanut oil	8.3
	Borgou	22	2	9	0	0	0	0	0.03	0.1	Sorghum	0.5
CAMEROON	**Duala**	29	2	7	1	3	0	0	0.1	0.2	Other vegetable oil	3.0
	North	17	4	24	1	6	1	6	0.7	0.7	Other vegetable oil	9.2
MALI	**Bamako**	27	4	15	3	11	1	4	0.3	0.4	Millet	4.8
	Sikasso	21	6	29	3	14	0	0	0.4	0.4	Peanuts	2.9
NIGERIA	**Lagos**	29	3	10	1	3	1	3	0.2	0.3	Palm oil	5.3
	Kano	23	4	17	2	9	1	4	0.5	0.6	Peanut oil	8.7

* LB: lower-bound scenario where the concentration of non-detected analyte is zero and the concentration of detected but non-quantified analyte is the limit of detection. UB: upper-bound scenario where the concentration of non-detected analyte is the limit of detection and the concentration of detected but non-quantified analyte is the limit of quantification; ** Samples of the rainy season were collected in October 2017 and samples of the dry season were collected in February 2018.

2.4. Ochratoxin A (OTA)

OTA was detected in 10% of tested composite samples (Table 6). Six percent (6%) of all tested samples exceeded 1 µg/kg OTA, including maize (13%), wheat (pasta 50%) and peanut oil (50%). Only three samples contained OTA concentrations exceeding Codex standards applying to unprocessed wheat, barley or rye (5 µg/kg): sorghum (Sikasso: 5.6 µg/kg), rice (Borgou: 6.3 µg/kg) and palm oil (Benin Littoral: 35.4 µg/kg).

There is currently no standard regulating OTA in edible oils, rice and sorghum.

Table 6. Occurrence and concentration of ochratoxin A (μg/kg wet weight) by core food and by study centre.

CORE FOOD		N	n > LOD	% > LOD	n > 1μg/kg	% > 1μg/kg	n > 5μg/kg	% > 5μg/kg	Mean Conc. *		Max Conc.	
									LB	UB	Season **	UB
Palm oil		4	1	25	1	25	1	25	8.9	8.9	Rainy	35.4
Rice		16	5	31	4	25	1	6	0.9	0.9	Dry	6.3
Sorghum		10	2	20	2	20	1	10	0.8	0.9	Rainy	5.6
Maize		16	2	13	2	13	0	0	0.2	0.2	Rainy	1.4
Peanut oil		2	1	50	1	50	0	0	1.2	1.3	Rainy	2.5
Pasta		2	1	50	1	50	0	0	0.5	0.6	Rainy	1.1
Other core foods		144	7	5	0	0	0	0	0.03	0.1	Rainy	0.8
TOTAL		194	19	10	11	6	3	2	0.4	0.4	-	35.4
CENTRE		**N**	**n > LOD**	**% > LOD**	**n > 1 μg/kg**	**% > 1 μg/kg**	**n > 5 μg/kg**	**% > 5 μg/kg**	**Mean Conc. ***		**Max Conc.**	
									LB	**UB**	**Core food**	**UB**
BENIN	Littoral	26	5	19	5	19	1	4	1.6	1.6	Palm oil	35.4
	Borgou	22	3	14	3	14	1	5	0.4	0.5	Rice	6.3
CAMEROON	Duala	29	3	10	3	10	0	0	0.04	0.1	Cassava fresh	0.7
	North	17	3	18	3	18	0	0	0.2	0.3	Rice	2.0
MALI	Bamako	27	1	4	1	4	0	0	0.1	0.2	Maize	1.4
	Sikasso	21	1	5	1	5	1	5	0.3	0.4	Sorghum	5.6
NIGERIA	Lagos	29	0	0	0	0	0	0	0	0.1	ND	0.1
	Kano	23	2	9	3	13	0	0	0.2	0.3	Rice	2.6

* LB: lower-bound scenario where the concentration of non-detected analyte is zero and the concentration of detected but non-quantified analyte is the limit of detection. UB: upper-bound scenario where the concentration of non-detected analyte is the limit of detection and the concentration of detected but non-quantified analyte is the limit of quantification; ** Samples of the rainy season were collected in October 2017 and samples of the dry season were collected in February 2018.

2.5. Citrinin (CIT)

CIT was detected in 19% of all samples (Table 7), including maize (63%), sorghum (70%) and rice (38%). The only available citrinin standard (EU) applies to food supplements based on rice fermented by red yeast (2000 µg/kg). Ten percent (10%) of tested samples had CIT concentrations of 5 µg/kg or more and four maize composite samples exceeded 100 µg/kg (25% of maize samples and 2% of all samples).

Table 7. Occurrence and concentration of total citrinin (µg/kg wet weight) by core food and by study centre.

CORE FOOD		N	n > LOD	% > LOD	n > 5 µg/kg	% > 5 µg/kg	n > 100 µg/kg	% > 100 µg/kg	Mean Conc. *		Max Conc.	
									LB	UB	Season **	UB
Maize		16	10	63	9	56	4	25	76.4	76.8	Rainy	416.5
Sorghum		10	7	70	4	40	0	0	5.5	6.3	Rainy	18.2
Rice		16	6	38	3	19	0	0	2.8	3.2	Dry	18.0
Other core foods		152	14	9	4	3	0	0	0.3	1.0	Rainy	7.4
TOTAL		194	37	19	20	10	4	2	7.0	7.7	Rainy	416.5
CENTRE		N	n > LOD	% > LOD	n > 5 µg/kg	% > 5 µg/kg	n > 100 µg/kg	% > 100 µg/kg	Mean Conc. *		Max Conc.	
									LB	UB	Core food	UB
BENIN	Littoral	26	4	15	3	12	2	8	19.0	19.7	Maize	372.3
	Borgou	22	9	41	7	32	1	5	21.9	22.5	Maize	416.5
CAMEROON	Duala	29	4	14	2	7	1	3	5.7	6.5	Maize	123.6
	North	17	4	24	1	6	0	0	2.2	3.0	Maize	31.9
MALI	Bamako	27	2	7	0	0	0	0	0.1	0.9	Maize/Sorghum	2.5
	Sikasso	21	4	19	2	10	0	0	1.5	2.1	Sorghum	17.0
NIGERIA	Lagos	29	6	21	3	10	0	0	2.6	3.2	Maize	55.9
	Kano	23	4	17	2	9	0	0	3.4	4.0	Maize	61.4

* LB: lower-bound scenario where the concentration of non-detected analyte is zero and the concentration of detected but non-quantified analyte is the limit of detection. UB: upper-bound scenario where the concentration of non-detected analyte is the limit of detection and the concentration of detected but non-quantified analyte is the limit of quantification; ** Samples of the rainy season were collected in October 2017 and samples of the dry season were collected in February 2018.

2.6. Foods Contaminated by Other Regulated Mycotoxins

2.6.1. Zearalenone (ZEN)

ZEN was detected in 6% of samples and never exceeded EU standards of 100 µg/kg for maize intended for direct human consumption. However, the three composite samples containing the highest ZEN concentrations were collected in the same study centre (Duala): maize (wet season: 7.6 µg/kg; dry season: 97.0 µg/kg) and cassava having undergone a drying process prior to being prepared as consumed (dry season: 7.6 µg/kg).

There is currently no Codex standard for ZEN in foods.

2.6.2. Deoxynivalenol (DON)

DON was also detected in 6% of composite samples, including in (1) bread samples (detection rate: 100%) with a mean concentration of 68.8 µg/kg (min: 31.9 µg/kg; max: 134.6 µg/kg), (2) in 100% of pasta prepared as consumed (mean LB–UB: 9.8–14.3 µg/kg). This is inferior to Codex standards applying to DON cereal-based foods for children (200 µg/kg) and for wheat, maize and barley flour, meal, semolina and flakes (1000 µg/kg).

2.6.3. Ergot Alkaloids

Twelve ergot alkaloids were detected in foods processed from wheat (5 of 6 bread samples), with a mean concentration 62.4 of µg/kg, ranging from non-detected to 165.7 µg/kg, for the sum of ergocornine (1.4%), ergocorninine (0.9%), ergocristine (21.0%), ergocristinine (6.8%), ergocryptine (7.6%), ergocryptinine (2.2%), ergometrine (14.4%), ergometrinine (0.5%), ergosin (21.0%), ergosinine (1.4%), ergotamin (21.7%) and ergotaminine (1.0%). There is no Codex standard for ergot alkaloids, and, to the best of our knowledge, the only available standard (EU) is 0.5 g/kg for the sum of ergot alkaloids in unprocessed cereals, except for maize and rice.

2.7. Non-Detected Mycotoxins of Health and Economic Significance

T2 and HT2 toxins, patulin and diacetoxyscirpenol were never detected in this present study.

2.8. Remarks on a Selection of Other Secondary Fungal, Bacterial and Plant Metabolites

2.8.1. *Aspergillus fumigatus* Metabolites in Palm Oil

The presence of 11 *Aspergillus fumigatus* metabolites was observed in palm oil composites only. Bisdethiomethylgliotoxin was detected in three of four samples, with a mean (LB–UB) concentration of 117.7–118.0 µg/kg. Tryptoquivaline was detected in three of four samples (mean: 81.6–81.8 µg/kg). Gliotoxin was detected in three of four samples (mean: 36.6–36.2 µg/kg). Helvolvic acid was detected in two of four samples (mean 27.2–29.3 µg/kg). Fumigaclavin was detected in three of four samples (mean: 16.6–16.9 µg/kg). Fumagillin was detected in one of four samples (mean: 10.9–13.4 µg/kg). Methylsulochrin was detected in four of four samples (mean: 8.54 µg/kg). Pyripyropene A was detected in one of four samples (mean: 5.7–5.8 µg/kg). Fumitremorgin was detected in three of four samples (mean: 3.6–3.7 µg/kg). Pseurotin A was detected in two of four samples (mean: 2.1–2.8 µg/kg). Pyripyropene D was detected in one of four samples (mean 0.3–0.5 µg/kg). Little is known about these substances, which are not likely to represent a threat to consumer at these concentrations. Their presence, however, reveals that *Aspergillus fumigatus*, a human pathogen, may thrive in the palm oil production chain at some point between the palm tree and final production. Therefore, it represents a risk to value chain operators, if not to consumers [30].

2.8.2. Cereulide in Smoked Fish

The bacterial metabolite cereulide was only detected five times (2.6%) in 194 samples, but was quantified in three of six or 50% of smoked fish samples. Mean (LB–UB) cereulide concentration in smoked fish was 0.8–0.9 µg/kg, and the maximum concentration was 2.5 µg/kg. The two other composites containing cereulide concentrations above the detection limit of 0.19 µg/kg were beef (2.0 µg/kg) and palm oil (0.7 µg/kg).

2.8.3. Cyanogenic Glucosides in Cassava

Following TDS methodology, all samples were prepared as consumed, but a distinction was made between cassava samples having undergone size reduction, fermentation and drying processes (e.g., cossets or gari using dehydration as preservation and toxins reduction means) before preparation, including rehydration (cassava dry) [31], and other cassava samples (cassava fresh).

Exposure to cyanogenic glucosides, such as linamarin and lotaustralin, may cause serious motor neuron diseases, called konzo [32–35].

A seasonal pattern was observed, with higher concentrations of both linamarin and lotaustralin in fresh cassava during the dry season ($p < 0.05$), which was already reported by previous studies on the matter [36].

While linamarin concentrations ranged from below LD (2.3 µg/kg) to 317 mg/kg wet weight (mean: 134 mg/kg) in cassava fresh samples, it was quantified between 0.15 mg/kg and 18 mg/kg (mean: 2.8 mg/kg) in cassava dry composite samples (1:47 ratio).

Similarly, lotaustralin ranged from 0.04 mg/kg to 0.66 mg/kg (mean: 0.16 mg/kg) in cassava dry and from below LD (1.3 µg/kg) to 18 mg/kg (mean: 6.1 mg/kg) in cassava fresh (1:26 ratio).

Overall, linamarin and lotaustralin were less concentrated in dry cassava samples ($p > 0.05$) than in fresh cassava. The wide range of cyanogenic glucoside concentrations in dry cassava composites (max/min ratio of 120:1 in the case of linamarin and 510:1 for lotaustralin) may be explained by different processing practices, such as the use of the wetting method in cassava flour [37,38], although we were not able to verify these aspects from information requested during the collection of samples.

Four composite samples of cassava fresh were collected in each country during the wet season and again during the dry season or harmattan. Surprisingly, neither linamarin, nor lotaustralin were detected above LD (2.3 and 1.3 µg/kg wet weight, respectively) in samples collected in Nigeria, whereas concentrations varied from 93 to 101 mg/kg (wet season) and from 198 to 317 mg/kg (dry season) in Benin, Cameroon and Mali. We have not figured out the reason of this Nigeria-specific pattern, which may include different cassava varieties or cultivars [39] as well as different cooking methods [40–42].

2.8.4. Low Contaminated Core Foods

We observed relatively low or no occurrence of mycotoxins and other toxins in foods prepared from fresh yam without dehydration processes, in rice and in traditional, soft and fermented drinks, as well as in sugar, onion, garlic, and eggs.

2.9. Secondary Metabolites Profile

Figure 1 shows 62 of the most frequently occurring metabolites out of 164 analytes, on the basis of detection in our samples.

More than a third (36%) of detected metabolites are unspecific to any fungi genera and might also be of plant origin.

Among most prevalent metabolites, six were detected in more than 50% of samples:

- asperglaucide (141 samples or 73% of 194 samples);
- cyclo(L-pro-L-val) (138 samples or 71%);
- cyclo(L-pro-L-tyr) (123 samples or 63%);
- flavoglaucin (105 samples or 54%);
- emodin (103 samples or 53%); and
- tryptophol (99 samples or 51%).

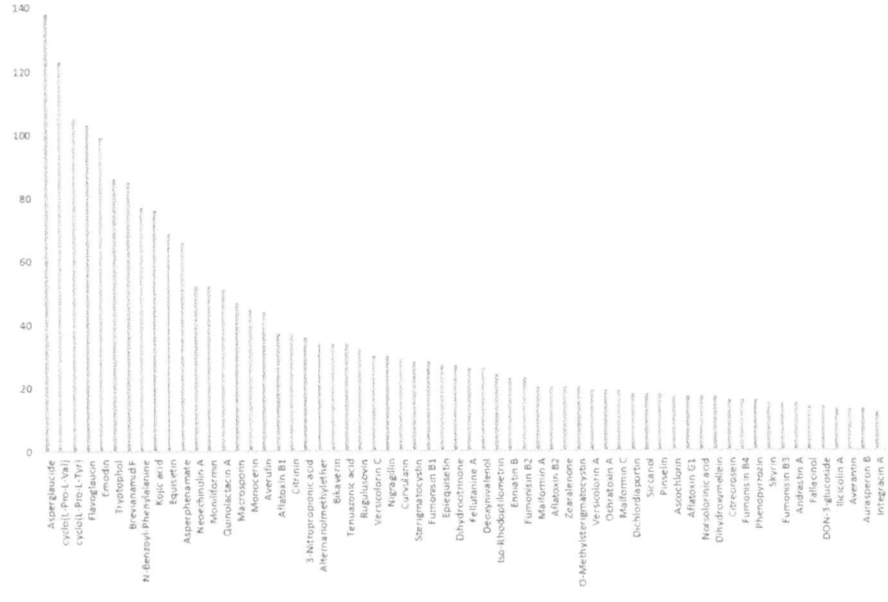

Figure 1. Most frequently detected secondary metabolites in Sub-Saharan Africa Total Diet Study (SSA-TDS) composite samples.

Crude red palm oil composite samples from Cotonou (Benin), Lagos and Kano (Nigeria) showed a higher number of metabolite concentrations above LOD than any other sample (n: 3; mean: 88; min–max: 83–98 metabolites), as noted in Table 8.

Table 8. Range of metabolites detected by core food composite samples.

Range Number of Analytes (min-max)	Composite Samples (n)	CORE FOODS
(51–98)	3	Palm oil
(21–50)	46	Maize, dried tubers, sorghum, peanuts, bread, various oils
(11–20)	62	Beans, dried cassava, rice, millet, smoked fish, onion and garlic, fermented drinks
(6–10)	45	Onion and garlic, meat, tubers, dairy products, rice, traditional soft drinks
(1–5)	36	Fresh tubers, sugar, onion and garlic, rice, eggs
0	2	Onion and garlic
(0–98)	194	TOTAL

All analytical data including quality control checks are enclosed in Table S1.

3. Discussion

First of all, the fact that 164 metabolites were detected in typical African foods does not mean that all of them represent a threat to human health. As of now, in the case of many analytes, the lack of knowledge on their toxicity and their combined (and potentially synergistic) effect with other substances limits our interpretation.

It does, however, represent a contribution to knowledge which may be used when new toxicological data with regards to some of these metabolites will be available. Therefore, the uniqueness of the multi-analyte LC–MS/MS approach used in this study, which enabled the occurrence characterization of a wide range of toxins and other fungal, plant and bacterial secondary metabolites, needs to be emphasized.

In the rest of this discussion, we found it relevant to focus on mycotoxins of public health and international trade significance.

The prevalence of mycotoxins in maize and peanut samples, though, has often been highlighted in previous surveys [43].

Worldwide, several total diet studies from various countries have included mycotoxins, including France [44–46], Canada [47], Lebanon [48], Vietnam [49], and China [50].

In Africa, studies of urinary biomarkers [51–53], surveys of food commodities [54,55], and the analysis foods prepared as consumed [56], have contributed to the rise in attention of the public health community to the threat that mycotoxins represent.

Unsurprisingly, the high concentration of the sum of AFB1, AFB2, AFG1 and AFG2 in African foods as consumed is probably the most significant public health and trade outcome of this multi-mycotoxin analysis compared with other regions of the world [57]. The fact that peanut oil may contain high AF_{tot} concentrations has only recently been described [58]. Peanut oil, peanut and maize are, therefore, likely to contribute significantly to AFB1 exposure, which will be used to characterize the risk of hepatocellular carcinoma.

The presence of fumonisins in staple foods such as maize, with concentrations below the Codex standard of 2 mg/kg, does not guarantee safety for our study populations. The Joint Expert Committee on Food Additives and Contaminants (JECFA) noted in the 83rd session [14], that the current worldwide exposure estimate was established with occurrence data belonging to countries of the WHO European region, and there was no available information on fumonisin levels in maize from the African, Eastern Mediterranean and South-East Asia regions. The JECFA also noted [14] that the interaction between AFB1, a compound with known genotoxic properties, and fumonisins, which have the potential to induce regenerative cell proliferation, is a concern. The completion of the dietary exposure assessment to the sum of FB1, FB2, FB3 and FB4 with the data presented in this paper will also result in conclusions with regards to the adequacy of protective levels of current Codex tolerances in the context of Africa (manuscript in preparation).

Surprisingly, the presence of a high concentration of AF precursor sterigmatocystin, not only in peanut oil but also in cottonseed oil and palm oil, was noted. In contrast, AF was never detected in cottonseed oil but detected in only one of four palm oil samples (0.5 µg/kg AFB1 in a red palm oil composite from sub-samples collected in the Littoral of Benin). This may be due to the production of STC by non-aflatoxigenic *Aspergilli*, such as *Aspergillus nidulans* [59], as well as other fungi genera [60]. The fact that we quantified STC in millet and sorghum composite samples is consistent with recent findings in sorghum [55]. Sorghum and millet, therefore, also qualify as potential STC dietary exposure contributors, noting that typical Sub-Saharan-Sahelian diets largely rely on these cereals [28].

The fact that citrinin was most concentrated in maize means that maize is likely to be a major contributor to CIT dietary exposures in centres where (1) maize CIT concentrations were high and (2) maize is consumed in large amounts.

We would like to bring forward the absence of Codex standards for mycotoxins in edible oils and, in light of occurrence data submitted in this paper, the need for surveillance of mycotoxin contamination levels in edible oils. The presence of (1) OTA in one palm oil with 97 other secondary metabolites, (2) high AF_{tot} concentrations in peanut oil, and (3) the presence of STC in cotton seed oil supports the need for an elaboration in the Codex code of practice for the production of safe edible oil.

Results of the risk characterization (manuscript in preparation) using this occurrence data and adequate food consumption data will clarify to what extent edible oils, as well as other core foods, contribute to the total dietary exposure to mycotoxins in Africa.

Mycotoxin exposure risk mitigation measures include growth prevention of toxin-producing fungi via biocontrol [61] in the field, good post-harvest practices [62] and mycotoxin degradation [63].

As human co-exposure to natural toxins through typical African foods is currently inevitable, national food safety authorities need to ensure that risk assessments are carried out properly to safeguard human health and to maintain international trade.

As demonstrated by the current study, AFB1, FB1, STC and many other secondary fungal metabolites are frequent co-contaminants in many foods (such as maize and sorghum) that threaten human health. Populations in North Cameroon and from Benin (where multiple toxins, including ABF1, FB1 and STC, have been detected within the TDS) may suffer repeated simultaneous exposure to natural toxins. In a recent study [15], the combined effects of various toxins at realistic concentrations were further investigated and revealed additive, antagonistic or synergistic effects. The results have confirmed that combinations of toxins may pose a considerable risk to human health. Clearly, further research is needed to understand the mechanics of toxicological interactions in order to effectively protect public health. Moreover, more TDSs in other locations of Benin, Cameroon, Mali and Nigeria, as well as in other countries belonging to Sub-Saharan Africa need to be carried out to better document actual dietary exposure levels to natural toxins in this region.

4. Conclusions

At this stage of the SSA-TDS project, the first ever multi-centric total diet study carried out in Africa, we have detected 164 secondary metabolites. However, our main results with regards to the occurrence of regulated mycotoxins in eight study centres are as follows:

- Mean AF_{tot} concentrations exceed EU and Codex tolerances applying to peanuts. Similar AF_{tot} levels were quantified in peanut oil (although no Codex or EU standards are currently available for edible oils), as well as in maize samples (aflatoxins in maize are not currently regulated by Codex).
- The TDS approach allowed for the capture of seasonal variations of the AF_{tot} contamination pattern in maize, which contains higher concentrations in samples collected during the rainy season.
- The geographic component of the AF_{tot} contamination pattern was suggested by variations in the mean AF_{tot} concentrations among study centres, which was also observed between two study centres from the same country (Duala versus North Cameroon).

Due to the systematic approach applied to this study, we consider these data fit for the completion of chronic dietary exposure assessment of mycotoxins, for which a health based guidance value is available (e.g., Tolerable Daily Intake (TDI) or end point for genotoxic carcinogenic substances using the margin of exposure approach). We will then be able to take into consideration food consumption data, at the household level, for eight population groups. We expect maize, peanut and peanut oil to contribute to most of the dietary exposure to AFB1. Likewise, we expect maize to contribute highly to FUM_{tot} and CIT dietary exposure. However, other core foods, in which lower mycotoxin concentrations were estimated, especially highly-consumed staple foods, may also significantly contribute to households' total dietary exposure.

Although Codex maximum limits were not exceeded in the case of FUM_{tot} and OTA, a household dietary exposure assessment will enable risk characterization of the investigated population groups. From this exercise, we will be able to conclude whether currently available Codex mycotoxin standards are sufficiently protective to African consumers.

The dietary exposure assessment of our study populations (manuscript in preparation) will provide guidance to risk managers from Benin, Cameroon, Mali and Nigeria for the identification of national priorities to the consumer protection agenda. We can nonetheless readily address our recommendations to risk managers based on AF_{tot} occurrence data referencing Codex standards only. It will indeed be beneficial for health and trade if national food safety authorities, with the support of their technical and financial partners, draft and implement a road-map and mobilize adequate resources taking the following into consideration:

- Food commodity value chain structures and organization;
- Prevention of field contamination by toxin-producing fungi; and
- Post-harvest practices with emphasis on hygiene, drying and storage conditions.

This will reduce the occurrence and concentrations of mycotoxins in African foods.

To date, these observations about STC occurrence in maize and in oils are new findings which were not reported or highlighted by the last JECFA evaluation of mycotoxins (2016) due to a lack of data at the time of the assessment.

Mitigation measures from Codex Alimentarius may include the updating of current codes of practices and standards and the elaboration of new ones to contribute to the reduction of natural toxins occurrence. This is in an effort to effectively safeguard African consumers' health and food quality.

5. Experimental

5.1. Sample Selection and Preparation of Foods as Consumed

Food consumption data were derived from household budget surveys generated by National Statistics Authorities, from Benin, Cameroon, Mali and Nigeria and gathering a total of 72,979 households. Core foods of each study centre were selected based on the relative importance of their mean consumption [28], so as to cover at least 90% of the mean total diet in grams per adult male equivalent per day (g/AME/d).

Each core food was sampled through available representation criteria [64] (such as market share or the origins of the food) using 12 subsamples of equal size, prepared as consumed and pooled into composites, which underwent laboratory tests. The subsamples were prepared individually according to recipe books [65–68]. These references are considered as representative of the diet of the study populations and were, therefore, selected by the representatives of national competent authorities. These recipe books allow the identification of the processes used in the preparation of the foods, especially cooking time and temperature. The actual recipes were, however, not prepared as each composite sample only contained one core food or ingredient. The inedible parts were removed at the preparation stage, as a typical consumer would do. Distilled water instead off tap water was used to prepare food as consumed to avoid contamination. The quantity of water added during the cooking process of each of the 12 subsamples was measured by weighing the food at each stage of the preparation process.

Two seasons were captured [69] for five main food groups, which cover staple foods and most of the mean total diet by weight (i.e., cereals, tubers, legumes, vegetables and fruits):

- The rainy season in October 2017; and
- The dry season, or harmattan, in February 2018.

Other food groups were collected during the rainy season only (i.e., nuts and seeds, dairy, oils, beverages and miscellaneous).

Among 335 composite samples, 194 consisted of foods which may be stored in conditions allowing for the growth of moulds and, consequently, are likely to comprise mycotoxins. Those 194 composite samples were selected for mycotoxin analysis. Samples were frozen and shipped by air in coolers with dry ice, within a timeframe never exceeding 24 hours, from the kitchen laboratory (Benin, Cameroon, Mali and Nigeria) to the testing laboratory (Austria).

5.2. Reagents and Chemicals

LC gradient grade methanol and acetonitrile, as well as MS grade ammonium acetate and glacial acetic acid (p.a.), were purchased from Sigma Aldrich (Vienna, Austria). A Purelab Ultra system (ELGA LabWater, Celle, Germany) was used for further purification of reverse osmosis water.

Standards of fungal and bacterial metabolites were obtained either as gifts from various research groups or from the following commercial sources: Romer Labs®Inc. (Tulln, Austria), Sigma-Aldrich (Vienna, Austria), BioAustralis (Smithfiled, Australia), AnalytiCon Discovery (Potsdam, Germany), Fermentek (Jerusalem, Israel), Iris Biotech GmbH (Marktredwitz, Germany), Enzo Life Sciences Europe (Lausanne, Switzerland) and LGC Promochem GmbH (Wesel, Germany). Stock solutions of each analyte were prepared by dissolving the solid substance in acetonitrile, acetonitrile/water 1:1 (v/v), methanol, methanol/water 1:1 (*v*/*v*) or water. Thirty-four combined working solutions were prepared by mixing the stock solutions of the corresponding analytes for easier handling and were stored at −20 °C. The final working solution was freshly prepared prior to spiking experiments through mixing of the combined working solutions.

5.3. Laboratory Sample Preparation

Twenty millilitres (20 mL) of extraction solvent (acetonitrile/water/acetic acid 79:20:1, *v*/*v*/*v*) were added to 5 g of sample. The samples were extracted for 90 minutes using a GFL 3017 rotary shaker (GFL, Burgwedel, Germany) and subsequently centrifuged for two minutes at 3000 rpm (radius 15 cm) on a GS-6 centrifuge (Beckman Coulter Inc., Fullerton, CA, USA). The extracts were diluted (1:1) with dilution solvent (acetonitrile/water/acetic acid 20:79:1, *v*/*v*/*v*). After appropriate mixing, 5 μL of the diluted extract was injected into the LC-MS/MS system without further pre-treatment.

5.4. LC-MS/MS Parameters

Metabolite analysis was carried out using a 1290 Series HPLC System (Agilent, Waldbronn, Germany) coupled to a QTrap 5500 LC-MS/MS System (Applied Biosystems SCIEX, Foster City, CA, USA) equipped with Turbo Ion Spray electrospray ionization source, as described earlier [70]. Chromatographic separation was performed at 25 °C on a Gemini® C_{18}-column (150 × 4.6 mm i.d., 5 μm particle size) equipped with a C_{18} 4 × 3 mm i.d. security guard cartridge (Phenomenex, Torrance, CA, USA). Confirmation of positive metabolite identification was carried out by two instances of scheduled multiple reaction monitoring (MRMs) which yielded 4.0 identification points according to the European Commission decision 2002/657 [71].

In order to further decrease the limits of detection (LODs) for aflatoxin B1 and ochratoxin A, larger aliquots of 20 μL of the diluted extracts (previously fortified with the related ^{13}C-labelled internal standards) were re-analysed using the QTrap 6500 LC-MS/MS system while keeping all other method parameters constant.

5.5. Quantification and Quality Control

Quantification was performed using external calibration based on serial dilution of a multi-analyte stock solution. Results were corrected using apparent recoveries that were determined for each of the investigated matrices by spiking experiments. The accuracy of the method is verified on a continuous basis by participation in a proficiency testing scheme organized by BIPEA (Gennevilliers, France) with a current success rate (i.e., a z-score between −2 and 2) of >94% of the >900 results submitted.

Supplementary Materials: The following are available online at http://www.mdpi.com/2072-6651/11/1/54/s1. Table S1: Raw Analytical Data.

Author Contributions: The first draft was produced by L.I. and M.S. and was then reviewed by R.K., A.A., and J.-C.L. Technical support was provided by J.-C.L., P.V., and B.L.B. National coordination, including sample collection, was assured by A.A., A.Z.K., S.E.H., and S.E. Food preparation was supervised by A.D.O., C.S.K.J.K., Y.K.D., and L.I. M.S. and R.K. supervised laboratory tests. The views expressed in this publication are those of the authors and do not necessarily reflect the views and policies of the Food and Agriculture Organization of the United Nations.

Funding: The project was funded under grant STDF/PG/303 and the authors are thankful to Kenza le Mentec and Marlynne Hopper of the Standard and Trade Development Facility (STDF), the donor institution.

Acknowledgments: We would like to remember the late Marie Madeleine Gimou, the initiator of this study. Many thanks also to FAO staff (Renata Clarke, Markus Lipp, Caroline Merten, Blaise Ouattara, Jean Kamanzi, Sekou Hebie and Alex Nyarko) who supported the total diet study at various stages of its submission and its implementation. The CPC management, as well as the various heads of national coordinating institutions of the other participating countries, ABSSA (Benin), ANSSA (Mali) and NAFDAC (Nigeria), contributed to the success of this project. The scientific committee members, who provide guidance and validation of the methodology with their valuable experience of implementing total diet studies, are Katie Egan, Peter Fürst, Thierry Guérin, Adam Probert, Siswanto Siswanto and Christina Tlustos. We are extremely grateful for their support.

Conflicts of Interest: The authors declare that there is no conflict of interest.

References

1. CDC (Centers for Disease Control and Prevention). Outbreak of Aflatoxin Poisoning—Eastern and Central Provinces, Kenya, January–July 2004. Available online: https://www.cdc.gov/nceh/hsb/chemicals/pdfs/mmwr5334p790.pdf (accessed on 2 November 2018).
2. Lewis, L.; Onsongo, M.; Njapau, H.; Schurz-Rogers, H.; Luber, G.; Kieszak, S.; Nyamongo, J.; Backer, L.; Dahiye, A.M.; Misore, A.; et al. Aflatoxin contamination of commercial maize products during an outbreak of acute aflatoxicosis in eastern and central Kenya. Kenya Aflatoxicosis Investigation Group. *Environ. Health Perspect.* **2005**, *113*, 1763–1767. [CrossRef] [PubMed]
3. IARC (International Agency for Research on Cancer). 2012. Available online: https://monographs.iarc.fr/ENG/Monographs/%E2%80%A6/mono100F-23.pdf (accessed on 2 November 2018).
4. Theumer, M.G.; Henneb, Y.; Khoury, L.; Snini, S.P.; Tadrist, S.; Canlet, C.; Puel, O.; Oswald, I.P.; Audebert, M. Genotoxicity of aflatoxins and their precursors in human cells. *Toxicol. Lett.* **2018**, *287*, 100–107. [CrossRef] [PubMed]
5. Rheeder, J.P.; Marasas, W.F.O.; Thiel, P.G.; Sydenham, E.W.; Shephard, G.S.; van Schalkwyk, D.J. Fusarium moniliforme and fumonisins in corn in human esophageal cancer in Transkei. *Phytopathology* **1992**, *82*, 353–357. [CrossRef]
6. Sun, G.; Wang, S.; Hu, X.; Su, J.; Huang, T.; Yu, J.; Tang, L.; Gao, W.; Wang, J.S. Fumonisin B1 contamination of home-grown corn in high-risk areas for esophageal and liver cancer in China. *Food Addit. Contam.* **2007**, *24*, 181–185. [CrossRef] [PubMed]
7. Etzel, R.A. Reducing malnutrition: Time to consider potential links between stunting and mycotoxin exposure? *Pediatrics* **2014**, *134*, 4–6. [CrossRef] [PubMed]
8. Smith, L.E.; Prendergast, A.J.; Turner, P.C.; Mbuya, M.N.; Mutasa, K.; Kembo, G.; Stoltzfus, R.J. Sanitation Hygiene Infant Nutrition Efficacy (SHINE) Trial Team. The Potential Role of Mycotoxins as a Contributor to Stunting in the SHINE Trial. *Clin. Infect. Dis.* **2015**, *61*, S703–S709. [CrossRef]
9. McMillan, A.; Renaud, J.B.; Burgess, K.M.N.; Orimadegun, A.E.; Akinyinka, O.O.; Allen, S.J.; Miller, J.D.; Reid, G.; Sumarah, M.W. Aflatoxin exposure in Nigerian children with severe acute malnutrition. *Food Chem. Toxicol.* **2018**, *111*, 356–362. [CrossRef] [PubMed]
10. Chen, C.; Mitchell, N.J.; Gratz, J.; Houpt, E.R.; Gong, Y.; Egner, P.A.; Groopman, J.D.; Riley, R.T.; Showker, J.L.; Svensen, E.; et al. Exposure to aflatoxin and fumonisin in children at risk for growth impairment in rural Tanzania. *Environ. Int.* **2018**, *115*, 29–37. [CrossRef]
11. De Onis, M.; Branca, F. Childhood stunting: A global perspective. *Matern. Child. Nutr.* **2016**, *12*, 12–26. [CrossRef]
12. Khlangwiset, P.; Shephard, G.S.; Wu, F. Aflatoxins and growth impairment: A review. *Crit. Rev. Toxicol.* **2011**, *41*, 740–755. [CrossRef]
13. De Ruyck, K.; De Boevre, M.; Huybrechts, I.; De Saeger, S. Dietary mycotoxins, co-exposure, and carcinogenesis in humans: Short review. *Mutat. Res. Rev. Mutat. Res.* **2015**, *766*, 32–41. [CrossRef] [PubMed]
14. World Health Organization. *Evaluation of Certain Food Contaminants, 83rd Report of the Joint FAO/WHO Expert Committee on Food Additives*; WHO Technical Report Series 1002; WHO Press, World Health Organization: Geneva, Switzerland, 2016.
15. Meneely, J.P.; Hajšlová, J.; Krska, R.; Elliott, C.T. Assessing the combined toxicity of the natural toxins, aflatoxin B(1), fumonisin B(1) and microcystin-LR by high content analysis. *Food Chem. Toxicol.* **2018**, *121*, 527–540. [CrossRef] [PubMed]
16. European Commission. Rapid Alert System on Food and Feed. 2016. Available online: https://ec.europa.eu/food/safety/rasff_en (accessed on 6 April 2018).
17. European Commission. Commission Regulation (EC) No 1881/2006 of 19 December 2006 Setting Maximum Levels for Certain Contaminants in Foodstuffs. 2006. Available online: https://eur-lex.europa.eu/legal-content/ (accessed on 6 April 2018).
18. *Codex Alimentarius. General Standard for Contaminants in Food and Feed (CODEX STAN 193-1995)*; Adopted in 1995; Revised in 1997, 2006, 2008, 2009; Amended in 2010, 2012, 2013, 2014, 2015; 2015.
19. Pennington, J.A.T. Revision of the total diet study food list and diets. *J. Am. Diet. Assoc.* **1983**, *82*, 166–173. [PubMed]

20. Egan, S.K.; Bolger, P.M.; Carrington, C.D. Update of US FDA's Total Diet Study food list and diets. *Expo. Sci. Environ. Epidemiol.* **2007**, *17*, 573–582. [CrossRef] [PubMed]
21. Sirot, V.; Volatier, J.L.; Calamassi-Tran, G.; Dubuisson, C.; Menard, C.; Dufour, A.; Leblanc, J.C. Core food of the French food supply: Second Total Diet Study. *Food Addit. Contam. Part A* **2009**, *26*, 5623–5639. [CrossRef] [PubMed]
22. EFSA (European Food Safety Agency). Overview of the procedures currently used at EFSA for the assessment of dietary exposure to different chemical substances. *EFSA J.* **2011**, *9*, 2490. [CrossRef]
23. EFSA (European Food Safety Agency). Towards a harmonised Total Diet Study approach: A guidance document. *EFSA J.* **2011**, *9*, 2450. [CrossRef]
24. Hulin, M.; Bemrah, M.N.; Nougadere, A.; Volatier, J.L.; Sirot, V.; Leblanc, J.C. Assessment of infant exposure to food chemicals: The French Total Diet Studydesign. *Food Addit. Contam. Part A* **2014**, *31*, 1226–1239. [CrossRef]
25. Gimou, M.M.; Charrondiere, U.R.; Leblanc, J.C.; Pouillot, R. Dietary exposure to pesticide residues in Yaounde: The Cameroonian total diet study. *Food Addit. Contam.* **2008**, *25*, 458–471. [CrossRef]
26. Gimou, M.M.; Charrondiere, U.R.; Leblanc, J.C.; Pouillot, R.; Noel, L.; Guerin, T. Concentration of 25 elements in foodstuffs in Yaounde: The Cameroonian total diet study. *J. Food Compos. Anal.* **2014**, *34*, 39–55. [CrossRef]
27. FAO (Food and Agriculture Organization of the United Nations). Total Diet Study as a Tool to Assess Chemical Contamination in Foods. Application in Sub-Saharan Africa. 2014. Available online: http://www.fao.org/fileadmin/user_upload/agns/pdf/Highlights/SubAfricaHighlight-LR.pdf (accessed on 6 April 2018).
28. Ingenbleek, L.; Jazet, E.; Dzossa, A.D.; Adebayo, S.B.; Ogungbangbe, J.; Dansou, S.; Diallo, Z.J.; Kouebou, C.; Adegboye, A.; Hossou, E.; et al. Methodology design of the regional Sub-Saharan Africa Total Diet Study in Benin, Cameroon, Mali and Nigeria. *Food Chem. Toxicol.* **2017**, *109*, 155–169. [CrossRef] [PubMed]
29. Díaz Nieto, C.H.; Granero, A.M.; Zon, M.A.; Fernández, H. Sterigmatocystin: A mycotoxin to be seriously considered. *Food Chem. Toxicol.* **2018**, *118*, 460–470. [CrossRef]
30. Russell, K.; Broadbridge, C.; Murray, S.; Waghorn, D.; Mahoney, A. Gardening can seriously damage your health. *Lancet* **2008**, *371*, 2056. [CrossRef]
31. Codex Alimentarius. *CAC/RCP 73-2013: Code of Practice for the Reduction of Hydrocyanic Acid (HCN) in Cassava and Cassava Products*; 2013.
32. Cliff, J.; Muquingue, H.; Nhassico, D.; Nzwalo, H.; Bradbury, J.H. Konzo and continuing cyanide intoxication from cassava in Mozambique. *Food Chem. Toxicol.* **2011**, *49*, 631–635. [CrossRef] [PubMed]
33. Newton, C.R. Cassava, konzo, and neurotoxicity. *Lancet Glob. Health* **2017**, *5*, e853–e854. [CrossRef]
34. Boivin, M.J.; Okitundu, D.; Makila-Mabe, B.; Sombo, M.T.; Mumba, D.; Sikorskii, A.; Mayambu, B.; Tshala-Katumbay, D. Cognitive and motor performance in Congolese children with konzo during 4 years of follow-up: A longitudinal analysis. *Lancet Glob. Health* **2017**, *5*, e936–e947. [CrossRef]
35. Kashala-Abotnes, E.; Okitundu, D.; Mumba, D.; Boivin, M.J.; Tylleskär, T.; Tshala-Katumbay, D. Konzo: A distinct neurological disease associated with food (cassava) cyanogenic poisoning. *Brain Res. Bull.* **2018**, *5*. [CrossRef]
36. Banea-Mayambu, J.P.; Tylleskär, T.; Gitebo, N.; Matadi, N.; Gebre-Medhin, M.; Rosling, H. Geographical and seasonal association between linamarin and cyanide exposure from cassava and the upper motor neurone disease konzo in former Zaire. *Trop. Med. Int. Health* **1997**, *2*, 1143–1151. [CrossRef]
37. Banea, J.P.; Howard Bradbury, J.; Mandombi, C.; Nahimana, D.; Denton, I.C.; Kuwa, N.; Tshala Katumbay, D. Effectiveness of Wetting Method for Control of Konzo and Reduction of Cyanide Poisoning by Removal of Cyanogens from Cassava Flour. *Food Nutr. Bull.* **2014**, *35*, 28–32. [CrossRef]
38. Nhassico, D.; Bradbury, J.H.; Cliff, J.; Majonda, R.; Cuambe, C.; Denton, I.C.; Foster, M.P.; Martins, A.; Cumbane, A.; Sitoe, L.; et al. Use of the wetting method on cassava flour in three konzo villages in Mozambique reduces cyanide intake and may prevent konzo in future droughts. *Food Sci. Nutr.* **2015**, *4*, 555–561. [CrossRef]
39. Jørgensen, K.; Bak, S.; Busk, P.K.; Sørensen, C.; Olsen, C.E.; Puonti-Kaerlas, J.; Møller, B.L. Cassava plants with a depleted cyanogenic glucoside content in leaves and tubers. Distribution of cyanogenic glucosides, their site of synthesis and transport, and blockage of the biosynthesis by RNA interference technology. *Plant Physiol.* **2005**, *139*, 363–374. [CrossRef] [PubMed]

40. Frazzoli, C.; Mazzanti, F.; Achu, M.B.; Pouokam, G.B.; Fokou, E. Elements of kitchen toxicology to exploit the value of traditional (African) recipes: The case of Egusi Okra meal in the diet of HIV+/AIDS subjects. *Toxicol. Rep.* **2017**, *4*, 474–483. [CrossRef] [PubMed]
41. Obueh, H.; Kolawole, E. Comparative Study on the Nutritional and Anti-Nutritional Compositions of Sweet and Bitter Cassava Varieties for Garri Production. *J. Nutr. Health Sci.* **2016**, *3*, 3–7. [CrossRef]
42. Babalola, O.O. Cyanide Content of Commercial Gari from Different Areas of Ekiti State, Nigeria. *J. Nat. Sci. Res.* **2014**, *2*, 58–60.
43. Caballero, B.; Finglas, P.; Toldra, F. *Encyclopedia of Food and Health*, 1st ed.; Academic Press: Cambridge, MA, USA, 2015.
44. Leblanc, J.C.; Tard, A.; Volatier, J.L.; Verger, P. Estimated dietary exposure toprincipal food mycotoxins from the first French Total Diet Study. *Food Addit. Contam.* **2005**, *22*, 652–672. [CrossRef] [PubMed]
45. Sirot, V.; Fremy, J.M.; Leblanc, J.C. Dietary exposure to mycotoxins and health risk assessment in the second French total diet study. *Food Chem. Toxicol.* **2013**, *52*, 1–11. [CrossRef]
46. Fleury, S.; Rivière, G.; Allès, B.; Kesse-Guyot, E.; Méjean, C.; Hercberg, S.; Touvier, M.; Bemrah, N. Exposure to contaminants and nutritional intakes in a French vegetarian population. *Food Chem. Toxicol.* **2017**, *109*, 218–229. [CrossRef] [PubMed]
47. Tam, J.; Pantazopoulos, P.; Scott, P.M.; Moisey, J.; Dabeka, R.W.; Richard, I.D. Application of isotope dilution mass spectrometry: Determination of ochratoxin A in the Canadian Total Diet Study. *Food Addit. Contam. Part A Chem. Anal. Control Expo. Risk Assess.* **2011**, *28*, 754–761. [CrossRef]
48. Raad, F.; Nasreddine, L.; Hilan, C.; Bartosik, M.; Parent-Massin, D. Dietary exposureto aflatoxins, ochratoxin A and deoxynivalenol from a total diet study in an adult urban Lebanese population. *Food Chem. Toxicol.* **2014**, *73*, 35–43. [CrossRef]
49. Huong, B.T.M.; Tuyen, L.D.; Tuan, D.H.; Brimer, L.; Dalsgaard, A. Dietary exposure toaflatoxin B(1), ochratoxin A and fuminisins of adults in Lao Cai province, Vietnam: A total dietary study approach. *Food Chem. Toxicol.* **2016**, *98*, 127–133. [CrossRef]
50. Qiu, N.N.; Lyu, B.; Zhou, S.; Zhao, Y.F.; Wu, Y.N. The contamination and dietary exposure analysis for seven mycotoxins in the Fifth Chinese Total Diet Study. *Zhonghua Yu Fang Yi Xue Za Zhi* **2017**, *51*, 943–948. [CrossRef] [PubMed]
51. Abia, W.A.; Warth, B.; Sulyok, M.; Krska, R.; Tchana, A.; Njobeh, P.B.; Turner, P.C.; Kouanfack, C.; Eyongetah, M.; Dutton, M.; et al. Bio-monitoring of mycotoxin exposure in Cameroon using a urinary multi-biomarker approach. *Food Chem. Toxicol.* **2013**, *62*, 927–934. [CrossRef] [PubMed]
52. Ediage, E.N.; Diana Di Mavungu, J.; Song, S.; Sioen, I.; De Saeger, S. Multimycotoxin analysis in urines to assess infant exposure: A case study in Cameroon. *Environ. Int.* **2013**, *57–58*, 50–59. [CrossRef]
53. Ezekiel, C.N.; Warth, B.; Ogara, I.M.; Abia, W.A.; Ezekiel, V.C.; Atehnkeng, J.; Sulyok, M.; Turner, P.C.; Tayo, G.O.; Krska, R.; et al. Mycotoxin exposure in rural residents in northern Nigeria: A pilot study using multi-urinary biomarkers. *Environ. Int.* **2014**, *66*, 138–145. [CrossRef] [PubMed]
54. Misihairabgwi, J.M.; Ezekiel, C.N.; Sulyok, M.; Shephard, G.S.; Krska, R. Mycotoxincontamination of foods in Southern Africa: A 10-year review (2007–2016). *Crit. Rev. Food Sci. Nutr.* **2017**, *11*, 1–16. [CrossRef]
55. Ssepuuya, G.; Van Poucke, C.; Ediage, E.N.; Mulholland, C.; Tritscher, A.; Verger, P.; Kenny, M.; Bessy, C.; De Saeger, S. Mycotoxin contamination of sorghum and its contribution to human dietary exposure in four sub-Saharan countries. *Food Addit. Contam. Part A* **2018**, *35*, 1384–1393. [CrossRef] [PubMed]
56. Abia, W.A.; Warth, B.; Ezekiel, C.N.; Sarkanj, B.; Turner, P.C.; Marko, D.; Krska, R.; Sulyok, M. Uncommon toxic microbial metabolite patterns in traditionally home-processed maize dish (fufu) consumed in rural Cameroon. *Food Chem. Toxicol.* **2017**, *107*, 10–19. [CrossRef] [PubMed]
57. Marin, S.; Ramos, A.J.; Cano-Sancho, G.; Sanchis, V. Mycotoxins: Occurrence, toxicology, and exposure assessment. *Food Chem. Toxicol.* **2013**, *60*, 218–237. [CrossRef]
58. Shephard, G.S. Aflatoxins in peanut oil: Food safety concerns. *World Mycotoxin J.* **2018**, *11*, 149–158. [CrossRef]
59. Molnár, Á.P.; Németh, Z.; Fekete, E.; Flipphi, M.; Keller, N.P.; Karaffa, L. Analysis of the Relationship between Alternative Respiration and Sterigmatocystin Formation in Aspergillus nidulans. *Toxins* **2018**, *10*, 168. [CrossRef] [PubMed]

60. Uzma, F.; Mohan, C.D.; Hashem, A.; Konappa, N.M.; Rangappa, S.; Kamath, P.V.; Singh, B.P.; Mudili, V.; Gupta, V.K.; Siddaiah, C.N.; et al. Endophytic Fungi-Alternative Sources of Cytotoxic Compounds: A Review. *Front. Pharmacol.* **2018**, *9*, 309. [CrossRef] [PubMed]
61. Mwakinyali, S.E.; Ding, X.; Ming, Z.; Tong, W.; Zhang, Q.; Li, P. Recent development of aflatoxin contamination biocontrol in agricultural products. *Biol. Control* **2019**, *128*, 31–39. [CrossRef]
62. Codex Alimentarius. *CAC/RCP 55-2004: Code of Practice for the Prevention and Reduction of Aflatoxin Contamination in Peanuts*; 2004.
63. Peng, Z.; Chen, L.; Zhu, Y.; Huang, Y.; Hu, X.; Wu, Q.; Nüssler, A.K.; Liu, L.; Yang, W. Current major degradation methods for aflatoxins: A review. *Trends Food Sci. Technol.* **2018**, *80*, 155–166. [CrossRef]
64. Tsukakoshi, Y. Sampling variability and uncertainty in total diet studies. *Analyst* **2011**, *136*, 533–539. [CrossRef] [PubMed]
65. Vinakpon-Gbaguidi, V. *Saveurs du Benin et de la Sous-Région*; Nouvelle Presse Publications: Benin, 2003.
66. Nya-Njike, P. *L'Art Culinaire Camerounais*; L'Harmattan: Paris, France, 1998.
67. Gautier, L.; Mallet, J.F. *Le Vrai Goût du Mali. Collection Gastronomie*; Herme: Paris, France, 2006.
68. Madubike, F. *All Nigerian Recipes Cookbook*; 2013; Mass Market Paperback; ISBN-13:978-8461617548.
69. Elegbede, F.C.; Papadopoulos, A.; Kolbaum, A.E.; Turrini, A.; Mistura, L.; Lindtner, O.; Sirot, V. TDS exposure project: How and when to consider seasonality in a total diet study? *Food Chem. Toxicol.* **2017**, *105*, 119–126. [CrossRef] [PubMed]
70. Malachová, A.; Sulyok, M.; Beltrán, E.; Berthiller, F.; Krska, R. Optimization and validation of a quantitative liquid chromatography-tandem mass spectrometric method covering 295 bacterial and fungal metabolites including all regulated mycotoxins in four model food matrices. *J. Chromatogr. A* **2014**, *1362*, 145–156. [CrossRef]
71. European Commission. Commission Decision 2002/657/EC of 12 August 2002 implementing Council Directive 96/23/EC concerning the performance of analytical methods and the interpretation of results (Text with EEA relevance) (notified under document number C(2002) 3044). *Off. J. Eur. Communities* **2002**, *L221*, 8–36.

toxins

MDPI

Article

Variation of *Fusarium* Free, Masked, and Emerging Mycotoxin Metabolites in Maize from Agriculture Regions of South Africa

Theodora Ijeoma Ekwomadu [1,*], Toluwase Adeseye Dada [2], Nancy Nleya [2], Ramokone Gopane [1], Michael Sulyok [3] and Mulunda Mwanza [2]

1 Department of Biological Sciences, Faculty of Natural and Agricultural Sciences, Mafikeng Campus, North-West University, Mmabatho 2735, South Africa; ramokone.gopane@gmail.com

2 Department of Animal Health, Faculty of Natural and Agricultural Sciences, Mafikeng Campus, North-West University, Mmabatho 2735, South Africa; adedad@gmail.com (T.A.D.); dangwa@gmail.com (N.N.); mulunda.mwanza@nwu.ac.za (M.M.)

3 Department of Agro Biotechnology (IFA-Tulln), University of Natural Resources and Life Sciences Vienna (BOKU), A-3430 Tulln, Austria; michael.sulyok@boku.ac.at

* Correspondence: 23115394@nwu.ac.za

Received: 11 October 2019; Accepted: 8 January 2020; Published: 28 February 2020

Abstract: The presence of mycotoxins in cereal grain is a very important food safety issue with the occurrence of masked mycotoxins extensively investigated in recent years. This study investigated the variation of different *Fusarium* metabolites (including the related regulated, masked, and emerging mycotoxin) in maize from various agriculture regions of South Africa. The relationship between the maize producing regions, the maize type, as well as the mycotoxins was established. A total of 123 maize samples was analyzed by a LC-MS/MS multi-mycotoxin method. The results revealed that all maize types exhibited a mixture of free, masked, and emerging mycotoxins contamination across the regions with an average of 5 and up to 24 out of 42 investigated *Fusarium* mycotoxins, including 1 to 3 masked forms at the same time. Data obtained show that fumonisin B_1, B_2, B_3, B_4, and A_1 were the most prevalent mycotoxins and had maximum contamination levels of 8908, 3383, 990, 1014, and 51.5 μg/kg, respectively. Deoxynivalenol occurred in 50% of the samples with a mean concentration of 152 μg/kg (max 1380 μg/kg). Thirty-three percent of the samples were contaminated with zearalenone at a mean concentration of 13.6 μg/kg (max 146 μg/kg). Of the masked mycotoxins, DON-3-glucoside occurred at a high incidence level of 53%. Among emerging toxins, moniliformin, fusarinolic acid, and beauvericin showed high occurrences at 98%, 98%, and 83%, and had maximum contamination levels of 1130, 3422, and 142 μg/kg, respectively. Significant differences in the contamination pattern were observed between the agricultural regions and maize types.

Keywords: *Fusarium*; maize; masked mycotoxins; LC-MS/MS; agricultural regions; South Africa

Key Contribution: The most important finding of this study was the environmental influence as well as the maize type on the occurrence of *Fusarium* and their free, masked, and emerging mycotoxins. However, significant differences in contamination pattern were observed between the agricultural regions and maize types.

1. Introduction

Mycotoxins can adversely affect human and animal health condition, productivity, economics, and trade [1–3]. The United Nations' Food and Agricultural Organization (FAO) made an estimate that there was significant contamination of about twenty-five percent of the world's food crops with mycotoxins leading to annual loss in the range of one million tons [4]. Recently, studies suggest that the percentage of contaminated cereals is much higher at 72% [5]. The difference may be due in part, to what levels are considered as contamination, in addition to advances in detection and monitoring [6]. Recently, it has come to be clearer that in mycotoxin-contaminated products, various structurally-related compounds produced during plant metabolism or during after food processing can co-occur with the parent toxins [7]. These mycotoxin derivatives had a very different chemical behavior including polarity and solubility, compared to the precursor and thus, can easily escape routine analyses [8]. Since they are undetectable by conventional analytical techniques because of their altered structures, there is thus generally an underestimation of the mycotoxin load. Also, despite their chemical alteration, coupled with the fact that they are generally not regulated by legislation, they may be considered as being masked [6]. Furthermore, it has to be highlighted that masked mycotoxins can be "unmasked" again in the digestive tract of animals and humans, releasing the parent compound with its toxicological effects again. A similar situation exists with emerging mycotoxins: toxicological data are scarce which makes it difficult to set up regulations and maximum tolerated limits to protect humans and animals from potential health risks. On the other hand, "emerging mycotoxins" generally represents a group of chemically altered mycotoxins which have no regulations up till date. Studies employing LC–MS/MS for structure elucidation provide insights about these newly discovered metabolites [9]. Common emerging *Fusarium* mycotoxins are eniantins (ENNs), beauvericin (BEA), moniliformin (MON), fusaproliferin (FP), fusidic acid (FA), culmorin (CUL), and butenolide (BUT). Jestoi [10] published an extensive review regarding this diverse set of chemical compounds.

Conjugated or masked mycotoxins first came to the attention of public health officials, when animals fed with apparently low mycotoxin contaminated feed, showed high severity of mycotoxicosis. The unanticipated high toxicity was ascribed to the presence of undetected, conjugated forms of mycotoxins [8].

Historically, Gareis et al., [11] for the first time, used the term 'masked mycotoxins' and it refers to the products that are formed when plants metabolize mycotoxins, as part of their natural defense system. These secondary metabolites are not detectable by conventional techniques because their structure has been altered in the plant, nor are they regulated. The metabolites are so-called masked as they become toxic again as soon as they cleave off their sugar molecule in the intestine of the humans and animals. The term conventional applies to the analytical detection methods that have previously or initially been developed for specific mycotoxins only. Then in 2013, researchers revisited the masked mycotoxin topic again and Berthiller et al. [8] made a clear definition of what masked mycotoxins are. The term "masked mycotoxins" is now widely accepted. However, in 2014, Michael Rychlik and his research group came up with a comprehensive definition to include all modified forms of mycotoxins as well as masked mycotoxins as "modified mycotoxins" [12]. Actually, the latter is the umbrella term of all mycotoxins that are modified by some sort of process (for example, food processing). Masked mycotoxins are sort of, part of this definition, but only entail the molecules that are formed by the plants.

The likelihood of mobilization of mycotoxins that interact with metabolically active plants in the field is the issue. As *Fusarium* infection usually occurs in the field (in contrast to *Aspergillus* and *Penicillium* infections), the *Fusarium* mycotoxins (deoxynivalenol, zearalenone, fumonisins, nivalenol, fusarenon- X, T-2 toxin, HT-2 toxin) are the most prominent target for conjugation [8]. Although, transformation of other mycotoxins e.g., ochratoxin A, patulin, and destruxins, by plants has also been described. Specifically, deoxynivalenol-3-glucoside (D3G), zearalenone-14-glucoside (Z14G), and zearalenone-14-sulphate (Z14S) are the most commonly found masked mycotoxins occurring in food commodities [8]. Currently, only glucoside and sulfate conjugates of DON, ZON, T-2, and HT-2 have been proven to occur in naturally infected cereals such as maize, wheat, and barley [13]. However, many researchers have carried out studies on the occurrence of free and masked *Fusarium* mycotoxins in cereal-based food and feed commodities (Table 1).

Table 1. Occurrence data of glucoside and sulfate conjugates (masked mycotoxins) and their respective free forms identified in cereal grains and food/feed products.

Mycotoxin	Food/Feed Commodity	% Positive Samples	Range (μg/kg)	References
ZON	Maize	83	59–1071	[14]
	Maize	-	<LOQ-15,700	[15]
	Maize	-	56	[16]
	Maize	37%	<50–196	[17]
	Maize	90	0–135	[18]
	Maize	-	14	[19]
	Maize	-	10.2	[20]
	Maize	92.5	247.8	[21]
	Maize silage(feed)	79.4	20-11	[22]
	Barley	5.9	<LOQ-17	[23]
	Barley	67	2–31	[24]
	Beans	90	185.2	[21]
	Wheat	46.7	<LOQ-234	[23]
	Wheat	83.3	12–109	[14]
	Wheat	47.5	1–100	[24]
	Peanut	57	70	[25]
	Oat	41.9	<LOQ-675	[26]
	Oat	66.7	13–85	[14]
	Oat	100	5–15	[24]
	Cocoa seeds	23.5	24.2–83.6	[27]
	Bread	83.3	19–53	[14]
	Breakfast cereals	-	4	[28]
	Corn flakes	83.3	34–90	[14]
	Malt	56	102–2213	[29]
	Traditional brewed beer	45	2.6–426	[30]
ZON-14G	Maize	17	274	[14]
	Maize	-	<LOQ-9750 (total)	[14]
	Barley	17.6	<LOQ-9.6	[23]
	Wheat	6.7	<LOQ-0.6	[23]
	Wheat	0	-	[14]
	Oats	3.2	<LOQ	[23]
	Oats	0	-	[14]
	Bread	33.3	20–20	[14]
	Corn flakes	0	-	[14]
ZON-16G	Barley	23.5	<LOQ	[23]
	Wheat	6.7	<LOQ-2.8	[23]
	Oat	58.1	<LOQ-7.9	[23]
ZON-14S	Maize	17	51	[14]
	Maize	-	<LOQ-9750 (total)	[15]
	Maize silage(feed)	42.3	2–4318	[22]
	Barley	8.8	<LOQ-26.1	[23]
	Wheat	33.3	11	[14]
	Wheat	40.0	<LOQ-22.5	[23]
	Oats	16.7	12	[14]
	Oats	29.0	<LOQ-220	[23]
	Bread	16.7	24	[14]
	Corn flakes	0	-	[14]

Table 1. *Cont.*

Mycotoxin	Food/Feed Commodity	% Positive Samples	Range (μg/kg)	References
DON	Maize	100	255–5245	[14]
	Maize	100	90–680	[31]
	Maize	100	32–2246	[32]
	Maize	90	74–1382	[33]
	Maize	22	9.6–745.1	[34]
	Maize	100	1,469	[20]
	Maize	100	0.3–4374	[32]
	Maize	-	380	[16]
	Maize silage (feed)	71.8	1.5–13,488	[22]
	Barley	82.4	LOQ-1180	[23]
	Barley	100	<60	[32]
	Barley	83	54–1602	[24]
	Whole wheat	-	8000	[28]
	Wheat	83	11–1265	[13]
	Wheat	97.6	LOQ-5510	[23]
	Wheat	66.7	16–150	[14]
	Wheat	100	540–5080	[31]
	Wheat	100	46–2683	[32]
	Wheat	46.5	25–2975	[24]
	Wheat	46.1	LOQ-297	[35]
	Wheat	75	LOQ-10,130	[36]
	Wheat	-	40–490	[37]
	Wheat	68	up to 302	[38]
	Rice	23.8	107.9	[39]
	Durum wheat	100	1750	[36]
	Rye	100	<50	[32]
	Oat	100	2690	[23]
	Oat	16.7	46	[14]
	Oat	81.8	62–2216	[32]
	Oat	100	67–149	[24]
	Bread	66.7	20–102	[14]
	Corn flakes	16.7	207	[14]
	Breakfast cereals	28.6	28.6	[40]
	Snacks	61.8	61.8	[40]
	Flours	72.3	72.3	[40]
	Wheat flour	97.2	1.3–825.9	[41]
	Wheat flour	100	78.9–325.8	[35]
	Swine feed	93.8	50–931	[33]
	Poultry feed	93.3	157–1231	[33]
	Feed	99	124–2352	[42]
	Lager beer	100	1.6–6.4	[43]
	Beer	90	LOQ-35.9	[44]

Table 1. *Cont.*

Mycotoxin	Food/Feed Commodity	% Positive Samples	Range (μg/kg)	References
DON-3G	Maize	100	36–1003	[14]
	Maize	67	<LOQ-70	[31]
	Maize	100	<35	[32]
	Maize	80	14–121	[33]
	Maize	-	<LOQ-1100	[15]
	Maize silage(feed)	63.0	1.0–3159	[22]
	Barley	73.5	<LOQ-1300	[23]
	Barley	0	-	[32]
	Barley	29	43–277	[24]
	Wheat	83.3	<LOQ-922	[23]
	Wheat	16.7	18	[14]
	Wheat	100	59–200	[31]
	Wheat	83.3	43–737	[32]
	Wheat	27.3	40-356	[24]
	Wheat	27	16-138	[45]
	Wheat	31	<LOQ	[35]
	Wheat	75	100-1230	[36]
	Durum wheat	94	LOQ-850	[46]
	Rye	0	-	[32]
	Oat	87.1	<LOQ-6600	[23]
	Oat	100	28–97	[14]
	Oat	45.5	162–287	[32]
	Oat	0	-	[24]
	Bread	83.3	26–29	[14]
	Corn flakes	50.0	24–28	[14]
	Breakfast cereals	85.7	19–66	[40]
	Snacks	82.4	11–94	[40]
	Flours	68.2	5–72	[40]
	Wheat flour	33.4	0.2–15.7	[41]
	Swine feed	93.8	6–80	[33]
	Poultry feed	93.3	30–107	[33]
	Beer	95	LOQ-27.5	[44]
NIV	Barley	73.5	<LOQ-302	[23]
	Wheat	43.3	<LOQ-74.0	[23]
	Oat	71.0	<LOQ-4940	[23]
	Rice	-	97	[47]
	Wheat flour	100	LOQ-140.6	[35]
NIV-3G	Barley	61.8	<LOQ- 65.3	[23]
	Wheat	43.3	<LOQ-33.6	[23]
	Oat	16.1	<LOQ-58.3	[23]
HT-2	Barley	35.3	<LOQ-39.5	[23]
	Barley	100	23–233	[48]
	Barley	100	3–213	[49]
	Wheat	100	19–96	[48]
	Wheat	63.3	<LOQ-39.5	[23]
	Wheat	77.8	26–85	[50]
	Oat	100	11–187	[48]
	Oat	74.2	<LOQ-1830	[23]
	Oat	88.9	21–851	[50]

Table 1. *Cont.*

Mycotoxin	Food/Feed Commodity	% Positive Samples	Range (µg/kg)	References
HT-2G	Barley	52.9	<LOQ-38.5	[23]
	Barley	94.4	0.6–162.8	[49]
	Wheat	53.3	<LOQ-15.9	[23]
	Oat	16.1	<LOQ-300	[23]
T-2	Barley	100	41–160	[48]
	Barley	100	1–154	[49]
	Wheat	100	17–76	[48]
	Wheat	55.6	11–23	[50]
	Oat	100	31–165	[48]
	Oat	88.9	10–377	[50]
	Maize	36	7.5–29	[17]
T-2G	Barley	50	0.1–14.5	[50]

Though toxicological data are scarce since masked mycotoxins represent an emerging condition, studies highlight the potential threat to consumer safety from these substances. However, the possible hydrolysis of masked mycotoxins back to their parent toxin during food/feed processing or during mammalian digestion raises concerns. To ensure the safety of agricultural products (food safety), there is a need for the identification and determination of mycotoxins and their masked forms in order to assess possible effects on consumers. Furthermore, since there are very few investigations on the impact of climatic differences on mycotoxins variation in the different agriculture regions of South Africa, a comprehensive study was imperative.

2. Results and Discussion

2.1. General Mycotoxin Occurrence Data

Mycotoxins produced by *Fusarium* species are found on grains cultivated on every continent. The susceptibility of maize to *Fusarium* fungi and mycotoxins contamination is well documented [51–53]. All the maize samples analyzed in this study were contaminated with an average of 5 to 24 out of 42 *Fusarium* mycotoxins, including 0 to 3 masked forms at the same time. The summary data in Table 2 highlights the relevance of fumonisin B1, B2, B3, B4, and A1 in the samples with 98%, 91%, 80%, 82%, and 54% of 123 samples contaminated with maximum contamination levels of 8908, 3383, 990, 1014, and 51.5 µg/kg respectively. Fumonisin B1 was the most common mycotoxin in the maize samples. Fumonisins commonly occur in maize while in the field, predominantly when they are cultivated in warmer regions [54].

Table 2. Summary statistics of occurrence of investigated 42 *Fusarium* free mycotoxins, masked and emerging mycotoxin metabolites in 123 maize samples on LC-MS/MS.

	Concentration (μg/kg)			Number of Positive Samples Across ARs [a]		N [b] (123)	P [c] %	LOQ [d]	LOD [e]	Rec [f] (%)
Metabolite Group	**Minimum**	**Maximum**	**Mean**	**ER (n = 58)**	**WR (n = 65)**					
Free *Fusarium* Mycotoxins										
Fumonisin B_1	12.6	8908	596	58	63	121	98	8	2.4	75.0
Fumonisin B_2	7.9	3383	221	55	57	112	91	7	2.1	79.0
Fumonisin B_3	<LOQ	990	85.6	44	54	98	80	7	2.1	75.0
Fumonisin B_4	<LOQ	1014	81.6	45	56	101	82	7	2.1	75.0
Fumonisin A_1	<LOQ	51.5	6.1	28	38	66	54	2.1	0.6	75.0
Fumonisin Total	**20.5**	**14,347**	**990**	**n/a**	**n/a**	**n/a**	**n/a**	**n/a**	**n/a**	**n/a**
Zearalenone	<LOQ	146	13.6	19	22	41	33	0.6	0.2	54.0
HT-2 toxin	40.2	40.2	40.2	0	1	1	0.8	6.4	1.9	96.8
T-2 toxin	148.0	148	148	0	1	1	0.8	2.4	0.8	102.0
Deoxynivalenol	8.2	1380	152	35	26	61	50	-	1.2	85.0
Nivalenol	7.7	35.7	14.2	4	10	14	11	3.8	1.1	80.0
Masked *Fusarium* Mycotoxins										
DON-3Glucoside	2.43	112	15.0	31	34	65	53	-	0.8	92.0
ZON-sulphate	11	146	14.0	9	13	22	18	n.a	-	75.0
Hydrolysed FB_1	0.7	28.0	3.9	12	19	31	25	0.7	0.2	100.0
HT-2 Glucoside	31.2	31.2	31.2	0	1	1	0.8	5.7	1.7	71.9
Emerging Mycotoxins and Other Less Well-Known *Fusarium* Metabolites										
Monoacetoxyscirpenol	20.9	20.9	20.9	0	1	1	0.8	5.2	1.6	89.0
Diacetoxyscirpenol	4.4	5.0	4.7	0	2	2	1.7	0.5	0.2	81.8
Neosolaniol	4.5	4.5	4.5	0	1	1	0.8	4.6	1.4	83.3
Culmorin	13.3	465	90.0	7	11	18	15	5.5	1.6	76.0
15Hydroxyculmorin	<LOQ	2022	181	24	25	49	40	20.8	6.2	55.7
5Hydroxyculmorin	<LOQ	578	167	12	6	18	15	50	15	102.0
Moniliformin	<LOQ	1130	219	57	63	120	98	5	1.5	100.0
Beauvericin	<LOQ	142	7.2	56	51	107	87	0.03	0.01	100.0
Enniatin B	0.1	4.9	2.5	0	2	2	1.7	0.04	0.01	81.0
Enniatin B_1	0.13	3.0	1.6	0	2	2	1.7	0.11	0.03	100.0
Enniatin B_2	<LOQ	<LOQ	<LOQ	0	1	1	0.8	0.57	0.2	91.2
Aurofusarin	<LOQ	5470	296	46	43	89	72	3.7	1.1	79.8
Bikaverin	<LOQ	651	72.0	38	44	82	67	55	16	79.6
Butenolid	<LOQ	214	48.9	22	13	35	28	10.8	3.2	61.0
Epiequisetin	<LOQ	18.9	5.0	7	12	19	15	2	0.6	70.3
Equisetin	<LOQ	129	19.4	14	16	30	24	2.3	0.7	79.9
Apicidin	2.9	15.4	9.1	1	1	2	1.6	0.65	0.2	108.0
Deoxyfusapyron	<LOQ	53.0	11.1	7	16	23	19	2.7	0.8	103.0
Fusapyron	<LOQ	18.0	11.1	14	24	38	31	2.4	0.7	106.0
Fusaric acid	57.9	195	85.2	14	10	24	20	30	10	91.4
Fusarinolic acid	<LOQ	3422	506	57	63	120	98	30	10	89.5
7Hydroxypestalotin	<LOQ	16.5	7.5	8	17	25	20	2.9	0.9	100.0
Acuminatum B	<LOQ	219	33.0	4	8	12	9.8	30	10	105.0
Acuminatum C	26.5	204	98.8	4	3	7	5.7	30	10	62.4
Chlamydospordiol	2.1	5.1	7.3	1	1	2	1.7	0.3	0.1	96.5
Chlamydosporol	27.0	26.9	26.9	0	1	1	0.8	1.1	0.3	94.0
Chrysogin	<LOQ	7.7	4.4	6	42	48	30	1.9	0.6	99.2
Siccanol	34.6	252	64.3	43	48	91	74	-	-	86.9

[a] Contamination across agricultural regions (ARs), eastern region (ER), western region (WR). [b] Total number of contaminated samples (N). [c] Percentage of total number of samples contaminated (P). [d] LOQ: limit of quantification, <LOQ: less than LOQ. [e] LOD: limit of detection [s/n = 3:1] expressed as μg/kg sample. [f] Recovery: calculated from spiking experiment of maize samples. n/a: Not applicable.

Contamination of maize by fumonisins was considered an important risk factor in human oesophageal cancer in the former Transkei region of South Africa and Santa Catarina State, Brazil [55,56]. Fumonisin B1 has also been implicated in the development of neural tube defects in babies of mothers consuming fumonisin-contaminated maize, especially in certain regions of South Africa, China, and Italy [57,58].

Deoxynivalenol occurred in 50% of the samples with a mean concentration of 152 μg/kg (max 1380 μg/kg). Deoxynivalenol DON causes different unspecific symptoms, for example vomiting and diarrhea, also causes loss of weight and feed refusal in livestock, and hence is known as vomitoxin [59].

The presence of DON in maize has often been associated with samples originating from temperate regions such as northern Europe and North America [60]. However, reports emerging from tropical countries, specifically from South Africa, continue to reveal the occurrence of DON in maize and maize products [61]. It has also been reported at a lower concentration than in this study on maize from Burkina Faso at a mean of 31.4 μg/kg, Mozambique at a range of 116–124 μg/ kg [62] and in Nigerian maize at a range of 11–479 μg/kg [63] and at a mean of 225 μg/kg [64].

Furthermore, according to Ediage et al. [65], DON was detected in the urine of toddlers (1.5–5 years) from Cameroon, which also affirms its presence in food commodities originating from sub-Saharan Africa. However, although half of the analyzed samples contained DON, the maximum level at 1380 μg/kg did not exceed the maximum allowable levels for DON in unprocessed maize as was set by South African Regulation in 2016 at 2000 μg/kg.

Forty-one samples were contaminated with zearalenone at a mean concentration of 13.6 μg/kg. The occurrence of ZON in agricultural commodities has not been thoroughly investigated in sub-Saharan Africa. It was first reported on South African maize and subsequently recovered from maize and other commodities elsewhere on the continent [66]. The maize samples analyzed in this study showed a 33% occurrence rate for ZON, with none of the samples exceeding the maximum level of 350 μg/kg for unprocessed maize products according to European commission [67]. Similarly, it was also reported by Boutigny et al. [68] at 33% occurrence in naturally infected field-grown maize samples. Meyer et al. [69] also reported that less than 10% of the South African commercial maize samples were contaminated with ZON. The detected maximum concentration from the samples was 145.6 μg/kg (Table 2). Compared to other reports on the occurrence of ZON in maize samples from other parts of Africa, the levels detected in this study were lower than levels earlier reported by Adejumo et al. [34] in maize samples from Nigeria. Zearalenone was also reported in maize samples from Cameroon with a mean concentration of 68 μg/kg and a maximum concentration of 309 μg/kg [70]. However, the relatively low level of ZON observed in this study somewhat supports the notion that ZON is perhaps a persistent yet minor contaminant of foods/feeds in South Africa [66,69,71] but its significance is in its oestrogenic potential to mammals.

Occurrence of HT-2 and T-2 in the samples was at a very low level at 0.8% (1 sample) at maximum concentration of 40.2 and 148.0 μg/kg, respectively. The EU permissible levels of mycotoxins by the European Commission have recommended the limit of 15–1000 μg/kg for the sum of HT-2 and T-2 toxins in various matrices [72]. The detected maximum concentration of HT-2 and T-2 from the samples did not exceed the recommended limit. Meanwhile, HT-2 toxin and T-2 are two related compounds that may be synthesized by several *Fusarium* species. Their presence in cereal grain has been well documented, with several reports originating from the cold climate of northern Europe [73,74]. Nevertheless, few reports from Africa indicated; 1–8% incidence rates for HT-2 in Nigerian cereals [64] and 25% rate in Tanzanian maize for HT-2, with range of 15–25 ppb [75]. Also, South African Grain Laboratory (SAGL) reported detection of HT-2 and T-2 only in one maize sample, with the levels 72 and 232 μg/kg, respectively [76].

Nivalenol occurred only in 11% of the samples at a contamination range of 7.7–35.7 μg/kg. Occurrence of NIV in maize has previously been reported in Nigerian samples, although at a higher incidence rate (54%) [63] and higher contamination range of 163–271 μg/kg [64]. Nivalenol is known to be immunosuppressive and also as a protein inhibitor.

Of the masked mycotoxins, DON-3-glucoside occurred at highest incidence rate of 53% than the other masked forms detected in this study. Among emerging toxins, moniliformin, fusarinolic acid, and beauvericin showed high occurrence being found in 98%, 98%, and 83% of samples, respectively. High incidences of these toxins in maize which serve as a staple food in South Africa is an important cause for concern. From available literature, little or no appreciable study has been done on the occurrence of these mycotoxins in food and food products in South Africa and neglecting them increases the risk of exposure to humans and animals. The occurrences of these emerging mycotoxins produced by

Fusarium specie, have been reported in food crops which represents an important problem in some parts of the world [77,78].

Furthermore, the high incidence of these emerging toxins should not be taken for granted, since moniliformin is known to be cytotoxic to many mammalian systems [10]. Beauvericin is also known to be genotoxic to human lymphocytes [79]. Other detected metabolites included emerging mycotoxins and other less well-known Fusarium metabolites such as enniatins (B, B1, and B2), monoacetoxyscirpenol, diacetoxyscirpenol, equisetin, epi-equisetin, aurofusarin, apicidin, neosolaniol, culmorin, 5- & 15-hydroxyculmorin, bikerverin, butenolide, chlamydospordiol, chlamydosporol, chrysogin, fusaric acid, fusapyron, deoxyfusapyron, 7-hydroxypestalotin, acuminatum (B and C) fusarinolic acid, and fusarin C. The co-occurrence of these emerging mycotoxins with other major mycotoxins and many other *Fusarium* metabolites of unknown toxicity is a source of concern. Beauvericin, like enniatins, is a cyclodepsipeptide that has antibiotic, insecticidal, and cytotoxic properties presumably related to their ionophoric properties [80].

2.2. Distribution of Fusarium (Free and Masked) Mycotoxins across the Agriculture Regions of South Africa

Mycotoxins are among the food-borne risks that are dependent upon climatic conditions. Indeed, the ability of fungi to produce mycotoxins is largely influenced by temperature, relative humidity, insect attack, and stress conditions of the plants [81]. The main effect due to agriculture regions (AR) on the distribution of *Fusarium* free and masked mycotoxins in the maize samples is presented in Table 2.

All the 42 *Fusarium* toxins and metabolites investigated in the maize samples across the agricultural regions were detected and quantified except for the emerging toxin, enniantin B2, which was only detected in 2% of the samples from the western region but was not quantified because the values were below the limit of quantification (<LOQ). Of the major mycotoxins, HT-2 was not detected at all in the eastern region but was quantified only in 2% of the maize from the western region at 40.2 μg/kg concentration. Detection of HT-2 in the warm western region and not the cold eastern region as would be expected is contrary as it was reported that HT-2 was previously considered a problem in colder European climates [13]. Of the fumonisin Bs, FB1 occurred at the most frequency rate than FB2, FB3, and FB4. Fumonisin B1 was the most contaminant mycotoxin, occurring at a mean concentration of 752 ± 1469 μg/kg from the warm western region and of 440 ± 514 μg/kg in the cold eastern region with only 3% (2 samples) not contaminated. A similar trend was also observed when the sum of the fumonisins was considered, for instance, the highest fumonisin (FB1 + FB2 + FB3) concentrations were detected in warm western region and this is consistent with Munkvold [82] and De La Campa et al. [83] who reported optimum temperatures of 30 and 32 °C for fumonisin production. This observation could be correlated to the high mycotoxins production potentials of *Fusarium* fungi in warmer climates [84]. In general, fumonisin B1 was detected in 98% of all the samples analyzed, irrespective of the maize type and AR, ranging from 12.6 to 8908 μg/kg with the mean concentration of 596 μg/kg (Table 2).

Fumonisin B^2 was the second most contaminant mycotoxin, across the ARs and generally, fumonisin A1 was quantified in 32% of all the maize samples analyzed and in 22% of the samples, it was detected but below the limit of quantification (<LOQ).

Deoxynivalenol was more prevalent in the eastern region than in the western region, having 60% and 40% of the samples all detected and quantified respectively (Table 2). This was in line with what was reported by Chilaka et al. [64] who observed the highest incidence rate of DON in maize samples from the DS zone in Nigeria, (an agricultural region) characterized by a lower temperature and higher average annual. Zearalenone, nivalenol, HT-2, and T-2 were also among the major *Fusarium* mycotoxins, detected and quantified in the two regions. Nivalenol, HT-2, and T-2 were detected in higher concentration in the WR while ZON occurred at lower concentration in the same region, occurring higher in the ER. This observation could be because ZON is known to be produced in somewhat cool environments compared to other mycotoxins [9,22]. The highly toxic mycotoxin, T-2 toxin, has until recently not been recorded in South Africa. T-2 toxin is most commonly produced by *F.* sporotrichioides, a fungus well-adapted to survive in colder countries [85]. Some T-2-producing

Fusarium species have occasionally been isolated from wheat with *Fusarium* head blight (FHB) and maize with *Fusarium* ear rot (FER) symptoms in South Africa [86]. The presence of T-2 toxin in local maize grain, was reported by the SAGL, [76] and it has been shown to have a relationship with *F. verticillioides* and *F. graminearum* [87,88].

Masked mycotoxins detected in the samples include DON-3-glucoside, zearalenone-sulphate, hydrolysed fumonisin B1, and HT-2-glucoside. Deoxynivalenol-3-glucoside (DON-3-G) and HT-2-glucoside had the highest and least occurrence respectively across the two regions. Although there were no significant differences in their distribution across the agriculture regions, but to our knowledge, there seem to be no available data on the occurrence of some of these masked mycotoxins in South Africa. Hence, this is the first report of zearalenone-sulfate and HT-2-glucoside in South African maize. Although toxicological data are still limited, the occurrence of masked mycotoxins is expected to add substantially to the overall mycotoxins load and toxicity. This invariably will increase the toxic health effects by these masked mycotoxins, which may be either direct or indirect through hydrolysis, or released from the matrix during digestion into the free mycotoxins [89].

Moreover, monoacetoxyscirpenol, diacetoxyscirpenol, neosolaniol, enniatins (B and B1) were not detected in maize samples from the eastern regions but had low incidences of 2%, 3%, 2%, 3%, and 3% respectively in the western region. This could be explained by the agroclimatic differences in the agriculture regions that favored the accumulation of these emerging mycotoxins in the warmer and drier climate of the western maize regions.

2.3. *Agricultural Region and Maize Type on Fusarium Mycotoxin Distribution and Accumulation on Maize*

Mycotoxin occurrence and distribution is influenced by different factors such as crop species, climatic, and environmental conditions of a given region. The interaction between agricultural region and maize type on the concentration of *Fusarium* mycotoxins on maize samples is presented in Table 3.

From the statistical analyses, the concentration of the major mycotoxins; fumonisin Bs (B1, B2, B3, B4) and fumonisin A1 in white maize samples collected from the western region (WR) was significantly higher ($p < 0.05$) than that of yellow maize from the same region, as well as on white and yellow maize from the eastern region (ER). The mean values were 1023 ± 1698, 377 ± 645, 146 ± 197, 140 ± 207, and 10.7 ± 12 µg/kg respectively. The fact that the white maize samples of the western region (WR) had significantly higher mean levels of fumonisin can be explained partly by high mycotoxins production potentials of *Fusarium* fungi in warmer climates as reported by [84]. Then, Munkvold [82] and De La Campa et al. [83] also reported optimum temperatures of 30 and 32 °C for fumonisin production, which is what is obtained in the western region. Furthermore, differences in plant genotype cannot be ruled out as it has been reported that white maize is a better substrate for fumonisin production than yellow maize [90].

Deoxynivalenol concentration in yellow maize samples from the western region was significantly higher ($p < 0.05$) than that of the white maize samples from the same region and on the white and yellow maize samples from the eastern region, with the mean value being 367 ± 503 µg/kg (Table 3). Occurrence of DON at significantly higher concentration in the western region could be explained by high mycotoxins production potentials of *Fusarium* fungi in warmer climates as reported by [84]. The higher DON concentration in yellow maize than the white maize could be due to variation in localization of pigments in yellow maize kernel, which might have a prominent role in the actual degree of resistance to *Fusarium* infection/mycotoxin accumulation [91].

Table 3. Occurrence of *Fusarium* mycotoxins in white and yellow maize from the agriculture regions.

Free *Fusarium* Mycotoxins	ER		WR		LSD (<0.05)
	WM (Mean ± SD)	YM (Mean ± SD)	WM (Mean ± SD)	YM (Mean ± SD)	
	Concentration (μg/kg)				
Fumonisin B_1	607 ± 579	273 ± 362	1023 ± 1698	482 ± 634	407.26
Fumonisin B_2	200 ± 189	102 ± 121	377 ± 645	203 ± 248	156.06
Fumonisin B_3	80.9 ± 60	43.5 ± 35	146 ± 197	7.2 ± 59	4.87
Fumonisin B_4	71.6 ± 53	39.7 ± 32	140 ± 207	75.1 ± 74	52.61
Fumonisin A_1	4.2 ± 2	3.7 ± 1	10.7 ± 12	5.8 ± 3	4.37
Zearalenone	21.0 ± 47	9.8 ± 13	4.4 ± 9	19.2 ± 27	17.64
Deoxynivalenol	121 ± 186	51.0 ± 35	71.0 ± 71	367 ± 503	136.80
Nivalenol	10.2 ± 0	15.3 ± 4	15.8 ± 9	15.5 ± 6	6.79
T-2	<LOD	0	148 ± 0	0	-
HT-2	0	0	31.2 ± 0	0	-
Masked *Fusarium* Mycotoxins					
DON-3-glucoside	18.6 ± 25	6.7 ± 4	8.7 ± 11	25.9 ± 36	13.73
Zearalenone-Sulphate	2.78 ± 48	1.14 ± 97	7.36 ± 54	9.5 ± 71	22.30
Hydrolysed Fumonisin B_1	3.8 ± 3	1.8 ± 1	6.5 ± 7	3.6 ± 4	3.13
HT-2-glucoside	0	0	31.2 ± 0	0	-
Emerging Mycotoxins and Other Less Well-Known *Fusarium* Metabolites					
Monoacetoxyscirpenol	0	0	20.9 ± 0	0	-
Diacetoxyscirpenol	0	0	4.7 ± 1	0	-
Neosolaniol	0	0	4.8 ± 0	0	-
Culmorin	62.4 ± 25	159 ± 177	22.7 ± 7	116 ± 102	101.47
15Hydroxyculmorin	179 ± 148	79.2 ± 55	55.5 ± 21	410 ± 597	190.36
5Hydroxyculmorin	111 ± 16	58.7 ± 349	167 ± 0	331 ± 0	233.20
Moniliformin	184 ± 178	165 ± 144	195 ± 227	331 ± 327	96.21
Beauvericine	2.3 ± 5	4.4 ± 8	7.8 ± 27	14.2 ± 26	9.24
Enniatin B	0	0	5.0 ± 0	0.1 ± 0	-
Enniatin B_1	0	0	2.98	0.13	-
Enniatin B_2	0	0	<LOQ	<LOQ	-
Aurofusarin	130 ± 188	283 ± 577	127 ± 127	643 ± 1490	351.51
Bikaverin	56.1 ± 0	156 ± 0	0	76.3 ± 16	75.81
Butenolid	31.2 ± 14	27.6 ± 11	19.3 ± 9	118 ± 97	42.67
Epiquisetin	1.5 ± 0	11.8 ± 7	4.4 ± 2	2.2 ± 0	4.13
Equisetin	4.2 ± 2	50.7 ± 56	17.8 ± 10	4.7 ± 1	28.94
Apicidin	15.4 ± 0	0	2.9 ± 0	0	-
Deoxyfusapyron	3.3 ± 0	5.0 ± 2	30.1 ± 23	6.1 ± 3	21.74
Fusapyron	3.3 ± 0	5.0 ± 8	30.1 ± 4	6.1 ± 5	21.74
Fusaric Acid	0	109 ± 46	133 ± 26	99.4 ± 22	22.70
Fusarinolic Acid	530 ± 387	493 ± 281	557 ± 590	446 ± 293	161.10
Hydrolysed Fumonisin B_1	3.8 ± 3	1.8 ± 1	6.5 ± 7	3.6 ± 4	3.13
7Hydroxypestalotin	8.5 ± 4	5.8 ± 1	6.8 ± 3	9.2 ± 4	2.76
Acuminatum B	0	0	27 ± 6	103 ± 82	94.90
Acuminatum C	58.4 ± 25	164 ± 41	129 ± 0	44.8 ± 18	45.80
Chlamydospordiol	2.1 ± 0	-	5.1 ± 0	-	-
Chlamydosporal	-	-	26.9 ± 0	-	-
Chrysogin	5.6 ± 6	3.8 ± 2	3.6 ± 2	4.5 ± 2	2.01
Siccanol	62.0 ± 51.2	42.3 ± 26.2	61.7 ± 59.3	91.0 ± 55.3	24.9

AR—agriculture region, ER—eastern region, WR—western region, Qty—percentage quantified, <LOD—percentage below limit of detection, <LOQ—percentage below limit of quantification, YM—yellow maize, WM—white maize, LSD—least significant difference, SD—standard deviation.

Zearalenone and nivalenol concentrations showed no significant difference ($p > 0.05$) amongst white and yellow maize samples collected from the western region, as well as white and yellow maize from the eastern region. In addition, white maize from the eastern region has the highest mean concentration of 21.0 ± 47 µg/kg. This is in line with the finding of Ediage et al. [92], who noted no significant difference in zearalenone accumulation in maize samples from Cameroon, irrespective of geographical location.

The concentration of the masked mycotoxins DON-3-glucoside in yellow maize samples from the western region was significantly higher ($p < 0.05$) than that of the white maize samples from the same region and with the yellow maize samples from the eastern region. A similar trend was also observed with the parent toxin deoxynivalenol in this study. This can also have the same explanation as is the case with DON accumulation.

DON-3-glucoside concentration in white maize samples from the eastern region had no statistically significant difference ($p > 0.05$) with the yellow maize samples from the same region and with white maize samples from the western region. In addition, yellow maize from the western region had the highest concentration with mean value of 25.9 ± 36 µg/kg.

Zearalenone sulfate concentrations had no significant differences ($p > 0.05$) amongst white and yellow maize samples collected from the western region, white and yellow maize from the eastern region. A similar trend was also observed with the parent mycotoxin, zearalenone, in this study.

The concentration of hydrolyzed fumonisin B1 in white maize samples collected from the western region was significantly higher ($p < 0.05$) than that of yellow maize from the eastern region. The same trend was observed with other fumonisin parent mycotoxins. Also, hydrolyzed fumonisin B1 in white maize samples collected from the western region had no significant difference ($p > 0.05$) with that of the yellow maize samples from the same region and also with the white maize samples from the eastern region. HT-2-glucoside occurred only on white maize from the western region at mean contamination level of 31.2 ± 0 µg/kg.

Of the emerging toxins, the concentration of 15-Hydroxyculmorin, moniliformin, aurofusarin, and butenolid on yellow maize from the western region was significantly higher ($p < 0.05$) than that of the white maize samples from the same region, as well as white and yellow maize from the eastern region. The mean values were 410 ± 597, 331 ± 327, 643 ± 1490, 118 ± 97 µg/kg respectively. Bikaverin, epiquisetin and equisetin concentrations in yellow maize samples from the eastern region was significantly higher ($p < 0.05$) than that of the white maize samples from the same region and with white and yellow maize samples from the western region. The concentrations of deoxyfusapyron and fusaric acid in white maize samples collected from the western region (WR) is significantly higher ($p < 0.05$) than that of yellow maize from the same region, white and yellow maize from the eastern region. Fusarinolic acid and chrysogin concentrations had no significant difference ($p > 0.05$) amongst white and yellow maize samples collected from the western region, white and yellow maize from the eastern region. Then, beauvericin and acuminatum B concentrations in yellow maize samples collected from the western region is significantly higher ($p < 0.05$) than that of yellow and white maize from the eastern region and had no significant difference ($p > 0.05$) with white maize samples from the western region.

3. Conclusions and Recommendations

The analysis showed that maize types were contaminated with a mixture of both free, masked, and emerging mycotoxins across the maize producing regions of South Africa. All the maize samples analyzed had an average of 5 up to 24 out of 42 mycotoxins, including 1 to 3 masked forms at the same time. This study has also shown that there is higher risk of *Fusarium* mycotoxins exposure, especially fumonisin Bs, with consumption of maize grown in the western than with the eastern agriculture regions of South Africa. White maize samples from the western region (WR) had significantly higher mean levels of fumonisins. It also showed that there is no significant difference in the occurrence of the masked toxins across the agriculture regions. Although toxicological data are still limited, the occurrence or presence of masked mycotoxins will add substantially to the overall mycotoxin load and toxicity. However, studies highlight the potential threat to consumer safety from these substances.

4. Materials and Methods

4.1. Chemicals and Standards

The reagents and chemicals used were of analytical grade (LC gradient grade and MS grade), and were obtained from Merck and Co, Sigma or Microsep, Pretoria, South Africa, except others from Sigma-Aldrich (Vienna, Austria), Merck (Darmstadt, Germany) and VWR (Leuven, Belgium). Mycotoxin standards were obtained from various research groups or purchased from various commercial sources (Romer Labs®Inc. (Tulln, Austria), AnalytiCon Discovery (Potsdam, Germany), Bio Australis (Smithfield, Australia). Water was purified successfully by reverse osmosis using Purite™ water purification technology Suez, UK from LASEC, South Africa.

4.2. Sampling Site Description

The agriculture regions are the eastern region (ER), which is situated in the higher rainfall and cooler areas and the western region (WR), which is located in the drier and warmer areas. The ER consists of Gauteng province and the eastern Free State where the mean maximum temperatures ranged from 24 to 27 °C while the WR consists of the northwest province and western Free State, where the average temperature ranged from 29 to 32 °C. The geography and climatic conditions of the regions have previously been documented [68].

4.3. Sampling and Sample Preparation

A total of 123 maize samples harvested during the 2015/2016 season were collected from randomly selected silo sites in the northwest, Gauteng, and Free state province in the two agriculture regions. From the silos located at the western region (WR), 65 composite samples were collected, comprising of 44 white maize and 21 yellow maize. From the eastern region (ER) silo locations, 58 composite samples were collected, comprising of 30 white maize and 28 yellow maize. For each maize sample, at least 10 incremental samples of 100 g each were taken and combined, making approximately 1 kg according to EC 401/2006, in sterile zip lock polythene bags, well labelled and transported to the laboratory. The samples were thoroughly mixed and milled using a sterile high-speed laboratory blender (IKA M20, Merck, Sarstedt, Nümbrecht, Germany) and packaged in sealed sterile plastic bags to avoid contamination. A cleaning and decontamination routine of the equipment was performed using 70% methanol, after each milling practice. Samples were stored prior to analysis at 4 °C in the freezer. The milled sample material was vigorously homogenized with a spatula before weighing.

4.4. *Fusarium* Mycotoxins Analysis

4.4.1. Sample Preparation and Cleanup for LC-MS/MS Multi-mycotoxin Analyses

This was done according to Sulyok et al. [93], sample preparation was rather simple and eludes any clean-up. Briefly, 5 g representative amount of ground maize kernels was weighed into a 50 mL polypropylene tube (Sarstedt, Nümbrecht, Germany) and extracted for 90 min at 180 rpm on a GFL 3017 rotary shaker (GFL, Burgwedel, Germany) with 20 mL of extraction solvent, (acetonitrile/water/acetic acid 79:20:1, v/v/v). Subsequently, the extracts were centrifuged for 2 min at 3000 rpm (radius 15 cm) on a GS-6 centrifuge (Beckman Coulter Inc., Fullerton, CA, USA). The raw extracts were transferred into glass vials using Pasteur pipettes, and 350 µL aliquots were diluted in the same volume (1/1) with dilution solvent, (acetonitrile/water/acetic acid 20:79:1, v/v/v) to adjust the solvent strength. After appropriate mixing, 50 µL of the diluted extract was analyzed by LC-MS/MS without further pre-treatment.

4.4.2. LC-MS/MS Multi-mycotoxin Measurement Parameters

Analyses for the *Fusarium* toxins was performed by liquid chromatography–tandem mass spectrometry (LC-MS/MS) multi-mycotoxin method, at the Centre for Analytical Chemistry, Department of Agro biotechnology (IFA-Tulln), University of Natural Resources and Life Sciences, Vienna, Austria. The analysis was performed according to the methods described by Sulyok et al. [93] and Malachová et al. [94] with slight modifications.

Briefly, a QTrap 4000 LC-MS/MS System (Applied Biosystems, Foster City, CA) was equipped with a turbo ion spray electrospray ionization (ESI) source and an 1100 Series HPLC System (Agilent, Waldbronn, Germany). Chromatographic separation of the analytes was performed at 25 °C on a Gemini® C_{18} column, 150 × 4.6 mm i.d., 5 μm particle size, equipped with a C_{18} 4 × 3 mm i.d. security guard cartridge (Phenomenex, Torrance, CA, US), using (eluent A) methanol/water/acetic acid 10:89:1 (v/v/v) or (eluent B), methanol/water/acetic acid 97:2:1. Both eluents contained 5 mM ammonium acetate. After an initial time of 2 min at 100% A, the proportion of B was increased linearly to 100% within 12 min, followed by a hold-time of 3 min at 100% B and 4 min column re-equilibration at 100% A. The injection volume of 50 μL and flow rate of 1 mL min^{-1} was used.

The ESI-MS/MS source temperature operated at 550 °C, in the multiple reaction monitoring (MRM) mode, both in positive and negative polarities in two separate chromatographic runs per sample by scanning two fragmentation reactions per analyte. Further MS parameters were as follows: curtain gas 10 psi (69 kPa of max. 99.5% nitrogen); ion source gas 1 (sheath gas) 50 psi (345 kPa of nitrogen); ion source gas 2 (drying gas) 50 psi (345 kPa of nitrogen); ion spray voltage −4000 V and +4000 V respectively, collision-activated dissociation gas (nitrogen) high.

4.4.3. Method Validation

LC-MS/MS multi-mycotoxins method was validated in terms of linearity, apparent recovery (AP), limit of detection (LOD), limit of quantification (LOQ), (Table 1), using blank matrices of maize, (401/2006/EC, 2006). Apparent recoveries of the analytes were calculated by spiking five different samples that were not contaminated with mycotoxins with a multi-analyte standard. The spiked samples (0.25 g) each were left overnight in the dark at room temperature for evaporation of the solvent to establish equilibrium between the analytes and the sample was then extracted with 1 mL of extraction solvent as described above. The corresponding peak areas of the spiked samples were used for estimation of the apparent recovery by comparison with a standard of the same concentration prepared by dilution in pure solvent:

$$\text{RA\%} = 100 \times \frac{\text{Peak area spiked samples}}{\text{Peak area liquid standards}}$$

Limits of detection (LOD) and limit of quantification (LOQ) were calculated from the signal to noise ratios (S/N) of the respective multiple reaction monitoring (MRM) chromatograms deriving from analysis of the spiked samples: LOD = 3 × S/N and LOD = 10 × S/N, respectively.

4.4.4. Data Analysis

The data collected were subjected to analyses of variance (ANOVA), as outlined for factorial arrangement in a completely randomized design (CRD) using GenStat Release 10.3DE [95] statistical software. The means were compared using Fisher's least significant difference (F-LSD) at 5% probability level. The data were not log-transformed, only the geometric means were taken. Microsoft Office Excel software version 2016 was used to determine the frequencies, range, and percentages.

Author Contributions: Conceptualization, M.M., and T.I.E.; Methodology and validation, M.S., T.I.E., and M.M.; Formal analysis, T.I.E., T.A.D. and N.N.; Supervision, M.M., and R.G.; Original manuscript preparation, T.I.E.; Review and editing, M.M., and M.S. All authors have read and agreed to the published version of the manuscript.

Funding: This research was funded by The National Research Foundation of South Africa (NRF), Grant number; 105882 Health and Welfare Sector, Education and Training (HWSETA) PhD bursary, 2017-2018. and APC was funded by The North-west University South Africa (NWU).

Acknowledgments: The authors wish to acknowledge the North-west University South Africa (NWU), The National Research Foundation of South Africa (NRF), and The Health and Welfare Sector, Education and Training South Africa (HWSETA).

Conflicts of Interest: The authors declare no conflict of interest.

References

1. Smith, J.E.; Solomons, G.; Lewis, C.; Anderson, J.G. Role of mycotoxins in human and animal nutrition and health. *Nat.Toxins* **1995**, *3*, 187–192. [CrossRef] [PubMed]
2. Wild, C.P.; Gong, Y.Y. Mycotoxins and human disease: A largely ignored global health issue. *Carcinogenesis* **2009**, *31*, 71–82. [CrossRef] [PubMed]
3. Ekwomadu, T.I.; Mwanza, M. A decade of mycotoxins research in Africa: A review. In *Mycotoxins: Occurrence, Toxicology and Management Strategies*; Cesar, R., Ed.; Nova Science Publishers: New York, NY, USA, 2015.
4. Smith, J.E.; Solomons, G.; Lewis, C.; Anderson, J.G. *Mycotoxins in Human Nutrition and Health*; Chapter 9; European Commission: Luxembourg, Luxembourg, 1994.
5. Streit, E.; Naehrer, K.; Rodrigues, I.; Schatzmayr, G. Mycotoxin occurrence in feed and feed raw materials worldwide: Long-term analysis with special focus on Europe and Asia. *J. Sci. Food Agric.* **2013**, *93*, 2892–2899. [CrossRef]
6. Berthiller, F.; Maragos, C.M.; Dall'Asta, C. (Eds.) *Chapter 1: Introduction to Masked Mycotoxins*; RSC Publishing: London, UK, 2016; pp. 1–13.
7. Galaverna, G.; Dallsta, C.; Mangia, M.A.; Dossena, A.; Marchelli, R. Masked mycotoxins: An emerging issue for food safety. *Czech J. Food Sci.* **2009**, *27*, 89–92. [CrossRef]
8. Berthiller, F.; Crews, C.; Dall'Asta, C.; Saeger, S.D.; Haesaert, G.; Karlovsky, P.; Oswald, I.P.; Seefelder, W.; Speijers, G.; Stroka, J. Masked mycotoxins: A review. *Mol. Nutr. Food Res.* **2013**, *57*, 165–186. [CrossRef]
9. Krska, R. How does climate change impact on the occurrence and the determination of natural toxins. In Proceedings of the 7th International Symposium on Recent Advances in Food Analysis, Prague, Czech Republic, 3–6 November 2015.
10. Jestoi, M. Emerging *Fusarium* Mycotoxins Fusaproliferin, Beauvericin, Enniatins, And Moniliformin—A Review. *Crit. Rev. Food Sci. Nutr.* **2008**, *48*, 21–49. [CrossRef]
11. Gareis, M.; Bauer, J.; Thiem, J.; Plank, G.; Grabley, S.; Gedek, B. Cleavage of Zearalenone-Glycoside, a "Masked" Mycotoxin, during Digestion in Swine. *J. Vet. Med. Ser. B* **1990**, *37*, 236–240. [CrossRef]
12. Rychlik, M.; Humpf, H.U.; Marko, D.; Dänicke, S.; Mally, A.; Berthiller, F.; Klaffke, H.; Lorenz, N. Proposal of a comprehensive definition of modified and other forms of mycotoxins including "masked" mycotoxins. *Mycotoxin Res.* **2014**, *30*, 197–205. [CrossRef]
13. Bryła, M.; Waśkiewicz, A.; Ksieniewicz-Woźniak, E.; Szymczyk, K.; Edrzejczak, R.J. Modified fusarium mycotoxins in cereals and their products—Metabolism, occurrence, and toxicity: An updated review. *Molecules* **2018**, *23*, 1–34.
14. De Boevre, M.; Di Mavungu, J.D.; Maene, P.; Audenaert, K.; Deforce, D.; Haesaert, G.; Eeckhout, M.; Callebaut, A.; Berthiller, F.; Van Peteghem, C.; et al. Development and validation of an LC-MS/MS method for the simultaneous determination of deoxynivalenol, zearalenone, T-2-toxin and some masked metabolites in different cereals and cereal-derived food. *Food Addit. Contam. Part A* **2012**, *29*, 819–835. [CrossRef]
15. De Boevre, M.; Landschoot, S.; Audenaert, K.; Maene, P.; Di Mavungu, D.; Eeckhout, M.; Haesaert, G.; De Saeger, S. Occurrence and within field variability of *Fusarium* mycotoxins and their masked forms in maize crops in Belgium. *World Mycotoxin J.* **2014**, *7*, 91–102. [CrossRef]
16. Ngoko, Z.; Daoudou, I.H.; Kamga, P.T.; Mendi, S.; Mwangi, M.; Bandyopadhyay, R.; Marasas, W.F.O. Fungi and mycotoxins associated with food commodities in Cameroon. *J. App. Biosci.* **2008**, *6*, 164–168.
17. Afolabi, C.G.; Ekpo, E.J.A.; Bandyopadhyay, R. Maize contamination by zearalenone and T-2 toxin and human exposure in Nigeria. *Mycotoxins* **2013**, *63*, 143–149. [CrossRef]
18. Chilaka, C.A.; de Kock, S.; Phoku, J.Z.; Mwanza, M.; Egbuta, M.A.; Dutton, M.F. Fungal and mycotoxin contamination of South African commercial maize. *J. Food Agric. Environ.* **2012**, *10*, 296–303.
19. Zinedine, A.; Brera, C.; Elakhdari, S.; Catano, C.; Debegnach, F.; Angelini, S.; De Santis, B.; Faid, M.; Benlemlih, M.; Minardi, V.; et al. Natural occurrence of mycotoxins in cereals and spices commercialized in Morocco. *Food Control* **2006**, *17*, 868–874. [CrossRef]

20. Mohale, S.; Medina, A.; Rodríguez, A.; Sulyok, M.; Magan, N. Mycotoxigenic fungi and mycotoxins associated with stored maize from different regions of Lesotho. *Mycotoxin Res.* **2013**, *29*, 209–219. [CrossRef]
21. Mulunda, F. Mycotoxins occurrence in selected staple food in main markets from Lubumbashi, Democratic Republic of Congo. *J. Food Agric. Environ.* **2013**, *11*, 51–55.
22. Kovalsky, P.; Kos, G.; Nährer, K.; Schwab, C.; Jenkins, T.; Schatzmayr, G.; Sulyok, M.; Krska, R. Co-Occurrence of Regulated, Masked and Emerging Mycotoxins and Secondary Metabolites in Finished Feed and Maize-An Extensive Survey. *Toxins* **2016**, *8*, 363. [CrossRef]
23. Nathanail, A.V.; Syvähuoko, J.; Malachová, A.; Jestoi, M.; Varga, E.; Michlmayr, H.; Adam, G.; Sieviläinen, E.; Berthiller, F.; Peltonen, K. Simultaneous determination of major type A and B trichothecenes, zearalenone and certain modified metabolites in Finnish cereal grains with a novel liquid chromatography-tandem mass spectrometric method. *Anal. Bioanal. Chem.* **2015**, *407*, 4745–4755. [CrossRef]
24. Bryła, M.; Waśkiewicz, A.; Podolska, G.; Szymczyk, K.; Jędrzejczak, R.; Damaziak, K.; Sułek, A. Occurrence of 26 mycotoxins in the grain of cereals cultivated in Poland. *Toxins* **2016**, *8*, E160. [CrossRef]
25. Njobeh, P.B.; Dutton, M.F.; Koch, S.H.; Chuturgoon, A.A.; Stoev, S.D.; Mosonik, J.S. Simultaneous occurrence of mycotoxins in human food commodities from Cameroon. *Mycotoxin Res.* **2010**, *26*, 47–57. [CrossRef] [PubMed]
26. Wu, Q.; Kuča, K.; Humpf, H.U.; Klímová, B.; Cramer, B. Fate of deoxynivalenol and deoxynivalenol-3-glucoside during cereal-based thermal food processing: A review study. *Mycotoxin Res.* **2017**, *33*, 79–91. [CrossRef] [PubMed]
27. Egbuta, M.A.; Chilaka, C.A.; Phoku, J.Z.; Mwanza, M.; Dutton, M.F. Co-contamination of Nigerian cocoa and cocoa-based powder beverages destined for human consumption by mycotoxins. *Stud. Ethno Med.* **2013**, *7*, 187–194. [CrossRef]
28. Mashinini, K.; Dutton, M. The Incidence of Fungi and Mycotoxins in South Africa Wheat and Wheat-Based Products. *J. Environ. Sci. Heal Part B Pestic.* **2006**, *41*, 285–296. [CrossRef]
29. Nkwe, D.O.; Taylor, J.E.; Siame, B.A. Fungi, Aflatoxins, Fumonisin Bl and Zearalenone Contaminating Sorghum-based Traditional Malt, Wort and Beer in Botswana. *Mycopathologia* **2005**, *160*, 177–186. [CrossRef]
30. Odhav, B.; Naicker, V. Mycotoxins in South African traditionally brewed beers. *Food Addit. Contam.* **2002**, *19*, 55–61. [CrossRef]
31. Berthiller, F.; Dall'Asta, C.; Schuhmacher, R.; Lemmens, M.; Adam, G.; Krska, A.R. Masked mycotoxins: Determination of a deoxynivalenol glucoside in artificially and naturally contaminated wheat by liquid chromatography-tandem mass spectrometry. *J. Agric. Food Chem.* **2005**, *53*, 3421–3425. [CrossRef]
32. Rasmussen, P.H.; Nielsen, K.F.; Ghorbani, F.; Spliid, N.H.; Nielsen, G.C.; Jørgensen, L.N. Occurrence of different trichothecenes and deoxynivalenol- 3-β-D-glucoside in naturally and artificially contaminated Danish cereal grains and whole maize plants. *Mycotoxin Res.* **2012**, *28*, 181–190. [CrossRef]
33. Zhao, Z.; Rao, Q.; Song, S.; Liu, N.; Han, Z.; Hou, J.; Wu, A. Simultaneous determination of major type B trichothecenes and deoxynivalenol-3-glucoside in animal feed and raw materials using improved DSPE combined with LC-MS/MS. *J. Chromatogr. B* **2014**, *963*, 75–82. [CrossRef]
34. Adejumo, T.O.; Hettwer, U.; Karlovsky, P. Occurrence of Fusarium species and trichothecenes in Nigerian maize. *Int. J. Food Microbiol.* **2007**, *116*, 350–357. [CrossRef]
35. Trombete, F.; Barros, A.; Vieira, M.; Saldanha, T.; Venâncio, A.; Fraga, M. Simultaneous Determination of Deoxynivalenol, Deoxynivalenol-3-Glucoside and Nivalenol in Wheat Grains by HPLC-PDA with Immunoaffinity Column Cleanup. *Food Anal. Methods* **2016**, *9*, 2579–2586. [CrossRef]
36. Jin, Z.; Zhou, B.; Gillespie, J.; Gross, T.; Barr, J.; Simsek, S.; Bruggeman, R.; Schwartz, P. Production of deoxynivalenol (DON) and DON-3-glucoside during the malting of Fusarium infected hard red spring wheat. *Food Control* **2018**, *85*, 6–10. [CrossRef]
37. Ayalew, A.; Fehrmann, H.; Lepschy, J.; Beck, R.; Abate, D. Natural occurrence of mycotoxins in staple cereals from Ethiopia. *Mycopathologia* **2006**, *162*, 57–63. [CrossRef] [PubMed]
38. Muthomi, J.W.; Ndung'u, J.K.; Gathumbi, J.; Mutitu, E.; Wagacha, M. The occurrence of Fusarium species and mycotoxins in Kenyan wheat. *Crop Prot.* **2008**, *27*, 1215–1219. [CrossRef]
39. Makun, H.A.; Dutton, M.F.; Njobeh, P.B.; Mwanza, M.; Kabiru, A.Y. Natural multi-occurrence of mycotoxins in rice from Niger State, Nigeria. *Mycotoxin Res.* **2011**, *27*, 97–104. [CrossRef]

40. Malachova, A.; Dzuman, Z.; Veprikova, Z.; Vaclavikova, M.; Zachariasova, M.; Hajslova, J. Deoxynivalenol, Deoxynivalenol-3-glucoside, and Enniatins: The Major Mycotoxins Found in Cereal-Based Products on the Czech Market. *J. Agric. Food Chem.* **2011**, *59*, 12990–12997. [CrossRef]
41. Li, F.; Jiang, D.; Zhou, J.; Chen, J.; Li, W.; Zheng, F. Mycotoxins in wheat flour and intake assessment in Shandong province of China. *Food Addit. Contam. Part B* **2016**, *9*, 170–175. [CrossRef]
42. Njobeh, P.B.; Dutton, M.F.; Åberg, A.T.; Haggblom, P. Estimation of multi-mycotoxin contamination in South African compound feeds. *Toxins* **2012**, *4*, 836–848. [CrossRef]
43. Mbugua, S.K.; Gathumbi, J.K. The contamination of Kenyan lager beers with Fusarium mycotoxins. *J. Inst. Brew.* **2004**, *110*, 227–229. [CrossRef]
44. Zachariasova, M.; Hajslova, J.; Kostelanska, M.; Poustka, J.; Krplova, A.; Cuhra, P.; Hochel, I. Deoxynivalenol and its conjugates in beer: A critical assessment of data obtained by enzyme-linked immunosorbent assay and liquid chromatography coupled to tandem mass spectrometry. *Anal. Chim. Acta* **2008**, *625*, 77–86. [CrossRef]
45. Bryła, M.; Ksieniewicz-Woźniak, E.; Waśkiewicz, A.; Szymczyk, K.; Jędrzejczak, R. Natural occurrence of nivalenol, deoxynivalenol, and deoxynivalenol-3-glucoside in polish winter wheat. *Toxins* **2018**, *10*, 81. [CrossRef] [PubMed]
46. Palacios, S.A.; Erazo, J.G.; Ciasca, B.; Lattanzio, V.M.; Reynoso, M.M.; Farnochi, M.C.; Torres, A.M. Occurrence of deoxynivalenol and deoxynivalenol-3-glucoside in durum wheat from Argentina. *Food Chem.* **2017**, *230*, 728–734. [CrossRef] [PubMed]
47. Serrano, A.B.; Font, G.; Ruiz, M.J.; Ferrer, E. Co-occurrence and risk assessment of mycotoxins in food and diet from Mediterranean area. *Food Chem.* **2012**, *135*, 423–429. [CrossRef] [PubMed]
48. Veprikova, Z.; Vaclavikova, M.; Lacina, O.; Dzuman, Z.; Zachariasova, M.; Hajslova, J. Occurrence of mono- and di-glycosylated conjugates of T-2 and HT-2 toxins in naturally contaminated cereals. *World Mycotoxin J.* **2012**, *5*, 231–240. [CrossRef]
49. Lattanzio, V.M.T.; Ciasca, B.; Terzi, V.; Ghizzoni, R.; McCormick, S.P.; Pascale, M. Study of the natural occurrence of T-2 and HT-2 toxins and their glucosyl derivatives from field barley to malt by high-resolution Orbitrap mass spectrometry. *Food Addit. Contam. Part A* **2015**, *32*, 1647–1655. [CrossRef]
50. Lattanzio, V.; Visconti, A.; Haidukowski, M.; Pascale, M. Identification and characterization of new Fusarium masked mycotoxins, T2 and HT2 glycosyl derivatives, in naturally contaminated wheat and oats by liquid chromatography-high-resolution mass spectrometry. *J. Mass Spectrom.* **2012**, *47*, 466–475. [CrossRef]
51. Marasas, W.F.O. Discovery and Occurrence of the Fumonisins: A Historical Perspective. *Environ. Health Perspect.* **2001**, *109*, 239–243.
52. Fandohan, P.; Hell, K.; Marasas, W.F.O.; Wingfield, M.J. Infection of maize by Fusarium species and contamination with fumonisin in Africa. *Afr. J. Biotechnol.* **2003**, *2*, 646–665.
53. Ekwomadu, T.I.; Gopane, R.E.; Mwanza, M. Occurrence of filamentous fungi in maize destined for human consumption in South Africa. *Food Sci. Nutr.* **2018**, *6*, 884–890. [CrossRef]
54. Bhat, R.; Rai, R.V.; Karim, A.A. Mycotoxins in Food and Feed: Present Status and Future Concerns. *Compr. Rev. Food Sci. Food Saf.* **2010**, *9*, 57–81. [CrossRef]
55. Peraica, M.; Radic, B.; Lucic, A.; Pavlovic, M. Toxic effects of mycotoxins in humans. *Bull. World Health Org.* **1999**, *77*, 754–763. [PubMed]
56. Bennett, J.W.; Klich, M. Mycotoxins. *Clin. Microbiol. Rev.* **2003**, *16*, 497–516. [CrossRef] [PubMed]
57. Bhat, R.V.; Vasanthi, S. Mycotoxin food safety risk in developing countries. Food safety in food security and food trade. Vision 2020 for Food. *Agric. Environ. Focus* **2003**, *10*, 1–2.
58. Marasas, W.F.O.; Riley, R.T.; Hendricks, K.A.; Stevens, V.L.; Sadler, T.W.; Gelineau-van Waes, J.; Missmer, S.A.; Cabrera, J.; Torres, O.; Gelderblom, W.C.A.; et al. Fumonisins Disrupt Sphingolipid Metabolism, Folate Transport, and Neural Tube Development in Embryo Culture and In Vivo: A Potential Risk Factor for Human Neural Tube Defects among Populations Consuming Fumonisin-Contaminated Maize. *J. Nutr.* **2004**, *134*, 711–716. [CrossRef]
59. da Rocha, M.E.B.; Freire, F.D.C.O.; Maia, F.E.F.; Guedes, M.I.F.; Rondina, D. Mycotoxins and their effects on human and animal health. *Food Control* **2014**, *36*, 159–165. [CrossRef]
60. Perkowski, J.; Plattner, R.D.; Goliński, P.; Vesonder, R.F.; Chelkowski, J. Natural occurrence of deoxynivalenol, 3-acetyl-deoxynivalenol, 15-acetyl-deoxynivalenol, nivalenol, 4, 7-dideoxynivalenol, and zearalenone in Polish wheat. *Mycotoxin Res.* **1990**, *6*, 7–12. [CrossRef]

61. Shephard, G.S.; van der Westhuizen, L.; Katerere, D.R.; Herbst, M.; Pineiro, M. Preliminary exposure assessment of deoxynivalenol and patulin in South Africa. *Mycotoxin Res.* **2010**, *26*, 181–185. [CrossRef]
62. Warth, B.; Parich, A.; Atehnkeng, J.; Bandyopadhyay, R.; Schuhmacher, R.; Sulyok, M.; Krska, R. Quantitation of mycotoxins in food and feed from Burkina Faso and Mozambique using a modern LC-MS/MS multitoxin method. *J. Agric. Food Chem.* **2012**, *60*, 9352–9363. [CrossRef]
63. Adetunji, M.; Atanda, O.; Ezekiel, C.N.; Sulyok, M.; Warth, B.; Beltrán, E.; Krska, R.; Obadina, O.; Bakare, A.; Chilaka, C.A. Fungal and bacterial metabolites of stored maize (*Zea mays*, L.) from five agro-ecological zones of Nigeria. *Mycotoxin Res.* **2014**, *30*, 89–102. [CrossRef]
64. Chilaka, C.A.; De Boevre, M.; Atanda, O.O.; De Saeger, S. Occurrence of Fusarium mycotoxins in cereal crops and processed products (Ogi) from Nigeria. *Toxins* **2016**, *8*, 342. [CrossRef]
65. Njumbe Ediage, E.; Diana Di Mavungu, J.; Song, S.; Sioen, I.; De Saeger, S. Multimycotoxin analysis in urines to assess infant exposure: A case study in Cameroon. *Environ. Int.* **2013**, *58*, 50–59. [CrossRef] [PubMed]
66. Placinta, C.M.; D'mello, J.P.F.; Macdonald, A.M.C. A review of worldwide contamination of cereal grains and animal feed with Fusarium mycotoxins. *Anim. Feed Sci. Technol.* **1999**, *78*, 21–37. [CrossRef]
67. European Commission. Commission Regulation (EC) No 401/2006 laying down the methods of sampling and analysis for the official control of the levels of mycotoxins in foodstuffs. *Off. J. Eur. Union* **2006**, *70*, 12–34.
68. Boutigny, A.L.; Beukes, I.; Small, I.; Zühlke, S.; Spiteller, M.; Van Rensburg, B.J.; Flett, B.; Viljoen, A. Quantitative detection of *Fusarium* pathogens and their mycotoxins in South African maize. *Plant Pathol.* **2012**, *61*, 522–531. [CrossRef]
69. Meyer, H.; Skhosana, Z.D.; Motlanthe, M.; Louw, W.; Rohwer, E. Long term monitoring (2014–2018) of multi-mycotoxins in South African commercial Maize and wheat with a locally developed and validated LC-MS/MS method. *Toxins* **2019**, *11*, E271. [CrossRef]
70. Abia, W.A.; Warth, B.; Sulyok, M.; Krska, R.; Tchana, A.N.; Njobeh, P.B.; Dutton, M.F.; Moundipa, P.F. Determination of multi-mycotoxin occurrence in cereals, nuts and their products in Cameroon by liquid chromatography tandem mass spectrometry (LC-MS/MS). *Food Control* **2013**, *31*, 438–453. [CrossRef]
71. Changwa, R.; Abia, W.; Msagati, T.; Nyoni, H.; Ndleve, K.; Njobeh, P. Multi-mycotoxin occurrence in dairy cattle feeds from the gauteng province of South Africa: A pilot study using UHPLC-QTOF-MS/MS. *Toxins* **2018**, *10*, 294. [CrossRef]
72. European Union (2013/165/EU CR). Recomendations on the presence of T-2 and HT-2 toxin in cereals and cereal products. *Off. J. Eur. Union.* **2013**, *56*, 12–15.
73. Van Der Fels-Klerx, H.J.; Klemsdal, S.; Hietaniemi, V.; Lindblad, M.; Ioannou-Kakouri, E.; Van Asselt, E.D. Mycotoxin contamination of cereal grain commodities in relation to climate in North West Europe. *Food Addit. Contam. Part A* **2012**, *29*, 1581–1592. [CrossRef]
74. Nordkvist, E.; Häggblom, P. Fusarium mycotoxin contamination of cereals and bedding straw at Swedish pig farms. *Anim. Feed Sci. Technol.* **2014**, *198*, 231–237. [CrossRef]
75. Kamala, A.; Ortiz, J.; Kimanya, M.; Haesaert, G.; Donoso, S.; Tiisekwa, B.; De Meulenaer, B. Multiple mycotoxin co-occurence in maize grown in three agro-ecological zones of Tanzania. *Food Control* **2015**, *54*, 208–215. [CrossRef]
76. South African Grain Laboratory. South African Maize Crop Quality Report 2012/2013. Available online: https://sagl.co.za/maize/maize-reports/ (accessed on 1 July 2019).
77. Kosiak, B.; Torp, M.; Skjerve, E.; Thrane, U. The Prevalence and Distribution of *Fusarium* species in Norwegian Cereals: A Survey. *Acta Agric. Scand. Sect. B Soil Plant Sci.* **2003**, *53*, 168–176.
78. Jestoi, M.; Rokka, M.; Yli-Mattila, T.; Parikka, P.; Rizzo, A.; Peltonen, K. Presence and concentrations of the Fusarium-related mycotoxins beauvericin, enniatins and moniliformin in Finnish grain samples. *Food Addit. Contam.* **2004**, *21*, 794–802. [CrossRef]
79. Çelik, M.; Aksoy, H.; Yilmaz, S. Evaluation of beauvericin genotoxicity with the chromosomal aberrations, sister-chromatid exchanges and micronucleus assays. *Ecotoxicol. Environ. Saf.* **2010**, *73*, 1553–1557. [CrossRef] [PubMed]
80. Juan, C.; Ritieni, A.; Mañes, J. Occurrence of Fusarium mycotoxins in Italian cereal and cereal products from organic farming. *Food Chem.* **2013**, *141*, 1747–1755. [CrossRef] [PubMed]
81. Miraglia, M.; Marvin, H.J.P.; Kleter, G.A.; Battilani, P.; Brera, C.; Coni, E.; Cubadda, F.; Croci, L.; De Santis, B.; Dekkers, S. Climate change and food safety: An emerging issue with special focus on Europe. *Food Chem. Toxicol.* **2009**, *47*, 1009–1021. [CrossRef] [PubMed]

82. Munkvold, G.P. Epidemiology of Fusarium Diseases and their Mycotoxins in Maize Ears. *Eur. J. Plant Pathol.* **2003**, *109*, 705–713. [CrossRef]
83. De La Campa, R.; Hooker, D.C.; Miller, J.D.; Schaafsma, A.W.; Hammond, B.G. Modeling effects of environment, insect damage, and Bt genotypes on fumonisin accumulation in maize in Argentina and the Philippines. *Mycopathologia* **2005**, *159*, 539–552. [CrossRef]
84. Shephard, G.S.; Thiel, P.G.; Stockenström, S.; Sydenham, E.W. Worldwide survey of fumonisin contamination of corn and corn-based products. *J. AOAC Int.* **1996**, *79*, 671–687. [CrossRef]
85. Leslie, J.F.; Summerell, B.A. (Eds.) *The Fusarium Laboratory Manual*; Blackwell Publishing: Ames, IA, USA, 2006.
86. Beukes, I.; Rose, L.J.; Shephard, G.S.; Flett, B.C.; Viljoen, A. Mycotoxigenic Fusarium species associated with grain crops in South Africa—A review. *S. Afr. J. Sci.* **2017**, *113*, 1–12.
87. Patience, J.F.; Ensley, S.M.; Hoyer, S. *Mycotoxin Contamination of Corn: What it is, What it does to Pigs and What can be done about it*; Iowa Pork Industry Center Fact Sheets Paper: Ames, IA, USA, 2010.
88. Matny, O.N. Screening of Mycotoxin Produced by *Fusarium verticillioides* and *F. proliferatum* in Culture Media. *Asian J. Agric. Rural Dev.* **2014**, *4*, 36–41.
89. De Boevre, M.; Graniczkowska, K.; De Saeger, S. Metabolism of modified mycotoxins studied through in vitro and in vivo models: An overview. *Toxicol. Lett.* **2015**, *233*, 24–28. [CrossRef] [PubMed]
90. Rheeder, J.P.; Sydenham, E.W.; Marasas, W.F.O.; Thiel, P.G.; Shephard, G.S.; Schlechter, M.; Stockenstrom, S.; Cronje, D.W.; Viljoen, J.H. Fungal infestation and mycotoxin contamination of South African commercial maize harvested in 1989 and 1990. *S. Afr. J. Sci.* **1995**, *91*, 127–131.
91. Bernardi, J.; Stagnati, L.; Lucini, L.; Rocchetti, G.; Lanubile, A.; Cortellini, C.; De Poli, G.; Busconi, M.; Marocco, A. Phenolic profile and susceptibility to *Fusarium* infection of pigmented maize cultivars. *Front. Plant Sci.* **2018**, *9*, 1–13. [CrossRef] [PubMed]
92. Njumbe Ediage, E.; Hell, K.; De Saeger, S. A comprehensive study to explore differences in mycotoxin patterns from agro-ecological regions through maize, peanut, and cassava products: A case study, Cameroon. *J. Agric. Food Chem.* **2014**, *62*, 4789–4797. [CrossRef] [PubMed]
93. Sulyok Michael Krska, R.; Schuhmacher, R. A liquid chromatography/tandem mass spectrometric multi-mycotoxin method for the quantification of 87 analytes and its application to semi-quantitative screening of moldy food samples. *Anal. Bioanal. Chem.* **2007**, *389*, 1505–1523. [CrossRef]
94. Malachová, A.; Sulyok, M.; Beltrán, E.; Berthiller, F.; Krska, R. Optimization and validation of a quantitative liquid chromatography–tandem mass spectrometric method covering 295 bacterial and fungal metabolites including all regulated mycotoxins in four model food matrices. *J. Chromatogr. A* **2014**, *1362*, 145–156. [CrossRef]
95. International VSN. *GenStat for Windows*; VSN International: Hemel Hempstead, UK, 2011.

toxins

MDPI

Article

Multi-Mycotoxin Occurrence and Exposure Assessment Approach in Foodstuffs from Algeria

Choukri Khelifa Mahdjoubi [1,2], Natalia Arroyo-Manzanares [3], Nisserine Hamini-Kadar [2], Ana M. García-Campaña [1], Kihel Mebrouk [2] and Laura Gámiz-Gracia [1,*]

1 Department of Analytical Chemistry, Faculty of Sciences, University of Granada, 18071 Granada, Spain; choukri_khelifa@hotmail.com (C.K.M.); amgarcia@ugr.es (A.M.G.-C.)

2 Department of Biology, Faculty of Natural and Life Science, University of Oran 1, 31100 Oran, Algeria; hamini.kadar@yahoo.fr (N.H.-K.); kihalm@gmail.com (K.M.)

3 Department of Analytical Chemistry, Faculty of Chemistry, University of Murcia, 30003 Murcia, Spain; natalia.arroyo@um.es

* Correspondence: lgamiz@ugr.es; Tel.: +34-9-5824-8594

Received: 9 February 2020; Accepted: 17 March 2020; Published: 19 March 2020

check for updates

Abstract: A survey on 120 cereal samples (barley, maize, rice and wheat) from Algerian markets has been carried out to evaluate the presence of 15 mycotoxins (ochratoxin A, deoxynivalenol, fumonisin B1 and B2, T-2 and HT-2 toxins, zearalenone, fusarenon X, citrinin, sterigmatocystin, enniatins A, A1, B and B1, and beauvericin). With this purpose, a QuEChERS-based extraction and ultra-high performance liquid chromatography coupled to tandem mass spectrometry (UHPLC-MS/MS) were used. Analytical results showed that 78 cereal samples (65%) were contaminated with at least one toxin, while 50% were contaminated with three to nine mycotoxins. T-2 toxin, citrinin, beauvericin and deoxynivalenol were the most commonly found mycotoxins (frequency of 50%, 41.6%, 40.8% and 33.3%, respectively). Fumonisins (B1 + B2), enniatins B and B1, deoxynivalenol and zearalenone registered high concentrations (289–48878 µg/kg, 1.2–5288 µg/kg, 15–4569 µg/kg, 48–2055 µg/kg and 10.4–579 µg/kg, respectively). Furthermore, concentrations higher than those allowed by the European Union (EU) were observed in 21, 8 and 1 samples for fumonisins, zearalenone and deoxinivalenol, respectively. As a conclusion, the high levels of fumonisins (B1 + B2) in maize and deoxynivalenol, zearalenone and HT-2 + T-2 toxins in wheat, represent a health risk for the average adult consumer in Algeria. These results pointed out the necessity of a consistent control and the definition of maximum allowed levels for mycotoxins in Algerian foodstuffs.

Keywords: Mycotoxins; cereals; UHPLC-MS/MS; exposure assessment; Algeria

Key Contribution: The study describes the incidence of 15 mycotoxins in 120 Algerian cereal samples (barley, maize, rice and wheat grains) by UHPLC-MS/MS. The results have pointed out the high co-occurrence of mycotoxins, as well as the high concentration (above the maximum allowed concentration) of some mycotoxins legislated in the EU in those cereals, posing a risk for consumers. These results highlight the necessity of establishing maximum levels for mycotoxins in Algeria.

1. Introduction

Mycotoxins are secondary metabolites of low molecular weight produced by a variety of fungi (mainly *Aspergillus, Fusarium* and *Penicillium* spp.). They present diverse chemical structures, having different biological effects on animals and humans, such as teratogenicity, carcinogenicity, mutagenicity, immunotoxicity or neurotoxicity [1,2]. These toxic compounds are often naturally occurring in the food chain and, therefore, result in human exposure either by direct consumption of contaminated crops, plants, seeds and fruits, or indirectly through ingestion of food derived from exposed animals (meat, eggs or milk) [3–5].

More than 400 mycotoxins have been reported so far, being aflatoxins (AFs), trichothecenes, fumonisins, zearalenone (ZEN) and ochratoxin A (OTA) the most representatives. However, other mycotoxins are becoming a global issue of high concern due to their extensive incidence in food and feed, including emerging *Fusarium* toxins, such as beauvericin (BEA) and enniatins (ENNs) [6–8].

Humans and animals are more frequently exposed to multiple mycotoxins than to a single one [9–11], producing an increasing concern about the health hazard. The combined effects of mycotoxins have been reported in several studies [12–15]. However, the actual health risk assessment is limited to their single effects, since there is not enough information available about the nature of the observed effects or the relative potencies of each mycotoxin in the mixture [16].

Due to the toxic effects of these compounds, the EU [17,18] and many countries have set maximum levels or recommendations in foodstuffs intended to human consumption (see Table 1). Moreover, for the most well-documented toxins, tolerable daily intakes (TDI) or provisional maximum tolerable daily intakes (PMTDI) have been established by the Food and Agriculture Organization (FAO)/World Health Organization (WHO) Joint Expert Committee on Food Additives (JECFA) [19,20].

Table 1. Maximum tolerable levels set for some mycotoxins in cereals.

Mycotoxin	Commodity	Maximum Level (μg/kg)
	European Union [17,18]	
Aflatoxin B1	Cereals and cereal products	2
Aflatoxins [a]	Cereals and cereal products	4
	Maize	10
Ochratoxin A	Unprocessed cereals	5
Zearalenone	Unprocessed maize	350
	Unprocessed cereals other than maize	100
Fumonisins B1 + B2	Unprocessed maize	4000
Deoxynivalenol	Unprocessed cereals other than durum wheat, oats, maize	1250
	Unprocessed durum wheat, oats and maize	1750
T-2 + HT-2	Unprocessed barley and maize	200
	Unprocessed oats	1000
	Unprocessed wheat, rye and other cereals	100
	United States of Americ [21,22]	
Aflatoxin B1	All food crops	15
Fumonisins B1 + B2 + B3	Unprocessed maize	4000
	Canada [23]	
Deoxynivalenol	Unprocessed wheat	2000
	Japan [24,25]	
Aflatoxin B1	All food crops	10
Deoxynivalenol	Unprocessed wheat	1100 [b]
	China [26]	
Aflatoxin B1	Unprocessed maize	20
	Rice (brown rice)	10
	Unprocessed wheat barley, other cereals	5
Deoxynivalenol	Maize, barley, wheat, other cereals	1000
Ochratoxin A	Unprocessed cereals	5
Zearalenone	Unprocessed wheat and maize	60

Table 1. *Cont.*

Mycotoxin	Commodity	Maximum Level (μg/kg)
	Morocco [27]	
Aflatoxin B1	Cereals and cereal products	2
Aflatoxins [a]	Cereals and cereal products	4
Ochratoxin A	Unprocessed cereals	5
Deoxynivalenol	Unprocessed durum wheat, oats and maize	1750
Zearalenone	Unprocessed maize	200

[a] AFs: Sum of AFB1 + AFB2 + AFG1 + AFG2; [b] Provisional maximum level.

Studies show that mycotoxins are ubiquitous contaminants [28,29]. It is estimated that they are present in approximately 25% of cereals consumed worldwide and a recent study suggests that this percentage could be even higher [30]. In general, mycotoxin contamination is higher when climatic conditions are favourable. Algeria is a North African country whose climate is characterised by high temperatures and high relative humidity in some areas that seems to stimulate the toxigenic moulds growth and toxinogenesis, in which cereals and cereal-based products represent a staple food for the population [31]. Moreover, no applicable norms concerning mycotoxin contamination of cereal (local or imported) have been set in Algeria. Indeed, a large amount of cereal commercialized in Algeria is imported and little is known about toxin contamination. Therefore, in order to the health risks associated with mycotoxin exposure different preventive approaches have been used. These include enforcement of legislation, good agricultural practices and the monitoring of mycotoxin contamination. But for this, it is essential to establish evidence and data on the presence of mycotoxins and exposure levels. Accurate exposure data on mycotoxins is an important input in risk assessment and management efforts as well as in the establishment of appropriated legislation for the monitoring and control of mycotoxin exposure in food [32].

In order to quantify the concentration of these hazards in different commodities, reliable and accurate analytical methods that allow their unambiguous identification and accurate quantification at low concentration are needed. In this sense, liquid chromatography (LC) or ultra-high performance LC (UHPLC) coupled to tandem mass spectrometry (MS/MS) have become the techniques of choice for the determination of multiple mycotoxins in food and feed [33–35]. In addition, alternative sample treatment methods, such as QuEChERS (acronym of Quick, Easy, Cheap, Effective, Rugged and Safe), are being increasingly applied to the analysis of mycotoxins, due to their feasibility, flexibility, versatility, low cost and rapidity [36].

Within this context, the main objective of this work is the validation of an analytical method based on a simple QuEChERS-based extraction and UHPLC–MS/MS for the determination of 15 mycotoxins, namely: OTA, deoxynivalenol (DON), fumonisin B1 (FB1) and B2 (FB2), T-2 and HT-2 toxins, ZEN, fusarenon X (F-X), citrinin (CIT), sterigmatocystin (STE), enniatin A (ENNA), A1 (ENNA1), B (ENNB) and B1 (ENNB1) and BEA. Secondly, the analysis of cereals samples (barley, maize, rice and wheat) collected from different Algerian markets will help to estimate the potential contribution to the dietary exposure of Algerian consumers.

2. Results

2.1. Method Validation

The selected analytical method was validated for each cereal in terms of linearity, matrix effect, recovery, precision, limits of detection (LODs) and limits of quantification (LOQs). The results are summarised in the supplementary material (Table S1).

Method linearity was assessed by spiking blank samples at five concentration levels (processed in duplicate). All calibration curves showed a good linearity, with coefficients (R^2) higher than 0.98 in all the cases. LODs and LOQs were determined as the concentration of analyte giving a signal to noise ratio (S/N) equal to 3 and 10, respectively. In all cases, LOQs were lower than maximum permitted or recommended concentrations established by the EU for those mycotoxins in cereals (FBs, DON, OTA, ZEN, T-2 and HT-2 toxin) [17,18].

Matrix effect was evaluated at the following concentration levels: OTA and STE: 25 μg/kg; CIT: 100 μg/kg; FB1, FB2, T-2, HT-2 and ZEN: 250 μg/kg; ENNB, ENNB1, ENNA, ENNA1 and BEA: 400 μg/kg; DON: 1000 μg/kg and F-X: 2500 μg/kg. Matrix effect was calculated as follow:

ME = 100 × (signal of spiked extract − signal of standard solution)/signal of standard solution

A strong ion suppression was observed for all the selected compounds in the four matrices under study, ranging from −77.8% to −18.2 % for barley, −76.5% to −15.6% for maize, −80.4% to −18.2% for rice and −76.9% to −14.3% for wheat (see supplementary material, Figure S1). As a consequence, calibration curves in matrix were used.

The efficiency of the extraction process was evaluated by recovery studies, spiking blank samples at the same levels used in the matrix effect study. Each sample was processed in triplicate and injected three times. The ratio of peak areas of the samples spiked before extraction and the extracts spiked after extraction was used to calculate the recovery. The average recovery values were: 87.5% for barley, 84.8 % for maize, 88.5 % for rice and 86.2 % for wheat.

Intra-day (repeatability) and inter-day precision (intermediate precision) were evaluated and expressed as relative standard deviation (%RSD). Spiked blank samples (processed and injected three times) at the same concentration levels mentioned above for the matrix effect and recovery studies were used. For the intra-day precision study, samples were analysed on the same day, while the inter-day precision was estimated through samples analysed on three consecutive days. The relative standard deviation (%RSD) for intra-day and inter-day precision were lower than 14% and 23%, respectively, for all the mycotoxins and matrix combination (see supplementary material Table S2). These values were all in a permitted range by European Commission [37].

2.2. *Mycotoxins Occurrence Data*

A total of 120 samples comprising barley (n = 30), maize (n = 30), rice (n = 30) and wheat (n = 30) were evaluated for the occurrence of mycotoxins (OTA, DON, FB1, FB2, T-2, HT-2, ZEN, F-X, CIT, STE, ENNA, ENNA1, ENNB, ENNB1 and BEA). Table 2 presents the occurrence, concentration range and mean concentration of each mycotoxin in positive cereal samples that is, considering only mycotoxins with a concentration above the LOQ.

Moreover, 78 out of 120 samples (65%) evidenced at least one mycotoxin above the LOQ, and 13 out of 15 mycotoxins included in the study were found in some of the analysed samples, being the exceptions OTA and STE (not detected). Overall, T-2, CIT, BEA and DON were the most commonly found mycotoxins with a global incidence of 50%, 40.8%, 38.3% and 33.3%, respectively. However, the maximum concentration value was found for FB1 in maize (42,143 μg/kg).

The results obtained for each mycotoxin (or group of mycotoxins) are commented below.

2.2.1. Occurrence of Trichothecenes

Regarding the distribution of the studied trichothecenes (HT-2, T-2, DON and F-X), T-2 was the most frequently found, being present in 100% samples of maize and wheat at concentrations ranging from 16.6 to 47.2 μg/kg, being the mean concentrations (considering only positive samples) 24.9 μg/kg in maize and 21.8 μg/kg in wheat. HT-2 was present only in 7 wheat samples (23%) at concentrations from 8.4 to 36.7 μg/kg (mean value 18.1 μg/kg). None of the samples exceeded the maximum recommended concentration for these toxins in non-processed cereals (ranging from 100–1000 μg/kg for the sum of T-2 + HT-2) [17].

DON was found in 27 wheat samples (90%) and 13 maize samples (43.3%), with mean concentrations of positives samples of 588 μg/kg and 632 μg/kg, respectively. The highest concentration found for DON was 2055 μg/kg, corresponding to one maize sample that exceeded the maximum permitted concentration for DON established by the EU in maize (1750 μg/kg) [18]. As an example, a chromatogram of a wheat sample contaminated with DON is shown in supplementary material (Figure S2A).

Table 2. Summary of mycotoxins found in the analysed cereal samples.

	Barley (n = 30)				Maize (n = 30)				Rice (n = 30)				Wheat (n = 30)			
	I (%) [a]	Mean [b]	LOD-LOQ [c]	Range [d]	I (%) [a]	Mean [b]	LOD-LOQ [c]	Range [d]	I (%) [a]	Mean [b]	LOD-LOQ [c]	Range [d]	I (%) [a]	Mean [b]	LOD-LOQ [c]	Range [d]
Analytes		(μg/kg)				(μg/kg)				(μg/kg)				(μg/kg)		
FB1	nd	nd	–	–	29 (96.6)	14,812		289–42,143	nd	nd	–	–	nd	nd	–	–
FB2	nd	nd	–	–	27 (90)	2789	–	27.5–8603	nd	nd	–	–	nd	nd	–	–
HT-2	nd	nd	9	–	nd	nd	–	–	nd	nd	–	–	7 (23)	18.1	3	8.4–36.7
T-2	nd	nd	–	–	30 (100)	24.9	–	24.6–25.7	nd	nd	–	–	30 (100)	21.8	–	16.6–47.2
DON	nd	nd	–	–	13 (43)	632	–	47.6–2055	nd	nd	–	–	27 (90)	588	1	68.3–1363
ZEN	nd	nd	2	–	7 (23.3)	109	–	20.4–579	6 (20)	9.9	17	8.6–15.5	19 (63.3)	102	9	9.6–295
F-X	3 (10)	190	2	142–284	24 (80)	281	1	177–477	nd	nd	–	–	3 (10)	152	7	139–159
OTA	nd	nd	–	–	nd	nd	–	–	nd	nd	–	–	nd	nd	–	–
CIT	9 (30)	26.2	2	10.9–52.0	25 (83.3)	32.7	–	8.6–273	nd	nd	23	–	15 (50)	16.8	14	9.8–32.3
STE	nd	nd	–	–	nd	nd	–	–	nd	nd	–	–	nd	nd	–	–
BEA	nd	nd	–	–	25 (83.3)	3.8	4	0.85–31.4	nd	nd	–	–	21 (70)	155.4	–	2.8–486
ENNA	nd	nd	–	–	nd	nd	–	–	nd	nd	–	–	7 (23.3)	28.3	–	8.4–87.6
ENNA1	nd	nd	–	–	3 (10)	56.4	–	11.5–103	nd	nd	–	–	14 (46.7)	107	3	4.0–395
ENNB	nd	nd	–	–	nd	nd	–	–	nd	nd	–	–	18 (60)	1668	–	1.2–5288
ENNB1	nd	nd	–	–	2 (6)	60.9	–	15.0–107	nd	nd	–	–	21 (70)	469	–	19.5–4569
Total	**12 (40)**				**30 (100)**				**6 (20)**				**30 (100)**			

[a] Incidence of samples ≥ LOQ (% of samples ≥ LOQ), [b] Mean value for samples ≥ LOQ, [c] Number of samples ≥ LOD and ≤ LOQ, [d] minimum value—maximum value, n: Number of samples; nd: Not detected.

F-X was determined in 3 samples of barley (10%), 24 of maize (80%) and 3 of wheat (10%). The maximum content of F-X was found in a maize sample (477 μg/kg).

In addition, the incidence of T-2 toxin in positive samples was higher than the incidence of DON in the same samples. Similar to our finding, Bouafifssa et al. [38] reported higher levels of T-2 and HT-2 compared to DON in Moroccan pasta, with contamination levels from 4 to 419 μg/kg and 4 to 50 μg/kg for T-2 and HT-2, respectively. The higher incidence of DON in maize and wheat compared to other cereals was also reported by Pleadin et al. [39] in 181 cereal samples from Croatia, where DON was found in 71%, 65%, 53% and 21% samples of maize, wheat, barley and oat, with mean concentrations of 1565, 223, 342 and 145 μg/kg, respectively. However, lower incidence and concentrations of DON, HT-2, T-2 and F-X were found in cereal samples from Italy, where mean levels in positive samples were 20.1, 4.8, 0.3 and 36.7 μg/kg, for DON, HT-2, T-2 and F-X respectively [40].

2.2.2. Occurrence of Zearalenone

ZEN was found in 7 samples of maize (23.3%), 6 of rice (20%) and 19 of wheat (63.3%), being the mean concentrations of positive samples 109 μg/kg in maize, 9.9 μg/kg in rice and 102 μg/kg in wheat. One sample of maize (579 μg/kg) and 7 samples of wheat (with concentrations up to 295 μg/kg) exceeded the maximum permitted levels established in the EU for ZEN (350 μg/kg for maize and 100 μg/kg for other unprocessed cereals) [18].

The incidence of ZEN in our study was similar to those reported in other studies [41,42], but the contamination levels were lower than other values from literature. For instance, ZEN was found at concentrations up to 1399 μg/kg in cereals from Nigeria [43] and up to 15,700 μg/kg in maize samples from Belgium, where mean level of ZEN in the analysed samples was 2180 μg/kg [44]. However, other studies reported lower ZEN concentrations with an average of 12 and 14 μg/kg in Italian and Moroccan cereals, respectively [40,45].

2.2.3. Occurrence of Fumonisins

FB1 and FB2 were found only in maize samples, with high concentrations and incidence rate. Thus, FB1 was present in 29 samples (96.6%) at concentrations from 289 up to 42,143 μg/kg, whereas FB2 was quantified in 27 samples (90%) at concentrations from 27.5 to 8603 μg/kg. The mean concentrations for positives samples were 14,812 and 2789 μg/kg for FB1 and FB2, respectively. Considering the sum of fumonisins (FB1 + FB2), it was in the range of 289–48,878 μg/kg. A total of 21 samples (70%) showed concentrations of (FB1 + FB2) above the maximum allowed level established by the EU (4000 μg/kg for unprocessed maize) [18] suggesting the high exposure of the population to these toxins.

According to the analyses, fumonisins were not detected in barley, rice and wheat samples. Such a trend was also observed by Ghali et al. in cereals from Tunisia, reporting the highest levels for fumonisins in maize samples at an incidence rate of 52% [46]. This confirms that the risk of fumonisin contamination of wheat, barley and rice is rather low due to the known tendency of the *Fusarium* spp. producing fumonisins (*F. verticillioides* and *F. proliferatum*) to infect maize [47]. This result is in agreement with other previous studies: thus, a high fumonisin incidence was reported in Nigerian maize-based products with concentrations ranging from 74 to 22064 μg/kg [43], in maize from South Africa, reporting concentrations up to 53863 μg/kg for FB1 [48], and from Ethiopia, where 77% samples of maize contaminated with fumonisins at concentrations between 25–4500 μg/kg, were attributed to *F. verticillioides* [49].

2.2.4. Occurrence of Citrinin

Remarkably, 40.8% of the total analysed samples were contaminated by CIT; it was present in 9 (30%), 25 (83.3%) and 15 (50%) samples of barley, maize and wheat, respectively, showing mean levels of 26.2 μg/kg in barley, 32.7 μg/kg in maize and 16.8 μg/kg in wheat samples. The highest concentration of CIT was found in a maize sample (273 μg/kg). CIT was not detected in rice samples.

These results may be explained by the susceptibility of the analysed cereals (barley, maize and wheat) to CIT-producing fungi (*Aspergillus* and *Penicillium* spp.) and to the influence of climatic conditions such as substrate composition, temperature and water activity (aw) to enhance CIT production, especially during storage, as it is well-known that mycotoxin production is modulated by environmental factors [50].

2.2.5. Occurrence of Emerging Mycotoxins

Only maize and wheat samples were contaminated with BEA and ENNs, while these emerging mycotoxins were not detected in rice and barley. ENNB1 was the most frequent ENN (2 samples (6.6%) of maize and in 21 samples (70%) of wheat) with concentrations from 15.0–107 µg/kg and 19.5–4569 µg/kg for maize and wheat, respectively. ENNA1 was quantified in 3 samples (10%) of maize and 14 samples (46.7) of wheat, with mean concentrations for positive samples of 56.4 µg/kg and 107 µg/kg, respectively. ENNA and ENNB were found in 7 (23.3%) and 18 samples (60%) of wheat, respectively. The contamination levels varied between 8.4 and 87.6 µg/kg for ENNA and from 1.2 to 5288 µg/kg for ENNB, with mean values for positive samples of 28.3 and 1668 µg/kg, respectively. ENNA and ENNB were not detected in maize. A chromatogram of a wheat sample contaminated with ENNB1 is shown in supplementary material (Figure S2B).

Concerning BEA, 25 samples (83.3%) of maize and 21 samples (70%) of wheat were positives, with concentrations between 0.85–31.4 µg/kg and 2.8–486 µg/kg in maize and wheat samples, respectively.

Other previous works including determination of ENNs and BEA in cereals also showed the high incidence of these mycotoxins. Thus, a high incidence of BEA (80%) was recently reported in Serbian maize with levels ranging from 8 to 129 µg/kg [7]. Moreover, a study on the occurrence of emerging mycotoxins in Spanish cereals showed high incidence of ENNs (73.4%), wherein ENNA1 was the most frequent emerging mycotoxins, with the highest concentrations (33.3–814 mg/kg) [51]. Furthermore, the high contamination of BEA and ENNs was also reported in cereals from Italy [40] and up to 800 mg/kg of ENNB1 were found in a wheat-based cereal sample from Morocco [8]. In contrast with our results, Oueslati et al. [52] did not detect BEA in maize and wheat from Tunisia.

2.2.6. Co-Occurrence of Mycotoxins in Analysed Samples

The co-occurrence of mycotoxins as well as the main combinations found in the analysed cereal samples were evaluated (see supplementary material Table S3). Among the positive samples, 50% (all maize (n = 30) and wheat (n = 30) samples) were found to be contaminated with more than one mycotoxin. The most frequent co-occurrence was the combination of 5 mycotoxins for maize and 8 mycotoxins for wheat. Moreover, different combinations were observed, depending on the cereal, the most frequent combinations being: (FB1 + FB2 + T-2 + F-X + CIT + BEA) and (FB1 + FB2 + T-2 + F-X + BEA) in maize, and (DON + T-2 + ZEN + ENNA1 + ENNB + ENNB1) in wheat samples. The highest number of mycotoxins occurring simultaneously was nine in 2 maize samples (DON + FB1 + FB2 + T-2 + ZEN + F-X + CIT + BEA + ENNA1) and 1 wheat sample (DON + HT-2 + T-2 + ZEN + CIT + BEA + ENNA1 + ENNB + ENNB1).

The co-occurrence of mycotoxins in cereals has been studied previously, especially in the Mediterranean area [38,40,52]. In agreement with our results, in wheat grains from Morocco, 51% samples were contaminated with 2–6 mycotoxins [53], whereas that at least one mycotoxin was present in the 65% cereal-derived samples from Spain [54]. Recently, a study performed in Italy showed that 81% cereal samples were contaminated with more than one mycotoxin and the most frequent co-occurrence was with DON, F-X, ENNB and ENNA1 [40]. A summary of the results obtained in this study and other previously reported occurrence studies are presented in Table 3.

Table 3. Occurrence of mycotoxins in cereals from different surveys.

	Barley (n = 30)			Maize (n = 30)			Rice (n = 30)			Wheat (n = 30)			Analytical Method	LOQ (µg/kg)	Ref.
	I (%) [a]	Range [c] (µg/kg)	Mean [b] (µg/kg)	I (%) [a]	Range [c] (µg/kg)	Mean [b] (µg/kg)	I (%) [a]	Range [c] (µg/kg)	Mean (µg/kg)	I (%) [a]	Range [c] (µg/kg)	Mean [b] (µg/kg)			
FB1	nd	–	nd	29 (96.6)	289–42,143	14812	nd	–	nd	nd	–	nd	Ultra-high performance liquid chromatography coupled to tandem mass spectrometry (UHPLC-MS/MS)	2.6–4.8	This work
FB2	nd	–	nd	27 (90)	27.5–8603	2789	nd	–	nd	nd	–	nd		2.2–10	
HT-2	nd	–	nd	nd	–	nd	nd	–	nd	7 (23)	8.4–36.7	18.1		2.8–9.9	
T-2	nd	–	nd	30 (100)	24.6–25.7	24.9	nd	–	nd	30 (100)	16.6–47.2	21.8		2.3–4.4	
DON	nd	–	nd	13 (43)	47.6–2055	632	nd	–	nd	27 (90)	68.3–1363	588		4.2–4.8	
ZEN	nd	–	nd	7 (23.3)	20.4–579	109	6 (20)	8.6–15.5	9.9	19 (63.3)	9.6–295	102		4.3–9.7	
F-X	3 (10)	142–284	190	24 (80)	177–477	281	nd	–	nd	3 (10)	139–159	152		90–174	
OTA	nd	–	nd	nd	–	nd	nd	–	nd	nd	–	nd		20–92	
CIT	9 (30)	10.9–52.0	26.2	25 (83.3)	8.6–273	32.7	nd	–	nd	15 (50)	9.8–32.3	16.8		8.4–23	
STE	nd	–	nd	nd	–	nd	nd	–	nd	nd	–	nd		0.6–1.3	
BEA	nd	–	nd	25 (83.3)	0.85–31.4	3.8	nd	–	nd	21 (70)	2.8–486	155.4		0.6–1.3	
ENNA	nd	–	nd	nd	–	nd	nd	–	nd	7 (23.3)	8.4–87.6	28.3		0.5–1.2	
ENNA1	nd	–	nd	3 (10)	11.5–103	56.4	nd	–	nd	14 (46.7)	4.0–395	107		1.4–2.7	
ENNB	nd	–	nd	nd	–	nd	nd	–	nd	18 (60)	1.2–5288	1668		1.2–3.8	
ENNB1	nd	–	nd	2 (6)	15.0–107	60.9	nd	–	nd	21 (70)	19.5–4569	469		2.6–4.4	
	Barley (n = 9)			Oat (n = 7)			Rye (n = 11)			Wheat (n = 57)			Analytical Method	LOQ (µg/kg)	Ref.
	I (%) [a]	Range [c] (µg/kg)	Mean [b] (µg/kg)	I (%) [a]	Range [c] (µg/kg)	Mean [b] (µg/kg)	I (%) [a]	Range [c] (µg/kg)	Mean (µg/kg)	I (%) [a]	Range [c] (µg/kg)	Mean [b] (µg/kg)			
DON	1 (11)	up to 35.5		4 (57)	10.3–83	29.9	5 (45.5)	16.5–79.6	23.23	16 (28)	9.6–99.6	10.96	LC–MS/MS	10	[40]
3-AcDON	nd	–	nd	1 (14.2)	5.23	5.24	nd	–	nd	nd	–	nd		15	
15-AcDON	nd	–	nd	nd	–	nd	nd	–	nd	2 (3.5)	10.8–29.13	0.64		15	
FUS-X	4 (44)	27.5–47.3	18.43	3 (42.8)	26–75	23	5 (45.5)	42.4–70.2	28.52	14 (24)	12.50–102	18.44		15	
NIV	3 (33)	21.7–106	25.15	4 (57)	45.5–50.4	27.13	2 (18)	33.9–34.4	56.9	11 (19)	12–106	8.86		15	
DAS	nd	–	nd	nd	–	nd	nd	–	nd	nd	–	nd		10	
NEO	nd	–	nd	nd	–	nd	nd	–	nd	nd	–	nd		5	
HT-2	nd	–	nd	nd	–	nd	3 (27.2)	6.98–50.3	5.34	3 (5.2)	6.78–60.10	4.44		10	
T-2	nd	–	nd	nd	–	nd	nd	–	nd	2 (3.5)	7.14–17.8	0.39		5	
ZEN	2 (22)	11.15	11.16	nd	–	nd	nd	–	nd	5 (8.7)	2.35–27.15	12.17		5	
α -ZEN	nd	–	nd	nd	–	nd	nd	–	nd	nd	–	nd		5	
β -ZEN	nd	–	nd	nd	–	nd	nd	–	nd	nd	–	nd		5	
BEA	nd	–	nd	4 (57)	7.2–41	8.8	5 (45.5)	8.9–16.5	2.72	5 (8.7)	9.6–35	12.8		15	
ENNB	nd	–	nd	3 (42.8)	5.5–97	2.8	6 (54.4)	6.7–45	5.8	16 (28)	5.5–97	20.2		15	
ENNB1	2 (22)	5.5–7.3	1.4	nd	–	nd	nd	–	nd	2 (3.5)	5.47–33.1	0.43		15	

Table 3. *Cont.*

	Barley (n = 9)			Oat (n = 7)			Rye (n = 11)			Wheat (n = 57)			Analytical Method	LOQ (µg/kg)	Ref.
	I (%) [a]	Range [c] (µg/kg)	Mean [b] (µg/kg)	I (%) [a]	Range [c] (µg/kg)	Mean [b] (µg/kg)	I (%) [a]	Range [c] (µg/kg)	Mean (µg/kg)	I (%) [a]	Range [c] (µg/kg)	Mean [b] (µg/kg)			
ENNB4	8 (88.9)	6.6–60	15.7	4 (57)	20–284.2	50.7	4 (36.3)	23.4–74	13.8	18 (31)	5.7–110.2	38.44		15	
ENNA	nd	-	nd	nd	-	nd	4 (36.3)	7.8–9.8	7.1	6 (10)	8.4–29.8	1.56		10	
ENNA1	nd	-	nd	2 (28.5)	9–45.5	8.7	nd	-	nd	11 (19)	5.3–55	0.74		15	
	Barley (n = 5)			Sorghum (n = 3)			Processed cereals (n = 13)			Wheat (n = 34)			Analytical Method	LOQ (µg/kg)	Ref.
	I (%) [a]	Range [c] (µg/kg)	Mean [b] (µg/kg)	I (%) [a]	Range [c] (µg/kg)	Mean [b] (µg/kg)	I (%) [a]	Range [c] (µg/kg)	Mean (µg/kg)	I (%) [a]	Range [c] (µg/kg)	Mean [b] (µg/kg)			
AFB1	nd	-	nd	66.6	14.4–79.9	46.7	nd	-	nd	nd	-	nd	UHPLC–MS/MS	1	[52]
AFG2	4 (80)	23.1–52.4	35.5	3 (100)	13–36.8	24.6	nd	-	nd	4 (11)	5.2–8.7	6.6		1	
HT-2	3 (60)	4.9–11	7.6	nd	-	nd	nd	-	nd	4 (11)	5.3–7.1	5.8		5	
FB1	1 (20)	up to 63.1	na	2 (66.6)	6.4–120	na	nd	-	nd	nd	-	nd		5	
FB2	nd	-	nd	1 (33.3)	61.5	61.5	1 (7.6)	6.4	6.4	1 (3)	8.7	8.7		5	
OTA	nd	-	nd	nd	-	nd	1 (7.6)	5	5	nd	-	nd		5	
	Barley (n = 4)			Maize (n = 28)			Rice (n = 1)			Wheat (n = 21)			Analytical Method	LOQ (µg/kg)	Ref.
	I (%) [a]	Range [c] (µg/kg)	Mean [b] (µg/kg)	I (%) [a]	Range [c] (µg/kg)	Mean [b] (µg/kg)	I (%) [a]	Range [c] (µg/kg)	Mean (µg/kg)	I (%) [a]	Range [c] (µg/kg)	Mean [b] (µg/kg)			
ENNA	nd	-	nd	nd	-	nd	I (%)[a]	Range[b] (µg/kg)	Mean[c] (µg/kg)	nd	-	nd	LC-DAD	6	[51]
ENNA1	na	up to 361,570	148,160	na	up to 813,010	167,700	nd	-	nd	na	up to 634,850	225370		4	
ENNB	1 (25)	up to 21,370	21370	na	up to 6310	4470	1(100)	up to 814,410	814420	nd	-	nd		4	
ENNB1	1 (25)	up to 4340	4340	na	up to 21,370	21370	1(100)	up to 7950	7950	nd	-	nd		5	
BEA	2 (50)	up to 6940	4870	6 (21.4)	up to 9310	5720	nd	-	nd	9 (42)	up to 3500	2300		5	
FUS	nd	-	nd	na	up to 2470	2470	1(100)	up to 11780	11780	na	up to 6630	3120		6	
	Maize (n =136)			Sorghum (n =110)			Millet (n = 87)			Ogi (n = 30)			Analytical Method	LOQ (µg/kg)	Ref.
	I (%) [a]	Range [c] (µg/kg)	Mean [b] (µg/kg)	I (%) [a]	Range [c] (µg/kg)	Mean [b] (µg/kg)	I (%) [a]	Range [c] (µg/kg)	Mean (µg/kg)	I (%) [a]	Range [c] (µg/kg)	Mean [b] (µg/kg)			
FB1	(65)	up to 8222	541	(8)	up to 78	64	(9)	up to 18.172	2333	(93)	up to 1903	590	LC–MS/MS	16.4–20	[43]
FB2	(54)	up to 2885	376	(2)	up to 55	48	(13)	up to 3892	609	(87)	up to 1283	472		22.6–24.2	
FB3	(43)	up to 445	117	(2)	up to 46	38	nd	-	nd	(77)	up to 37	121		28	
DON	(16)	up to 225	99	(3)	up to 119	100	(13)	up to 543	151	(13)	up to 74	61		14–24	
15-AcDON	nd	-	nd	(2)	up to 44	39	(1)	up to 11	11	nd	-	nd		8.8–14	
DON-3G	nd	-	nd	(23)	up to 63	24	nd	-	nd	(17)	up to 44	30		7.54–30.6	

Table 3. *Cont.*

	Maize (*n* =136)			Sorghum (*n* =110)			Millet (*n* = 87)			Ogi (*n* = 30)			Analytical Method	LOQ (μg/kg)	Ref.
	I (%) [a]	Range [c] (μg/kg)	Mean [b] (μg/kg)	I (%) [a]	Range [c] (μg/kg)	Mean [b] (μg/kg)	I (%) [a]	Range [c] (μg/kg)	Mean (μg/kg)	I (%) [a]	Range [c] (μg/kg)	Mean [b] (μg/kg)			
ZEN	(1)	up to 65	65	(1)	up to 38	38	(14)	up to 1399	419	(3)	up to 39	39		6.5–7.7	
ZEN-14G	(9)	up to 24	21	(3)	up to 22	19	(6)	up to 34	23	(3)	up to 31	31		9.2–10.2	
α -ZEN	(1)	up to 20	20	(3)	up to 33	33	nd	-	nd	(7)	up to 22	20		10–14	
β-ZEN	(2)	up to 21	20	(1)	up to 21	21	(1)	up to 39	39	(10)	up to 20	19		14.4–16	
HT-2	(1)	up to 20	20	(8)	up to 31	20	(5)	up to 36	36	(3)	up to 13	13		13	
NIV	(2)	up to 271	206	nd	-	nd	nd	-	nd	(7)	up to 160	148		175–162.6	
FUS-X	(1)	up to 154	154	nd	-	nd	nd	-	nd	(7)	up to 133	137		41.2–147.2	
DAS	(13)	up to 8	3	(18)	up to 16	5	(29)	up to 25	5	nd	-	nd		0.64–1	
	Barley (*n* = 34)			Maize (*n* = 63)			Oats (*n* = 33)			Wheat (*n* = 51)			Analytical Method	LOQ (μg/kg)	Ref.
	I (%) [a]	Range [c] (μg/kg)	Mean [b] (μg/kg)	I (%) [a]	Range [c] (μg/kg)	Mean [b] (μg/kg)	I (%) [a]	Range [c] (μg/kg)	Mean (μg/kg)	I (%) [a]	Range [c] (μg/kg)	Mean [b] (μg/kg)			
DON	(53)	74–228	342	(71)	215–2942	1565	(21)	34–201	145	(65)	115–278	223	ELISA-UV	20.5	[39]
ZEN	(9)	5–68	32	(78)	10–611	187	(6)	4–43	17	(69)	7–107	56		2.1	
FUS-X	(15)	25–121	44	(90)	37–4434	1756	(6)	25–31	28	(39)	28–203	66		24.5	
T-2	(32)	5–26	13	(57)	5–42	24	(18)	5–10	7	(25)	6–18	9		4.1	
	Barley (*n* = 20)			Maize (*n* = 20)			Wheat (*n* =20)						Analytical Method	LOQ (μg/kg)	Ref.
	I (%) [a]	Range [c] (μg/kg)	Mean [b] (μg/kg)	I (%) [a]	Range [c] (μg/kg)	Mean [b] (μg/kg)	I (%) [a]	Range [c] (μg/kg)	Mean (μg/kg)						
OTA	8(40)	up to 0.8	0.17	8 (40)	up to 7.2	1.08	8 (40)	up to 1.73	0.42				HPLC-FLD	0.02	[45]
ZEN	na	-	na	3 (15)	up to 17	14	na	-	na					10	
FB1	na	-	na	10 (50)	up to 5960	1930	na	-	na					60	
	Maize based products (*n* = 17)			Rice based products (*n* = 9)			Wheat based products (*n* =7)						Analytical Method	LOQ (μg/kg)	Ref.
	I (%) [a]	Range [c] (μg/kg)	Mean [b] (μg/kg)	I (%) [a]	Range [c] (μg/kg)	Mean [b] (μg/kg)	I (%) [a]	Range [c] (μg/kg)	Mean (μg/kg)						
ENNA	nd	-	nd	nd	-	nd	nd	-	nd				LC-DAD	6	[8]
ENNA1	na	423,600	113,000	na	61,400	55,100	1 (14)	up to 46,900	46,900					4	
ENNB	nd	-	nd	1 (11)	1050	1050	nd	-	nd					4	
ENNB1	1 (5.8)	20,100	20100	1 (11)	600	600	na	up to 79500	79500					5	
BEA	nd	-	nd	nd	-	nd	nd	-	nd					5	
FUS	nd	-	nd	1 (11)	3900	3900	nd	-	nd					6	

[a] Incidence of samples ≥LOQ (% of samples ≥LOQ); [b] minimum value – maximum; [c] Mean value for samples ≥LOQ; *n*: Number of samples; nd: Not detected; na: Not available; DAD: diode array detector; 3-acetyl-deoxynivalenol (3-AcDON), 15-acetyl-deoxynivalenol (15-AcDON), neosolaniol (NEO), diacetoxyscirpenol (DAS), α-zearalenol (α-ZEN), β-zearalanol (β-ZEN), fusaproliferin (FUS), zearalenone-14-glucoside (ZEN-14G), fumonisin B3 (FB3), deoxynivalenol-3-glucoside (DON-3G), enniatin B4 (ENNB4), Nivalenol (NIV).

These results demonstrated that it is not unusual to find cereals contaminated with several mycotoxins, and evidenced the human exposure to multiple mycotoxins [9,11–13]. Therefore, these findings point out the necessity of more toxicity studies that consider co-exposure to multiple mycotoxins, to detect possible synergism and additive effects and its consequent potential impact for public health.

2.3. Exposure Estimates

The dietary exposure to the studied mycotoxins was evaluated by calculating the probable daily intake (PDI), which combines mycotoxins analysis data obtained from the analysed samples with the food consumption of the adult population with a body weight of 60 kg [55].

The PDI (μg/kg per body weight (bw)/day) of each mycotoxin was calculated using the following equation [55]:

$$\text{PDI} = (\text{Cm} \times \text{K})/\text{bw} \quad (1)$$

where C_m is the mean content of a mycotoxin in the cereal (μg/kg); K is the average consumption of the commodity (g/day) and bw is the body weight used for adult population.

Once the PDI had been calculated, the health risk characterization of each mycotoxin (% of relevant TDI) was estimated as the ratio of PDI to TDI (μg/kg bw/day) for each mycotoxin as follows:

$$\%\text{TDI} = (\text{PDI}/\text{TDI}) \times 100 \quad (2)$$

The PMTDI or TDI and the provisional tolerable weekly intake (PTWI) set by both the FAO/WHO JECFA and the Scientific Committee on Food (SCF), were used as reference doses [19,20]. Data on consumption of barley (36 g/day), maize (44 g/day), rice (8 g/day) and wheat (502 g/day) by Algerian population were mainly obtained from FAO statistical study [56]. The results obtained are summarised in Table 4.

Table 4. Results of the probable daily intake (PDI) assessment of the studied mycotoxins.

Analytes	TDI (μg/kg bw/day)	Barley (*n* = 30)		Maize (*n* = 30)		Rice (*n* = 30)		Wheat (*n* = 3)	
		PDI (μg/kg bw/day)	% TDI	PDI (μg/kg bw/day)	% TDI	PDI (μg/kg bw/day)	% TDI	PDI (μg/kg bw/day)	% TDI
FB1		0		10.86		0		0	
FB2		0		2.05		0		0	
Sum_{FBs}	2	0	0	12.91	645.4	0	0	0	0
HT-2		0		0		0		0.15	
T-2		0		0.02		0		0.18	
$Sum_{HT2\text{-}T2}$	0.1	0	0	0.02	18.26	0	0	0.33	333.8
DON	1	0	0	0.46	46.35	0	0	4.92	491.9
ZEN	0.25	0	0	0.08	31.97	$1.3 \cdot 10^{-3}$	0.53	0.85	341.4
F-X		0.11		0.21		0		1.27	
CIT		0.02		0.02		0		0.14	
BEA		0		$2.8 \cdot 10^{-3}$		0		1.30	
ENNA		0		0		0		0.24	
ENNA1		0		0.04		0		0.90	
ENNB		0		0		0		13.96	
ENNB1		0		0.04		0		3.92	

From the PDI values, it can be concluded that maize and wheat samples represent an important dietary exposure of mycotoxins for the Algerian population, with a dietary exposure range of 2.8×10^{-3}–12.91 (μg/kg bw/day) and 0.14–13.96 (μg/kg bw/day) in maize and wheat, respectively. In these samples, the obtained values for FB1 (10.86 μg/kg bw/day) in maize and ENNB (13.96 μg/kg bw/day) in wheat were the highest contribution for the PDI of mycotoxins for the Algerian population.

The exposure assessment was evaluated for all the mycotoxins with TDI available, namely DON, FB1+ FB2, ZEN, T-2 + HT-2 (see Table 4). The results showed that Algerian consumers present at a high risk of exposure to the sum of fumonisins (FB1 + FB2) through maize consumption (%TDI of 645.4), and to DON, ZEN and the sum of (HT-2 + T-2) through wheat consumption with %TDI of 491.9, 341.4 and 333.8, respectively. These values are several-hundred fold higher than the values established by JECFA for FB1 + FB2 (PMTDI of 2 μg/kg bw/day), DON (PMTDI of 1 μg/kg bw/day), ZEN (TDI of 0.25 μg/kg bw/day) and for the sum of HT-2 and T-2 toxins (TDI of 0.1 μg/kg bw/day).

These findings suggest that the intake of mycotoxins from analysed maize and wheat samples represent a high health risk for the average adult consumers in Algeria and pointed towards the necessity for a consistent control over these contaminants.

These results are globally in line with other studies from African countries; for instance, a study performed in Tanzania reported that fumonisin exposures to adult individuals in 38% of the households exceeded the provisional maximum tolerable daily intake (PMTDI) of 2 μg/kg bw based on the fumonisins (FB1 + FB2) concentration of up to 11 mg/kg in the maize grains [57]. Similar results were found for fumonisins (FB1 + FB2 + FB3) dietary exposure of 12.4 μg/kg bw/day in Nigerian infants and 8.2 μg/kg bw/day in children by maize intake with %TDI of 622.2 and 414.8, which were 311 and 207 times higher than the tolerable daily intake of 2 μg/kg bw/day [58].

Moreover, high attention should be devoted to the health risk scenarios for consumers in the case of co-exposures to multiple mycotoxins. However, the effect of co-occurrence of mycotoxins have not been well understood yet. These effects could be additive, synergistic and can vary with dose, exposure time, and toxicological end point [9].

3. Conclusions

In this study, a LC-MS/MS analytical method for determination of 15 mycotoxins in cereals has been applied to provide data on the occurrence of these hazards in barley, maize, rice and wheat from Algeria. Moreover, the risk associated with the exposure to mycotoxins through intake of cereal products has been estimated.

Results on mycotoxins occurrence showed that 65% of the samples were contaminated with at least one mycotoxin. The maximum acceptable levels established in EU where exceeded in 21 samples (70%) for fumonisins (4000 μg/kg of FB1 + FB2 in maize), 8 samples for ZEN (100 μg/kg for maize and 350 for other unprocessed cereals) and 1 sample for DON (1750 μg/kg for wheat and maize) [18]. Co-contamination was observed in 50% of the analysed samples (all maize and wheat samples). Among them, some samples were contaminated with up to 9 mycotoxins. Due to these high levels of contamination, this study concluded that the Algerians are at high risk of exposure to the sum of fumonisins (FB1 + FB2) through the consumption of maize, and to DON, ZEN and the sum of (HT-2 +T-2) through the consumption of wheat. Therefore, preventive approaches to curtail health risks associated with mycotoxins exposures are needed, which requires, at first instance, government intervention. The analytical results of this survey should encourage Algerian authorities to introduce allowed maximum limits of mycotoxins in cereals. In our opinion, continuous monitoring of these mycotoxins with a higher number of samples and other susceptible foodstuffs (such as cereal-based products, nuts or dried fruits) are recommended to assess the situation, at least for such a time until proper regulatory limits are set. On the other hand, the focus should also be directed towards reduction and control of mycotoxins producing fungi in the food chain.

4. Materials and Methods

4.1. Reagents and Materials

Methanol (MeOH) and acetonitrile (MeCN) of LC-MS grade, and ammonium formate were supplied by VWR International Eurolab S.L. (Barcelona, Spain). Formic acid eluent additive for LC–MS was obtained from Sigma Aldrich (St. Louis, MO, USA). Magnesium sulphate ($MgSO_4$), sodium chloride (NaCl) and sodium citrate were purchased from Panreac Química (Barcelona, Spain), while disodium hydrogen citrate sesquihydrate was supplied by Merck (Darmstadt, Germany). Ultrapure water was obtained from a Milli-Q Plus system (Millipore Bedford, MA, USA).

Mycotoxin standard solutions (10 mg/L in MeCN) of OTA, STE, F-X, DON, CIT, ZEN, FB1, FB2, T-2 and HT-2 were purchased from Techno Spec (Barcelona, Spain). Individual standards (powder) of ENNA, ENNA1, ENNB, ENNB1 and BEA were obtained from Sigma Aldrich and stock solutions were prepared at 1000 mg/L in MeCN. Multi-mycotoxins intermediate working solutions in MeCN (1 mg/L of OTA and STE; 2 mg/L of CIT; 10 mg/L of FB1, FB2, T-2, HT-2 and ZEN; 100 mg/L of DON, ENNA, ENNA1, ENNB, ENNB1 and BEA and 1000 mg/L of F-X) were prepared by combining suitable aliquots of each individual standard stock solution. These solutions were stored at −20 °C.

Nylon syringe filters (13 mm, 0.22 μm, from VWR) were used for filtration of extracts prior to the injection into the chromatographic system.

4.2. Instruments and Equipment

UHPLC-MS/MS analyses were performed in an Agilent 1290 Infinity LC (Agilent Technologies, Waldbronn, Germany) coupled to an API 3200 triple quadrupole mass spectrometer (AB Sciex, Darmstadt, Germany) with electrospray ionization (ESI). The chromatographic separation was performed using an Agilent Zorbax Eclipse Plus RRHD C18 column (50 × 2.1 mm, 1.8 μm). Analyst software (Version 1.6.3, AB Sciex, Darmstadt, Germany) was used for acquisition and data analysis.

During the sample treatment, an evaporator System (System EVA-EC, from VLM GmbH, Bielefeld, Germany), a vortex-2 Genie (Scientific Industries, Bohemia, NY, USA), a universal 320R centrifuge (Hettich ZENtrifugen, Tuttlingen, Germany), and a kitchen blender were used.

4.3. Samples

A total of 120 cereal samples (barley, maize, rice and wheat) destined for human consumption were randomly purchased from different local markets in three areas of the western region of Algeria—Aint Temouchent, Oran and Tiaret—during the year 2018 (see Table 5). In order to obtain representative samples, several sub-samples were taken from each batch, being thoroughly mixed to achieve a final 1-kg sample. Finally, the samples were grinded, homogenized and stored in a dark and dry place until analysis.

Table 5. Sampling information.

Sample Type	Selected Areas	No. of Markets	No. of Samples	Origin
Barley	Aint Temouchent	5	10	**OAIC**
	Oran	7	10	
	Tiaret	8	10	
			Total = 30	
Maize	Aint Temouchent	7	10	**Imported**
	Oran	6	10	
	Tiaret	8	10	
			Total = 30	
Rice	Aint Temouchent	10	10	**Imported**
	Oran	7	10	
	Tiaret	10	10	
			Total = 30	
Wheat	Aint Temouchent	6	10	**NG**
	Oran	9	10	
	Tiaret	8	10	
			Total = 30	

OAIC: Office Algérien interprofessionnel des céréales, NG: not given.

4.4. Mycotoxins Extraction Procedure

A previous method was used for the extraction of mycotoxins in the different samples [59]. Briefly, 2 g of grounded sample were weighed in a polypropylene centrifuge tube (50 mL), 8 mL of water was added, and the mixture was vortexed for 10 s. Subsequently, 10 mL of 5% formic acid in MeCN was added to the tube, shaking by vortex for 2 min. Then, 4 g of $MgSO_4$, 1 g of NaCl, 1 g of sodium citrate and 0.5 g of disodium hydrogen citrate sesquihydrate were added and the tube was shaken vigorously for 1 min. After centrifugation at 4500 rpm (3722× *g*) for 5 min, 2 mL of the upper supernatant layer was transferred to a 4-mL vial, evaporated to dryness under a gentle stream of nitrogen, and reconstituted to a final volume of 1 mL with a mixture of MeOH:water (50:50, *v/v*). The samples were filtered before injection and the 15 mycotoxins were determined by UHPLC-MS/MS. A graphical scheme of the procedure is shown in Figure 1.

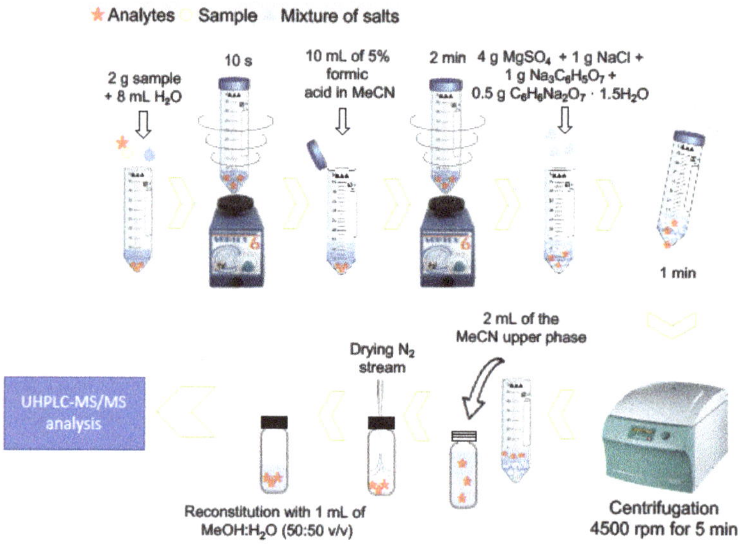

Figure 1. Graphical scheme of the extraction procedure.

4.5. *Ultra-High Performance Liquid Chromatography Coupled to Tandem Mass Spectrometry (UHPLC-MS/MS) Analysis*

Chromatographic analyses were performed using a gradient elution with water (phase A), and MeOH (phase B), both containing 0.3% formic acid and 5 mM ammonium formate, at a flow rate of 0.4 mL/min. The gradient elution program was as follows: 0–1 min, 5% B; 4 min, 50% B; 5 min, 80% B; 5.5 min, 90% B; 5.7 min, 5% B; 5.7–8 min, 5% B. The temperature of the column was kept at 35 °C and the injection volume was 5 µL.

The electrospray ionization was carried out in in the positive mode and the acquisition was performed under multiple reaction monitoring (MRM) conditions. The ionization source parameters were set as follows: source temperature: 500 °C; curtain gas (nitrogen): 30 psi; ion spray voltage: 5000 V; and GAS 1 and GAS 2 (both of them nitrogen): 50 psi. The applied cone voltages and collision energies for each mycotoxin are summarized in the supplementary material (Table S4). In all cases, the most abundant product ion was used for quantification, while the second one was used for confirmation.

Supplementary Materials: The following are available online at http://www.mdpi.com/2072-6651/12/3/194/s1, Figure S1: Matrix effect (%) for each studied mycotoxin extracted from barley, maize, rice and wheat samples (concentration levels: OTA and STE: 25 µg/kg; CIT: 100 µg/kg; FB1, FB2, T-2, HT-2 and ZEN: 250 µg/kg; ENNB, ENNB1, ENA, ENA1 and BEA: 400 µg/kg, DON: 1000 µg/kg and FUS-X: 2500 µg/kg); Figure S2: UHPLC-MS/MS MRM chromatogram of two samples of positive wheat contaminated with (A) DON (1362 µg/kg) and (B) ENNB1 (419 µg/kg); Table S1: Performance characteristics of the UHPLC-MS/MS method for each mycotoxin in barley, maize, rice and wheat samples; Table S2: Recovery (%R), intra-day precision (%RSD_r) and inter-day precision (%RSD_R) for barley, maize, rice and wheat samples (n = 9); Table S3: Co-occurrence of analysed mycotoxins in wheat and maize samples; Table S4: Monitored ions of the target analytes and MS/MS parameters.

Author Contributions: Conceptualization, L.G.-G., A.M.G.-C., C.K.M., N.H.-K. and K.M.; methodology, L.G.-G. and N.A.-M.; validation, C.K.M. and N.A.-M.; formal analysis, C.K.M. and N.A.-M.; investigation, C.K.M., N.A.-M. and L.G.-G.; resources, A.M.G.-C.; writing—original draft preparation, C.K.M. and N.A.-M.; writing—review and editing, L.G.-G. and A.M.G.-C.; visualization, C.K.M. and L.G.-G.; supervision, A.M.G.-C. and L.G.-G.; project administration, A.M.G.-C.; funding acquisition, A.M.G.-C. All authors have read and agreed to the published version of the manuscript.

Funding: This research was funded by SPANISH MINISTRY OF ECONOMY AND COMPETITIVENESS and EUROPEAN REGIONAL DEVELOPMENT FUND (MINECO/FEDER, UE) (Project ref: AGL2015-70708-R, MINECO/FEDER, UE). Choukri Khelifa Mahdjoubi would like to thanks the University of Granada for ERASMUS Mundus Doctoral Fellowship.

Conflicts of Interest: The authors declare no conflict of interest.

References

1. Ostry, V.; Malir, F.; Toman, J.; Grosse, Y. Mycotoxins as human carcinogens—The IARC Monographs classification. *Mycotoxin Res.* **2017**, *33*, 65–73. [CrossRef] [PubMed]
2. Zain, M.E. Impact of mycotoxins on humans and animals. *J. Saudi Chem. Soc.* **2011**, *15*, 129–144. [CrossRef]
3. De Boevre, M.; Diana Di Mavungu, J.; Landschoot, S.; Audenaert, K.; Eeckhoutn, M.; Maene, P.; Haesaert, G.; De Saeger, S. Natural occurrence of mycotoxins and their masked forms in food and feed products. *World Mycotoxin J.* **2012**, *5*, 207–219. [CrossRef]
4. Otto, K.; Santini, A.; Oancea, S. Recent aflatoxin survey data in milk and milk products: A review. *Int. J. Dairy Technol.* **2017**, *70*, 1–12. [CrossRef]
5. Mikušová, P.; Šrobárová, A.; Sulyok, M.; Santini, A. *Fusarium* fungi and associated metabolites presence on grapes from Slovakia. *Mycotoxin Res.* **2013**, *29*, 97–102. [CrossRef]
6. Gruber-Dorninger, C.; Novak, B.; Nagl, V.; Berthiller, F. Emerging mycotoxins: Beyond traditionally determined food contaminants. *J. Agric. Food Chem.* **2017**, *65*, 7052–7070. [CrossRef]
7. Jajić, I.; Dudaš, T.; Krstovic, S.; Krska, R.; Sulyok, M.; Bagi, F.; Savić, Z. Emerging *Fusarium* mycotoxins fusaproliferin, beauvericin, enniatins, and moniliformin in Serbian maize. *Toxins* **2019**, *11*, 357. [CrossRef]
8. Mahnine, N.; Meca, G.; Elabidi, A.; Fekhaoui, M.; Saoiabi, A.; Font, G.; Mañes, J.; Zinedine, A. Further data on the levels of emerging *Fusarium* mycotoxins enniatins (A, A1, B, B1), beauvericin and fusaproliferin in breakfast and infant cereals from Morocco. *Food Chem.* **2011**, *124*, 481–485. [CrossRef]
9. Grenier, B.; Oswald, I.P. Mycotoxin co-contamination of food and feed: Meta-analysis of publications describing toxicological interactions. *World Mycotoxin J.* **2011**, *4*, 285–313. [CrossRef]
10. Lee, H.J.; Ryu, D. Worldwide occurrence of mycotoxins in cereals and cereal-derived food products: Public health perspectives of their co-occurrence. *J. Agric. Food Chem.* **2017**, *65*, 7034–7051. [CrossRef]
11. Assunção, R.; Silva, M.J.; Alvito, P. Challenges in risk assessment of multiple mycotoxins in food. *World Mycotoxin J.* **2016**, *9*, 791–811. [CrossRef]
12. Alassane-Kpembi, I.; Schatzmayr, G.; Taranu, I.; Marin, D.; Puel, O.; Oswald, I.P. Mycotoxins co-contamination: Methodological aspects and biological relevance of combined toxicity studies. *Crit. Rev. Food Sci. Nutr.* **2017**, *57*, 3489–3507. [CrossRef] [PubMed]
13. De Ruyck, K.; De Boevre, M.; Huybrechts, I.; De Saeger, S. Dietary mycotoxins, co-exposure, and carcinogenesis in humans: Short review. *Mutat. Res. Rev. Mutat. Res.* **2015**, *766*, 32–41. [CrossRef] [PubMed]
14. Smith, M.C.; Madec, S.; Coton, E.; Hymery, N. Natural co-occurrence of mycotoxins in foods and feeds and their in vitro combined toxicological effects. *Toxins* **2016**, *8*, 94. [CrossRef] [PubMed]
15. Tavares, A.M.; Alvito, P.; Loureiro, S.; Louro, H.; Silva, M.J. Multi-mycotoxin determination in baby foods and in vitro combined cytotoxic effects of aflatoxin M1 and ochratoxin A. *World Mycotoxin J.* **2013**, *6*, 375–388. [CrossRef]
16. Wan, L.Y.; Turner, P.C.; El-Nezami, H. Individual and combined cytotoxic effects of *Fusarium* toxins (deoxynivalenol, nivalenol, zearalenone and fumonisins B1) on swine jejunal epithelial cells. *Food Chem. Toxicol.* **2013**, *57*, 276–283. [CrossRef]
17. European Commission (EC). Commission recomendations of 27 March 2013 on the presence of T-2 and HT-2 toxin in cereals and cereal products. *Off. J. Eur. Union. L* **2013**, *91*, 12–15.
18. European Commission (EC). Commission regulation (EC) No 1881/2006 of 19 December 2006 setting maximum levels for certain contaminants in foodstuffs. *Off. J. Eur. Union.* **2006**, *364*, 5–24.

19. JEFCA. Joint FAO/WHO Standards Programme Codex Committee on Contaminants in Food. Working Document for Information and Use in Discussions Related to Contaminants and Toxins in the GSCTFF. 2018. Available online: http//www.fao.org/fao-who-codexalimentarius/sh-proxy/en/?lnk=1&url=https%253A%252F%252Fworkspace.fao.org%252Fsites%252Fcodex%252FMeetings%252FCX-735-12%252FWD%252Fcf12_INF01x.pdf (accessed on 6 December 2019).
20. JEFCA. Joint FAO/WHO Expert Committee on Food Additives. Summary and Conclusion of Eighty-Third Meeting of the JEFCA. 2016. Available online: http://www.fao.org/3/BQ821E/bq821e.pdf (accessed on 6 December 2019).
21. U.S. Food and Drug Administration. Compliance Policy Guide Sec. 555.400 Foods—Adulteration with Aflatoxin. Available online: https://www.fda.gov/Food/GuidanceRegulation/GuidanceDocumentsRegulatoryInformation/ChemicalContaminantsMetalsNaturalToxinsPesticides/default.htm (accessed on 20 February 2020).
22. U.S. Food and Drug Administration. Guidance for Industry. Fumonisin Levels in Human Foods and Animal Feeds. Available online: https://www.fda.gov/Food/GuidanceRegulation/GuidanceDocumentsRegulatoryInformation/ChemicalContaminantsMetalsNaturalToxinsPesticides/default.htm (accessed on 20 February 2020).
23. Canadian Food Inspection Agency. RG-8 Regulatory Guidance. Contaminants in Feed (Formerly RG-1, Chapter 7), Section 1: Mycotoxins in Livestock Feed. Available online: http://www.inspection.gc.ca/animals/feeds/regulatory-guidance/rg8/eng/1347383943203/1347384015909 (accessed on 20 February 2020).
24. Food and Agricultural Materials Inspection Centre (FAMIC). Aflatoxin. Available online: http://www.famic.go.jp/ffis/oie/obj/hc_aflatoxin.pdf (accessed on 20 February 2020).
25. Food and Agricultural Materials Inspection Centre (FAMIC). Regulation Value of Pesticides, Heavy Metals and Mycotoxins (Administrative Guideline). Available online: http://www.famic.go.jp/ffis/feed/r_safety/r_feeds_safety22.html (accessed on 20 February 2020).
26. China Food Safety National Standard. Maximum Levels of Mycotoxins in Food, Released 17 March 2017, Implemented 17 September 2017. Available online: https://www.fas.usda.gov/data/china-china-releases-standard-maximum-levels-mycotoxins-foods (accessed on 20 February 2020).
27. Bulletin Officiel. Arrêté Conjoint du Ministre de L'agriculture et de la Pêche Maritime et du Ministre de la Santé n° 1643-16 du 23 Chaabane 1437 (30 Mai 2016) Fixant les Limites Maximales Autorisées des Contaminants dans les Produits Primaires et les Produits Alimentaires. Available online: http://www.onssa.gov.ma/fr/images/reglementation/transversale/ARR.1643-16.FR.pdf (accessed on 20 February 2020).
28. Santini, A.; Meca, G.; Ritieni, A. Fusaproliferin, beauvericin, and enniatins: Occurrence in food—A review. *World Mycotoxin J.* **2012**, *5*, 71–81. [CrossRef]
29. Santini, A.; Raiola, A.; Meca, G.; Ritieni, A. Aflatoxins, ochratoxins, trichothecenes, patulin, fumonisins and beauvericin in finished products for human consumption. *J. Clin. Toxicol.* **2015**, *5*, 265. [CrossRef]
30. Eskola, M.; Kos, G.; Elliott, C.T.; Hajšlová, J.; Mayar, S.; Krska, R. Worldwide contamination of food-crops with mycotoxins: Validity of the widely cited 'FAO estimate' of 25%. *Crit. Rev. Food Sci. Nutr.* **2019**, *3*, 1–17. [CrossRef] [PubMed]
31. Tantaoui-Elaraki, A.; Riba, A.; Oueslati, S.; Zinedine, A. Toxigenic fungi and mycotoxin occurrence and prevention in food and feed in northern Africa–A review. *World Mycotoxin J.* **2018**, *11*, 385–400. [CrossRef]
32. Van Egmond, H.P.; Schothorst, R.C.; Jonker, M.A. Regulations relating to mycotoxins in food. *Anal. Bioanal. Chem.* **2007**, *389*, 147–157. [CrossRef] [PubMed]
33. Berthiller, F.; Cramer, B.; Iha, M.H.; Krska, R.; Lattanzio, V.M.T.; MacDonald, S.; Malone, R.J.; Maragos, C.; Solfrizzo, M.; Stranska-Zachariasova, M.; et al. Developments in mycotoxin analysis: An update for 2016–2017. *World Mycotoxin J.* **2018**, *11*, 5–31. [CrossRef]
34. Santini, A.; Ferracane, R.; Meca, G.; Ritieni, A. Overview of analytical methods for beauvericin and fusaproliferin in food matrices. *Anal. Bioanal. Chem.* **2009**, *395*, 1253–1260. [CrossRef]
35. Santini, A.; Ferracane, R.; Somma, M.C.; Somma, A.; Aragon, A.; Ritieni, A. Multitoxin extraction and detection of thricotecenes in cereals: An improved LC-MS/MS approach. *J. Sci. Food Agric.* **2009**, *89*, 1145–1153. [CrossRef]
36. Arroyo-Manzanares, N.; Huertas-Pérez, J.F.; García-Campaña, A.M.; Gámiz-Gracia, L. Review of sample treatments and the state-of-the-art of analytical techniques for mycotoxins in food. In *Analysis of Food Toxins and Toxicants*; Wong, Y.-C., Lewis, R.J., Eds.; John Wiley & Sons Ltd.: New York, NY, USA, 2017; pp. 51–102. [CrossRef]

37. European Commission (EC). Commission regulation (EC) No 401/2006 of 23 February 2006 laying down the methods of sampling and analysis for the official control of the levels of mycotoxins in foodstuffs. *Off. J. Eur. Union.* **2006**, *L70*, 12–34.
38. Bouafifssa, Y.; Manyes, L.; Rahouti, M.; Mañes, J.; Berrada, H.; Zinedine, A.; Fernández-Franzón, M. Multi-occurrence of twenty mycotoxins in pasta and a risk assessment in the Moroccan population. *Toxins* **2018**, *10*, 432. [CrossRef]
39. Pleadin, J.; Vahčić, N.; Perši, N.; Ševelj, D.; Markov, K.; Frece, J. *Fusarium* mycotoxins' occurrence in cereals harvested from Croatian fields. *Food Control* **2013**, *32*, 49–54. [CrossRef]
40. Juan, C.; Ritieni, A.; Mañes, J. Occurrence of *Fusarium* mycotoxins in Italian cereal and cereal products from organic farming. *Food Chem.* **2013**, *141*, 1747–1755. [CrossRef]
41. Zaied, C.; Zouaoui, N.; Bacha, H.; Abid, S. Natural occurrence of zearalenone in Tunisian wheat grains. *Food Control* **2012**, *25*, 773–777. [CrossRef]
42. Savi, G.D.; Piacentini, K.C.; Rocha, L.O.; Carnielli-Queiroz, L.; Furtado, B.G.; Scussel, R.; Zanoni, E.T.; Machado de Ávila, R.A.; Corrêa, B.; Angioletto, E. Incidence of toxigenic fungi and zearalenone in rice grains from Brazil. *Int. J. Food Microbiol.* **2018**, *270*, 5–13. [CrossRef] [PubMed]
43. Chilaka, C.A.; De Boevre, M.; Atanda, O.O.; De Saeger, S. Occurrence of *Fusarium* mycotoxins in cereal crops and processed products (Ogi) from Nigeria. *Toxins* **2016**, *8*, 342. [CrossRef] [PubMed]
44. De Boevre, M.; Landschoot, S.; Audenaert, K.; Maene, P.; Diana Di Mavungu, J.; Eeckhout, M.; Haesaert, G.; De Saeger, S. Occurrence and within field variability of *Fusarium* mycotoxins and their masked forms in maize crops in Belgium. *World Mycotoxin J.* **2014**, *7*, 91–102. [CrossRef]
45. Zinedine, A.; Brera, C.; Elakhdari, S.; Catano, C.; Debegnach, F.; Angelini, S.; De Santis, B.; Faid, M.; Benlemlih, M.; Minardi, V.; et al. Natural occurrence of mycotoxins in cereals and spices commercialized in Morocco. *Food Control* **2006**, *17*, 868–874. [CrossRef]
46. Ghali, R.; Ghorbel, H.; Hedilli, A. *Fumonisin* determination in Tunisian foods and feeds. ELISA and HPLC methods comparison. *J. Agric. Food Chem.* **2009**, *57*, 3955–3960. [CrossRef]
47. Bryła, M.; Waśkiewicz, A.; Podolska, G.; Szymczyk, K.; Jędrzejczak, R.; Damaziak, K.; Sułek, A. Occurrence of 26 mycotoxins in the grain of cereals cultivated in Poland. *Toxins* **2016**, *8*, 160. [CrossRef]
48. Phoku, J.Z.; Dutton, M.F.; Njobeh, P.B.; Mwanza, M.; Egbuta, M.A.; Chilaka, C.A. *Fusarium* infection of maize and maize-based products and exposure of a rural population to fumonisin B1 in Limpopo Province, South Africa. *Food Addit. Contam. Part A* **2012**, *29*, 1743–1751. [CrossRef]
49. Tsehaye, H.; Brurberg, M.B.; Sundheim, L.; Assefa, D.; Tronsmo, A.; Tronsmo, A.M. Natural occurrence of *Fusarium* species and fumonisin on maize grains in Ethiopia. *Eur. J. Plant Pathol.* **2017**, *147*, 141–155. [CrossRef]
50. Houbraken, J.; De Vries, R.P.; Samson, R.A. Modern taxonomy of biotechnologically important *Aspergillus* and *Penicillium* species. *Adv. Appl. Microbiol.* **2014**, *86*, 199–249. [CrossRef]
51. Meca, G.; Zinedine, A.; Blesa, J.; Font, G.; Mañes, J. Further data on the presence of *Fusarium* emerging mycotoxins enniatins, fusaproliferin and beauvericin in cereals available on the Spanish markets. *Food Chem. Toxicol.* **2010**, *48*, 1412–1416. [CrossRef] [PubMed]
52. Oueslati, S.; Romero-González, R.; Lasram, S.; Garrido-Frenich, A.; Martínez-Vidal, J.L. Multi-mycotoxin determination in cereals and derived products marketed in Tunisia using ultra-high performance liquid chromatography coupled to triple quadrupole mass spectrometry. *Food Chem. Toxicol.* **2012**, *50*, 2376–2381. [CrossRef] [PubMed]
53. Blesa, J.; Moltó, J.C.; El Akhdari, S.; Mañes, J.; Zinedine, A. Simultaneous determination of *Fusarium* mycotoxins in wheat grain from Morocco by liquid chromatography coupled to triple quadrupole mass spectrometry. *Food Control* **2014**, *46*, 1–5. [CrossRef]
54. Rodríguez-Carrasco, Y.; Ruiz, M.J.; Font, G.; Berrada, H. Exposure estimates to *Fusarium* mycotoxins through cereals intake. *Chemosphere* **2013**, *93*, 2297–2303. [CrossRef] [PubMed]
55. International Programme on Chemical Safety (IPCS). Principles and Methods for the Risk Assessment of Chemicals in Food. In *Environmental Health Criteria 240*; WHO: Stuttgart, Germany, 2009; Available online: https://www.who.int/foodsafety/publications/chemical-food/en/ (accessed on 6 December 2019).
56. FAO Statistics Division (FAOSTAT). Food and Agricultural Commodities Production. Available online: http://faostat.fao.org/site/339/default.aspx (accessed on 10 January 2019).

57. Kimanya, M.; De Meulenaer, B.; Tiisekwa, B.; Ndomondo-Sigonda, M.; Kolsteren, P. Human exposure to fumonisins from home grown maize in Tanzania. *World Mycotoxin J.* **2008**, *1*, 307–313. [CrossRef]
58. Adetunji, M.C.; Atanda, O.O.; Ezekiel, C.N. Risk assessment of mycotoxins in stored maize grains consumed by infants and young children in Nigeria. *Children* **2017**, *4*, 58. [CrossRef]
59. Arroyo-Manzanares, N.; Huertas-Pérez, J.F.; García-Campaña, A.M.; Gámiz-Gracia, L. Simple methodology for the determination of mycotoxins in pseudocereals, spelt and rice. *Food Control* **2014**, *36*, 94–101. [CrossRef]

toxins

MDPI

Article

Occurrence of Mycotoxins in Swine Feeding from Spain

Natalia Arroyo-Manzanares [1,2], Vicente Rodríguez-Estévez [3], Plácido Arenas-Fernández [1], Ana M. García-Campaña [1] and Laura Gámiz-Gracia [1,*]

1 Department Analytical Chemistry, Faculty of Sciences, University of Granada, Campus Fuentenueva s/n, 18071 Granada, Spain; natalia.arroyo@um.es (N.A.-M.); placi_94_miercoles@hotmail.com (P.A.-F.); amgarcia@ugr.es (A.M.G.-C.)

2 Department Analytical Chemistry, Faculty of Chemistry, Regional Campus of International Excellence "Campus Mare Nostrum", University of Murcia, E-30100 Murcia, Spain

3 Department Animal Production, Faculty of Veterinary, Campus Univ. Rabanales, University of Córdoba, 14071 Córdoba, Spain; vrestevez@uco.es

* Correspondence: lgamiz@ugr.es

Received: 17 May 2019; Accepted: 13 June 2019; Published: 15 June 2019

Abstract: A survey including 228 pig feed samples from Spain has been developed, exploring the occurrence of 19 mycotoxins (aflatoxins B1, B2, G1 and G2, ochratoxin A, fumonisins B1 and B2, citrinin, zearalenone, deoxynivalenol, fusarenon X, sterigmatocystin, T-2 toxin, HT-2 toxin, enniatins A, A1, B and B2, and beauvericin). The samples were analysed by solid-liquid extraction followed by liquid chromatography coupled with fluorescence or mass spectrometry detection. Enniatin B was found in 100% of the samples (up to 1200 μg/kg) and beauvericin in more than 90%. Moreover, 40% of samples were contaminated with more than five mycotoxins. This high occurrence is insurmountable and surpasses all previous studies, probably due to the inclusion of emerging mycotoxins, scarcely explored. The majority of the samples (96.9%) were in accordance with EU regulations, which do not address emerging mycotoxins or co-occurrence. These results show that in order to ensure mycotoxin absence, emerging mycotoxins should always be considered.

Keywords: feed; pig; mycotoxins; liquid chromatography; fluorescence detection; mass spectrometry; solid-liquid extraction; co-occurrence

Key Contribution: A total of 19 mycotoxins have been determined in 228 pig feed samples from Spain by solid-liquid extraction followed by HPLC-FLD (aflatoxins) and LC-MS/MS. The results show the great incidence of emerging mycotoxins as well as the co-occurrence of up to five mycotoxins in a significant number of samples.

1. Introduction

The European Union (EU) ensures the safety of foods by setting maximum levels of different contaminants including mycotoxins, which are toxic secondary metabolites produced by certain fungi (as *Aspergillus*, *Fusarium* or *Penicillium*) that can contaminate food during harvesting, processing or storage. Among them, there are well-known hazards such as aflatoxins (AFs), considered as carcinogenic to humans by the International Agency for Research on Cancer [1] or ochratoxin A (OTA) and fumonisin B1, considered as possibly carcinogenic to humans [2,3]. Currently, several mycotoxins are included in the EU legislation and maximum contents have been established in different raw materials and food commodities. However, despite the effort to control these food contaminants, every year mycotoxins are found among the "top ten" hazards reported annually by the Rapid Alert System for Food and Feed (RASFF), aflatoxins being the most frequently reported [4]. Moreover, there are still some mycotoxins without maximum content allowed in regulation, but with some evidence of adverse effects on human health. These are the so-called "emerging mycotoxins", including some *Fusarium* toxins such as enniatins (ENNs) and beauvericin (BEA).

In addition to food, it is also important to control contamination in feeds, as they are the first link in the food-chain, having influence not only on the health of animals but also on humans consuming animal-derived products. To deal with feed safety, the EU has established regulatory levels for aflatoxin B1 (AFB1) and guidelines for deoxynivalenol (DON), zearalenone (ZEA), OTA, fumonisin B1 and B2 (FB1 and FB2), and T-2 and HT-2 toxins in different raw materials and feed, including pig diets (Table 1) [5–7]. Moreover, the European Food Safety Authority (EFSA) has published different scientific opinions about the risks to animal health related to the presence of different mycotoxins [8]. One of these opinions about the risks to human and animal health, related to the presence of BEA and ENNs in food and feed, concluded that in the absence of toxicological data for most livestock animals, research studies on their adverse effects are required [9].

Table 1. Maximum content (AFB1) or recommended levels for mycotoxins in feed (only those concerning swine) [5–7].

Mycotoxin	Product Intended for Pig Feed	Maximum/Guidance Value (μg kg^{-1})
AFB1	Feed materials	20
	Complementary and complete feed	10
	Compound feed for pigs (except piglets)	20
	Compound feed for piglets	5
DON	Maize by-products	12,000
	Cereals and cereal products with the exception of maize by-products	8000
	Maize by-products	12,000
	Complementary and complete feedingstuffs for pigs	900
ZEA	Cereals and cereal products with the exception of maize by-products	2000
	Maize by-products	3000
	Complementary and complete feedingstuffs for piglets and gilts (young sows)	100
	Complementary and complete feedingstuffs for sows and fattening pigs	250
OTA	Cereals and cereal products	250
	Complementary and complete feedingstuffs for pigs	50
FB1+FB2	Maize and maize products	60,000
	Complementary and complete feedingstuffs for pigs	5000
T-2+HT-2	Oat milling products (husks)	2000
	Other cereal products	500
	Compound feed	250

Given their relatively high susceptibility to *Fusarium* toxins and the high content of cereals in their diet, pigs should be given the highest priority. Thus, several studies have shown the toxic effect of mycotoxins on pig health, including a modulation of the immune response, resulting in an increase in susceptibility and severity of infectious diseases, and a reduction in vaccine efficacy, whilst also having an indirect effect on animal productivity [10]. Especially abundant are the number of studies about the pathological effects of *Fusarium* toxins on pig reproduction [11]; these include abortion, embryonic and foetal death, increased number of oestrus repetitions, failure of induction programs with PGF2α and increased number of stillbirths and splaylegged piglets. What is more, ZEA can produce hyperestrogenism [12] and tail necrosis in suckling piglets [13]. The effects of DON on gut function [14,15], and DON and ZEA on colon microbiota, have also been reported [16]. Moreover, diets co-contaminated with AFs and FBs negatively affected the growth of piglets, despite the absence of pathology and despite the absence of clinical signs [17].

Different studies and reviews have reported the occurrence of mycotoxins in raw materials and feed, showing that multi contamination is very frequent and seems to be normal and not exceptional [18–27]. However, most of these studies are focused on regulated mycotoxins and only a few studies included emerging mycotoxins [28–30]. Nevertheless, EFSA has concluded that further data on the co-occurrence of different mycotoxins, including BEA and ENNs with other toxins in feed, and the probable combined effects are needed [9].

The aim of this work was to evaluate the occurrence of nineteen different mycotoxins, including *Fusarium* emerging mycotoxins, in pig feed from Spain. With this purpose, 228 samples were collected during 2017 from different farms and analysed using high performance liquid chromatography with post-column photochemical derivatisation and fluorescence detection (HPLC-FLD) for quantification of AFs and ultra-high performance liquid chromatography with tandem mass spectrometry (UHPLC-MS/MS) for the rest, while solid-liquid extraction was used as sample treatment.

2. Results and Discussion

Ocurrence of Mycotoxins in Pig Feed Samples

Among the 19 mycotoxins under study, the most commonly found (considering samples with concentrations above the LOQs) were: ENNB (100% samples), BEA (93.4%), ENNB1 (53.5%), FB1 (50.0%) and ENNA1 (40.8%). Besides that, none of the samples presented concentrations above the LOD for AFG2 and OTA. Figure 1 summarises the results. This high mycotoxin occurrence surpasses all previous studies of occurrence; i.e., Streit et al. [18] found 72% positive samples for at least one mycotoxin when analysing 17,316 samples of feed and feed raw materials from all over the world to study contamination with AFs, OTA, ZEA, DON and fumonisins. However, other regional and national studies of occurrence have found higher occurrences for some mycotoxins above the LOD; i.e., Li et al. [31] detected DON and ZEA at percentages of 97 and 100% in samples of complete feeds randomly collected from 15 pig farms in the Beijing region (China), and Ma et al. [24] found AFB1, ZEA and DON at percentages from 96.4 to 100% also in samples of complete feeds from different provinces of China. The maximum incidence reached in the present study is probably due to the high number of mycotoxins analysed, including emerging mycotoxins, scarcely explored in other studies. This fact shows that for both a correct diagnosis and in order to ensure its absence, it is not enough to analyse only the most common mycotoxins.

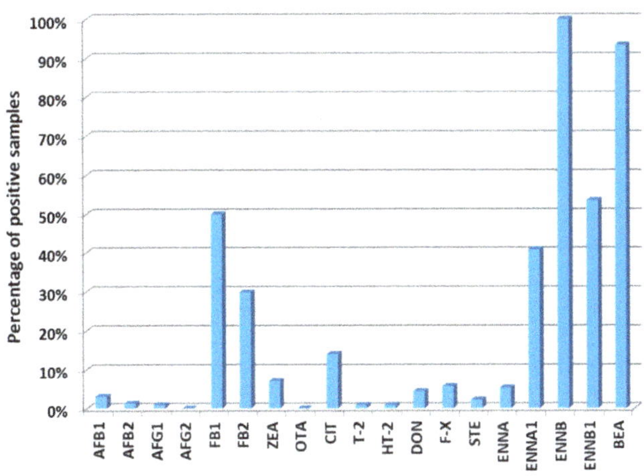

Figure 1. Percentage of positive samples of each mycotoxin.

Concerning the concentrations found, the contents of the two mycotoxins with higher incidence ranged from 2.0 to 1222 μg kg^{-1} for ENNB, and from 1.7 to 747 μg kg^{-1} for BEA in positive samples (100% and 93.4% respectively). However, the mycotoxin detected at the highest level was FB1 with a concentration of 3959 μg kg^{-1}. A summary of all the results obtained is shown in Table 2. As can be seen from the high values of the %RSD, a great difference in the content of the different mycotoxins was observed among the samples.

The EU has established recommended guidance values on raw materials, complementary and complete feeding stuff with differences associated to age of animal and phase of production. Focusing the attention on those mycotoxins with maximum permitted or recommended levels, it can be concluded that all the samples fulfil the requirements about maximum content of AFB1, as well as the guidelines for OTA, FB1+FB2, T-2+HT-2 toxins, and DON. However, several samples (3.1%) showed contents of ZEA above the recommended levels: one sample of maize (7681 μg kg^{-1}); five samples of compound feed for piglets (ranging from 125 to 956 μg kg^{-1}), and one for fattening pigs (290 μg kg^{-1}). In a recent study, it has been proven that growing pigs (six-week-old and 19 kg BW at the beginning) fed ad libitum with feed containing 80 μg kg^{-1} ZEA for four weeks reduced body weight gain, daily feed intake, feed conversion rate, and the serum levels of immunoglobulin IgG and IgM, and total antioxidants, showing microscopic lesions in kidneys [32]. This study shows the need to reduce the levels of tolerance for ZEA in feeds for pigs.

These results are in accordance with other survey studies carried out in Europe, where *Fusarium* toxins, such as DON, ZEA, and FBs, were the most frequently found mycotoxins in feed [23].

Another aspect of concern was the high co-occurrence of mycotoxins. Thus, 69% of analysed samples contained between 3 and 5 mycotoxins (Figure 2), the most common combinations being those involving co-occurrence of different emerging mycotoxins: ENNB + BEA (12.7%); ENNB + ENNB1 + BEA (8.3%) and ENNA1 + ENNB + ENNB1 + BEA (10.5%). Moreover, the co-occurrence of more than one regulated mycotoxin (AFs, OTA, FBs, DON, ZEA, and T-2 + HT-2) was detected in 19 samples (8.3% of all samples). The natural co-occurrence of mycotoxins in feeds is already known and several surveys have reported this everywhere [21]; this can be explained by these diets being usually made up of multiple cereal and other grain sources; and mycotoxin frequency and levels in feed depend on these raw materials and, especially, on their geographic origin [21,24,31]; i.e., a study of 17,316 samples of feed and feed raw materials from all over the world showed 38% samples co-contaminated [18]. Although several reviews of existing data and of the literature on worldwide mycotoxin contamination of food and feed are available, the impact of the different raw materials used on feed formulation has not been widely studied. Concerning cereals and derived cereal product samples, 127 mycotoxin combinations were described by Smith et al., showing that 70%, 24% and 6% of the studies concerning the effect of co-occurrence of mycotoxins were binary, ternary and quaternary or quinary mixtures, respectively [33]. However, this level of study is not proportional to the co-occurrence found in the present study with 13.16%, 22.37% and 46.49% for binary, ternary and quaternary or quinary mixtures, respectively. Nevertheless, the current regulations do not consider the mycotoxin combination effects and the maximum allowed or recommended levels are fixed for a single mycotoxin. These results should be a matter of concern, as additive or synergistic effect (still not well-known) can increase the toxicity of mycotoxins. Hence, guidelines should not only be individually set for each mycotoxin but also for especially concerning combinations, particularly for those including emerging mycotoxins such as moniliformin, BEA or ENNs [22]. This fact has also been pointed out in a recent review compiling in vitro experimental data on mycotoxins combined toxicity [33]. Besides, more stringent mycotoxin limits of tolerance should be established, especially when there is a risk of chronic ingestion of regular contaminants of pig feed [32].

Table 2. Summary of the occurrence of the studied mycotoxins: number and percentage of positive samples, mean concentration value of positive samples, median, 1st and 3rd quartile of positives samples, RSD% and minimum and maximum found concentrations.

	Nº Positive Samples	Incidence (%)	Mean (µg kg^{-1})	%RSD	Min (µg kg^{-1})	Max (µg kg^{-1})	0.25th Quantile (µg kg^{-1})	0.50th Quantile (µg kg^{-1})	0.75th Quantile (µg kg^{-1})	Nº Non-Compliant Samples [a]
AFB1	7	3.07	0.94	90	0.29	2.91	0.44	0.57	1.17	0
AFB2	3	1.32	0.60	55	0.28	1.06	0.28	0.47	1.06	—
AFG1	2	0.88	0.33	33	0.22	0.44	—	0.33	—	—
AFG2	0	0	-	-	-	-	—	—	—	—
FB1	114	50.00	403	133	4.09	3959	99.1	209	541	0
FB2	68	29.82	184	101	3.63	961	61.9	133	221	
ZEA	16	7.02	741	244	101	7681	126	137.0	463	7
OTA	0	0	-	-	-	-	—	—	—	0
CIT	32	14.04	147	80	10.7	512	71.0	109.0	197	—
T-2	2	0.88	31.9	12	28.0	35.9	—	31.9	—	0
HT-2	2	0.88	117	4.8	112	123	—	117	—	
DON	10	4.39	237	51	153	555	164	182	263	0
F-X	13	5.70	291	32	65.1	821	94.1	212	428	—
STE	5	2.19	104	115	11.2	308	11.3	12.8	243	—
ENNA	12	5.26	9.82	170	3.29	64.9	3.52	3.9	8.07	—
ENNA1	93	40.79	19.0	128	4.54	140	6.65	10.5	20.9	—
ENNB	228	100	118	137	2.01	1222	14.9	54.3	165	—
ENNB1	122	53.51	34.3	141	2.58	247	6.53	15.1	38.2	—
BEA	213	93.42	20.7	270	1.71	747	4.93	8.73	19.2	—

[a] According to EU directives and recommendations [5–7].

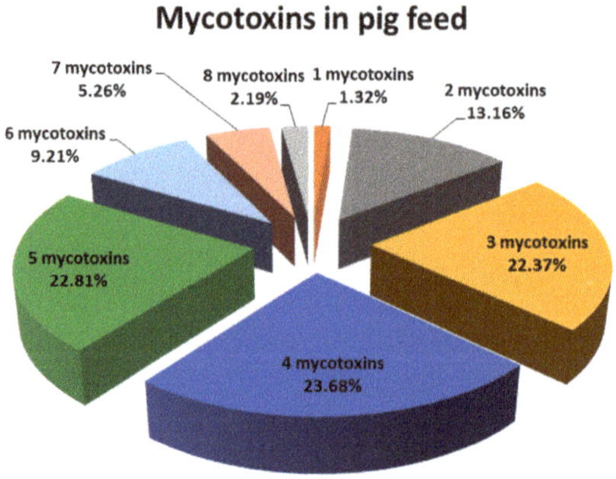

Figure 2. Frequency and co-occurrence of different mycotoxins in feed samples.

In order to elucidate which factors were the most relevant, one-way ANOVA studies were carried out for total content of mycotoxins as well as for each mycotoxin individually, considering different factors.

- *PRESENTATION.* ANOVA showed that no significant differences for total mycotoxin content (flour: 467 ± 549 μg kg^{-1}; pellet: 404 ± 360 μg kg^{-1}; *p*-value: 0.4741) existed between both presentations, however higher concentration of sterigmatocystin (STE) was found in pellet (flour: 0.193 ± 1.50 μg kg^{-1}; pellet: 11.3 ± 53.1; *p*-value: 0.0053).
- *SAMPLING POINT.* This factor was significant for total mycotoxin content (silo: 436 ± 607 μg kg^{-1}, feeder: 244 ± 160 μg kg^{-1}, sack: 677 ± 554 μg kg^{-1}, bulk: 3091 ± 4075 μg kg^{-1}, *p*-value: 0.0000). In addition, significant differences were obtained for citrinin (CIT) (silo: 11 ± 31 μg kg^{-1}, hopper: 22 ± 50 μg kg^{-1}, sack: 70 ± 135 μg kg^{-1}, bulk: 0 ± 0 μg kg^{-1}, *p*-value: 0.0022), DON (silo: 14 ± 56 μg kg^{-1}, hopper: 0 ± 0 μg kg^{-1}, sack: 0 ± 0 μg kg^{-1}, bulk: 93 ± 207 μg kg^{-1}, *p*-value: 0.0074), FB1 (silo: 150 ± 353 μg kg^{-1}, hopper: 12 ± 38 μg kg^{-1}, sack: 323 ± 399 μg kg^{-1}, bulk: 1278 ± 1462 μg kg^{-1}, *p*-value: 0.0000), FB2 (silo: 0 ± 0 μg kg^{-1}, hopper: 45 ± 123 μg kg^{-1}, sack: 71 ± 105 μg kg^{-1}, bulk: 339 ± 362 μg kg^{-1}, *p*-value: 0.0000) and ZEA (silo: 41 ± 137 μg kg^{-1}, hopper: 0 ± 0 μg kg^{-1}, sack: 32 ± 132 μg kg^{-1}, bulk: 1297 ± 2855 μg kg^{-1}, *p*-value: 0.0001). In all the cases, higher concentrations were obtained for samples from bulk storage without container in warehouse except for CIT, where the highest concentrations were found in sack. The higher mycotoxin content could be explained because bulk storage without container is less hygienic and the feed is in contact with the warehouse floor. Sampling is usually the principal source of variation in toxin analysis, originating close to 90% of the error in some analyses [34]. Traceability is important and sampling of feed should cover from factory to feeder, including silos and warehouses.
- *TYPE OF ANIMAL.* These feeds have different proportions of raw materials to provide different nutrient levels; however, cereal grains are always the main component. ANOVA results in no significant differences among the feeds for the total mycotoxin content and for any isolated mycotoxin; the high proportion of cereals in all these feeds could explain the lack of differences. However, less contamination could be expected in piglet feed than in older pig feed, which could be explained by the lower level of maize used for piglets diet, maize being the main contributor to mycotoxin contamination [21].

3. Conclusions

A survey of nineteen mycotoxins (including emerging mycotoxins) in pig feed samples collected during 2017 from different farms and suppliers from Spain has been carried out. The methods of analysis were fully validated, using HPLC-FLD for AFs and UHPLC-MS/MS for the remaining mycotoxins as analytical techniques, while solid-liquid extraction was used as sample treatment. All the samples fulfilled the requirement about the maximum or recommended content of AFB1, OTA, FB1+FB2, T-2+HT-2 and DON. However, several samples showed contents of ZEA above the recommended levels.

This survey on mycotoxin occurrence in pig feed contains one of the most complete analyses, quantifying 19 different mycotoxins, including five emerging ones. Although compliance with EU regulations has been generally high, the study showed the high occurrence of emerging mycotoxins (100% contaminated samples), as well as the high co-occurrence of different mycotoxins in the same sample.

Livestock species respond differently to mycotoxin poisoning, so that clinical signs can be difficult to detect and a correct analysis of the food is the most appropriate way to diagnose and detect it. Moreover, the synergistic interaction of co-occurring mycotoxins is well-known, hence regulations for maximum levels should be set including some particular combinations, focusing on emerging mycotoxins such as ENNs. These results highlight the necessity for a reinforcement of quality control of feeds with continuous monitoring from feed mill to pig feeder.

4. Materials and Methods

4.1. Chemicals and Reagents

All reagents were of analytical reagent grade and solvents were HPLC grade. Acetonitrile (MeCN), methanol (MeOH) and ammonium formate were obtained from VWR BDH Prolabo (West Chester, PA, USA). Formic acid was supplied by Merck (Darmstadt, Germany). Sodium chloride, magnesium sulfate, sodium citrate, disodium hydrogen citrate sesquihydrate and sodium phosphate dihydrogen monohydrate were purchased from Panreac Química (Barcelona, Spain).

Ultrapure water used throughout the work was obtained from a Milli-Q water purification system (18.2 MΩ cm-1, Milli-Q Plus system, Millipore, Bedford, MA, USA).

AFs (AFB1, AFB2, AFG1, AFG2) were purchased from Sigma-Aldrich (St. Louis, MO, USA); OTA, DON, ZEA, CIT, T-2, HT-2, fusarenon-X (F-X), STE, FB1, FB2 and ENNs (ENNA, ENNB and ENNB1) were purchased from Techno Spec (Barcelona, Spain), while BEA and ENNA1 were supplied by VWR International Eurolab, S.L. (Barcelona, Spain). Individual stock standard solutions were prepared for all of them in MeCN and kept in the dark at −18 °C.

Different intermediate solutions were prepared due to the differences in the analytical method and sample treatment. The first solution was composed of ENNA, ENNA1, ENNB, ENNB1 and BEA at a concentration of 10 μg mL^{-1}. The second solution was composed of FB1 and FB2, DON, ZEA, CIT, OTA, T-2, HT-2, F-X and STE at a concentration of 1 μg mL^{-1} each mycotoxin. The third solution was composed of AFB1, AFB2, AFG1 and AFG2 at a concentration of 50 μg mL^{-1}. Working solutions were prepared by diluting the intermediate solutions in MeOH:water (50:50).

Syringe filters (25 mm, 0.2 μm nylon membrane, VWR, West Chester, PA, USA) were used for filtration of samples prior to the injection into the chromatographic system.

4.2. Instrumentation and Equipment

The determination of AFs was carried out using a modular HPLC system consisting of a quaternary low pressure gradient pump (Model PU-2089, Jasco, Tokyo, Japan); an autosampler with a 100 μL loop (Model AS-2055, Jasco, Tokyo, Japan); a C18 Kinetex separation column (150 mm × 4.6 mm, 2.6 μm) from Phenomenex (Torrance, CA, USA) placed in a column oven; a UV derivatization module, which consists of a photochemical reactor specific for the analysis of aflatoxins with a 254 nm lamp (LCTech, Dorfen, Germany) and finally, a fluorescence detector (Model FP 2020, Jasco, Tokyo, Japan) used to acquire the signals. ChromNAV software (1.09.03 version, Jasco, Tokyo, Japan) was used for data acquisition and processing of AFs.

In order to determine the remaining mycotoxins an Agilent 1290 Infinity LC system (Agilent Technologies, Waldbronn, Germany) equipped with a binary pump, an on-line degasser, autosampler and a column oven placing a Zorbax Eclipse Plus RRHD C18 column (50 mm × 2.1 mm, 1.8 µm) coupled to a triple quadrupole (QqQ) mass spectrometer API 3200 (AB SCIEX, Toronto, ON, Canada) with electrospray ionization (ESI) was used. Data were processed using the Analyst Software version 1.5 with schedule multiple reaction monitoring (MRM) TM Algorithm (AB SCIEX).

For sample treatment, an evaporator system (System EVA-EC, from VLM GmbH, Bielefeld, Germany), a vortex-2 Genie (Scientific Industries, Bohemia, NY, USA) and a universal 320R centrifuge (HettichZentrifugen, Tuttlingen, Germany) were used.

4.3. Samples

A total of 228 samples of pig feed were collected from February to August 2017 at different farms and manufacturing industries around Spain. Samples (approximately 500 g) were sent directly to the laboratory by the veterinarians responsible for farms or feed mills. Analytical personnel were not present to oversee the sampling. They included: 71 compound feed samples destined to fattening pigs, 42 to sows, 111 to piglets, 2 to gilts, and 2 samples of maize (grain given to the fattening pigs in two farms). The information of each sample comprised: animals to which it was destined, presentation (flour or pellet) and sampling point (silo, feeder, sack and bulk storage without container in warehouse). Samples were milled and homogenised using a standard grinder. Finally, all samples were stored at room temperature until analysis.

4.4. Sample Preparation

Two different extractions were carried out: one for determination of AFs by HPLC-FLD and the other for the rest of mycotoxins by UHPLC-MS/MS.

4.4.1. Aflatoxins

A previously reported method was adapted [35]. Briefly, 2 g of homogenised sample were placed into a 50-mL screw cap test tube with conical bottom. The sample was extracted with 10 mL of MeCN, shaking by vortex for 3 min. The sample was centrifuged at 4500 rpm for 5 min. Afterwards, 2 mL of the upper layer was pipetted into a vial and dried under a gentle stream of N2. Finally, the residue was dissolved with 1 mL of MeOH:water (50:50, *v/v*) and the solution thus obtained was filtered through a 0.22 µm nylon membrane filter before injection into the HPLC-FLD system.

4.4.2. Multimycotoxins, Enniatins and Beauvericin

Two g of homogenised sample and 8 mL of water were placed into a 50-mL screw cap test tube with conical bottom, shaken by vortex for 1 min. Then, 10 mL of solvent (MeCN, 5% formic acid) were added to the tube, shaken by vortex for 3 min, following by the addition of salts (4 g $MgSO_4$, 1 g NaCl), shaken vigorously for 2 min. After these steps, the tubes were centrifuged at 4500 rpm for 5 min. Finally, 2 mL of the organic layer was evaporated to dryness under a gentle stream of N_2. The residue was dissolved with 1 mL of MeOH:water (50:50, *v/v*). Each solution was filtrated with 0.22 µm membrane filter prior to the UHPLC-MS/MS analysis.

4.5. Chromatographic Separation and Detection

In order to achieve the lowest limits of quantification, determination of the different mycotoxins was carried out in three different analyses: AFs were determined by HPLC-FLD, multimycotoxins by UHLC-MS/MS and emerging mycotoxins (ENNs and BEA) in a different UHPLC-MS/MS analysis.

4.5.1. Determination of Aflatoxins by HPLC-FLD

The chromatographic separation was carried out according to Arroyo-Manzanares et al. [35]. The chromatographic separation was achieved using a C18 Kinetex separation column (150 mm × 4.6 mm, 2.6 μm). The mobile phase consisted of three different solvents: eluent A (water), eluent B (MeOH) and eluent C (MeCN) and a linear gradient elution was used as follows, keeping constant eluent B at 27%: 0–3 min: 0% C; 20 min: 13% C; and 21–23 min: 68% C, followed by an equilibration time of 10 min. The injection volume was 50 μL and the flow rate was 1 mL min^{-1}. The temperature of the column was kept constant at 30 °C. Excitation and emission wavelengths of the FLD for the determination of the aflatoxins were set at 365 nm and 460 nm, respectively. The FLD was working at gain x100.

4.5.2. Determination of Multimycotoxins by LC-MS/MS

Conditions similar to those reported in a previous paper were used [36]. The chromatographic separation of multimycotoxins (OTA, FB1, FB2, T-2, HT-2, STE, CIT, F-X, DON, ZEA) was carried out on a C18 Zorbax Eclipse Plus RRHD column (50 × 2.1 mm, 1.8 μm) and as mobile phase eluent A (5 mM ammonium formate aqueous solution, 0.3% formic acid), and eluent B (MeOH with 5 mM ammonium formate, 0.3% formic acid), under the following gradient conditions: 0 min: 5% B; 1 min: 50% B; 4 min: 80% B; 6 min: 90% B, 6.20 min: 5% B and 8 min: 5% B. The injection volume was 5 μL, the flow rate was 0.4 mL min^{-1} and the temperature of the column was 35 °C.

MRM conditions of the mass spectrometer for the determination of multimycotoxins are shown in Supplementary Materials (Table S1). The working conditions of the mass spectrometer were established in positive ESI mode under the following conditions: temperature of the source 500 °C; voltage of the ion spray 5 KV, collision gas (nitrogen) 5 psi; curtain gas (nitrogen) 30 psi; GAS 1 and GAS 2 (both of them nitrogen) 50 psi. In all circumstances, an ion was used as precursor and two ions were obtained from the fragmentation process: the most abundant one for quantification (Q) and the other for confirmation (I).

4.5.3. Determination of Enniatins and Beauvericin by LC-MS/MS

The same column (C18 Zorbax Eclipse Plus RRHD) was used for separation of the five emerging mycotoxins. However, the mobile phase did not include ammonium formate in its composition in order to avoid the formation of ammonium adducts. Thus, this mobile phase consisted of eluent A (0.3% formic acid aqueous solution), and eluent B (MeOH with 0.3% formic acid). The following linear gradient elution was used: 0 min: 70% B; 2 min: 90% B; 4 min: 90% B; 4.2 min: 70% B, and 6 min: 70% B. The flow rate was 0.4 mL min^{-1}. The injection volume was 5 μL and the column temperature was 35 °C.

MRM conditions of the mass spectrometer are shown in the Supplementary Materials (Table S1). The working conditions of the mass spectrometer were established in positive ESI mode, working under the conditions described for multimycotoxin determination.

4.6. Method Development

Optimisation of Sample Preparation

The extraction of AFs was performed according to a previously reported method [35], so no further optimisation was required.

Extraction of multimycotoxins and emerging mycotoxins was optimised, trying to obtain one single sample treatment suitable for the extraction of all mycotoxins. With this purpose, solid-liquid extraction with and without partitioning process were tested using 2 g of sample, as follows: (a) 10 mL MeCN with 5% formic acid; (b) 10 mL MeOH with 5% formic acid; (c) 8 mL H_2O + 10 mL MeCN with 5% formic acid + 4 g $MgSO_4$ + 1 g NaCl; (d) 8 mL H_2O + 10 mL MeCN with 5% formic acid + 4 g $MgSO_4$ + 1 g NaCl + 1 g sodium citrate + 0.5 g disodium hydrogen citrate sesquihydrate; (e) 8 mL 30 mM NaH_2PO_4 buffer pH 7.1 +10 mL MeCN with 5% formic acid + 4 g $MgSO_4$ + 1 g NaCl + 1 g sodium citrate + 0.5 g disodium hydrogen citrate sesquihydrate; and (f) 8 mL H_2O + 10 mL MeCN with 5% formic acid + 4 g $MgSO_4$, 1 g NaCl). The best results in terms of recovery were obtained with conditions described in (c).

Subsequently, the volume of extraction solvent (MeCN, 5% formic acid) was studied within 5 and 10 mL, 10 mL being chosen as optimum in view of the worst recoveries if lower extraction volumes were utilised. Afterwards, the volume of extract to dry was studied, choosing a volume of 2 mL as a compromise to achieve a short time of sample treatment and low limits of detection (LODs).

4.7. Method Validation

The validation of each method comprised the establishment of the calibration curves in the presence of matrix, evaluation of LODs and limits of quantification (LOQs), selectivity, precision (as interday and intraday precision at different concentration levels) and extraction recovery studies at different concentration levels. In all the cases, a sample of feed previously analysed and free of mycotoxins was chosen as the representative matrix.

4.7.1. Aflatoxins

As stated previously, this method had been previously validated for different kinds of feeds, including pig feeding [35]. The method provided good linearity in the concentration range 1–50 μg kg^{-1} (R^2 > 0.99), LOQs of 1 μg kg^{-1}, precisions expressed as RSD lower than 7.9% and 9.2% for intraday and interday studies, respectively, and recoveries higher than 85%.

4.7.2. Multimycotoxins

For the evaluation of linearity, calibration curves were obtained by spiking portions of a representative feed sample with different concentrations of analytes, depending on the analytical sensitivity for each mycotoxin. Samples were submitted to the sample treatment and analysed and the area of the Q ion was selected as analytical signal. LODs and LOQs were calculated as 3 × S/N and 10 × S/N, respectively. The results are shown in Table 3.

Table 3. Calibration curves, LODs and LOQs for multimycotoxin, enniatins and beauvericin determination.

	Calibration Curve	Linear Range (μg kg^{-1})	R^2	LOD (μg kg^{-1})	LOQ (μg kg^{-1})
DON	y = 3.5279x + 100.67	86.1–1000	0.998	26	86
F-X	y = 1.0892x − 8.111	135.9–2000	0.995	41	136
CIT	y = 395.99x − 13280	24.3–500	0.999	7.3	24
HT-2	y = 8.555x − 29.894	33.9–2000	0.998	10	34
FB1	y = 11.103x + 347.53	22.2–500	0.999	6.7	22
T-2	y = 40.038x + 1245.4	26.8–2000	0.995	8.1	27
ZEA	y = 29.141x − 758.98	99.0–500	0.998	30	99
OTA	y = 82.66x + 333.23	8.1–100	0.993	2.5	8.1
STE	y = 208.73x − 848.16	9.9–100	0.991	3.0	9.9
FB2	y = 13.57x + 112.34	38.5–500	0.999	12	38
ENNB	y = 987.45x − 2943.8	2.00–500	0.998	0.60	2.0
ENNB1	y = 1482.5x + 9430.2	2.53–500	0.999	0.76	2.5
BEA	y = 372.32x + 0.0047	1.72–500	0.998	0.52	1.7
ENNA1	y = 1092.9x + 3989.6	4.52–500	0.999	1.4	4.5
ENNA	y = 1578.8x + 3790.5	2.94–500	0.999	0.88	2.9

The matrix effect was evaluated by relative comparison of the analyte peak area of a spiked extract of sample and a standard solution at the same concentration. Ion suppression was observed for most of the analytes, so calibration in presence of matrix was mandatory for quantification. These studies, as well as intra and interday precision and recovery studies, were carried out at three different concentration levels (three samples, injected in triplicate). The results are shown in Supplementary Materials (Tables S2 and S3).

4.7.3. Enniatins and Beauvericin

A similar procedure was followed for validation of UHPLC-MS/MS determination of emerging mycotoxins. Calibration curves were obtained by spiking a feed sample with concentrations of analytes up to 500 μg kg^{-1}, submitted to the whole procedure and analysed. Table 3 summarises the results.

The matrix effect, precision and trueness were evaluated at three different concentration levels (three samples, injected in triplicate). Once again, ion suppression was observed for all the analytes, so calibration in presence of matrix was mandatory for quantification. The results are shown in Supplementary Materials (Tables S2 and S3).

4.8. Data Processing

One-way ANOVA studies were carried out for total content of mycotoxins as well as for each mycotoxin individually, considering different factors, namely: presentation (pellet or flour), sampling point (silo, feeder, sack and bulk storage without container in warehouse), and type of animal feed (piglet, gilts, sows and fattening pigs). When this information was not available, the sample was not included in the study. These studies were performed to determine statistical differences with a level of confidence of 95%.

PRESENTATION. A total of 226 samples were analysed, 183 presented as flour and 43 as pellet (n = 43). The two samples of maize (grain) were excluded from this study.

SAMPLING POINT. In this case, a total of 137 were processed, 77 samples were collected from a silo, 29 from a feeder, 25 from a sack and 6 were stored in bulk without container in warehouse.

TYPE OF ANIMAL. Another factor is the type of pig for which the compound feed is intended. In this sense, different types of feed (classified according to the necessities of the animal) can be distinguished: piglets (n = 111), fattening (n = 71), gilts (n = 2) and sows (n = 42). The two samples of maize (grain) were excluded from this study.

Supplementary Materials: The following are available online at http://www.mdpi.com/2072-6651/11/6/342/s1, Table S1: Monitored ions of multimycotoxins, enniatins and beauvericin, and MS/MS parameters; Table S2: Matrix effect (%ME), recovery (%R) and recovery precision (%RSD) for multimycotoxin enniatins and beauvericin determination (n = 9); Table S3: Precision study for multimycotoxin, enniatins and beauvericin determination at different concentration levels (n = 9).

Author Contributions: Conceptualization, L.G.-G., A.M.G.-C. and V.R.-E.; methodology, L.G.-G. and N.A.-M.; validation, L.G.-G. and N.A.-M.; formal analysis, N.A.-M.; investigation, P.A.-F. and N.A.-M.; resources, V.R.-E. and A.M.G.-C.; writing—original draft preparation, L.G.-G. and N.A.-M.; writing—review and editing, V.R.-E. and A.M.G.-C.; visualization, L.G.-G., V.R.-E. and N.A.-M.; supervision, A.M.G.-C. and L.G.-G.; project administration, A.M.G.-C.; funding acquisition, A.M.G.-C.

Funding: This research was funded by SPANISH MINISTRY OF ECONOMY AND COMPETITIVENESS and EUROPEAN REGIONAL DEVELOPMENT FUND (MINECO/FEDER, UE), grant number AGL2015-70708-R.

Conflicts of Interest: The authors declare no conflict of interest.

References

1. IARC Monographs Vol. 100F. Chemical Agents and Related Occupations. Aflatoxins. 2012. Available online: https://monographs.iarc.fr/wp-content/uploads/2018/06/mono100F-23.pdf (accessed on 31 March 2019).
2. IARC Monographs Vol. 56. Some Naturally Occurring Substances: Food Items and Constituents, Heterocyclic Aromatic Amines and Mycotoxins. Ochratoxin A. 1993. Available online: https://monographs.iarc.fr/wp-content/uploads/2018/06/mono56-18.pdf (accessed on 31 March 2019).

3. IARC Monographs Vol. 56. Some Naturally Occurring Substances: Food Items and Constituents, Heterocyclic Aromatic Amines and Mycotoxins. Toxins Derived from Fusarium moniliforme: fumonisin B1 and B2 and fusarin C. 1993. Available online: https://monographs.iarc.fr/wp-content/uploads/2018/06/mono56-15.pdf (accessed on 5 June 2019).
4. Rapid Alert System for Food and Feed (RASFF). Available online: https://ec.europa.eu/food/safety/rasff_en (accessed on 30 April 2019).
5. Regulation, E.C. Commission Regulation (EU) No 574/2011 of 16 June 2011 amending Annex I to Directive 2002/32/EC of the European Parliament and of the Council as regards maximum levels for nitrite, melamine. *Ambrosia spp. and carry-over of certain coccidiostats and histomonostats and consolidating Annexes I and II thereto. Off. J. Eur. Union L* **2011**, *L159*, 7–24.
6. European Commission. Commission Recommendation of 17 August 2006 on the presence of deoxynivalenol, zearalenone, ochratoxin A, T-2 and HT-2 and fumonisins in products intended for animal feeding. *Off. J. Eur. Union L* **2006**, *229*, 7–9.
7. European Commission. Commission Recommendation of 27 March 2013 on the presence of T-2 and HT-2 toxin in cereals and cereal products. *Off. J. Eur. Union L* **2013**, *91*, 12–15.
8. European Food Safety Authority (EFSA). Available online: http://www.efsa.europa.eu/en/topics/topic/mycotoxins (accessed on 30 April 2019).
9. EFSA Panel on Contaminants in the Food Chain (CONTAM). Scientific Opinion on the risks to human and animal health related to the presence of beauvericin and enniatins in food and feed. *EFSA J.* **2014**, *12*, 3802. [CrossRef]
10. Pierron, A.; Alassane-Kpembi, I.; Oswald, I.P. Impact of mycotoxin on immune response and consequences for pig health. *Anim. Nutr.* **2016**, *2*, 63–68. [CrossRef]
11. Kanora, A.; Maes, D. The role of mycotoxins in pig reproduction: A review. *Vet. Med.* **2009**, *54*, 565–576. [CrossRef]
12. Hennig-Pauka, I.; Koch, F.J.; Schaumberger, S.; Woechtl, B.; Novak, J.; Sulyok, M.; Nagl, V. Current challenges in the diagnosis of zearalenone toxicosis as illustrated by a field case of hyperestrogenism in suckling piglets. *Porcine Health Manag.* **2018**, *4*, 18–26. [CrossRef]
13. Van Limbergen, T.; Devreese, M.; Croubels, S.; Broekaert, N.; Michiels, A.; De Saeger, S.; Maes, D. Role of mycotoxins in herds with and without problems with tail necrosis in neonatal pigs. *Vet. Rec.* **2017**, *181*, 539–546. [CrossRef]
14. Ghareeb, K.; Awad, W.A.; Böhm, J.; Zebeli, Q. Impacts of the feed contaminant deoxynivalenol on the intestine of monogastric animals: Poultry and pig. *J. Appl. Toxicol.* **2015**, *35*, 327–337. [CrossRef]
15. Li, R.; Li, Y.; Su, Y.; Shen, D.; Dai, P.; Li, C. Short-term ingestion of deoxynivalenol in naturally contaminated feed alters piglet performance and gut hormone secretion. *Anim. Sci. J.* **2018**, *89*, 1134–1143. [CrossRef]
16. Reddy, K.E.; Jeong, J.Y.; Song, J.; Lee, Y.; Lee, H.J.; Kim, D.W.; Jung, H.J.; Kim, K.H.; Kim, M.; Oh, Y.K.; et al. Colon microbiome of pigs fed diet contaminated with commercial purified deoxynivalenol and zearalenone. *Toxins* **2018**, *10*, 347. [CrossRef] [PubMed]
17. Souza, C.F.; Da Silva, A.S.; Müller, L.K.F.; Baldissera, M.D.; Bottari, N.B.; Schetinger, M.R.C.; Santurio, J.M.; Gloria, E.M.; Machado, G.; Zanette, R.A.; et al. Changes of adenosinergic system in piglets fed a diet co-contaminated by mycotoxin and their effects on the regulation of adenosine. *Microb. Pathog.* **2018**, *114*, 328–332. [CrossRef] [PubMed]
18. Streit, E.; Naehrer, K.; Rodrigues, I.; Schatzmayr, G. Mycotoxin occurrence in feed and feed raw materials worldwide: Long-term analysis with special focus on Europe and Asia. *J. Sci. Food Agric.* **2013**, *93*, 2892–2899. [CrossRef] [PubMed]
19. Alonso, V.A.; Pereyra, C.M.; Keller, L.A.M.; Dalcero, A.M.; Rosa, C.A.R.; Chiacchiera, S.M.; Cavaglieri, L.R. Fungi and mycotoxins in silage: An overview. *J. Appl. Microbiol.* **2013**, *115*, 637–643. [CrossRef] [PubMed]
20. Zhang, Y.; Caupert, J. Survey of mycotoxins in U.S. distiller's dried grains with solubles from 2009 to 2011. *J. Agric. Food Chem.* **2012**, *60*, 539–543. [CrossRef] [PubMed]
21. Guerre, P. Worldwide mycotoxins exposure in pig and poultry feed formulations. *Toxins* **2016**, *8*, 350. [CrossRef] [PubMed]
22. Streit, E.; Schatzmayr, G.; Tassis, P.; Tzika, E.; Marin, D.; Taranu, I.; Tabuc, C.; Nicolau, A.; Aprodu, I.; Puel, O.; et al. Current situation of mycotoxin contamination and co-occurrence in animal feed-Focus on Europe. *Toxins* **2012**, *4*, 788–809. [CrossRef]

23. Romera, D.; Mateo, E.M.; Mateo-Castro, R.; Gómez, J.V.; Gimeno-Adelantado, J.V.; Jiménez, M. Determination of multiple mycotoxins in feedstuffs by combined use of UPLC–MS/MS and UPLC–QTOF–MS. *Food Chem.* **2018**, *267*, 140–148. [CrossRef]
24. Ma, R.; Zhang, L.; Liu, M.; Su, Y.T.; Xie, W.M.; Zhang, N.Y.; Dai, J.F.; Wang, Y.; Rajput, S.A.; Qi, D.S.; et al. Individual and combined occurrence of mycotoxins in feed ingredients and complete feeds in China. *Toxins* **2018**, *10*, 113. [CrossRef]
25. van der Fels-Klerx, H.J.; Adamse, P.; Punt, A.; van Asselt, E.D. Data analyses and modelling for risk based monitoring of mycotoxins in animal feed. *Toxins* **2018**, *10*, 54. [CrossRef]
26. Qian, M.; Yang, H.; Li, Z.; Liu, Y.; Wang, J.; Wu, H.; Ji, X.; Xu, J. Detection of 13 mycotoxins in feed using modified QuEChERS with dispersive magnetic materials and UHPLC–MS/MS. *J. Sep. Sci.* **2018**, *41*, 756–764. [CrossRef] [PubMed]
27. Gruber-Dorninger, C.; Jenkins, T.; Schatzmayr, G. Multimycotoxin screening of feed and feed raw materials from Africa. *World Mycotoxin J.* **2018**, *11*, 369–383. [CrossRef]
28. Changwa, R.; Abia, W.; Msagati, T.; Nyoni, H.; Ndleve, K.; Njobeh, P. Multimycotoxin occurrence in dairy cattle feeds from the Gauteng province of South Africa: A pilot study using UHPLC-QTOF-MS/MS. *Toxins* **2018**, *10*, 294. [CrossRef] [PubMed]
29. Abdallah, M.F.; Girgin, G.; Baydar, T.; Krska, R.; Sulyok, M. Occurrence of multiple mycotoxins and other fungal metabolites in animal feed and maize samples from Egypt using LC-MS/MS. *J. Sci. Food Agric.* **2017**, *97*, 4419–4428. [CrossRef] [PubMed]
30. Kovalsky, P.; Kos, G.; Nährer, K.; Schwab, C.; Jenkins, T.; Schatzmayr, G.; Sulyok, M.; Krska, R. Co-occurrence of regulated, masked and emerging mycotoxins and secondary metabolites in finished feed and maize–An extensive survey. *Toxins* **2016**, *8*, 363. [CrossRef] [PubMed]
31. Li, X.; Zhao, L.; Fan, Y.; Jia, Y.; Sun, L.; Ma, S.; Cheng, J.; Ma, Q.; Zhang, J. Occurrence of mycotoxins in feed ingredients and complete feeds obtained from the Beijing region of China. *J. Anim. Sci. Biotechnol.* **2014**, *5*, 37–44. [CrossRef] [PubMed]
32. Reddy, K.E.; Song, J.; Lee, H.J.; Kim, M.; Kim, D.W.; Jung, H.J.; Kim, B.; Lee, Y.; Yu, D.; Kim, D.W.; et al. Effects of high levels of deoxynivalenol and zearalenone on growth performance, and hematological and immunological parameters in pigs. *Toxins* **2018**, *10*, 114. [CrossRef] [PubMed]
33. Smith, M.C.; Madec, S.; Coton, E.; Hymery, N. Natural co-occurrence of mycotoxins in foods and feeds and their in vitro combined toxicological effects. *Toxins* **2016**, *8*, 94. [CrossRef] [PubMed]
34. Turner, N.W.; Bramhmbhatt, H.; Szabo-Vezse, M.; Poma, A.; Coker, R.; Piletsky, S.A. Analytical Methods for determination of mycotoxins: An update (2009–2014). *Anal. Chim. Acta* **2015**, *901*, 12–33. [CrossRef]
35. Arroyo-Manzanares, N.; Huertas-Pérez, J.F.; García-Campaña, A.M.; Gámiz-Gracia, L. Aflatoxins in animal feeds: A straightforward and cost-effective analytical method. *Food Control.* **2015**, *54*, 74–78. [CrossRef]
36. Arroyo-Manzanares, N.; Huertas-Pérez, J.F.; García-Campaña, A.M.; Gámiz-Gracia, L. Simple methodology for the determination of mycotoxins in pseudocereals, spelt and rice. *Food Control.* **2014**, *36*, 94–101. [CrossRef]

toxins

MDPI

Article

Combined (d)SPE-QuEChERS Extraction of Mycotoxins in Mixed Feed Rations and Analysis by High Performance Liquid Chromatography-High-Resolution Mass Spectrometry

Rocio Facorro [1,2], Maria Llompart [2] and Thierry Dagnac [1,*]

1 Galician Agency for Food Quality—Agronomic and Agrarian Research Centre (AGACAL-CIAM), Unit of Organic Contaminants, Apartado 10, 15080 A Coruña, Spain; rocio.facorro@rai.usc.es

2 Laboratory of Research and Development of Analytical Solutions (LIDSA), Department of Analytical Chemistry, Nutrition and Food Science, Faculty of Chemistry, E-15782 Campus Vida, Universidad de Santiago de Compostela, 15782 Santiago de Compostela, Spain; maria.llompart@usc.es

* Correspondence: thierry.dagnac@xunta.es; Tel.: +34-881-881-827

Received: 16 December 2019; Accepted: 13 March 2020; Published: 23 March 2020

check for updates

Abstract: The objective of this work was the development of a methodology capable of simultaneously determine 26 mycotoxins in mixed feed rations collected in 20 dairy farms. A sample preparation methodology based on a combination of (d)SPE and QuEChERS extractions was used. Liquid chromatography-high resolution mass spectrometry was employed for both identification and quantification purposes. To this respect, a powerful workflow based on data-independent acquisition, consisting of fragmenting all precursor ions entering the mass spectrometer in narrow m/z isolation windows (SWATH), was implemented. SWATH data file then contains all the information that would be acquired in a multitude of different experimental approaches in a single all-encompassing dataset. Analytical method performance was evaluated in terms of linearity, repeatability and matrix effect. Relative recoveries were also measured, giving values above 80% for most compounds. Matrix-matched calibration was carried out and enabled reaching the low ng mL^{-1} level for many mycotoxins. The observed matrix effect, in most cases suppressive, reached even values higher than 60%. The repeatability was also adequate, showing a relative standard deviation lower than 10%. All unified samples analyzed showed co-occurrence of two or more mycotoxins, recurrently zearalenone, fumonisin B1, and β-zearalenol, with an occurrence frequency ranging from 60% to 90%.

Keywords: Mycotoxins; mixed feed rations; QuEChERS; dispersive solid phase extraction; liquid chromatography; high-resolution mass spectrometry; data independent SWATH

Key Contribution: The optimized analytical methodology, based on QuEChERS with (d)SPE clean-up and HPLC-HRMS analysis by means of SWATH mode, allowed the identification and quantification of 26 mycotoxins in 97 mixed ration samples collected in 20 dairy farms (NW Spain). Fumonisins B1 and B2, zearalenone and its metabolite β-zearalenol, cyclopiazonic acid and enniatin B were the mycotoxins most frequently found. All samples showed co-occurrence of two or more of these mycotoxins, recurrently zearalenone and fumonisin B1. None of the samples showed concentrations higher than the maximum values established for feed intended for dairy cows.

1. Introduction

Mycotoxins are natural secondary metabolites produced by various filamentous fungi, mainly belonging to *Aspergillus*, *Penicillium*, and *Fusarium* genera. Around 400 of these metabolites are identified and classified, being the most predominant zearalenone, ochratoxin A, fumonisins and trichothecenes [1]. Mycotoxins can be produced during all stages; from pre-harvest, when plants are growing, to storage, transport or further processing [2].

Mycotoxins can cause some adverse effect on animal and human health. They can be carcinogenic agents (aflatoxin B_1 (AFB_1), fumonisins B1 and B2, (FB_1 and FB_2)), present estrogenic activity (zearalenone (ZEA)), or teratogenic (ochratoxin A, (OTA)) and immunotoxinogenic effects (deoxynivalenol, (DON)) [3]. In consequence, the presence and intake of mycotoxins can also strongly reduce animal productivity in farms. The severity of the damage depends on the compound, the degree of exposure and the possible co-occurrence of various mycotoxins. This is a frequent situation, such is the case with zearalenone and deoxynivalenol or zearalenone/deoxynivalenol/nivalenol/fumonisin B_1 [4].

Most studies concerning mycotoxins occurrence were carried out in food matrices, cereals [5–8] feedstuff [1,9,10], plant based beverages [11], and ruminant milk [12]. Nonetheless, some researchers also found traces of these toxins in human breast milk [13,14]. Due to mycotoxins stability to high temperatures and persistence during food preparation procedures [15], they present a risk to the whole food chain.

With the purpose of protecting consumers against mycotoxin intake, maximum levels of aflatoxins (AFB1, and sum of AFB1, AFB2, AFG1, and AFG2), sum of FB1 and FB2, DON, OTA, and ZEA, were established by the European Union for products intended for animal feeding [16,17]. In 2013, maximum recommended levels were set for the sum of toxins T-2 and HT-2 in cereals, as well [18].

Due to the hazard mycotoxins may cause, sensitive and selective analytical techniques are needed for their monitoring in food and feed. Besides, because of their frequent co-occurrence, the European Food Safety Authority (EFSA) recommends multianalyte methodologies. Nevertheless, finding an appropriate methodology, in terms of sample preparation, for a wide number of mycotoxins can be a tough task due to the structural variety and different chemical properties of the compounds.

Many sample treatment procedures for the simultaneous determination of multiple mycotoxins have been published. From classic approaches like solid liquid extraction (SLE) and liquid–liquid extraction (LLE) to the most recent ones, like modified Quick, Easy, Cheap, Effective, Rugged, and Safe (QuEChERS) extraction or matrix solid-phase dispersion (MSPD). Modified QuEChERS procedure is the sample pre-treatment of choice in many recent publications, due to its simplicity and cost-efficiency [8,9,19,20].

In the case of food and feed samples, clean-up steps are usually needed in order to reduce the matrix effect, especially in the case of liquid chromatography analyses. Solid phase extraction (SPE) and dispersive solid phase extraction (dSPE) are the most used as described in literature [19,21–23].

In terms of analysis, two different approaches are currently employed to control the presence of mycotoxins in food and feed. For screening and semi-quantitative analysis, enzyme-linked immune-sorbent assay (ELISA) and thin-layer chromatography (TLC) are the most frequently used. To provide accurate quantitative information and therefore reduce the number of possible false-positive results, chromatographic techniques such as gas chromatography (GC) and high-performance liquid chromatography (HPLC) coupled to ultraviolet (UV), fluorescence (FL), and mass spectrometry (MS) detectors are used [24,25].

HPLC coupled to tandem mass spectrometry (MS/MS) is considered the most reliable technique in order to achieve high specificity and sensitivity for the simultaneous determination of a large number of mycotoxins, reaching low limits of detection (LOD) and quantification (LOQ). Several studies using this technique for the analysis of multi-mycotoxins in food and feed commodities have been published [9,10,12,26,27].

More recently, high-resolution mass spectrometry (HRMS) has been implemented either in Data Dependent Acquisition mode or in Data Independent Acquisition mode. Besides providing high specificity and resolution due to the use of the accurate mass, HRMS allows for performing non-target screening. Quadrupole time-of-flight (QTOF) instruments offer the advantages of TOF detectors in terms of resolution and the possibility of acquiring product ion spectra working in MS/MS mode using accurate mass. This feature provides additional confirmatory parameters besides retention time and ion fragment ratio, i.e., exact mass, isotope pattern and spectral comparison against spectra libraries with smart confirmation criteria. Among the applications published employing quadrupole time-of-flight and Orbitrap® detectors [1,19,28,29] for the mycotoxin analysis in food and feed, the main part of them use Orbitrap technique.

Hence, the aim of this work was to develop a methodology that allowed the simultaneous analysis of multi-class mycotoxins in mixed rations. A QuEChERS extraction method was implemented, with the addition of a clean-up step based on solid phase extraction and dispersive solid phase extraction, in order to minimize the effect of the matrix. The analytical methodology was carried out using HPLC followed by triple quadrupole-time-of-flight detection based on a data independent workflow.

The method was optimized and validated for 26 mycotoxins produced by different genera. Several analytical parameters, such as linearity, repeatability, accuracy, matrix effect, measurement uncertainties, limits of detection, and recovery, were evaluated.

A monitoring of these mycotoxins was carried out using the proposed methodology analyzing 97 mixed ration samples collected in 20 dairy farms from Galicia (NW Spain) over one year. These rations may constitute the basis of the diet for dairy cows during 6 to 8 months of the year.

2. Results and Discussion

2.1. Chromatographic Analysis

The aim of this work was to develop an analytical methodology based on high performance liquid chromatography coupled to triple quadrupole-time of flight mass spectrometry for the simultaneous analysis of the 26 mycotoxins. Hence, the chromatographic conditions were optimized to achieve an efficient separation of the target compounds (see conditions in the Experimental section). The initial step was to consider the ionization of the target mycotoxins by injection of a standard containing all analytes, both in positive and negative mode. This allows for the selection of the most sensitive mode for each compound.

Two runs were needed for each analysis, since the methodology does not allow switching from one ionization mode to the other in such short time.

Different mobile phase configurations were evaluated. Being the solvents used always water as aqueous phase and methanol as organic phase, two different buffers were tested (ammonium formate and ammonium acetate, 3 mM), as well as the addition of an acidic component (formic acid and acetic acid, 0.1% v/v). The regression equations for matrix-matched calibration using each mobile phase configurations are specified in Tables S1 and S2. As shown, although the slopes obtained using ammonium acetate and ammonium formate did not present great differences, some frequently found compounds gave better responses using ammonium acetate rather than ammonium formate. Such is the case of deoxynivalenol acetylated derivatives (3+15-ADON), FB_2, cyclopiazonic acid (CPA) or alternariol (AOH) (examples in Figure 1). Besides, the addition of an acidic component (formic or acetic acid) caused a substantial signal suppression in almost all cases in comparison with the mobile phase without acid. Exceptionally, few compounds such as andrastin A (AND A), and toxins T-2 and HT-2 gave better responses with the addition of 0.1% of acid. Therefore, the mobile phase configuration selected was a combination of water (A) and methanol (B), both containing 3 mM ammonium acetate. The method performance will be then assessed with this mobile phase composition.

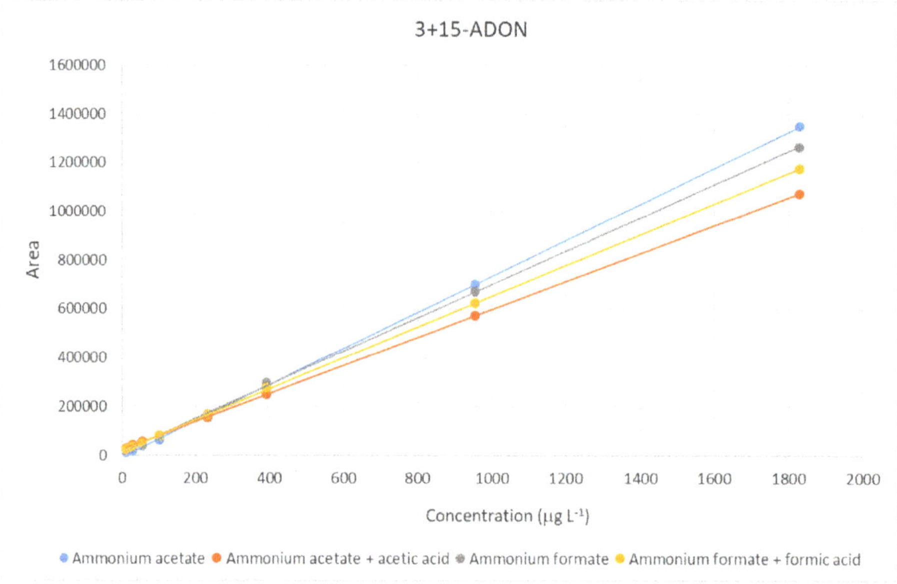

(a)

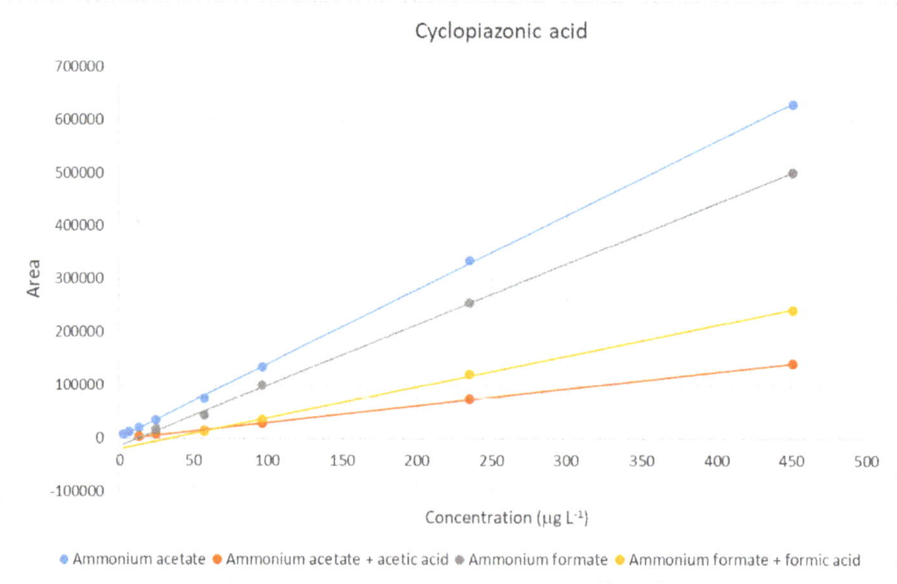

(b)

Figure 1. *Cont.*

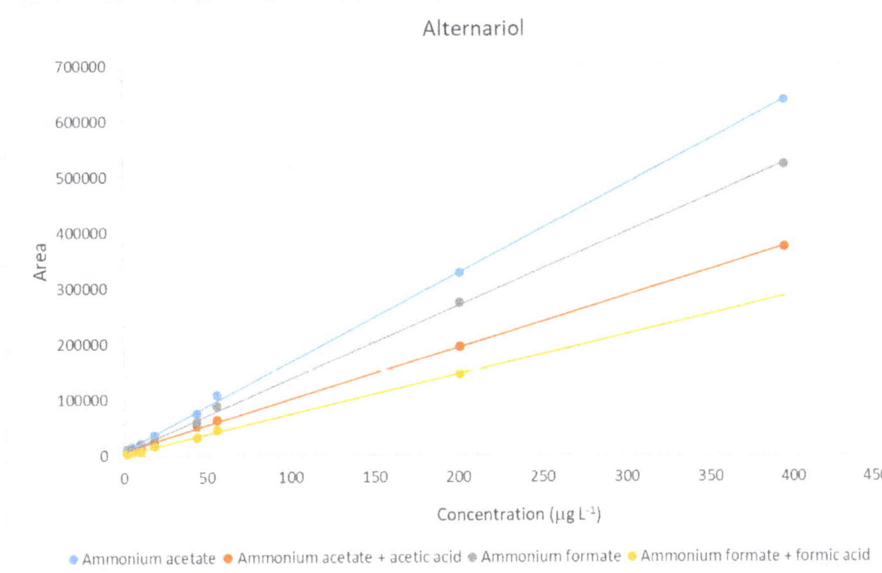

(c)

Figure 1. Response comparisons between the four mobile phase compositions: ammonium acetate 3mM, ammonium formate 3 mM and the addition of acetic or formic acid 0.1%, respectively for (**a**) 3+15 acetyldeoxynivalenol; (**b**) cyclopiazonic acid; and (**c**) alternariol.

The proposed chromatographic method offers a complete separation of all 26 mycotoxins and a good peak shape for quantification.

The only two exceptions were the isomers 3-DON and 15-DON for which the sum of the two metabolites was reported, as commonly done in previous published studies [29].

2.2. Method Performance Evaluation

Mixed ration is a difficult matrix to work with because of the complexity of its composition. Thus, to achieve a reliable detection and quantification of mycotoxins, and exhaustive analytical methodology is needed, beginning with a matrix effect evaluation. Most publications working with this kind of matrix emphasized the need of a matrix-matched calibration in order to properly evaluate the concentration of the analytes [1,21]. For this purpose, a comparison was made between a matrix-matched standard calibration and an external calibration prepared in MeOH. The concentration range values covered depended on the mycotoxin, but in all cases, the range was of two orders of magnitude (see individual ranges of concentration in Table 1) with eight concentration levels. Matrix-matched calibration was performed using the sample with the lowest analyte content found (blank sample), spiked in the same range of concentrations as for the external calibration. For most compounds, the signal showed a suppressive effect when working with matrix (Figure 2), showing a median value of −35%, with some compounds (e.g., HT-2 toxin and AOH) exhibiting over 70% of signal suppression. However, there were a few examples of signal enhancement, such as the case of FB_1 and FB_2. Because of this variance, further studies and quantification were carried out using the matrix-matched approach.

Table 1. HPLC-Triple time-of-flight (TOF) performance for the mycotoxin analysis: linearity, intra and inter-day precision, detection limits, recovery and matrix effect.

Compounds	R^2	Linear Range (ng mL^{-1})	Intra-Day Precision (%) (n = 3)	Inter-Day Precision (%) (n = 3)	LOQs (ng mL^{-1})	Recoveries				Matrix Effect (%)
						Concentration (ng mL^{-1})	Recovery (%) (% RSD)	Concentration (ng mL^{-1})	Recovery (%) (% RSD)	
3+15-ADON	0.9998	10.7–1828.3	2.4	5.8	24	339.6	85.8 (3.9)	95.4	90.5 (3.0)	−0.6
DON	0.9999	26.7–912.8	2.4	6.2	41.1	169.6	114.2 (0.8)	47.7	121.9 (8.7)	−22.4
ENN B	0.9999	40.6–905.4	2.5	5.9	46.8	168.2	88.8 (4.1)	47.6	132.6 (5.8)	33.9
ENN B_1	1.0000	13.5–479.7	1.1	1.4	24	170.6	58.3 (1.2)	47.9	67.4 (3.5)	−29.6
FB_1	0.9996	2.7–92.6	1.7	5.2	2.9	17.2	92.7 (2.2)	4.8	29.6 (6.6)	195.3
FB_2	0.9998	1.3–85.8	0.2	0.4	2.4	15.9	118.1 (0.3)	4.5	98.7 (4.2)	197.8
HT-2	0.9998	1.1–190.7	2.3	2.2	3.3	35.4	98.9 (0.4)	10.0	61.3 (2.0)	−69.8
MAC A	0.9998	2.6–87.9	3.1	6.0	4.2	16.3	104.0 (5.5)	4.6	70.4 (2.9)	−51.2
OTA	0.9991	1.6–36.4	4.8	5.2	2.0	6.8	71.4 (5.2)	1.9	54.8 (1.1)	−49.9
ROQ-C	0.9998	0.2–36.1	2.5	2.9	0.6	6.7	32.2 (2.5)	1.9	87.0 (8.0)	−31.0
STE	0.9996	0.5–35.9	4.7	5.1	1.2	6.7	80.9 (3.0)	1.9	76.3 (9.1)	−56.3
T-2	0.9999	0.4–9.1	0.9	1.1	0.5	1.7	155.2 (1.9)	0.6	85.7 (0.3)	−28.8
CPA	0.9998	2.6–450.5	1.1	3.8	7.0	83.7	53.7 (0.3)	23.5	83.4 (5.5)	−25.9
AND A[1]	0.9999	0.1–19.1	1.2	1.3	0.3	3.5	78.2 (2.5)	0.9	97.6 (3.0)	−53.0
AOH[1]	0.9996	2.0–394.7	3.7	5.6	3.5	73.1	66.7 (2.1)	18.6	84.2 (3.0)	−89.6
MPA[1]	0.9999	1.4–33.2	4.1	3.6	1.6	6.1	81.4 (1.8)	1.7	101.0 (1.1)	−34.6
PEN A	0.9993	1.0–191.3	1.4	4.0	2.1	35.4	37.8 (0.5)	9.0	56.5 (0.2)	−11.4
α-ZOL[1]	0.9999	2.5–249.2	4.9	9.4	5.7	90.8	95.8 (6.5)	23.2	90.5 (1.7)	−3.0
β-ZOL[1]	0.9999	2.6–256.4.4	4.8	5.2	5.7	93.4	77.9 (5.2)	23.8	108.2 (4.2)	−50.0
ZEA[1]	0.9999	0.5–100.7	4.6	4.8	1.2	18.7	83.4 (5.9)	4.8	103.7 (5.9)	−44.0
FUS X	0.9995	2.0–395.3	0.6	0.6	6	73.2	96.4 (0.7)	18.7	109.6 (0.9)	25.0
AF B_1	0.9998	0.1–103.5	4.4	10.4	0.3	21.3	103.3 (5.2)	11.4	57.8 (8.9)	−65.0
AF B_2	1.0000	0.4–25.9	4.7	4.4	1.2	5.3	105.2 (2.9)	2.9	62.5 (7.6)	−43.2
AF G_1	0.9998	0.1–103.5	6.8	10.1	0.3	21.3	96.5 (1.9)	11.4	49.2 (6.2)	−36.3
AF G_2	0.9998	0.4–25.9	4.6	4.3	1.2	5.3	118.3 (0.6)	2.9	135.5 (0.4)	−30.3

[1] negative mode.

Table 1 shows the determination coefficient (R^2) for all analytes using matrix-matched calibration. R^2 values were in all cases higher than 0.9991, showing an almost perfect linearity in the relation between the signals obtained and the concentration in the corresponding ranges. Both intra-day (n = 3) and inter-day (n = 3) precisions were calculated and expressed in terms of relative standard deviation (RSD) obtaining satisfactory values, ranging from 0.2% to 10.4%. The limits of quantification (LOQs) ranged from 0.3 to 50 ng mL^{-1}.

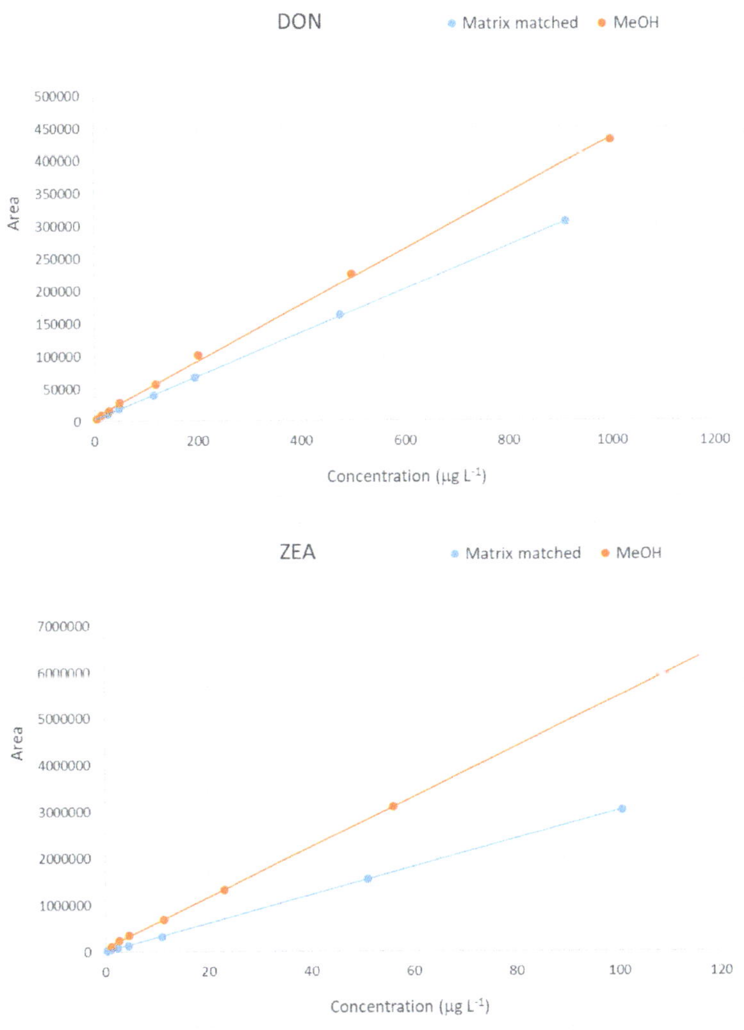

(a)

Figure 2. *Cont.*

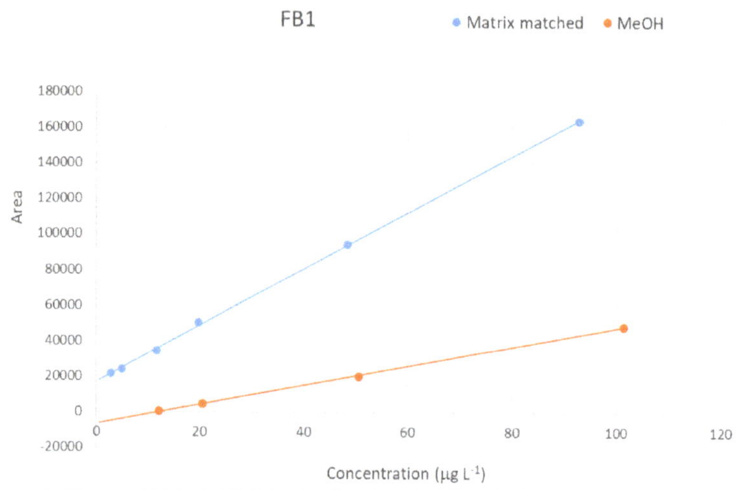

(b)

Figure 2. Matrix-matched calibration curves and external calibration. (**a**) Signal suppression: DON, ZEA; (**b**) signal enhancement: FB_1.

To evaluate the accuracy of the methodology proposed, a multitoxin reference material containing ZEA, DON and fumonisins was extracted. Mixed ration is a complex and heterogeneous matrix and there is no related reference material available; therefore, a maize based one was used. The extraction recoveries achieved were 78.6% in the case of DON and 98.6% for ZEA, endorsing the trueness of the proposed methodology.

To evaluate the whole sample preparation and chromatographic methodology for the determination of all 26 mycotoxins, recovery studies were carried out using the same matrix as in the matrix-matched calibration. The results, expressed in percentages, are shown in Table 1, as well as the relative standard deviation in each case. Most compounds gave recovery values between 70% and 120%, especially the most frequently found in this kind of matrix, DON, ZEA, and FB_1 and FB_2. Only two mycotoxins achieved recoveries below 60%, at the two spiking levels (Enniatin B1 (ENN B_1) and Penitrem A (PEN A)), but these analytes are not as recurrent and their contents in food/feed are not yet either regulated. The relative standard deviations were satisfactory since they were always lower than 9%.

ISO 21748:2017 provides recommendations [30] to assess the expanded measurement uncertainty (U). Thanks to the performance parameters given in Table 1, especially the intra-day and inter-day precisions, the relative expanded measurement uncertainty (% U) could be calculated. As shown in Table 2 for those 17 mycotoxins that were detected in the mixed ration samples, % U were lower than 15% in all cases, except for α-ZOL.

2.3. Application to Real Samples

Using the proposed methodology, 97 mixed ration samples collected in one year at 20 different dairy farms from Galicia (NW Spain) were analyzed. The criteria used for the correct identification of the analytes, in addition to the retention time, were the parameters given by the high-resolution mass spectrometry such as the exact mass accuracy, accepting only a 5 ppm error. The isotopic profile of the peak found as well as the formula proposed by the software for the mass obtained provided valuable additional information for the identification. Finally, contrasting the high resolution MS/MS spectra experimentally obtained with the available library spectra enabled a definitive confirmation.

Table 2. Concentration ranges, average, maximum and minimum concentrations (with relative expanded uncertainty, *U* (%)) and occurrence frequency of the 17 mycotoxins found in the 97 mixed ration samples.

Compounds	Concentration (ng g^{-1})				N° Positive	Occurrence Frequency (%)
	Average	*Max*	*Min*	*U* (%)		
3+15-ADON	426.9	770.0	214.1	*13.2*	33	33.7
DON	52.8	81.3	36.5	*14.1*	15	16.3
ENN B	322.0	1305.5	53.1	*13.4*	49	50.0
FB$_1$	578.8	1453.9	14.0	*11.6*	95	96.9
FB$_2$	160.6	481.1	13.3	*1.1*	64	65.3
MAC A	124.3	209.9	46.1	*13.6*	4	4.1
OTA	5.1	5.1	5.1	*11.0*	1	1.0
ROQ-C	36.9	59.3	19.6	*6.3*	4	4.1
STER	8.7	15.3	2.6	*10.9*	20	20.4
CPA	96.5	257.0	2.6	*8.7*	22	22.4
AND A	33.3	91.0	8.8	*2.8*	6	6.1
AOH	126.3	295.4	12.7	*12.4*	24	24.5
MPA	262.2	3151.4	8.2	*7.4*	19	19.4
PEN A	159.7	234.5	63.7	*9.1*	9	9.2
α-ZOL	32.5	50.3	25.3	*21.1*	28	28.6
β-ZOL	35.3	78.9	8.1	*11.0*	67	68.4
ZEA	10.2	23.2	3.5	*10.2*	90	91.8

Figure 3 summarizes the total of positives of each compound found in all samples. Fumonisins B_1 and B_2, zearalenone and its metabolite β-zearalenol, cyclopiazonic acid and enniatin B were the mycotoxins most frequently found. FB_1 was detected in 95 samples; ZEA in 90 samples; ENN B in 49 samples and DON in 15 samples. This results are in consonance with other studies published on mycotoxin occurrence in feed [8,20,31]. All samples exhibited co-occurrence of two or more mycotoxins, repeatedly zearalenone and fumonisin B_1, ZEA-FB_1-ENN B and ZEA-FB_1-DON. These co-occurrences are often found in feedstuff, as recent studies show [29,32]. However, the concentrations found in the mixed ration samples of the present study are much lower than those reported in these previous studies on maize based feed and silages [26,29,32].

Table 2 collects the concentration ranges and the average values in which the analyzed mycotoxins were present in the samples. ZEA, DON, FB_1, and FB_2 concentrations were always lower than the recommended values in feed intended for dairy cows; 0.5, 5, and 50 mg kg^{-1}, respectively. It should be noted that 3+15-ADON concentrations were one magnitude order higher than DON ones, with an occurrence frequency of 34%. As stated by EFSA in 2013 [33], it is important to collect more data on the incidence of these derivatives to better characterize their potential contribution to the total exposure to DON.

ENN B showed an average concentration of 0.32 mg kg^{-1} and reached a maximum content of 1.3 mg kg^{-1}. A particular attention should be paid to this cyclic depsipeptide since a recent EFSA scientific report, combining an in vivo toxicity and genotoxicity approach, concludes that ENN B poses a genotoxic hazard [34].

Finally, the screening of non-targeted mycotoxins conducted by implementing a retrospective approach with the same acquired data files and by contrasting HR MS/MS spectra with those of the mycotoxin library enabled to identify monocerin, aurofusarin, beauvericin, chlamydosporol, sambucinol, mevastatin, and two forms of enniantin, enniantins B4 and J1. Two anthraquinone derivatives, emodin and chrysophanol, were also identified.

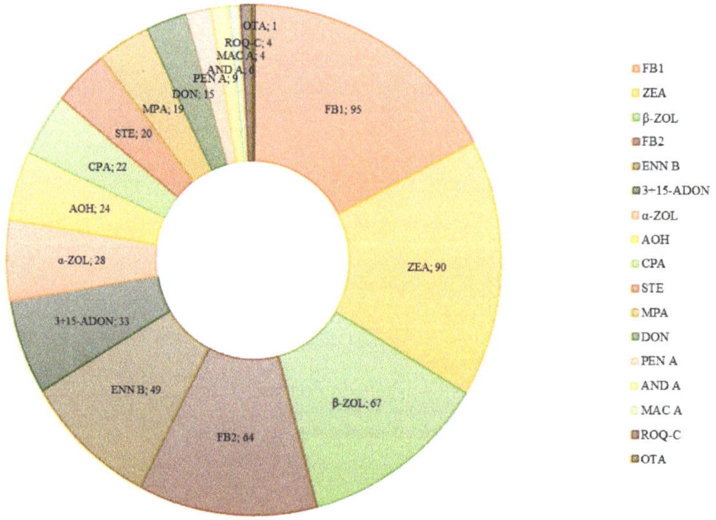

Figure 3. Total number of positives for each mycotoxin in the 97 samples analyzed.

3. Conclusions

The proposed analytical methodology based on QuEChERS with (d)SPE clean-up and HPLC-HRMS analysis allowed the screening and quantitation of 26 mycotoxins in mixed rations. The full method was successfully optimized, and its accuracy was confirmed using a reference material, as well. Recovery values were acceptable for most analytes, giving good results considering the complexity of the matrix for a multi-analyte methodology. Matrix-matched calibration was utilized to compensate for matrix effects, mostly suppressive, except from fumonisins.

The analytical method has been applied to the analysis of 97 samples. Fumonisins B_1 (97%) and B_2 (65%), zearalenone (92%), its metabolite β-zearalenol (68%) and enniatin B (50%) were the most frequently found mycotoxins. All samples presented co-occurrence of two or more mycotoxins, recurrently zearalenone and fumonisin B_1, amongst others. The use of SWATH^{TM} acquisition mode proved to be a very attractive approach for the unambiguous identification as well as for the accurate quantification of the target mycotoxins in complex feed samples. Additional screening based on a retrospective approach enabled the identification of non-targeted mycotoxins, most of them originated from *Fusarium* fungi.

4. Materials and Methods

4.1. Reagents and Standards

The studied compounds, their molecular formula, exact mass, CAS numbers and structure are summarized in Table 3. Standards of 3+15-ADON, DON, HT-2, OTA, ROQ-C, T-2, ZEA, AFB_1, AFG_1, AFB_2, AFG_2 were obtained from Biopure (Romer Labs, Tulln, Austria); ENN B, ENN B_1, STER, α-ZOL and β-ZOL from Sigma-Aldrich (Saint Louis, MO, USA); fumonisins FB_1 and FB_2 from Acros Organics (New Jersey, USA); MARC A and AND A from Santa Cruz Biotechnology (Dallas, TX, USA); CPA and MPA from Alpha Aesar (Ward Hill, MA, USA); AOH and PEN A from Cayman Chemical (Ann Harbor, MI, USA). Acetonitrile, water and methanol (both > 99.95% purity) were provided by Carlo Erba (Milan, Italy). Formic acid (99%), acetic acid glacial, ammonium formate and ammonium acetate were all LC-MS grade and were provided by Biosolve (Dieuze, France and Valkenswaard, The Netherlands).

Table 3. Target mycotoxins: molecular formula, exact mass, CAS number, molecular ion, retention time and structure.

Mycotoxins	Acronym	Molecular Formula	Exact Mass (m/z)	CAS Number	Molecular Ion	RT (min)	Structure
3-Acetyl deoxynivalenol	3-ADON	$C_{17}H_{22}O_7$	333.1365	50722-38-8	$[M+H]^+$	3.83	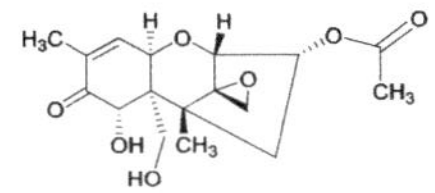
15-Acetyl deoxynivalenol	15-ADON	$C_{17}H_{22}O_7$	333.1365	88337-96-6	$[M+H]^+$	3.83	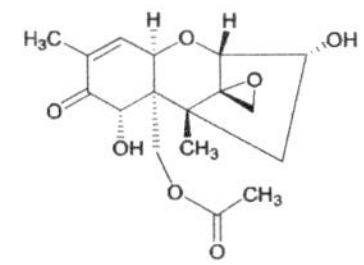
Deoxynivalenol	DON	$C_{15}H_{20}O_6$	295.1259	51481-10-8	$[M+H]^+$	2.24	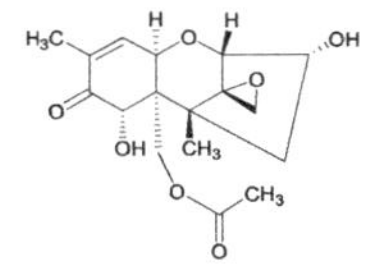
Enniatin B	ENN B	$C_{33}H_{57}N_3O_9$	639.4094	917-13-5	$[M+H]^+$	8.15	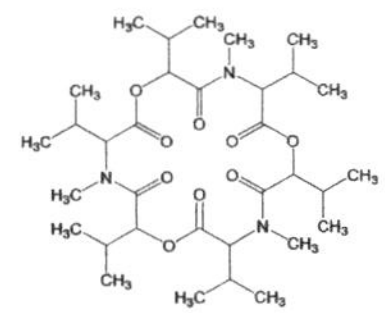
Enniatin B$_1$	ENN B$_1$	$C_{34}H_{59}N_3O_9$	653.4251	19914-20-6	$[M+H]^+$	8.30	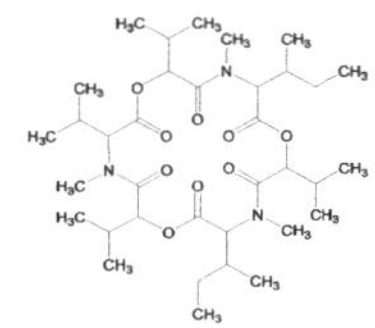

Table 3. *Cont.*

Mycotoxins	Acronym	Molecular Formula	Exact Mass (m/z)	CAS Number	Molecular Ion	RT (min)	Structure
Fumonisin B_1	FB_1	$C_{34}H_{59}NO_{15}$	721.3884	116355-83-0	$[M+H]^+$	5.20	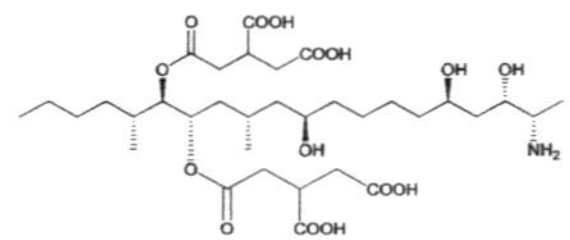
Fumonisin B_2	FB_2	$C_{34}H_{59}NO_{14}$	705.3935	116355-84-1	$[M+H]^+$	6.20	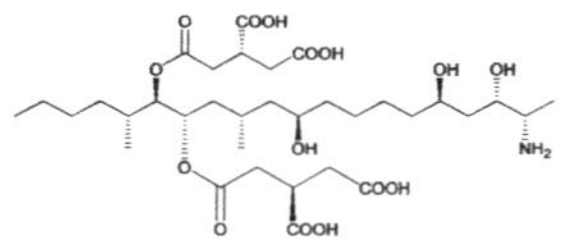
HT-2 toxin	HT-2	$C_{22}H_{32}O_8$	424.2097	26934-87-2	$[M+NH_4]^+$	5.70	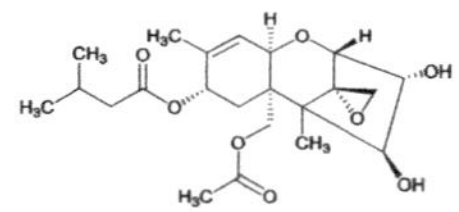
Marcfortine A	MAC A	$C_{28}H_{35}N_3O_4$	477.2627	75731-43-0	$[M+H]^+$	6.97	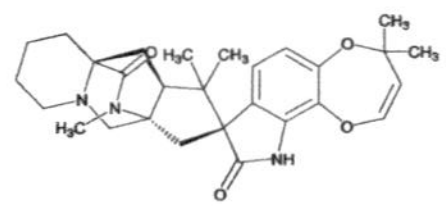
Ochratoxin A	OTA	$C_{20}H_{18}ClNO_6$	403.0822	303-47-9	$[M+H]^+$	5.36	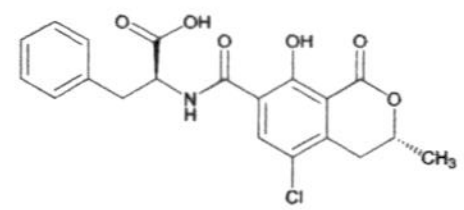
Roquefortine C	ROQ-C	$C_{22}H_{23}N_5O_2$	389.1851	58735-64-1	$[M+H]^+$	6.50	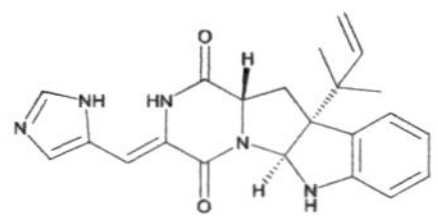

Table 3. *Cont.*

Mycotoxins	Acronym	Molecular Formula	Exact Mass (m/z)	CAS Number	Molecular Ion	RT (min)	Structure
Sterigmatocystin	STE	$C_{18}H_{12}O_6$	324.0633	10048-13-2	$[M+H]^+$	6.62	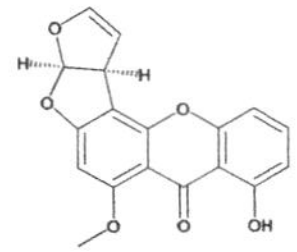
T-2 toxin	T-2	$C_{24}H_{34}O_9$	466.2202	21259-20-1	$[M+NH_4]^+$	6.16	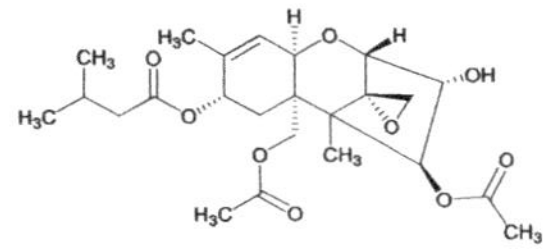
Cyclopiazonic acid	CPA	$C_{20}H_{20}N_2O_3$	336.1473	18172-33-3	$[M+H]^+$	5.00	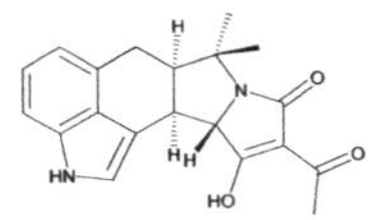
Andrastin A	AND A	$C_{28}H_{38}O_7$	486.2617	174232-42-9	$[M-H]^-$	5.95	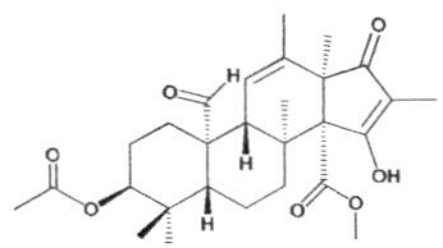
Alternariol	AOH	$C_{14}H_{10}O_5$	258.0528	641-38-3	$[M-H]^-$	5.64	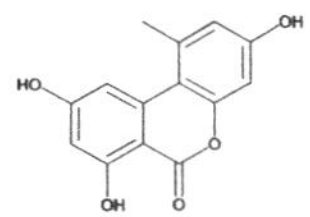
Mycophenolic acid	MPA	$C_{17}H_{20}O_6$	320.1259	24280-93-1	$[M-H]^-$	4.93	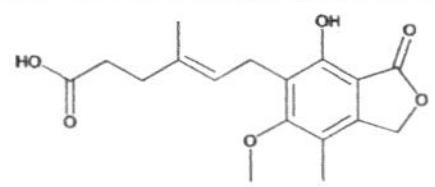

Table 3. *Cont.*

Mycotoxins	Acronym	Molecular Formula	Exact Mass (m/z)	CAS Number	Molecular Ion	RT (min)	Structure
Penitrem A	PEN A	$C_{37}H_{44}ClNO_6$	633.2857	12627-35-9	$[M+H]^+$	7.64	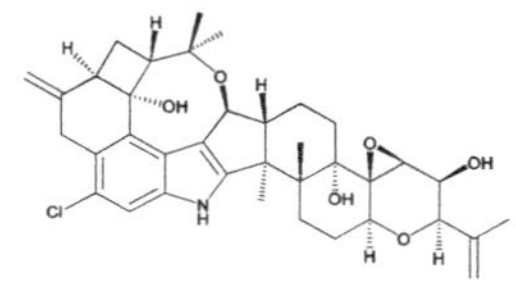
α-Zearalenol	α-ZOL	$C_{18}H_{24}O_5$	320.1623	36455-72-8	$[M+H]^+$	5.95	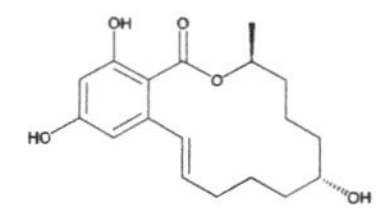
β-Zearalenol	β-ZOL	$C_{18}H_{24}O_5$	320.1623	71030-11-0	$[M+H]^+$	6.34	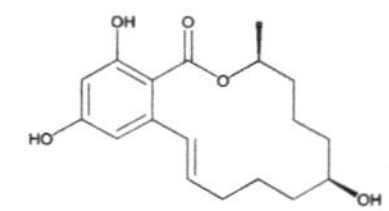
Zearalenone	ZEA	$C_{18}H_{22}O_5$	318.1467	17924-92-4	$[M+H]^+$	6.48	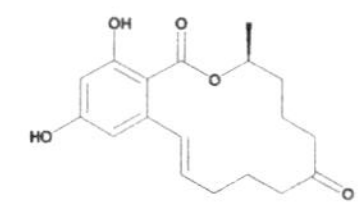
Fusarenon X	FUS X	$C_{17}H_{22}O_8$	354.1314	23255-69-8	$[M+H]^+$	3.03	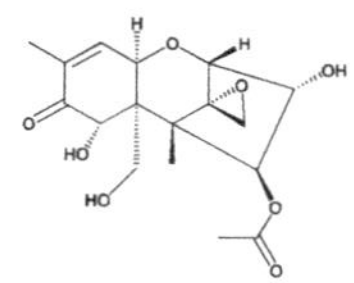

Table 3. *Cont.*

Mycotoxins	Acronym	Molecular Formula	Exact Mass (m/z)	CAS Number	Molecular Ion	RT (min)	Structure
Aflatoxin B_1	AF B_1	$C_{17}H_{12}O_6$	312.0633	1162-65-8	$[M+H]^+$	4.95	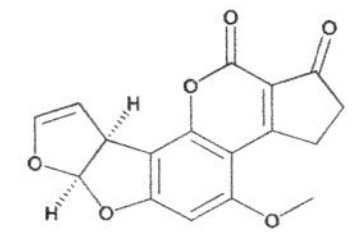
Aflatoxin B_2	AF B_2	$C_{17}H_{14}O_6$	314.0790	7220-81-7	$[M+H]^+$	4.75	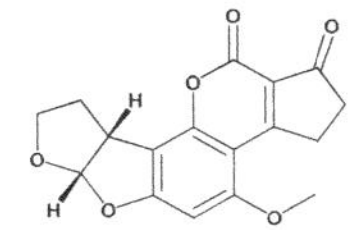
Aflatoxin G_1	AF G_1	$C_{17}H_{12}O_7$	328.0583	1165-39-5	$[M+H]^+$	4.54	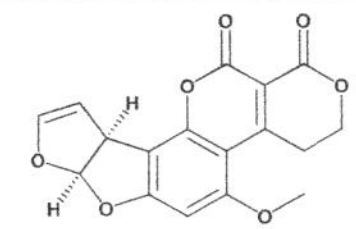
Aflatoxin G_2	AF G_2	$C_{17}H_{14}O_7$	330.0739	7241-98-7	$[M+H]^+$	4.33	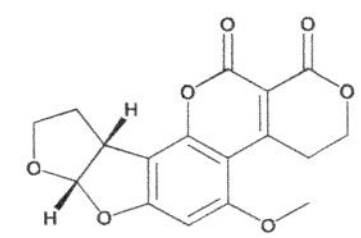

A multitoxin maize reference material (QCM7C1, Biopure, Romer Labs, Tulln, Austria) was used for checking the extraction method accuracy.

Individual stock solutions for each mycotoxin were prepared in acetonitrile. Further dilutions and mixtures were prepared in methanol. All solutions were stored at −20 °C.

4.2. Sampling and Sample Treatment

Mixed rations for dairy cows in Galicia usually rely on own forages and concentrates, and in recent years, the ensiled maize proportion is being increasing. Concentrates are mostly a complement of protein and energy to balance dairy cow rations. In Galicia, these concentrates are made with soya cake (30%) and rapeseed cake (20%) to increase protein, and with cereals, mainly maize (25%) and barley (15%), to achieve the cattle energy requirement.

97 mixed ration samples collected in 20 dairy farms between October 2017 and October 2018 in Galicia (NW Spain). Five hundred grams of the sample was dried for 48 h at 40 °C and then put in vacuum bags and stored at 4 °C until their analysis.

4.3. QuEChERS Extraction

Two grams of unifeed samples were weighed and placed in 50 mL falcon tubes with the addition of 10 mL of water and 10 mL of acetonitrile/formic acid (90:10% v/v). The tubes were placed in an automated shaker for 1 h at room temperature. Then, a mixture of 0.5 NaSesquihydrate/1g NaCitrate/1g NaCl/4g $MgSO_4$ (DisQuE^{TM} Pouch for CEN, Waters, Milford, USA) was added and the tube was shaken vigorously by hand for 1 min. After centrifugation (3398× *g* for 5 min), a portion of the supernatant was saved for subsequent clean-up.

4.4. Clean-Up Procedure

An Oasis PRiME HLB cartridge (3cc, 150 mg), Waters, Milford, MA, USA) was placed on a vacuum manifold. Without previous conditioning, after discarding 0.4 mL of supernatant, a 1 mL aliquot was passed-trough the cartridge and, the collected extract was then transferred to a 2 mL dSPE tube containing a mixture of sorbents (150 mg $MgSO_4$, 50 mg PSA, 30 mg C18, 30 mg Al-N) (Waters, Milford, USA). After centrifugation (2 min at 2360× *g*), a 500 µL aliquot was taken, evaporated under a gentle nitrogen stream, and reconstituted in 350 µL of methanol.

4.5. HPLC-QTOF Analysis

For LC-HRMS analysis, a HPLC system (Shimadzu Nexera X2) consisting of two high-pressure pumps (LC-30AD) and a SIL-30AC autosampler was used. HRMS detection was performed by a SCIEX (Ontario, Canada) TripleTOF® 5600+ equipped with a DuoSprayTM ion source and an Electrospray Ionization (ESI) probe. The chromatographic separation was achieved with a Kinetex bioZenTM Peptide XB-C18 (50 × 2.1 mm, 2.6 µm) column from Phenomenex (Torrance, CA, USA), kept at 40 °C with a CTO-30A column oven.

The mobile phase consisted of water (A) and methanol (B), both buffered with 3mM ammonium formate or ammonium acetate. Formic acid and acetic acid addition (0.1%) was also tested in combination with its respective buffer.

An 8-min gradient elution profile (10% B to 100% B) was employed reaching a total runtime of 15 min. The mobile phase flow-rate was kept to 0.25 mL mL^{-1} and the injection volume was set at 10 µL.

Regarding the HRMS conditions, the source temperature was set at 550 °C, the ion source gas at 50 (au, arbitrary units), the curtain gas at 30 (au), and the ion spray voltage floating at 5500 V (−4500 V in the negative mode).

The HRMS workflow consisted of a Full Scan, using 250 ms as accumulation time and 80 V (−80 V in negative mode) as the declustering potential in the ESI. Simultaneously, a data independent approach based on SWATH (Sequential Windowed Acquisition of All Theoretical Fragment Ion Mass Spectra) was performed. A wide mass range (80–850 Da) was divided in 30 mass windows with an accumulation time of 35 ms for each one. The cycle time was slightly higher than 1 s, enabling a perfect reconstruction of any chromatographic peak. The declustering potential was set to 80 V (−80 V in negative mode) and the collision energy was 40 V (−40 V in negative mode), with an energy spread of 20 V.

SWATH acquisition is a Data Independent Acquisition (DIA) strategy that delivers the complete picture of a sample. It provides the best method for performing sample analysis, identifying and quantifying every detectable analyte in a single chromatographic run.

In contrast to traditional mass spectrometry acquisition techniques that rely on Data Dependent Acquisition (DDA) strategies, SWATH acquisition is not "dependent" upon some pre-set criteria determined by the abundance of the compound. With SWATH acquisition, every detectable ion, irrespective of its concentration, is fragmented, identified and quantified to provide the high-resolution full MS and MS/MS picture for every peak. The data produced by SWATH then contains all the information that would be acquired in a multitude of different experimental approaches in a single all-encompassing dataset. Furthermore, SWATH acquisition cycle time does not increase as the number of target compounds increases.

In terms of post-acquisition, the complete set of data can serve as a digital archive of each sample, allowing full access to pull out anything that you may be interested in, without the need to re-inject your samples. Should new hypotheses arise in the future, this feature allows you to re-interrogate the sample data without re-analyzing the actual sample.

The system was operated by Analyst® 1.7.1 (SCIEX, Ontario, Canada) control software, PeakView® 2.2 (SCIEX, Ontario, Canada) processing software and MultiQuant® 3.0 (SCIEX, Ontario, Canada) quantitation software. A mycotoxin HR MS/MS spectral library (1.0), containing 288 compounds was used for identification and confirmation purposes.

As usual, mass accuracy is expressed in ppm, calculated as the relative difference between the measured mass and the actual mass of each analyte, as follows:

$$Accuracy\ (ppm) = \frac{(\text{Measured mass} - \text{Actual mass})}{\text{Actual mass}} * 1000000$$

4.6. Method Validation

The limit of quantification (LOQ) of each analyte was evaluated using the parameters of each matrix-matched calibration curve and the S/N obtained for the lowest concentration level of each analyte.

The matrix effect was calculated by subtracting the ratio of the slope values of the matrix-matched calibration curve and the external calibration curve to 1, and expressing it as a relative percentage (%) (See Table 1).

The mixed ration sample selected for recovery experiments was spiked in triplicate at two concentration levels (individual values indicated in Table 1). Aliquots of 2 g of fortified mixed ration samples were let interacting for 20 h at ambient temperature and protected from natural and artificial light exposure. Recovery values were given as the ratio of the signal obtained for these fortified samples and the response of the corresponding concentration level in the matrix-matched standards.

ISO 21748:2017 provides recommendations [30] to assess the expanded measurement uncertainty (U) which was calculated (see following equations) by multiplying the combined standard uncertainty (u_c) by a coverage factor (k). At a confidence level of 95%, this k factor was set to two. The standard uncertainty (u_c) formulae combined the variance of the inter-day precision (S^2_R) and the uncertainty associated with the bias ($u_{bias}{}^2$). The latter was calculated using the number of replicates (n), the number of different conditions (*p*), the variance of the intra-day precision (S^2r) and the variance of the inter-day precision ($S^2{}_R$).

$$ubias = \sqrt{\left(\frac{S^2{}_R - (1 - \frac{1}{n})S^2{}_r}{p}\right)}$$
$$uc = \sqrt{S^2{}_R + ubias2}$$
$$U = ky\ uc$$

Supplementary Materials: The following are available online at http://www.mdpi.com/2072-6651/12/3/206/s1, Table S1: Regression equations for matrix-matched calibrations of the target mycotoxins in positive mode, using different mobile phase compositions. Table S2: Regression equations for matrix-matched calibrations of some target mycotoxins in negative mode, using different mobile phase compositions.

Author Contributions: The following statements should be used: "conceptualization, T.D.; methodology, M.L. and T.D.; software; validation, R.F.; investigation, R.F., T.D. and M.L.; resources, T.D.; data curation; writing—original draft preparation, R.F.; writing—review and editing, R.F., T.D. and M.L.; visualization, R.F.; supervision, T.D. and M.L.; project administration, T.D. and M.L.; funding acquisition, T.D and M.L. All authors have read and agreed to the published version of the manuscript.

Funding: This work was supported by the CONECTA-PEME GALAC project IN852A 2016/103. This work was also supported by the project IGDC13-1E-2026 (Infrastructure Program, MINECO, Spain). The authors belong to the Galician Competitive Research Groups GPC 2017/04 and GPC IN607B 2019/13, and to the CRETUS Strategic Partnership (ED431E 2018/01). All these programs are co-funded by FEDER (UE).

Acknowledgments: Special thanks to Irmandiños for supplying the unifeed samples.

Conflicts of Interest: The authors declare no conflict of interest.

References

1. Romera, D.; Mateo, E.M.; Mateo-Castro, R.; Gómez, J.V.; Gimeno-Adelantado, J.V.; Jiménez, M. Determination of multiple mycotoxins in feedstuffs by combined use of UPLC–MS/MS and UPLC–QTOF–MS. *Food Chem.* **2018**, *267*, 140–148. [CrossRef]
2. Pitt, J.I.; Taniwaki, M.H.; Cole, M.B. Mycotoxin production in major crops as influenced by growing, harvesting, storage and processing, with emphasis on the achievement of Food Safety Objectives. *Food Control* **2013**, *32*, 205–215. [CrossRef]
3. Lee, H.J.; Ryu, D. Worldwide Occurrence of Mycotoxins in Cereals and Cereal-Derived Food Products: Public Health Perspectives of Their Co-occurrence. *J. Agric. Food Chem.* **2017**, *65*, 7034–7051. [CrossRef]
4. Escrivá, L.; Font, G.; Manyes, L. In vivo toxicity studies of fusarium mycotoxins in the last decade: A review. *Food Chem. Toxicol.* **2015**, *78*, 185–206. [CrossRef]
5. Arroyo-Manzanares, N.; De Ruyck, K.; Uka, V.; Gámiz-Gracia, L.; García-Campaña, A.M.; De Saeger, S.; Di Mavungu, J.D. In-house validation of a rapid and efficient procedure for simultaneous determination of ergot alkaloids and other mycotoxins in wheat and maize. *Anal. Bioanal. Chem.* **2018**, *410*, 5567–5581. [CrossRef] [PubMed]
6. Paschoal, F.N.; de Azevedo Silva, D.; de Souza, R.V.S.; de Oliveira, M.S.; Pereira, D.A.A.; de Souza, S.V.C. A Rapid Single-Extraction Method for the Simultaneous Determination of Aflatoxins B1, B2, G1, G2, Fumonisin B1, and Zearalenone in Corn Meal by Ultra Performance Liquid Chromatography Tandem Mass Spectrometry. *Food Anal. Methods* **2017**, *10*, 1631–1644. [CrossRef]
7. Nathanail, A.V.; Syvähuoko, J.; Malachová, A.; Jestoi, M.; Varga, E.; Michlmayr, H.; Adam, G.; Sieviläinen, E.; Berthiller, F.; Peltonen, K. Simultaneous determination of major type A and B trichothecenes, zearalenone and certain modified metabolites in Finnish cereal grains with a novel liquid chromatography-tandem mass spectrometric method. *Anal. Bioanal. Chem.* **2015**, *407*, 4745–4755. [CrossRef] [PubMed]
8. Bolechová, M.; Benešová, K.; Běláková, S.; Čáslavský, J.; Pospíchalová, M.; Mikulíková, R. Determination of seventeen mycotoxins in barley and malt in the Czech Republic. *Food Control* **2015**, *47*, 108–113. [CrossRef]
9. Qian, M.; Yang, H.; Li, Z.; Liu, Y.; Wang, J.; Wu, H.; Ji, X.; Xu, J. Detection of 13 mycotoxins in feed using modified QuEChERS with dispersive magnetic materials and UHPLC–MS/MS. *J. Sep. Sci.* **2018**, *41*, 756–764. [CrossRef]

10. Wang, R.G.; Su, X.O.; Cheng, F.F.; Wang, P.L.; Fan, X.; Zhang, W. Determination of 26 mycotoxins in feedstuffs by multifunctional clean-up column and liquid chromatography-tandem mass spectrometry. *Chin. J. Anal. Chem.* **2015**, *43*, 264–270. [CrossRef]
11. Miró-Abella, E.; Herrero, P.; Canela, N.; Arola, L.; Borrull, F.; Ras, R.; Fontanals, N. Determination of mycotoxins in plant-based beverages using QuEChERS and liquid chromatography–tandem mass spectrometry. *Food Chem.* **2017**, *229*, 366–372. [CrossRef] [PubMed]
12. Tsiplakou, E.; Anagnostopoulos, C.; Liapis, K.; Haroutounian, S.A.; Zervas, G. Determination of mycotoxins in feedstuffs and ruminant's milk using an easy and simple LC-MS/MS multiresidue method. *Talanta* **2014**, *130*, 8–19. [CrossRef]
13. Rubert, J.; León, N.; Sáez, C.; Martins, C.P.B.; Godula, M.; Yusà, V.; Mañes, J.; Soriano, J.M.; Soler, C. Evaluation of mycotoxins and their metabolites in human breast milk using liquid chromatography coupled to high resolution mass spectrometry. *Anal. Chim. Acta* **2014**, *820*, 39–46. [CrossRef] [PubMed]
14. Cherkani-Hassani, A.; Mojemmi, B.; Mouane, N. Occurrence and levels of mycotoxins and their metabolites in human breast milk associated to dietary habits and other factors: A systematic literature review, 1984–2015. *Trends Food Sci. Technol.* **2016**, *50*, 56–69. [CrossRef]
15. Marin, S.; Ramos, A.J.; Cano-Sancho, G.; Sanchis, V. Mycotoxins: Occurrence, toxicology, and exposure assessment. *Food Chem. Toxicol.* **2013**, *60*, 218–237. [CrossRef] [PubMed]
16. *Directive 2002/32/EC of the European Parliament and of the Council of 7 May 2002 on Undesirable Substances in Animal Feed and its Amendments*; European Parliament and of the Council: Brussels, Belgium, 2002.
17. *Commission Recommendation of 17 August 2006 on the Presence of Deoxynivalenol, Zearalenone, Ochratoxin A, T-2 and HT-2 and Fumonisins in Products Intended for Animal Feeding) (2006/576/EC)*; The Commission of The European Communities: Brussels, Belgium, 2006.
18. *2013/165/EU, C.R. Recomendations on the presence of T-2 and HT-2 Toxin in Cereals and Cereal Products*; Official Journal of the European Union: Brussels, Belgium, 2013.
19. Turner, N.W.; Bramhmbhatt, H.; Szabo-Vezse, M.; Poma, A.; Coker, R.; Piletsky, S.A. Analytical methods for determination of mycotoxins: An update (2009–2014). *Anal. Chim. Acta* **2015**, *901*, 12–33. [CrossRef]
20. Rubert, J.; Dzuman, Z.; Vaclavikova, M.; Zachariasova, M.; Soler, C.; Hajslova, J. Analysis of mycotoxins in barley using ultra high liquid chromatography high resolution mass spectrometry: Comparison of efficiency and efficacy of different extraction procedures. *Talanta* **2012**, *99*, 712–719. [CrossRef]
21. Dzuman, Z.; Zachariasova, M.; Lacina, O.; Veprikova, Z.; Slavikova, P.; Hajslova, J. A rugged high-throughput analytical approach for the determination and quantification of multiple mycotoxins in complex feed matrices. *Talanta* **2014**, *121*, 263–272. [CrossRef]
22. Zhou, J.; Xu, J.J.; Huang, B.F.; Cai, Z.X.; Ren, Y.P. High-performance liquid chromatographic determination of multi-mycotoxin in cereals and bean foodstuffs using interference-removal solid-phase extraction combined with optimized dispersive liquid–liquid microextraction. *J. Sep. Sci.* **2017**, *40*, 2141–2150. [CrossRef]
23. Martínez-Domínguez, G.; Romero-González, R.; Garrido Frenich, A. Determination of toxic substances, pesticides and mycotoxins, in ginkgo biloba nutraceutical products by liquid chromatography Orbitrap-mass spectrometry. *Microchem. J.* **2015**, *118*, 124–130. [CrossRef]
24. Bueno, D.; Istamboulie, G.; Muñoz, R.; Marty, J.L. Determination of Mycotoxins in Food: A Review of Bioanalytical to Analytical Methods. *Appl. Spectrosc. Rev.* **2015**, *50*, 728–774. [CrossRef]
25. Rahmani, A.; Jinap, S.; Soleimany, F. Qualitative and quantitative analysis of mycotoxins. *Compr. Rev. Food Sci. Food Saf.* **2009**, *8*, 202–251. [CrossRef]
26. Zhao, Z.; Liu, N.; Yang, L.; Deng, Y.; Wang, J.; Song, S.; Lin, S.; Wu, A.; Zhou, Z.; Hou, J. Multi-mycotoxin analysis of animal feed and animal-derived food using LC-MS/MS system with timed and highly selective reaction monitoring. *Anal. Bioanal. Chem.* **2015**, *407*, 7359–7368. [CrossRef] [PubMed]
27. Dagnac, T.; Latorre, A.; Lorenzo, B.F.; Llompart, M. Validation and application of a liquid chromatography-tandem mass spectrometry based method for the assessment of the co-occurrence of mycotoxins in maize silages from dairy farms in NW Spain. *Food Addit. Contam. Part A* **2016**, *33*, 1850–1863. [CrossRef] [PubMed]
28. Dzuman, Z.; Zachariasova, M.; Veprikova, Z.; Godula, M.; Hajslova, J. Multi-analyte high performance liquid chromatography coupled to high resolution tandem mass spectrometry method for control of pesticide residues, mycotoxins, and pyrrolizidine alkaloids. *Anal. Chim. Acta* **2015**, *863*, 29–40. [CrossRef] [PubMed]

29. Jensen, T.; de Boevre, M.; Preußke, N.; de Saeger, S.; Birr, T.; Sönnichsen, F.D. Evaluation of High-Resolution Mass Spectrometry for the Quantitative Analysis of Mycotoxins in Complex Feed Matrices. *Toxins* **2019**, *11*, 531. [CrossRef]
30. ISO 21748. *Guidance for the Use of Repeatability, Reproducibility and Trueness Estimates in Measurement Uncertainty Estimation*; International Organization for Standardization: Geneva, Switzerland, 2017.
31. Ediage, E.N.; Van Poucke, C.; De Saeger, S. A multi-analyte LC-MS/MS method for the analysis of 23 mycotoxins in different sorghum varieties: The forgotten sample matrix. *Food Chem.* **2015**, *177*, 397–404. [CrossRef]
32. Franco, L.T.; Petta, T.; Rottinghaus, G.E.; Bordin, K.; Gomes, G.A.; Oliveira, C.A.F. Co-occurrence of mycotoxins in maize food and maize-based feed from small-scale farms in Brazil: A pilot study. *Mycotoxin Res.* **2019**, *35*, 65–73. [CrossRef]
33. European Food Safety Authority. Deoxynivalenol in food and feed: Occurrence and exposure 1. *EFSA J.* **2013**, *11*, 3379.
34. Istituto Superiore di Sanità (ISS); Italian National Agency for New Technologies, Energy and Sustainable Economic Development (ENEA); French Agency for Food, Environmental and Occupational Health & Safety (ANSES). In vivo Toxicity and Genotoxicity of Auvericin and Enniatins. Combined Approach to Study In Vivo Toxicity and Genotoxicity of Mycotoxins Beauvericin (BEA) and Enniatin B (ENNB). *EFSA Supporting Publ.* **2018**, *15*, 1406E. [CrossRef]

toxins

MDPI

Article

Development of an UPLC-MS/MS Method for the Analysis of Mycotoxins in Rumen Fluid with and without Maize Silage Emphasizes the Importance of Using Matrix-Matched Calibration

Sandra Debevere [1,2], Siegrid De Baere [1], Geert Haesaert [3], Michael Rychlik [4], Veerle Fievez [2,†] and Siska Croubels [1,*,†]

1 Department of Pharmacology, Toxicology and Biochemistry, Faculty of Veterinary Medicine, Ghent University, Salisburylaan 133, 9820 Merelbeke, Belgium

2 Department of Animal Sciences and Aquatic Ecology, Faculty of Bioscience Engineering, Ghent University, Coupure links 653, 9000 Ghent, Belgium

3 Department of Plants and Crops, Faculty of Bioscience Engineering, Ghent University, Coupure Links 653, 9000 Ghent, Belgium

4 Chair of Analytical Food Chemistry, Technische Universität München, Maximus-von-Imhof-Forum 2, 85354 Freising, Germany

* Correspondence: siska.croubels@ugent.be; Tel.: +32-9-264-7345

† Shared last author.

Received: 30 June 2019; Accepted: 5 September 2019; Published: 7 September 2019

check for updates

Abstract: Ruminants are less susceptible to the effects of mycotoxins than monogastric animals as their rumen microbiota are claimed to degrade and/or deactivate at least some of these toxic compounds. However, the mycotoxin degradation is not well-known yet. For this, a sensitive, specific, and accurate analytical method is needed to determine mycotoxins in the rumen fluid. This study aims to develop and thoroughly validate an ultra-performance liquid chromatography tandem mass spectrometry method for the quantitative determination in the rumen fluid of some of the most relevant mycotoxins found in maize silage in Western Europe: deoxynivalenol (DON), nivalenol (NIV), zearalenone (ZEN), mycophenolic acid (MPA), roquefortine C (ROQ-C) and enniatin B (ENN B), as well as their metabolites deepoxy-deoxynivalenol (DOM-1), α-zearalenol (α-ZEL), β-zearalenol (β-ZEL), zearalanone (ZAN), α-zearalanol (α-ZAL) and β-zearalanol (β-ZAL). As feed is often present in the rumen fluid samples, the potential interaction of feed particles with the mycotoxin extraction and analysis was investigated. Extraction recovery and matrix effects were determined in the rumen fluid with and without maize silage. Differences in those parameters between rumen fluid alone and rumen fluid with maize silage highlight the importance of using matrix-matched calibration curves for the quantification of mycotoxins in rumen fluid samples. A cross-validation of the method with rumen fluid and maize silage demonstrates that this analytical method can be applied in research on rumen fluid samples to investigate the degradation of the reported mycotoxins by rumen microbiota if matrix-matched calibration is performed.

Keywords: mycotoxins; UPLC-MS/MS; rumen fluid; maize silage; matrix-matched

Key Contribution: A sensitive and accurate ultra-performance liquid chromatography tandem mass spectrometry (UPLC-MS/MS) method was developed and validated to determine mycotoxins and their metabolites in rumen fluid with or without addition of feed particles. This method can be applied in research to investigate the degradation of mycotoxins by rumen microbiota if matrix-matched calibration is performed.

1. Introduction

Mycotoxins are secondary fungal metabolites that are harmful to humans and/or animals. Monogastric animals, in particular pigs, are more susceptible to toxic effects of mycotoxins than ruminants as the symbiosis of ruminants with rumen microbiota allows (partial) mycotoxin degradation [1,2]. Despite this potential degradation, mycotoxin-associated subclinical health problems have been reported in high-productive dairy cows, reflected by vague and non-specific symptoms and periodic decrease in milk production [1]. The increased milk yield during the past decades is not only associated with a higher incidence of metabolic disorders and hence a lower detoxifying capacity of mycotoxins in the rumen [3], but also with a higher rumen passage rate and a higher proportion of highly fermentable maize silage in the ration of cows. Unfortunately, this maize silage is more vulnerable to contamination with multiple mycotoxins than e.g., grassland products [4–6].

In Belgium, which has a temperate climate, some of the most important mycotoxins found in maize silage are deoxynivalenol (DON), nivalenol (NIV), zearalenone (ZEN), mycophenolic acid (MPA), roquefortine C (ROQ-C), and enniatins, especially enniatin B (ENN B) [4,7–9] (Tables S1 and S2). However to determine their rumen degradation, a sensitive, specific, accurate and high-throughput multi-method allowing analysis of these mycotoxins and their metabolites in rumen fluid is essential.

Nowadays, ultra-performance liquid chromatography (UPLC) combined with tandem mass spectrometry (MS/MS) offers a powerful tool in the analysis of mycotoxins, as this method offers a high specificity and sensitivity and can detect low levels of mycotoxins in complex matrices. Hence, ultra-performance liquid chromatography tandem mass spectrometry (UPLC-MS/MS) has been used for the detection of several types of mycotoxins in different matrices such as food and feed [10–12], blood [13] and urine [14,15].

The analysis of mycotoxins in rumen fluid, however, is highly challenging because of the complex nature of the matrix, including partially degraded feed particles and its microbiota. First, an elaborate sample preparation before chromatographic analysis is crucial in the development of such a detection method to eliminate as much as possible the unwanted compounds that can interfere with the detection of the analytes. In addition, in case of multi-mycotoxin methods, it is difficult to develop an analytical method for the simultaneous determination of several mycotoxins, since they have different physicochemical properties. Hence, compromises have to be made in optimizing the sample preparation and MS parameters. Second, it has been described that bacteria and yeast cell walls, also present in rumen fluid, are able to bind mycotoxins such as DON, ZEN, aflatoxins and fumonisins leading to an inferior detection of free mycotoxins in the rumen fluid [16–24].

To the authors' knowledge, very few methods have been described to determine simultaneously several mycotoxins in rumen fluid and none of them included all mycotoxins that are often found in maize silage (DON, NIV, ZEN, MPA, ROQ-C, and ENN B). Kiessling et al. (1984) analyzed six different mycotoxins (aflatoxin B1, ochratoxin A, ZEN, T-2 toxin, diacetoxyscirpenol, and DON) in the rumen fluid [25]. However, to determine those mycotoxins from different classes, four different extraction procedures and detection methods were performed, which are time consuming and expensive. Gallo et al. (2015) determined simultaneously three *Penicillium roqueforti* mycotoxins (MPA, ROQ-C and *Penicillium roqueforti* toxin (PR toxin)) in the rumen fluid by means of HPLC-MS/MS [26]. Again, the extraction procedure was very time consuming as the purpose of the study was to distinguish between mycotoxins present in the supernatant of the rumen fluid and those present in the pellet of the rumen fluid, so two separate extractions were performed.

Therefore, the goal of this study is first to develop a high-throughput method for rumen fluid to extract the most prevalent and toxicological important mycotoxins found in maize silage (DON, NIV, ZEN, MPA, ROQ-C, and ENN B) and their metabolites (deepoxy-deoxynivalenol (DOM-1), α-zearalenol (α-ZEL), β-zearalenol (β-ZEL), zearalanone (ZAN), α-zearalanol (α-ZAL), and β-zearalanol (β-ZAL)) as these can be formed in the rumen [25,27–31]. Further, a highly specific and sensitive UPLC-MS/MS method to quantify these components in the rumen fluid was developed and validated. As rumen fluid is a complex matrix and mycotoxins can bind to feed or other particles present in rumen fluid, the impact of the matrix on extraction recovery and matrix effects was evaluated. As the importance of using matrix-matched calibration curves was evident, a cross-validation of the method with rumen fluid and maize silage as a matrix was performed to prove the usability of the method in the research on rumen fluid samples if matrix-matched calibration is used.

2. Results and Discussion

Several criteria were taken into account when developing the UPLC-MS/MS method. The sample preparation procedure had to be fast and straightforward to ensure an acceptable daily throughput (sample no. >140). Along with this, a run-time of ≤10 min per ionization mode (electrospray (ESI) positive (+) or negative (−)) was aimed for. However, simplicity and speed of analysis should not be reached at the expense of reliability, specificity, and sensitivity. Finally, a thorough method validation was performed prior to routine application. Special attention was paid to the potential interaction of feed particles, in this study on maize silage, with the extraction and analysis of mycotoxins, which is important to setup an appropriate calibration method.

2.1. Mycotoxin Selection

Mycotoxin concentrations in the rumen can be very low, certainly when kinetic degradation studies are performed as concentrations might gradually decrease. Hence, the developed UPLC-MS/MS method needs to be very sensitive to detect and quantify mycotoxins and their metabolites. As the sensitivity which can be reached by a multi-method also partially depends on the number of mycotoxins included, a selection of the most important mycotoxins found in maize silage in Western Europe was made. Sensitivity results are reported in Section 2.4.3. Limit of Quantification (LOQ) and Limit of Detection (LOD).

2.2. UPLC-MS/MS Method

Mass spectrometric parameters of both the precursor and product ions were tuned in their most sensitive ESI mode (+/−). For each compound, two specific product ions were selected from the mass spectrum after fragmentation (Table 1). The product ion with the highest intensity was selected for quantification, whereas the second most intense product ion was used for qualification.

Table 1. Overview of the multiple reaction monitoring (MRM) transitions and tandem mass spectrometry (MS/MS) parameters for the target analytes.

Analyte	Precursor Ion (*m/z*) [a]	Product Ions (*m/z*)	CV [b] (V)	CE [c] (eV)	RT [d] (min)
DON	297.2 $[M+H]^+$	248.9 231.0	20	10 12	4.06
$^{13}C_{15}$-DON	312.0 $[M+H]^+$	263.0 245.2	20	10 10	4.06
DOM-1	281.0 $[M+H]^+$	233.0 215.0	27	10 10	4.66
NIV	313.1 $[M+H]^+$	175.0	35	15	3.49
ENN B	640.2 $[M+H]^+$	196.2 213.8	70	20 22	8.04
$^{15}N_3$-ENN B	643.3 $[M+H]^+$	197.1 215.3	70	23 23	8.04
MPA	321.1 $[M+H]^+$	207.0 159.0	25	22 33	6.43

Table 1. *Cont.*

Analyte	Precursor Ion (*m/z*) [a]	Product Ions (*m/z*)	CV [b] (V)	CE [c] (eV)	RT [d] (min)
$^{13}C_{17}$-MPA	338.0 [M+H]$^+$	<u>320.0</u> 218.0	26	10 22	6.43
ROQ-C	390.2 [M+H]$^+$	<u>192.9</u> 322.0	32	25 20	5.94
$^{13}C_{22}$-ROQ-C	412.0 [M+H]$^+$	<u>201.0</u> 339.0	32	27 18	5.94
ZEN	317.3 [M−H]$^-$	<u>175.0</u> 131.0	15	25 30	4.95
$^{13}C_{18}$-ZEN	335.3 [M−H]$^-$	<u>185.1</u> 169.1	15	25 32	4.95
ZAN	319.2 [M−H]$^-$	<u>275.2</u> 205.1	20	20 22	4.90
α-ZEL	319.2 [M−H]$^-$	<u>275.2</u> 301.0	20	20 22	4.37
β-ZEL	319.2 [M−H]$^-$	<u>275.2</u> 301.0	20	20 22	3.99
α-ZAL	321.2 [M−H]$^-$	<u>277.1</u> 303.1	30	22 20	4.29
β-ZAL	321.8 [M−H]$^-$	<u>277.1</u> 303.1	30	22 20	3.92

[a] *m/z* = mass to charge ratio; [b] CV = cone voltage; [c] CE = collision energy; [d] RT = retention time. DON = deoxynivalenol, DOM-1 = deepoxy-deoxynivalenol, NIV = nivalenol, ENN B = enniatin B, MPA = mycophenolic acid, ROQ-C = roquefortine C, ZEN = zearalenone, ZAN = zearalanone, α-ZEL = α-zearalenol, β-ZEL = β-zearalenol, α-ZAL = α-zearalanol and β-ZAL = β-zearalanol. The underlined product ion is used for quantification.

Different columns, mobile phases, LC gradient programs and flow rates were evaluated in terms of resolution, peak intensity and shape, and signal-to-noise (S/N) ratio (data not shown). Four different reversed-phase columns were tested (A/ Acquity UPLC BEH C18, Waters; B/ Hypersil gold, ThermoScientific; C/ Acquity UPLC HSS T3 column, Waters; D/ Zorbax Eclipse Plus, Agilent). The best results were obtained with the Acquity UPLC® HSS T3 column (100 mm × 2.1 mm I.D., 1.8 μm particle size). Most mobile phases described in literature consist of a combination of water and methanol (MeOH) or acetonitrile (ACN) to which some additives can be added to improve chromatography and detection [32]. A better sensitivity was observed for DON when MeOH was used. For an optimal chromatographic separation of ZEN and its metabolites, ACN was needed [33]. Several concentrations of ammonium acetate and/or acetic acid or formic acid were tested. The mobile phases with their LC gradient program that showed the best results were finally chosen for the method are shown in Table 2. Because of the diverse physicochemical characteristics of the mycotoxins, it was not possible to analyze all mycotoxins in the same run. Hence, to detect mycotoxins in ESI− and ESI+ mode, the samples had to be analyzed by two subsequent analytical runs, within a total run-time of 17 min. Chromatographic separation of ZEN and its metabolites was performed in ESI− mode within a time frame of 7 min using a gradient elution of H_2O and ACN as mobile phases (MP); DON, DOM-1, NIV, ENN B, MPA, and ROQ-C were analyzed in ESI+ mode within a run-time of 10 min, using a gradient elution of 0.01% acetic acid (*v/v*) in H_2O and 0.01% acetic acid (*v/v*) in MeOH as MP. Figures 1 and 2 show the chromatograms of the mycotoxins present in a rumen fluid–buffer mixture after extraction, determined in ESI+ and ESI− mode, respectively. In comparison with the run-time of Gallo et al. (2015) to detect PR toxin, MPA, and ROQ-C in the rumen fluid by means of HPLC-MS/MS, the run-time for the positive ionization mode is 2 min shorter while 12 mycotoxins with various physicochemical properties were determined [26].

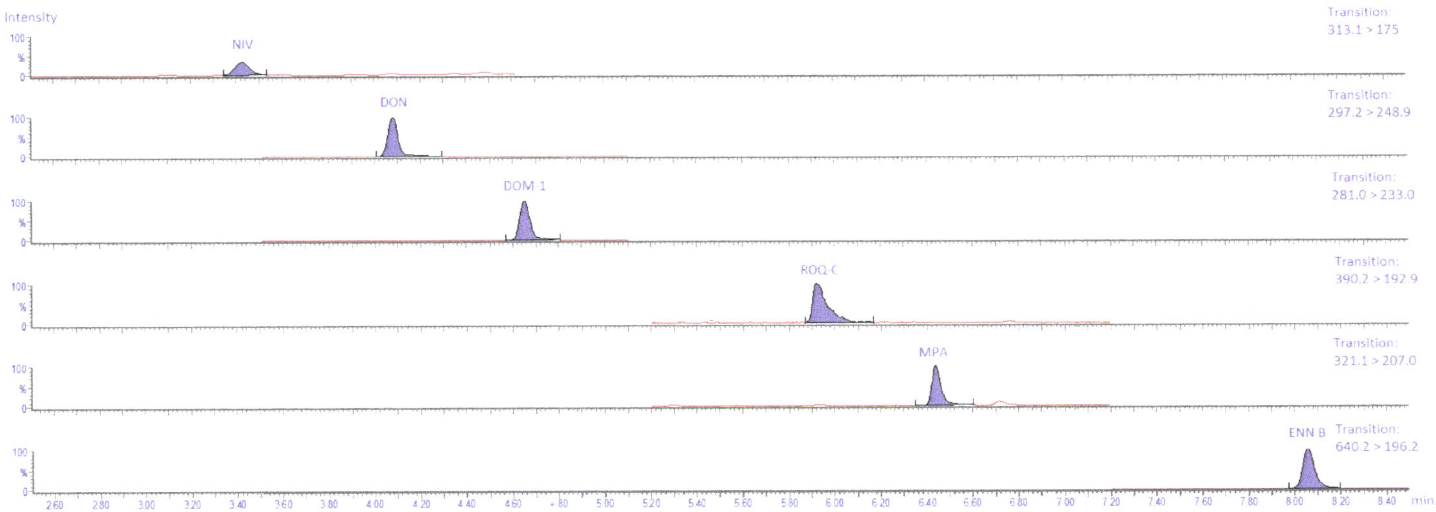

Figure 1. UPLC-MS/MS chromatograms of the analysis of nivalenol (NIV), deoxynivalenol (DON), deepoxy-deoxynivalenol (DOM-1), roquefortine C (ROQ-C), mycophenolic acid (MPA), and enniatin B (ENN B) in a rumen fluid–buffer mixture spiked at a level of 600 ng/mL (NIV), 120 ng/mL (DON), 122 ng/mL (DOM-1), 20 ng/mL (ROQ-C), 60 ng/mL (MPA), and 10 ng/mL (ENN B). These mycotoxins are detected in ESI+ mode. For each analyte, only the transition of the precursor ion to the product ion with the highest intensity (quantifier) is shown.

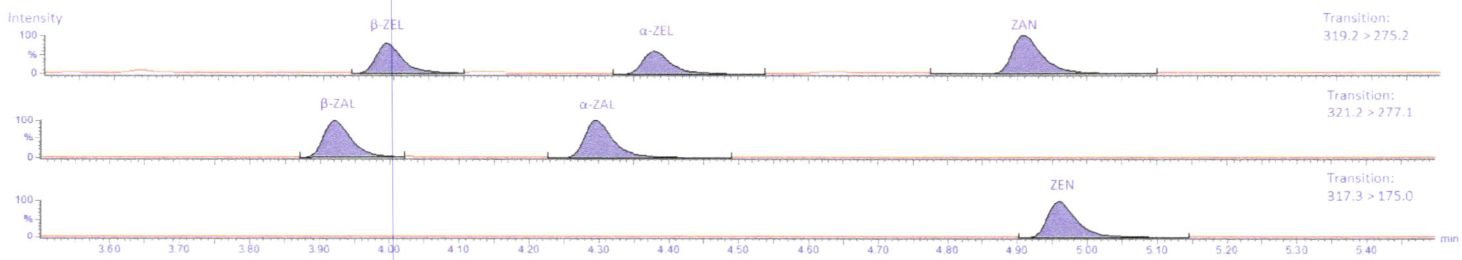

Figure 2. UPLC-MS/MS chromatograms of the analysis of zearalenone (ZEN) and its major metabolites in a rumen fluid–buffer mixture spiked at a level of 30 ng/mL of ZEN, zearalanone (ZAN), α-zearalenol (α-ZEL), β-zearalenol (β-ZEL), α-zearalanol (α-ZAL), and β-zearalanol (β-ZAL). These mycotoxins are detected in ESI− mode. For each analyte, only the transition of the precursor ion to the product ion with the highest intensity (quantifier) is shown.

Table 2. Overview of the UPLC gradient programs used for the mycotoxin ultra-performance liquid chromatography tandem mass spectrometry (UPLC-MS/MS) methods in ESI+ and ESI− mode.

Analyte(s) ESI+	Time (min)	MP A:B Ratio (v:v)
DON, $^{13}C_{15}$-DON, DOM-1, NIV, ENN B, $^{15}N_3$-ENN B, MPA, $^{13}C_{17}$-MPA, ROQ-C, $^{13}C_{22}$-ROQ-C	0.00–0.50	95:5
	0.50–5.50	Linear to 5:95
	5.50–7.50	5:95
	7.50–7.70	Linear to 95:5
	7.70–10.0	95:5
Analyte(s) ESI−	**Time (min)**	**MP C:D Ratio (v:v)**
ZEN, $^{13}C_{18}$-ZEN, ZAN, α-ZEL, β-ZEL, α ZAL, β ZAL	0.00–0.50	70:30
	0.50–3.50	Linear to 30:70
	3.50–4.90	30:70
	4.90–5.00	Linear to 70:30
	5.00–7.00	70:30

Note: ESI+: positive electrospray ionization mode; ESI-: negative ESI; MP: mobile phase. MP A = 0.01% acetic acid (v/v) in H_2O; MP B = 0.01% acetic acid (v/v) in MeOH; MP C = H_2O; MP D = ACN. DON = deoxynivalenol, DOM-1 = deepoxy-deoxynivalenol, NIV = nivalenol, ENN B = enniatin B, MPA = mycophenolic acid, ROQ-C = roquefortine C, ZEN = zearalenone, ZAN = zearalanone, α-ZEL = α-zearalenol, β-ZEL = β-zearalenol, α-ZAL = α-zearalanol and β-ZAL = β-zearalanol.

2.3. Sample Preparation and Extraction

The use of a simple and practical sample preparation procedure is desirable in order to reduce the time and cost of analysis. Theoretically, clean-up of samples can be kept to a minimum when analytical methods as sensitive and specific as UPLC-MS/MS are applied. However, the "dilute-and-shoot" principle was not considered in this study because of the complexity of the rumen fluid and the higher chance of clogging and pollution of the MS instrument. On the other hand, the more sophisticated extraction procedure with solid phase extraction (SPE) was not recommended as this seemed to be rather expensive and time-consuming, especially when large numbers of samples have to be analyzed. A liquid–liquid extraction (LLE) was chosen as an alternative and was a good compromise because of the simplicity of extraction, high throughput sample preparation procedure, and satisfactory sample clean-up. In the context of "green analytical chemistry" (GAC), ethyl acetate (EtAc) was chosen as the extraction solvent as this is a more environmentally friendly and cheaper alternative for ACN, which is often used in mycotoxin extraction protocols [34]. In addition, the simultaneous extraction of six mycotoxins and six mycotoxin metabolites minimizes the solvent consumption (250 µL PBS and 1.5 mL EtAc per sample) and chemical waste, and makes the chemical analysis more environmentally friendly [35]. In order to increase the extraction recovery, the volume of the organic extraction solvent was three times the volume of the aqueous phase (rumen fluid + buffer) as the distribution coefficient of the analytes between the organic and aqueous phase is dependent on their molar concentrations in the different phases. The combination of water and ACN as redissolution solvent at a 85/15 (v/v) ratio was crucial to obtain a good peak shape for all mycotoxins with their various physicochemical properties. Prior to UPLC-MS/MS analysis, the sample was passed through a filter to remove the remaining solid particles. For this step, several filters were tested. The Millex®-GV filter (0.22 µm, hydrophilic PVDF membrane, 13 mm) retained ROQ-C and ENN B and the Millex®-GN filter (0.20 µm, hydrophilic nylon membrane, 13 mm) retained MPA, ENN B, and ZEN and its metabolites. Only the Millex®-LG filter (0.20 µm, hydrophilic PTFE membrane, 4 mm) did not show any adsorption of mycotoxins.

2.4. Method Validation with Rumen Fluid–Buffer Mixture as Matrix

For this full method validation, a rumen fluid–buffer mixture was used as the matrix.

2.4.1. Linearity

Linear calibration curves were obtained for each component over the concentration ranges tested (see Table 3).

Table 3. Mycotoxin concentrations (limit of quantification (LOQ), medium and high concentration level) and concentration range in the rumen fluid of the calibration curve ($n = 10$) used for the in-house validation of the UPLC MS/MS method and validation results for linearity (r and GOF) and sensitivity (LOD (limit of detection) and LOQ; $n = 6$). For all components, the r and GOF met the acceptance criteria.

Analyte	Calibration Curve Range (ng/mL)	LOD (ng/mL)	LOQ (ng/mL)	Medium (ng/mL)	High (ng/mL)	*r*	GOF (%)
DON	0.45–180	0.05	0.45	12	120	0.9996	4.83
DOM-1	1.56–180	0.08	1.56	12	120	0.9995	4.67
NIV	36–600	5.43	36	120	600	0.9991	4.44
ENN B	0.39–15	<0.01	0.39	1.4	10	0.9982	6.36
MPA	0.6–90	0.17	0.60	6.0	60	0.9995	8.93
ROQ-C	0.1–30	<0.01	0.10	2.0	20	0.9996	5.29
ZEN	0.3–45	0.02	0.30	3.0	30	0.9993	7.82
ZAN	0.3–45	0.07	0.30	3.0	30	0.9997	5.02
α-ZEL	0.3–45	0.08	0.30	3.0	30	0.9990	6.24
β-ZEL	0.3–45	0.07	0.30	3.0	30	0.9938	7.78
α-ZAL	0.3–45	0.02	0.30	3.0	30	0.9983	4.82
β-ZAL	0.3–45	0.02	0.30	3.0	30	0.9931	11.15

Note: DON = deoxynivalenol, DOM-1 = deepoxy-deoxynivalenol, NIV = nivalenol, ENN B = enniatin B, MPA = mycophenolic acid, ROQ-C = roquefortine C, ZEN = zearalenone, ZAN = zearalanone, α-ZEL = α-zearalenol, β-ZEL = β-zearalenol, α-ZAL = α-zearalanol, and β-ZAL = β-zearalanol. Acceptability ranges: r = correlation coefficient ≥0.99; GOF = goodness-of-fit coefficient ≤20%. LOD: signal-to-noise ratio (S/N) = 3; LOQ value, lowest point of calibration curve and accuracy and precision within acceptability ranges [36].

The correlation coefficients (r) and goodness-of-fit coefficients (gof) met the acceptance criteria of ≥0.99 and ≤20%, respectively (Table 3).

The method of internal standardization was used to compensate for the analyte losses during sample preparation and for matrix effects during UPLC-MS/MS analysis. For DON, ENN B, MPA, ROQ-C, and ZEN, the corresponding isotopically labeled internal standard was used, respectively $^{13}C_{15}$-DON, $^{15}N_3$-ENN B, $^{13}C_{17}$-MPA, $^{13}C_{22}$-ROQ-C, and $^{13}C_{18}$-ZEN. Although the use of an IS for every single mycotoxin included in a multi-mycotoxin method is preferable, this is not feasible in practice because of financial limitations or lack of commercial availability. In this case, structurally related standards were also used, e.g., $^{13}C_{15}$–DON was chosen as IS not only for the analysis of DON, but also DOM-1 and NIV, and $^{13}C_{18}$-ZEN was chosen not only for the analysis of ZEN but also for its metabolites, as they have very similar physicochemical properties [10,33]. This is justified as it has been demonstrated during the method validation experiments that a reliable determination of these compounds was not impaired.

2.4.2. Accuracy and Precision

The within-run and between-run accuracy and precision were tested at three different concentration levels (limit of quantification (LOQ, see Section 2.4.3.), medium and high, see Table 3). The acceptability ranges for accuracy and within-run precision were met for all compounds at the specified levels according to VICH GL49 [36]. The between-run precision fell within the maximal allowed relative standard deviation ($RSD_{max} = 2^{(1-0.5\log Conc)}$) [36]. An overview of the results is given in Table 4.

Table 4. Validation results for within-run precision ($n = 6$) and between-run precision ($n = 2 \times 3$) with corresponding accuracy at low (LOQ), medium, and high concentration level. The different concentration levels for each mycotoxin are mentioned in Table 3. For all components, the acceptability ranges for accuracy and precision were met.

Analyte	Within-Run ($n = 6$)						Between-Run ($n = 2 \times 3$)					
	Accuracy (%)			Precision (RSD, %)			Accuracy (%)			Precision (RSD, %)		
	LOQ	Medium	High	LOQ	Medium	High	LOQ	Medium	High	LOQ	Medium	High
DON	19.4 ^{A}	−0.8 ^{C}	1.0 ^{C}	5.2 ^{A}	2.8 ^{C}	1.6 ^{D}	19.5 ^{A}	−1.6 ^{C}	0.4 ^{C}	5.2	2.5	1.8
DOM-1	3.7 ^{B}	−1.8 ^{C}	−5.9 ^{C}	5.5 ^{B}	5.2 ^{C}	9.2 ^{D}	2.6 ^{B}	−4.7 ^{C}	−2.9 ^{C}	5.2	5.8	6.6
NIV	4.2 ^{C}	−7.6 ^{C}	−7.2 ^{C}	5.0 ^{C}	7.6 ^{D}	7.7 ^{D}	1.8 ^{C}	−8.7 ^{C}	−6.1 ^{C}	6.8	7.8	6.5
ENN B	−14.0 ^{A}	3.6 ^{B}	−5.8 ^{C}	5.0 ^{A}	2.0 ^{B}	2.2 ^{C}	−14.0 ^{A}	1.1 ^{B}	−3.8 ^{C}	5.0	2.4	3.9
MPA	3.2 ^{A}	−8.2 ^{B}	5.2 ^{C}	5.3 ^{A}	2.9 ^{B}	4.5 ^{C}	4.5 ^{A}	−5.1 ^{B}	0.5 ^{C}	15.5	4.7	5.8
ROQ-C	4.7 ^{A}	−2.5 ^{B}	−4.9 ^{C}	4.5 ^{A}	2.0 ^{B}	1.3 ^{C}	6.8 ^{A}	−3.1 ^{B}	−4.5 ^{C}	4.2	2.4	2.5
ZEN	−5.6 ^{A}	2.0 ^{B}	−3.9 ^{C}	16.6 ^{A}	6.2 ^{B}	1.5 ^{C}	−0.3 ^{A}	−3.1 ^{B}	0.5 ^{C}	10.3	4.5	2.4
ZAN	3.6 ^{A}	−2.4 ^{B}	−6.5 ^{C}	11.1 ^{A}	5.2 ^{B}	3.9 ^{C}	-0.6 ^{A}	−0.5 ^{B}	−2.0 ^{C}	12.4	4.4	5.3
α-ZEL	−7.6 ^{A}	−2.4 ^{B}	−10.6 ^{C}	9.4 ^{A}	5.9 ^{B}	1.1 ^{C}	−9.2 ^{A}	−0.6 ^{B}	−5.1 ^{C}	12.4	5.9	8.2
β-ZEL	−4.6 ^{A}	−19.3 ^{B}	4.6 ^{C}	8.6 ^{A}	12.0 ^{B}	6.3 ^{C}	−2.8 ^{A}	−8.6 ^{B}	−5.8 ^{C}	8.2	13.4	14.5
α-ZAL	−21.3 ^{A}	−5.0 ^{B}	1.8 ^{C}	14.5 ^{A}	9.2 ^{B}	2.9 ^{C}	−13.5 ^{A}	−1.9 ^{B}	−5.5 ^{C}	16.9	10.1	9.6
β-ZAL	−6.7 ^{A}	−10.2 ^{B}	−9.9 ^{C}	16.4 ^{A}	12.0 ^{B}	10.9 ^{C}	−6.9 ^{A}	−5.3 ^{B}	−7.1 ^{C}	16.8	14.3	14.9

Note: DON = deoxynivalenol, DOM-1 = deepoxy-deoxynivalenol, NIV = nivalenol, ENN B = enniatin B, MPA = mycophenolic acid, ROQ-C = roquefortine C, ZEN = zearalenone, ZAN = zearalanone, α-ZEL = α-zearalenol, β-ZEL = β-zearalenol, α-ZAL = α-zearalanol, and β-ZAL = β-zearalanol. Acceptability ranges for accuracy: −50 to +20% for conc. <1 ng/mL (A in superscript), −30 to +10% for conc. ≥1 to 10 ng/mL (B in superscript), −20% to +10% for conc. ≥10 ng/mL (C in superscript), within-run precision: RSD_{max} = 30% for conc. <1 ng/mL (A in superscript), RSD_{max} = 25% for conc. ≥1 to 10 ng/mL (B in superscript), RSD_{max} = 15% for conc. ≥10 ng/mL to 100 ng/mL (C in superscript), RSD_{max} = 10% for conc. ≥100 ng/mL (D in superscript), between-run precision: $RSD_{max} = 2^{(1-0.5 \log Conc)}$.

2.4.3. Limit of Quantification (LOQ) and Limit of Detection (LOD)

The LOQ was set as the lowest concentration of the calibration curve that could be determined with an accuracy and precision within the acceptability ranges at the specified level. The LOQ ranged between 0.1 and 1.56 ng/mL, except for NIV that had an LOQ of 36 ng/mL (Table 3). The higher LOQ of NIV could be attributed to compromises that had to be made for a combined analysis of all analytes. Although NIV shows a better chromatographic response when determined in ESI- mode with mobile phases containing acetic acid [37], NIV is determined in ESI+ mode as adding acetic acid to the mobile phases led to a lower sensitivity for ZEN and its metabolites.

The LODs were theoretical and calculated based on the mean S/N ratios of the LOQ samples. Concentrations corresponding with a theoretical S/N ratio of 3 were set as the LOD and ranged from <0.01 to 0.17 ng/mL, except for NIV that had a LOD of 5.43 ng/mL.

In the study of Gallo et al. (2015) 15 mL of sample was used for mycotoxin determination with an LOD and LOQ of 4 and 10 ng/mL for both ROQ-C and MPA, which is much higher than the LOD of <0.01 ng/mL for ROQ-C and 0.28 ng/mL for MPA, and the LOQ of 0.1 ng/mL for ROQ-C and 0.6 ng/mL for MPA for the presented UPLC-MS/MS method that used only 250 μL of sample [26].

2.4.4. Signal Suppression/Enhancement (SSE) and Extraction Recovery (R_E)

The signal suppression/enhancement (SSE) was considered tolerable if the value ranged between 0.8 and 1.2 [38], with values outside this range indicating a strong matrix effect. SSE has been evaluated quantitatively and the results are shown in Table 5 (matrix A). As can be seen, there was a strong matrix effect for almost all analytes evaluated, which can be attributed to the complex matrix of rumen fluid. Extraction recoveries (R_E) were highest for DON and DOM-1 (>40%), around 20% for ZEN and its metabolites, around 15% for ENN B and ROQ-C and low for NIV and MPA (<10%) (see Table 5). These variations in R_E can be attributed to the different physicochemical properties of the different mycotoxins which e.g., results in differences in adsorption to particles present in rumen fluid. Such low R_E is hypothesized to be partially linked to enhanced adsorption/binding to (feed) particles as it has been described that bacteria and yeast cell walls could be involved in binding aflatoxins, DON, ZEN, and fumonisins, [16–24]. In order to minimize the variations because of R_E and SSE, the use of matrix-matched calibration curves based on samples, which had been subjected to the whole analytical procedure and the use of isotope labeled ISs are thus mandatory to quantify the selected mycotoxins and their metabolites in rumen fluid with an acceptable accuracy and precision. Despite the strong matrix effects and low R_E, very low LOD and LOQ values could be obtained because of the very sensitive UPLC-MS/MS analytical method.

2.4.5. Specificity

Prior to method validation, blank samples of the rumen fluid were analyzed to evaluate the presence/absence of mycotoxins. Small peaks could be detected at the elution zone of DON, DOM-1, ENN B, and ZEN, indicating that the rumen fluid already contained small amounts of these mycotoxins (DON: 0.45 ng/mL, DOM-1: 1.56 ng/mL, ENN B: 0.35 ng/mL, and ZEN: <LOQ) (Figures 3 and 4). However, these "background" mycotoxin levels could be compensated for by the use of a matrix-matched calibration approach.

Table 5. Extraction recovery (R_E) and signal suppression/enhancement (SSE) of mycotoxins in the rumen fluid–buffer mixture (Matrix A: RF) and in the rumen fluid–buffer mixture with 10 mg maize silage per mL (Matrix B: RF + MS). The results show that SSE and R_E differ between different matrices. Hence, matrix-matched calibration curves are needed when analyzing samples.

Analyte	Matrix A: RF		Matrix B: RF + MS	
	SSE (%)	R_E (%)	SSE (%)	R_E (%)
DON	79	42	55	44
NIV	68	9	60	8
ENN B	241	13	197	16
MPA	45	2	48	3
ROQ-C	69	16	62	16
ZEN	60	19	52	19
DOM-1	71	46	60	48
α-ZEL	60	20	53	21
β-ZEL	67	21	60	25
ZAN	64	19	53	21
α-ZAL	64	22	58	25
β-ZAL	68	23	63	29

Note: DON = deoxynivalenol, DOM-1 = deepoxy-deoxynivalenol, NIV = nivalenol, ENN B = enniatin B, MPA = mycophenolic acid, ROQ-C = roquefortine C, ZEN = zearalenone, ZAN = zearalanone, α-ZEL = α-zearalenol, β-ZEL = β-zearalenol, α-ZAL = α-zearalanol, and β-ZAL = β-zearalanol. SSE = $slope_{(spiked\ extract\ calibration\ curve)}/slope_{(standard\ calibration\ curve)}$, $R_E = slope_{(spiked\ calibration\ curve)}/slope_{(spiked\ extract\ calibration\ curve)}$.

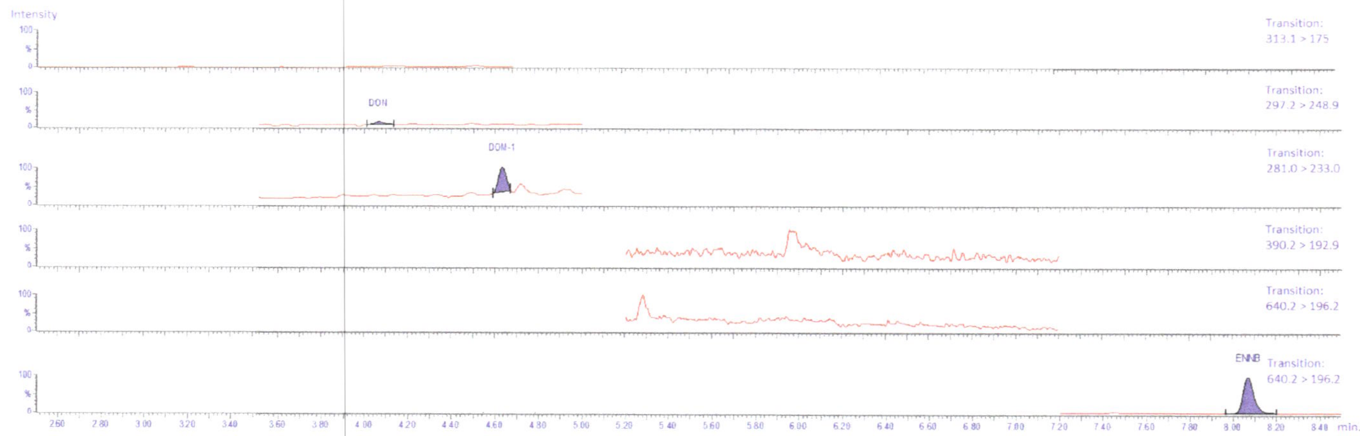

Figure 3. UPLC-MS/MS chromatograms of the analysis of nivalenol (NIV), deoxynivalenol (DON), deepoxy-deoxynivalenol (DOM-1), roquefortine C (ROQ-C), mycophenolic acid (MPA), and enniatin B (ENN B) in a rumen fluid–buffer mixture that was not spiked with mycotoxins. The rumen fluid was taken from a cow before morning feeding and showed traces of DON (0.45 ng/mL), DOM-1 (1.56 ng/mL), and ENN B (0.35 ng/mL). For each analyte, only the transition of the precursor ion to the product ion with the highest intensity (quantifier) is shown.

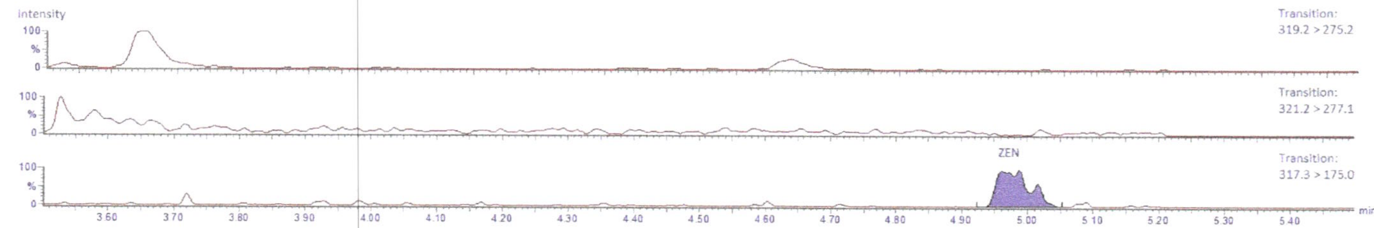

Figure 4. UPLC-MS/MS chromatograms of the analysis of zearalenone (ZEN) and its major metabolites zearalanone (ZAN), α-zearalenol (α-ZEL), β-zearalenol-(β-ZEL), α-zearalanol (α-ZAL), and β-zearalanol (β-ZAL) in a rumen fluid–buffer mixture that was not spiked with mycotoxins. The rumen fluid was taken from a cow before morning feeding and showed traces of ZEN (<LOQ). For each analyte, only the transition of the precursor ion to the product ion with the highest intensity (quantifier) is shown.

2.4.6. Carry-Over

No carry-over of analytes from one sample to another was observed, except for ROQ-C (0.3% carry over). This can be attributed to the flow-through-needle (FTN) design of the Acquity H-Class system. In case of ROQ-C analysis, carry-over could be overcome by the injection of two solvent samples after each validation/study sample with a suspected high concentration of ROQ-C [39].

2.5. Evaluation of the Impact of Maize Silage on the Extraction Recovery (R_E) and Signal Suppression/Enhancement (SSE)

As a substrate is always added to the rumen fluid to maintain fermentation during in vitro rumen incubations, and rumen fluid samples from in vivo studies also contain feed particles, feed particles can interact with the extraction and analysis of mycotoxins. Hence, the effect of feed particles, i.e., maize silage in this study, on R_E and SSE was investigated. The extraction recovery (R_E) in rumen fluid–buffer mixture including 10 mg maize silage per mL (RF + MS) was compared with the corresponding R_E in rumen fluid–buffer mixture without maize silage (RF). The results are shown in Table 5. SSE is mainly negatively influenced by the addition of maize silage in the samples, which was indeed expected as maize silage makes the matrix more complex. Moreover, the R_E is influenced by the addition of maize silage, which can be attributed to a change in solubility of the mycotoxins by addition of maize silage and/or the differences in adsorption to feed particles. Therefore, it is of utmost importance to prepare matrix-matched calibration curves.

2.6. Cross-Validation of the Method with Rumen Fluid and Maize Silage as Matrix

To prove that the described UPLC-MS/MS method can also be used with rumen fluid samples that also include substrate, the method was subjected to a cross-validation whereby 10 mg of maize silage was added per mL of rumen fluid–buffer mixture. Linearity, within-run accuracy and within-run precision were determined in the same way as the full method validation with rumen fluid–buffer mixture alone. The results are shown in Tables 6 and 7.

Table 6. Mycotoxin concentrations (limit of quantification (LOQ), medium and high concentration level) and concentration range of the calibration curve (n = 10) used for the in-house cross-validation of the UPLC MS/MS method with rumen fluid–buffer mixture and maize silage (10 mg/mL) and validation results for linearity (r and GOF) and sensitivity (LOD (limit of detection) and LOQ; n = 6). For all components, the r and GOF met the acceptance criteria.

Analyte	Calibration Curve Range (ng/mL)	LOD (ng/mL)	LOQ (ng/mL)	Medium (ng/mL)	High (ng/mL)	r	GOF (%)
DON	8–187	0.41	7.99	19.4	120	0.9998	1.65
DOM-1	6–180	0.35	6.00	12.0	120	0.9993	3.22
NIV	36–900	9.28	36	120	600	0.9974	6.64
ENN B	0.44–15.34	0.01	0.44	1.34	10	0.9985	5.81
MPA	1.20–90	0.42	1.20	6.0	60	0.9986	6.55
ROQ-C	0.2–30	0.01	0.20	2.0	20	0.9978	9.29
ZEN	1.06–45	0.04	1.06	3.5	30	0.9995	4.09
ZAN	0.3–45	0.10	0.30	3.0	30	0.9989	5.58
α-ZEL	0.6–45	0.15	0.60	3.0	30	0.9994	5.24
β-ZEL	0.6–45	0.08	0.60	3.0	30	0.9992	6.57
α-ZAL	1.5–45	0.04	1.50	3.0	30	0.9993	4.91
β-ZAL	0.3–45	0.06	0.30	3.0	30	0.9994	5.83

Note: DON = deoxynivalenol, DOM-1 = deepoxy-deoxynivalenol, NIV = nivalenol, ENN B = enniatin B, MPA = mycophenolic acid, ROQ-C = roquefortine C, ZEN = zearalenone, ZAN = zearalanone, α-ZEL = α-zearalenol, β-ZEL = β-zearalenol, α-ZAL = α-zearalanol, and β-ZAL = β-zearalanol. Acceptability ranges: r = correlation coefficient ≥0.99; GOF = goodness-of-fit coefficient ≤20%. LOD: signal-to-noise ratio (S/N) = 3; LOQ value, lowest point of calibration curve and accuracy and precision within acceptability ranges [36].

Table 7. Validation results of the in-house cross-validation of the UPLC MS/MS method with rumen fluid–buffer mixture and maize silage (10 mg/mL) for within-run precision (n = 6) with corresponding accuracy at low (LOQ), medium, and high concentration level. The different concentration levels for each mycotoxin are mentioned in Table 6. For all components, the acceptability ranges for accuracy and precision were met.

Analyte	Within-Run (n = 6)					
	Accuracy (%)			Precision (RSD, %)		
	LOQ	Medium	High	LOQ	Medium	High
DON	−0.3 ^{B}	0.1 ^{C}	−4.8 ^{C}	0.6 ^{B}	2.5 ^{C}	3.6 ^{D}
DOM-1	−7.9 ^{B}	−6.3 ^{C}	1.9 ^{C}	9.7 ^{B}	3.3 ^{C}	4.7 ^{D}
NIV	1.1 ^{C}	5.5 ^{C}	−1.0 ^{C}	5.6 ^{C}	2.8 ^{D}	8.0 ^{D}
ENN B	−2.0 ^{A}	4.5 ^{B}	−8.5 ^{C}	4.0 ^{A}	2.9 ^{B}	5.1 ^{C}
MPA	−1.1 ^{B}	−7.3 ^{B}	3.6 ^{C}	5.0 ^{B}	8.9 ^{B}	7.4 ^{C}
ROQ-C	−12.5 ^{A}	−1.0 ^{B}	−5.2 ^{C}	3.1 ^{A}	3.3 ^{B}	3.0 ^{C}
ZEN	−6.0 ^{B}	2.5 ^{B}	3.1 ^{C}	8.8 ^{B}	6.2 ^{B}	7.3 ^{C}
ZAN	−2.2 ^{A}	0.9 ^{B}	8.3 ^{C}	6.0 ^{A}	4.7 ^{B}	0.9 ^{C}
α-ZEL	1.9 ^{A}	−7.6 ^{B}	−5.2 ^{C}	4.7 ^{A}	2.9 ^{B}	6.5 ^{C}
β-ZEL	−1.4 ^{A}	−17.6 ^{B}	−0.7 ^{C}	8.2 ^{A}	10.0 ^{B}	7.8 ^{C}
α-ZAL	−10.9 ^{B}	−17.0 ^{B}	7.4 ^{C}	7.8 ^{B}	9.5 ^{B}	6.7 ^{C}
β-ZAL	−8.3 ^{A}	−16.8 ^{B}	−2.9 ^{C}	6.4 ^{A}	11.1 ^{B}	7.5 ^{C}

Note: DON = deoxynivalenol, DOM-1 = deepoxy-deoxynivalenol, NIV = nivalenol, ENN B = enniatin B, MPA = mycophenolic acid, ROQ-C = roquefortine C, ZEN = zearalenone, ZAN = zearalanone, α-ZEL = α-zearalenol, β-ZEL = β-zearalenol, α-ZAL = α-zearalanol, and β-ZAL = β-zearalanol. Acceptability ranges for accuracy: −50 to +20% for conc. <1 ng/mL (A in superscript), −30 to +10% for conc. ≥1 to 10 ng/mL (B in superscript), −20% to +10% for conc. ≥10 ng/mL (C in superscript), and within-run precision: RSD_{max} = 30% for conc. <1 ng/mL (A in superscript), RSD_{max} = 25% for conc. ≥1 to 10 ng/mL (B in superscript), RSD_{max} = 15% for conc. ≥10 ng/mL to 100 ng/mL (C in superscript), RSD_{max} = 10% for conc. ≥100 ng/mL (D in superscript).

Similar to the full method validation with rumen fluid–buffer mixture, the correlation coefficients (r) and goodness-of-fit coefficients (gof) met the acceptance criteria of ≥0.99 and ≤20%, respectively (Table 6). In addition, the within-run accuracy and precision at three different concentrations (LOQ, medium, and high) met the acceptability ranges for all compounds (Table 7). The LOQ ranged from 0.2 to 7.99 ng/mL, except for NIV that had a LOQ of 36 ng/mL. The LOD ranged from 0.01 to 0.42 ng/mL, except for NIV that had a LOD of 9.28 ng/mL. The LOQ and LOD values are somewhat higher compared to the LOQ and LOD values of the rumen fluid–buffer mixture alone, which can be explained by the more complex matrix. In addition, for the cross-validation, following concentrations were found in the blank sample with maize silage: DON: 7.39 ng/mL, ENN B: 0.34 ng/mL, and ZEN: 0.46 ng/mL, which could be due to a higher contamination of the rumen fluid used for the cross-validation, but also from the maize silage that is already contaminated with small amounts of mycotoxins. The relative high concentration of DON in the blank sample explains also the higher LOQ of DON compared to the full validation method without maize silage. However, these values are still very low compared to other methods. The method of Gallo et al. (2015) had an LOD and LOQ of 4 and 10 ng/mL for both ROQ-C and MPA, which is much higher than the LOD of 0.01 ng/mL for ROQ-C and 0.42 ng/mL for MPA, and the LOQ of 0.2 ng/mL for ROQ-C and 1.2 ng/mL for MPA for the cross-validation with rumen fluid and maize silage [26].

These results demonstrate that this method can be used for the quantitative determination of mycotoxins and their relevant metabolites in the rumen fluid samples if matrix-matched calibration curves are used. For in vitro rumen simulation studies for example, this implies that a calibration curve has to be prepared with rumen fluid which contains the same substrate as used in the study.

3. Conclusions

This study describes the development and in-house validation of a sensitive and specific UPLC-MS/MS method for the quantification of the most relevant mycotoxins and their possible metabolites in the rumen fluid. The sample extraction procedure is simple and straightforward and consists of a liquid–liquid extraction using EtAc. Furthermore, the UPLC-MS/MS analysis of each sample is split in two analytical runs, i.e., in ESI+ and ESI− mode, with a run-time of 10 and 7 min, respectively, resulting in a total analysis time of 17 min/sample allowing a high throughput. The method is successfully validated for all analytes of interest according to international guidelines and literature. Furthermore, as feed particles are often present in rumen fluid samples, the interaction of maize silage on SSE and R_E was demonstrated which emphasizes the importance of using matrix-matched calibration curves to allow a correct quantification of analytes. A cross-validation of the method with rumen fluid and maize silage as matrix demonstrates that this analytical method can be applied in research to investigate the degradation of the reported mycotoxins by rumen microbiota in rumen fluid samples if matrix-matched calibration curves are used.

4. Materials and Methods

4.1. Mycotoxin Selection

Mycotoxin selection was based on their prevalence in Belgium and their toxicological impact. In 2016, maize samples from 91 maize fields across Flanders (Belgium) were analyzed on mycotoxin contamination before ensiling. The mycotoxins most often found in these samples were NIV, DON, ZEN, and ENN B and were found in respectively 98.9%, 92.3%, 64.8%, and 42.9% of all samples (Table S1). Of those samples, 2% and 1% exceeded the European Union reference values for DON and ZEN, respectively. Hence, NIV, DON, ZEN, ENN B and their metabolites DOM-1, ZAN, α-ZEL, β-ZEL, α-ZAL, and β-ZAL were selected as pre-harvest mycotoxins. Additionally, mycotoxin contamination data of 21 maize silages and 100 silage samples were used and mycotoxins that had a minimum prevalence of 90% in non-moldy and/or moldy parts were also included. In addition to DON (in all samples), ZEN (in 90% of the maize silage samples), and ENN B (in 91% of silage samples), ROQ-C (in all moldy hot spots) and MPA (95% in non-moldy parts) were highly prevalent (Table S2). Although citrinin was also often present in maize silage (95%), this mycotoxin was not included in the UPLC-MS/MS method as this mycotoxin needs acidified extraction solvents in contrast to the other mycotoxins [40].

4.2. Rumen Fluid, Maize Silage, Mycotoxins, Chemicals, and Reagents

Rumen fluid was collected from three fistulated dairy cows (Institute for Agricultural, Fisheries and Food Research, EC2014/241) prior to the morning feeding and immediately transferred to thermos flasks before making the rumen fluid–buffer mixture (see Section 4.3). Sampling was done for the full method validation and another sampling was performed for the cross-validation.

Maize silage was obtained from Agrivet (Melle, Belgium) in May 2012 and lyophilized. Mycotoxin concentrations were determined at the Centre of Excellence in Mycotoxicology and Public Health, Department of Bioanalysis at Ghent University, Belgium. Only small amounts of NIV (201 μg/kg), DON (593 μg/kg), ZEN (68 μg/kg), and ROQ-C (12 μg/kg) were detected. The concentration of ENN B was below the cut-off value of 80 μg/kg.

The analytical standards of DON, NIV, MPA, ROQ-C, and ZEN were purchased from Fermentek (Jerusalem, Israel). The standards of DOM-1, ENN B, ZAN, α-ZEL, β-ZEL, α-ZAL, and β-ZAL were purchased from Sigma-Aldrich (Overijse, Belgium). The internal standards (IS) $^{13}C_{17}$-MPA, $^{13}C_{22}$-ROQ-C, and $^{13}C_{18}$-ZEN were purchased from Food Risk Management (Oostvoorne, The Netherlands). The IS $^{13}C_{15}$-DON was purchased from Sigma-Aldrich. The IS $^{15}N_3$-ENN B was kindly donated by the Chair of Analytical Food Chemistry at the Technical University of Munich (Freising, Germany) [41].

Methanol (MeOH), acetonitrile (ACN), ammonium acetate, and water (H_2O) were purchased from Biosolve (Valkenswaard, The Netherlands) and were of UPLC/MS grade. Acetic acid (AA), hydrochloric acid (HCl) 37%, ethyl acetate (EtAc) were purchased from Merck Millipore (Overijse,

Belgium) and were of analytical grade. Sodium chloride (NaCl), potassium dihydrogenphosphate (KH_2PO_4), and ammonium hydrogen carbonate (NH_4HCO_3) were purchased from VWR (Leuven, Belgium). The phosphate-buffered saline (PBS) powder packs were purchased from Thermo Fisher Scientific (Merelbeke, Belgium). Disodium hydrogen phosphate dodecahydrate ($Na_2HPO_4{\cdot}12H_2O$) and magnesium chloride hexahydrate ($MgCl_2{\cdot}6H_2O$) were from Carl Roth (Vienna, Austria). Sodium hydrogen carbonate ($NaHCO_3$) was purchased from Sigma-Aldrich. Carbon dioxide (CO_2) 100% was purchased from Air Liquide (Aalter, Belgium).

4.3. *Preparation of Standard Solutions and Rumen Fluid–Buffer Mixture*

Stock solutions of 1 mg/mL in ACN were prepared for DON, NIV, MPA, ROQ-C, ZEN, and ENN B and of 100 μg/mL in ACN for ZAN, α-ZEL, β-ZEL, α-ZAL, and β-ZAL. DOM-1 was already in solution upon purchase (50 μg/mL in ACN). An IS-stock solution of 5 μg/mL in ACN was prepared for $^{15}N_3$-ENN B. All other IS were already in solution upon purchase ($^{13}C_{15}$-DON: 25.3 μg/mL in ACN, $^{13}C_{17}$-MPA: 25.4 μg/mL in ACN, $^{13}C_{22}$-ROQ-C: 25 μg/mL in ACN, $^{13}C_{18}$-ZEN: 25.4 μg/mL in ACN). The standard stock solutions were used to make a combined working solution of all standards in ACN without IS (3 μg/mL for DON and DOM-1, 15 μg/mL for NIV, 0.25 μg/mL for ENN B, 1.5 μg/mL for MPA, 0.5 μg/mL for ROQ-C, and 0.75 μg/mL for ZEN, ZAN, α-ZEL, β-ZEL, α-ZAL, and β-ZAL). Serial dilutions (10×, 100×, and 1000×) of the combined working solutions were prepared. A combined working solution of all IS was prepared with a final concentration of 100 ng/mL for $^{13}C_{15}$-DON, $^{13}C_{17}$-MPA, and $^{13}C_{18}$-ZEN and 10 ng/mL for $^{15}N_3$-ENN B and $^{13}C_{22}$-ROQ-C. The standard and IS stock solutions were used for the preparation of matrix-matched calibrator and quality control (QC) samples. All stock and working solutions were stored at ≤ -15 °C.

As in vitro rumen simulations are performed in a buffered rumen environment, a rumen fluid–buffer mixture was also used for the method development. The buffer contained the following compounds per liter: 3.58 g $Na_2HPO_4{\cdot}12H_2O$, 1.55 g KH_2PO_4, 0.124 g $MgCl_2{\cdot}6H_2O$, 8.74 g $NaHCO_3$, and 1.00 g NH_4HCO_3. The buffer was saturated with CO_2 overnight and kept at 39 °C. The next day, fresh rumen fluid of three cows was mixed and sieved using a sieve with mesh width of 1 mm and added to the buffer at a rumen fluid/buffer ratio of 263.2 mL/1000 mL. The rumen fluid–buffer mixture was stored at ≤ -15 °C until calibrator, validation and quality control (QC) samples were prepared.

4.4. *Preparation of Calibrator, Validation and QC Samples*

Matrix-matched calibrator samples were prepared by adding the appropriate combined working solution to 250 μL rumen fluid–buffer mixture with or without maize silage to obtain a calibration range of 0.24–180 ng/mL of DON and DOM-1, 1.2–900 ng/mL of NIV, 0.02–15 ng/mL of ENN B, 0.12–90 ng/mL of MPA, 0.04–30 ng/mL of ROQ-C and 0.06–45 ng/mL of ZEN, ZAN, α-ZEL, β-ZEL, α-ZAL and β-ZAL. QC samples were prepared at medium and high concentration levels (Table 3). For DON and ZEN, the high concentration levels were based on the recommended maximum values in maize by-products formulated by the Commission of the European Communities [42]. As no recommended maximum values are available for the other mycotoxins, the high concentration levels of those mycotoxins were based on the worst case contamination levels in maize silage found in Belgium/the Netherlands described in literature [4,7–9].

4.5. *Mycotoxin Extraction*

The following procedure was applied to extract DON, DOM-1, NIV, ENN B, MPA, ROQ-C, ZEN, ZAN, α-ZEL, β-ZEL, α-ZAL, and β-ZAL from the matrix-matched calibrator, validation or QC samples: a 250-μL aliquot of the rumen fluid–buffer mixture was taken and transferred to a polypropylene 15-mL extraction tube, followed by the addition of 25 μL of the IS-mix working solution, 250 μL PBS and 1.5 mL of EtAc. The samples were vortex mixed and extracted on an overhead shaker (15 min). After centrifugation (3724× *g*, 5 min, 4 °C), the upper organic phase was collected and evaporated to dryness under a gentle nitrogen stream at ~50 °C. The dry residue was redissolved in 200 μL of

H_2O/ACN (85/15, *v*/*v*), vortex mixed, filtered using a 0.20 μm Millex®-LG PTFE filter (Merck Millipore, Overijse, Belgium), and collected in an autosampler vial.

4.6. UPLC-MS/MS Analysis

The UPLC system consisted of an Acquity H-Class Quaternary Solvent Manager and Flow-Through-Needle Sample Manager with temperature-controlled tray and column oven from Waters (Zellik, Belgium). Chromatographic separation of the analytes was achieved on an Acquity UPLC® HSS T3 column (100 mm × 2.1 mm I.D., particle size (dp): 1.8 μm) in combination with an Acquity HSS T3 Vanguard pre-column (5 mm × 2.1 mm I.D., dp: 1.8 μm) both from Waters.

Liquid chromatography parameters (mobile phase (MP) composition and gradient programs) were optimized for optimal chromatographic separation of all analytes. The MP for analysis of DON, $^{13}C_{15}$-DON, DOM-1, NIV, ENN B, $^{15}N_3$-ENN B, MPA, $^{13}C_{17}$-MPA, ROQ-C, and $^{13}C_{22}$-ROQ-C consisted of 0.01% AA in H_2O (A) and 0.01% AA in MeOH (B). The MP for the analysis of ZEN, $^{13}C_{18}$-ZEN and the metabolites ZAN, α-ZEL, β-ZEL, α-ZAL, and β-ZAL consisted of H_2O (A) and ACN (B). A gradient elution was performed as shown in Table 2. The flow-rate was set at 0.3 mL/min. The temperatures of the column oven and autosampler tray were set at 40 °C and 8 °C, respectively. The injection volume was 5 μL/sample.

The UPLC system was coupled to a Xevo® TQ-S MS/MS system, equipped with an ESI probe operating in both the positive and negative ionization mode (all from Waters, Zellik, Belgium). A divert valve was used and the UPLC effluent was directed to the mass spectrometer from 2.50 to 8.50 min for the ESI+ analysis and from 2.20 to 5.20 min for the ESI− analysis.

Mass spectrometric parameters were optimized by syringe infusion of working solutions of 1 μg/mL of each analyte (flow-rate: 10 μL/min) in combination with the mobile phase (50% A, 50% B, flow-rate: 0.3 mL/min). The following parameters were used: capillary voltage: 3.3 kV (ESI+) and 2.7 kV (ESI−), source offset: 40 V (ESI+) and 50 V (ESI−), desolvation temperature: 400 °C, desolvation gas: 600 L/h, cone gas: 150 L/h, source temperature: 150 °C, nebulizer pressure: 7.0 bar, LM resolution 1 and 2: 3.0, HM resolution 1 and 2: 15.0, ion energy 1: 0.8 (ESI+) and 0.5 (ESI−), ion energy 2: 1.0 (ESI+) and 0.5 (ESI−), collision gas flow: 0.20 mL/min (ESI+) and 0.25 mL/min (ESI−).

MS/MS acquisition was performed in the multiple reaction monitoring (MRM) mode for both the ESI+ and ESI− mode. In Table 1, an overview is given of compound specific parameters such as MRM transitions, cone voltage (CV), collision energy (CE), and retention time (RT). Dwell times were set at 10–50 msec/transition.

MassLynx version 4.1 software was used for data processing (Copyright© 2012, Waters, Zellik, Belgium).

4.7. In-House Method Validation

The developed UPLC-MS/MS method was validated in-house by a set of parameters that were in compliance with the recommendations as defined by the European Community [43] and with reference guidelines defined in other EU and FDA documents [36,44,45]. The following parameters have been evaluated for the full method validation with rumen fluid as matrix: linearity, within-run, and between-run accuracy, within-run and between-run precision, limit of quantification (LOQ), limit of detection (LOD), specificity, carry-over, extraction recovery and signal suppression/enhancement. For the cross-validation with rumen fluid + maize silage (10 mg maize silage per mL rumen fluid–buffer mixture) as matrix following parameters have been evaluated: linearity, within-run accuracy, within-run precision, limit of quantification (LOQ), and limit of detection (LOD).

Linearity. Calibration curve samples for the full validation and cross-validation were prepared by applying standard working solutions directly onto the homogenized blank samples, followed by a vortex mixing step. Mycotoxins from the calibration curve samples were extracted and quantified as mentioned in Sections 4.5 and 4.6. By using the same rumen fluid for the calibration curve as the samples to be analyzed, the initial concentration of mycotoxins can be determined by the standard addition method. By adding different concentrations of the mycotoxins to the rumen fluid samples, after measuring the final extract solutions, the observed signals are linearly regressed on the spiked amounts. The original unknown amount is estimated by the opposite of the abscissa intercept of the fitted linear curve. Linearity was evaluated by the correlation coefficients (r) and goodness- of-fit coefficients (g) that had to be ≥0.99 and ≤20%, respectively [43]. For a curve where no weighting is applied, the data at the high end of the calibration curve tend to dominate the calculation of the linear regression as the absolute variation is larger for higher concentrations. Hence, 1/x weighted calibration curves were used.

Accuracy and precision. Within-run accuracy and precision (repeatability) were determined by analysing six blank samples that were spiked at LOQ, medium, and high concentration levels on the same day. The between-run accuracy and precision were determined by analysing two blank samples that were spiked at LOQ, medium, and high concentration levels, repeated on three different days. The acceptance criteria for accuracy (i.e., the difference between spiked and measured concentration) were −30% to +10% for concentrations between 1 and 10 ng/mL and −20% to +10% for concentrations ≥10 ng/mL. For the precision, the relative standard deviation (RSD) had to be below $RSD_{max} = 2^{(1-0.5\log Conc)} \times 2/3$ for within-run precision and $2^{(1-0.5\log Conc)}$ for between-run precision [43,45].

Limit of quantification. The limit of quantification (LOQ) was the lowest concentration for which the method was validated with an accuracy and precision that fell within the recommended ranges. The LOQ was determined by analysing six spiked samples and was also the lowest point of the calibration curve.

Limit of detection. The limit of detection (LOD) was the lowest concentration that could be determined with a signal-to-noise ratio (S/N) of ≥3 and was calculated using samples spiked at the LOQ level.

Specificity. The specificity checks for interferences from endogenous compounds. The S/N ratio of a possible interfering peak in a blank sample ($n = 1$) had to be below the LOD of the analyte in the same elution zone.

Carry-over. The carry-over was evaluated by analyzing a solvent sample just after the highest calibrator sample. The eventual analyte concentration in the solvent sample had to be below the LOD.

Extraction recovery (R_E) and signal suppression/enhancement (SSE). These effects were determined according to Matuszewski et al. [46] and are visualized in Figure 5. Two types of matrix-matched calibration curves were prepared for each analyte by spiking the blank calibrator samples before (= spiked) and after (= spiked extract) extraction. In addition, one calibration curve was prepared using standard solutions (= standard). The extraction recovery (R_E) was determined by dividing the slopes of the resulting linear, spiked calibration curves by the related slopes of the calibration curves for spiked extracts. The signal suppression/enhancement (SSE) was evaluated by dividing the slopes of the calibration curves for spiked extracts by the related slopes of the calibration curves for standard solutions.

Figure 5. Determination of signal suppression/enhancement (SSE) and extraction recovery (R_E) in rumen fluid (RF) samples. Three types of calibration curves were prepared by adding mycotoxins (M) to: A/ solvent (standard), B/ rumen fluid before extraction (spiked), C/ rumen fluid samples after extraction (spiked extract). SSE was determined by dividing the slope of the calibration curve for spiked extract by the slope of the calibration curve for standard solutions. Extraction recovery was determined by dividing the slope of the spiked calibration curve by the slope of the calibration curve for spiked extract.

4.8. Evaluation of the Impact of Maize Silage on the Extraction Recovery (R_E) and Signal Suppression/Enhancement (SSE)

Ten milligram of maize silage was added per mL of rumen fluid–buffer mixture, R_E and SSE were determined as mentioned under Section 4.7 and compared with R_E and SSE values when using rumen fluid–buffer mixture alone.

Supplementary Materials: The following are available online at http://www.mdpi.com/2072-6651/11/9/519/s1, Table S1: Mycotoxins found in maize samples in Belgium, Table S2: Mycotoxins found in silage samples in Europe.

Author Contributions: Conceptualization, V.F. and S.C.; formal analysis, S.D.; investigation, S.D.; methodology, S.D.; resources, M.R.; supervision, S.D.B., V.F., and S.C.; validation, S.D.; writing—original draft, S.D; writing—review and editing, S.D.B., G.H., M.R., V.F., and S.C.

Funding: This research was funded by the government agency Flanders Innovation & Entrepreneurship, VLAIO-Belgium, grand number LA 140971.

Acknowledgments: The assistance of L. Decrop is greatly acknowledged.

Conflicts of Interest: The authors declare no conflict of interest. The funders had no role in the design of the study; in the collection, analyses, or interpretation of data; in the writing of the manuscript, or in the decision to publish the results

References

1. Fink-Gremmels, J. The role of mycotoxins in the health and performance of dairy cows. *Vet. J.* **2008**, *176*, 84–92. [CrossRef] [PubMed]
2. Upadhaya, S.D.; Park, M.A.; Ha, J.K. Mycotoxins and their biotransformation in the rumen: A review. *Asian-Australas. J. Anim. Sci.* **2010**, *23*, 1250–1260. [CrossRef]
3. Khafipour, E.; Li, S.; Plaizier, J.C.; Krause, D.O. Rumen microbiome composition determined using two nutritional models of subacute ruminal acidosis. *Appl. Environ. Microbiol.* **2009**, *75*, 7115–7124. [CrossRef] [PubMed]
4. Driehuis, F.; Spanjer, M.C.; Scholten, J.M.; Te Giffel, M.C. Occurrence of mycotoxins in maize, grass and wheat silage for dairy cattle in the Netherlands. *Food Addit. Contam. Part B* **2008**, *1*, 41–50. [CrossRef] [PubMed]
5. Zachariasova, M.; Dzuman, Z.; Veprikova, Z.; Hajkova, K.; Jiru, M.; Vaclavikova, M.; Zachariasova, A.; Pospichalova, M.; Florian, M.; Hajslova, J. Occurrence of multiple mycotoxins in European feedingstuffs, assessment of dietary intake by farm animals. *Anim. Feed Sci. Technol.* **2014**, *193*, 124–140. [CrossRef]
6. Valgaeren, B.; Théron, L.; Croubels, S.; Devreese, M.; De Baere, S.; Van Pamel, E.; Daeseleire, E.; De Boevre, M.; De Saeger, S.; Vidal, A.; et al. The role of roughage provision on the absorption and disposition of the mycotoxin deoxynivalenol and its acetylated derivatives in calves: From field observations to toxicokinetics. *Arch. Toxicol.* **2018**, *93*, 293–310. [CrossRef]
7. Tangni, E.K.; Pussemier, L.; Bastiaanse, H.; Haesaert, G.; Foucart, G.; Van Hove, F. Presence of mycophenolic acid, roquefortine C, citrinin and ochratoxin A in maize and grass silages supplied to dairy cattle in Belgium. *J. Anim. Sci. Adv.* **2013**, *3*, 598–612.
8. Driehuis, F.; Spanjer, M.C.; Scholten, J.M.; te Giffel, M.C. Occurrence of mycotoxins in feedstuffs of dairy cows and estimation of total dietary intakes. *J. Dairy Sci.* **2008**, *91*, 4261–4271. [CrossRef]
9. Wambacq, E.; Vanhoutte, I.; Audenaert, K.; De Gelder, L.; Haesaert, G. Occurrence, prevention and remediation of toxigenic fungi and mycotoxins in silage: A review. *J. Sci. Food Agric.* **2016**, *96*, 2284–2302. [CrossRef]
10. Jin, P.G.; Han, Z.; Cai, Z.X.; Wu, Y.J.; Ren, Y.P. Simultaneous determination of 10 mycotoxins in grain by ultra-high-performance liquid chromatography-tandem mass spectrometry using 13C15-deoxynivalenol as internal standard. *Food Addit. Contam. Part A* **2010**, *27*, 1701–1713. [CrossRef]
11. Ren, Y.; Zhang, Y.; Shao, S.; Cai, Z.; Feng, L.; Pan, H.; Wang, Z. Simultaneous determination of multi-component mycotoxin contaminants in foods and feeds by ultra-performance liquid chromatography tandem mass spectrometry. *J. Chromatogr. A* **2007**, *1143*, 48–64. [CrossRef] [PubMed]
12. Van Pamel, E.; Verbeken, A.; Vlaemynck, G.; De Boever, J.; Daeseleire, E. Ultrahigh-performance liquid chromatographic-tandem mass spectrometric multimycotoxin method for quantitating 26 mycotoxins in maize silage. *J. Agric. Food Chem.* **2011**, *59*, 9747–9755. [CrossRef] [PubMed]

13. Broekaert, N.; Devreese, M.; van Bergen, T.; Schauvliege, S.; De Boevre, M.; De Saeger, S.; Vanhaecke, L.; Berthiller, F.; Michlmayr, H.; Malachová, A.; et al. In vivo contribution of deoxynivalenol-3-β-d-glucoside to deoxynivalenol exposure in broiler chickens and pigs: Oral bioavailability, hydrolysis and toxicokinetics. *Arch. Toxicol.* **2017**, *91*, 699–712. [CrossRef] [PubMed]
14. Huybrechts, B.; Martins, J.C.; Debongnie, P.; Uhlig, S.; Callebaut, A. Fast and sensitive LC-MS/MS method measuring human mycotoxin exposure using biomarkers in urine. *Arch. Toxicol.* **2015**, *89*, 1993–2005. [CrossRef] [PubMed]
15. Vidal, A.; Claeys, L.; Mengelers, M.; Vanhoorne, V.; Vervaet, C.; Huybrechts, B.; De Saeger, S.; De Boevre, M. Humans significantly metabolize and excrete the mycotoxin deoxynivalenol and its modified form deoxynivalenol-3-glucoside within 24 h. *Sci. Rep.* **2018**, *8*, 1–11. [CrossRef]
16. Gallo, A.; Masoero, F. In vitro models to evaluate the capacity of different sequestering agents to adsorb aflatoxins. *Ital. J. Anim. Sci.* **2010**, *9*, 109–116. [CrossRef]
17. Niderkorn, V.; Morgavi, D.P.; Pujos, E.; Tissandier, A.; Boudra, H. Screening of fermentative bacteria for their ability to bind and biotransform deoxynivalenol, zearalenone and fumonisins in an in vitro model simulating corn silage. *Food Addit. Contam.* **2007**, *24*, 406–415. [CrossRef]
18. Peltonen, K.D.; El-Nezami, H.S.; Salminen, S.J.; Ahokas, J.T. Binding of aflatoxin B1 by probiotic bacteria. *J. Sci. Food Agric.* **2000**, *80*, 1942–1945. [CrossRef]
19. Oatley, J.T.; Rarick, M.D.; Ji, G.E.; Linz, J.E. Binding of aflatoxin B1 to Bifidobacteria in vitro. *J. Food Prot.* **2000**, *63*, 1133–1136. [CrossRef]
20. Karazhyan, R.; Shaker Sheyda, I.; Mehraban Sang Atash, M.; Tajalli, F.; Mojtahedi, M.; Sadegh, M. Effect of Saccharomyces cerevisiae yeast on ruminal detoxification of aflatoxin B1. *J. Vet. Res.* **2017**, *72*, 81–86.
21. El-Nezami, H.; Kankaanpää, P.; Salminen, S.; Ahokas, J. Physicochemical alterations enhance the ability of dairy strains of lactic acid bacteria to remove aflatoxin from contaminated media. *J. Food Prot.* **1998**, *61*, 466–468. [CrossRef] [PubMed]
22. El-Nezami, H.; Polychronaki, N.; Salminen, S.; Mykkänen, H. Binding rather than metabolism may explain the interaction of two food-grade Lactobacillus strains with zearalenone and its derivative α-zearalenol. *Appl. Environ. Microbiol.* **2002**, *68*, 3545–3549. [CrossRef] [PubMed]
23. El-Nezami, H.; Polychronaki, N.; Lee, Y.K.; Haskard, C.; Juvonen, R.; Salminen, S.; Mykkänen, H. Chemical moieties and interactions involved in the binding of zearalenone to the surface of Lactobacillus rhamnosus strains GG. *J. Agric. Food Chem.* **2004**, *52*, 4577–4581. [CrossRef] [PubMed]
24. Zhang, Y.G.; Liu, S.; Zhao, X.J.; Wang, N.; Jiang, X.; Xin, H.S.; Zhang, Y.G. Lactobacillus rhamnosus GG modulates gastrointestinal absorption, excretion patterns, and toxicity in Holstein calves fed a single dose of aflatoxin B1. *J. Dairy Sci.* **2018**, *102*, 1330–1340. [CrossRef] [PubMed]
25. Kiessling, K.H.; Pettersson, H.; Sandholm, K.; Olsen, M. Metabolism of aflatoxin, ochratoxin, zearalenone, and three trichothecenes by intact rumen fluid, rumen protozoa, and rumen bacteria. *Appl. Environ. Microbiol.* **1984**, *47*, 1070–1073. [PubMed]
26. Gallo, A.; Giuberti, G.; Bertuzzi, T.; Moschini, M.; Masoero, F. Study of the effects of PR toxin, mycophenolic acid and roquefortine C on in vitro gas production parameters and their stability in the rumen environment. *J. Agric. Sci.* **2015**, *153*, 163–176. [CrossRef]
27. Fink-Gremmels, J. Mycotoxins in cattle feeds and carry-over to dairy milk: A review. *Food Addit. Contam. Part A* **2008**, *25*, 172–180. [CrossRef]
28. Jeong, J.S.; Lee, J.H.; Simizu, Y.; Tazaki, H.; Itabashi, H.; Kimura, N. Effects of the *Fusarium* mycotoxin deoxynivalenol on in vitro rumen fermentation. *Anim. Feed Sci. Technol.* **2010**, *162*, 144–148. [CrossRef]
29. Dänicke, S.; Matthäus, K.; Lebzien, P.; Valenta, H.; Stemme, K.; Ueberschär, K.-H.; Razzazi-Fazeli, E.; Böhm, J.; Flachowsky, G. Effects of *Fusarium* toxin-contaminated wheat grain on nutrient turnover, microbial protein synthesis and metabolism of deoxynivalenol and zearalenone in the rumen of dairy cows. *J. Anim. Physiol. Anim. Nutr. (Berl.)* **2005**, *89*, 303–315. [CrossRef]
30. Seeling, K.; Dänicke, S.; Ueberschär, H.; Lebzien, P.; Flachowsky, G. On the effects of *Fusarium* toxin-contaminated wheat and the feed intake level on the metabolism and carry over of zearalenone in dairy cows. *Food Addit. Contam.* **2005**, *22*, 847–855. [CrossRef]
31. Kennedy, D.G.; Hewitt, S.A.; McEvoy, J.D.G.; Currie, J.W.; Cannavan, A.; Blanchflower, W.J.; Elliot, C.T. Zeranol is formed from *Fusarium* spp. toxins in cattle in vivo. *Food Addit. Contam.* **1998**, *15*, 393–400. [CrossRef] [PubMed]

32. Songsermsakul, E.; Razzazi-Fazeli, P. A review of recent trends in applications of liquid chromatography-mass spectrometry for determination of mycotoxins. *J. Liq. Chromatogr. Relat. Technol.* **2008**, *31*, 1641–1686. [CrossRef]
33. De Baere, S.; Osselaere, A.; Devreese, M.; Vanhaecke, L.; De Backer, P.; Croubels, S. Development of a liquid-chromatography tandem mass spectrometry and ultra-high-performance liquid chromatography high-resolution mass spectrometry method for the quantitative determination of zearalenone and its major metabolites in chicken and pig plasma. *Anal. Chim. Acta* **2012**, *756*, 37–48. [CrossRef] [PubMed]
34. Breidbach, A. A greener, quick and comprehensive extraction approach for LC-MS of multiple mycotoxins. *Toxins (Basel)* **2017**, *9*, 91. [CrossRef] [PubMed]
35. Gałuszka, A.; Migaszewski, Z.; Namieśnik, J. The 12 principles of green analytical chemistry and the SIGNIFICANCE mnemonic of green analytical practices. *Trends Anal. Chem.* **2013**, *50*, 78–84. [CrossRef]
36. Committee for Medicinal Products for Veterinary Use (CVMP). *VICH Topic GL49: Studies to Evaluate the Metabolism and Residue Kinetics of Veterinary Drugs in Food Producing Animals: Validation of Analytical Methods Used in Residue Depletion Studies*; Committee for Medicinal Products for Veterinary Use: London, UK, 2015.
37. Ye, J.; Wu, Y.; Wang, S.; Guo, Q.; Lu, M.; Wu, C.; Chen, Y. Simultaneous determination of 16 mycotoxins in cereals using an Agilent Triple Quadrupole LC/MS system and e-Method. *Agil. Technol.* **2017**, *2017*, 1–10.
38. Frenich, A.G.; Romero-González, R.; Gómez-Pérez, M.L.; Vidal, J.L.M. Multi-mycotoxin analysis in eggs using a QuEChERS-based extraction procedure and ultra-high-pressure liquid chromatography coupled to triple quadrupole mass spectrometry. *J. Chromatogr. A* **2011**, *1218*, 4349–4356. [CrossRef]
39. Rasmussen, R.R.; Storm, I.M.L.D.; Rasmussen, P.H.; Smedsgaard, J.; Nielsen, K.F. Multi-mycotoxin analysis of maize silage by LC-MS/MS. *Anal. Bioanal. Chem.* **2010**, *397*, 765–776. [CrossRef]
40. Molinié, A.; Faucet, V.; Castegnaro, M.; Pfohl-Leszkowicz, A. Analysis of some breakfast cereals on the French market for their contents of ochratoxin A, citrinin and fumonisin B1: Development of a method for simultaneous extraction of ochratoxin A and citrinin. *Food Chem.* **2005**, *92*, 391–400. [CrossRef]
41. Hu, L.; Rychlik, M. Biosynthesis of 15N3-labeled enniatins and beauvericin and their application to stable isotope dilution assays. *J. Agric. Food Chem.* **2012**, *60*, 7129–7136. [CrossRef]
42. European Commission. Commission Recommendation No 2006/576/EC of 17 August 2006 on the presence of deoxynivalenol, zearalenone, ochratoxin A, T-2 and HT-2 and fumonisins in products intended for animal feeding. *Off. J. Eur. Union* **2006**, *229*, 7–9.
43. European Commission. Commission Decision 2002/657/EC implementing Council Directive 96/23/EC concerning the performances of analytical methods and the interpretation of results. *Off. J. Eur. Communities* **2002**, *221*, 8–36.
44. Knecht, J.; Stork, G. Prozentuales und logarithmisches verfahren zur berechnung von eichkurven. *Z. Anal. Chem.* **1974**, *270*, 97–98. [CrossRef]
45. Heitzman, R. *Veterinary Drug Residues Residues in Food Producing Animals and Their Products: Reference Materials and Methods*; Blackwell Scientific Publications: Oxford, UK, 1996; Volume 636, ISBN 0-8412-3419-1.
46. Matuszewski, B.K.; Constanzer, M.L.; Chavez-Eng, C.M. Strategies for the assessment of matrix effect in quantitative bioanalytical methods based on HPLC-MS/MS. *Anal. Chem.* **2003**, *75*, 3019–3030. [CrossRef] [PubMed]

toxins

MDPI

Article

Characterization of Phase I and Glucuronide Phase II Metabolites of 17 Mycotoxins Using Liquid Chromatography—High-Resolution Mass Spectrometry

Irina Slobodchikova [1,2], Reajean Sivakumar [1], Md Samiur Rahman [1] and Dajana Vuckovic [1,2,*]

1 Department of Chemistry and Biochemistry, Concordia University, 7141 Sherbrooke Street West, Montreal, QC H4B 1R6, Canada

2 Centre for Biological Applications of Mass Spectrometry, Concordia University, 7141 Sherbrooke Street West, Montreal, QC H4B 1R6, Canada

* Correspondence: dajana.vuckovic@concordia.ca; Tel.: +1-514-848-2424 (ext. 3981)

Received: 15 June 2019; Accepted: 19 July 2019; Published: 24 July 2019

Abstract: Routine mycotoxin biomonitoring methods do not include many mycotoxin phase I and phase II metabolites, which may significantly underestimate mycotoxin exposure especially for heavily metabolized mycotoxins. Additional research efforts are also needed to measure metabolites in vivo after exposure and to establish which mycotoxin metabolites should be prioritized for the inclusion during large-scale biomonitoring efforts. The objective of this study was to perform human in vitro microsomal incubations of 17 mycotoxins and systematically characterize all resulting metabolites using liquid chromatography–high-resolution mass spectrometry (LC-HRMS). The results obtained were then used to build a comprehensive LC-MS library and expand a validated 17-mycotoxin method for exposure monitoring to screening of additional 188 metabolites, including 100 metabolites reported for the first time. The final method represents one of the most comprehensive LC-HRMS methods for mycotoxin biomonitoring or metabolism/fate studies.

Keywords: mycotoxins; metabolism; high-resolution mass spectrometry (HRMS); microsomal incubation; glucuronidation; human; biomonitoring

Key Contribution: This research provides the most comprehensive LC-MS library of mycotoxin metabolites fully characterized by HRMS to date to enable monitoring of 17 mycotoxins and 188 metabolites in plasma, urine or other biofluids of interest.

1. Introduction

Mycotoxins are toxic chemically diverse secondary metabolites produced by filamentous fungi. Their structural diversity can give rise to several adverse effects in humans and animals, such as carcinogenicity, immunosuppression, teratogenicity, nephrotoxicity, and hepatotoxicity [1]. The contamination of food and feed supply with low levels of mycotoxins is widespread, and includes commodities such as wine, apple juice, cereals, milk, coffee beans, maize, nuts, dried fruits, and meat products [2–7]. For example, a worldwide survey of more than 19,000 cereal and oilseed samples found that 72% were contaminated with one or more mycotoxins: aflatoxins (26%), deoxynivalenol (56%), ochratoxin A (25%), fumonisins (54%), and zearalenone (37%) [8]. In agreement with these findings, the most recent surveys of Canadian food supply showed 59% and 75% of the tested samples had detectable levels of at least one mycotoxin, with the most frequent incidence of deoxynivalenol [2,9]. Many other studies to date have also confirmed co-occurrence of multiple mycotoxins in food and feed samples [8,10,11], which in turn may lead to synergistic or antagonistic effects. Currently, the assessment of human mycotoxin exposure is primarily modelled from the measured/estimated levels of mycotoxins in the various foods and the calculated daily average food intake of various food groups to estimate population exposure and introduce regulations for food monitoring when appropriate. However, an individual's food consumption pattern depends on personal preferences. Thus, population-based food intake models can lead to the inaccurate estimation of human exposure to mycotoxins and, subsequently, higher health risk in some sub-populations. Vegetarian and non-vegetarian adult exposure to deoxynivalenol is one such example, whereby a recent U.K. study found ~2 × higher mean level of deoxynivalenol in vegetarians than in non-vegetarians [12]. Furthermore, the exceeded recommended tolerable daily intakes (32%) were found only in individuals belonging to the vegetarian group. Biomonitoring of adult and children urine samples in large-scale exposure studies also demonstrated that daily tolerable intake was exceeded for some mycotoxins [13–15].

To address the limitations of food-based models, direct human biomonitoring of biological fluids is proposed as an alternative approach to assess health risk [13–15]. However, this approach currently has several limitations. It requires high-throughout, multi-mycotoxin methods that have very low limits-of-detection in complex biological matrices such as blood and urine. Secondly, metabolic pathways have not been investigated thoroughly for all mycotoxins and key metabolites have not yet been prioritized for inclusion in routine biomonitoring [16]. Consequently, most of the existing analytical LC-MS methods used for the assessment of human exposure focus only on the detection of parent compounds. This can lead to significant underestimation of mycotoxin exposure. For example, a recent study of deoxynivalenol metabolism in humans confirmed the need for the inclusion of its metabolites in biomonitoring [17]. They showed that approximately 72.6% of total urinary deoxynivalenol (DON) was composed of its glucuronides, DON-15-glucuronide (15-Gluc-DON) and DON-3-glucuronide (3-Gluc-DON) and only 27.4% was present as free DON [17]. Thus, the measurement of DON only would underestimate DON exposure by ~4 ×. Other studies have also confirmed the importance of 15-Gluc-DON as a predominant glucuronide [15,18,19]. In general, biomonitoring methods should combine parent compounds and their predominant metabolite(s) in order to properly estimate exposure risk [20].

Currently, the majority of mycotoxin biomonitoring is performed using urine since it is non-invasive and accessible in relatively large volume. These methods can be divided into methods with and without β-glucuronidase treatment. B-glucuronidase catalyzes hydrolysis of conjugated mycotoxins, such as sulfate and glucuronide conjugates. Thus, the use of enzymatic hydrolysis can provide an appropriate alternative to direct metabolite monitoring for at least those mycotoxins which are predominately metabolized to phase II conjugated forms such as DON [17,21]. To date, such methods cover 7–11 mycotoxins [14,21–23]. The main disadvantages of β-glucuronidase treatment are: increases the cost per sample, requires longer processing time of about 16–18 h, and the additional step in sample preparation may give a rise to quantification errors. Multi-mycotoxin methods without β-glucuronidase treatment have been developed for 8–32 mycotoxins in urine [13,24–28], and for 8–27 mycotoxins in blood, serum or plasma [23,24,29,30]. However, these methods often include no or limited direct monitoring of mycotoxin metabolites.

Due to their toxicity, in vivo data on mycotoxin metabolism in humans after exposure is rare, with few exceptions [17]. Animal models have been used more frequently, but the interspecies differences in mycotoxin metabolism should be taken into consideration [31–33]. In vitro human liver microsomal incubations have been used extensively in the metabolism studies of mycotoxins, for example to obtain metabolic profile of T-2 and HT-2 toxins [32,33], or to study in detail the glucuronidation of the zearalenone group [34–36]. Human liver microsomes contain a variety of enzymes that are involved in both phase I and phase II toxin metabolism and reaction conditions can be easily controlled to generate the needed quantity of metabolites. The examples of phase I reactions are oxidation, reduction, dehalogenation, or hydrolysis and are catalyzed by several enzymes including cytochrome P450. Phase II reactions are conjugation reactions, for example with glucuronic acid, sulfate, glutathione and/or amino acids. High-resolution mass spectrometry (HRMS) provides an excellent analytical platform for the characterization and investigation of mycotoxin metabolites and putative biomarkers for further human biomonitoring [20]. The combination of HRMS and metabolic software can greatly speed up and expand the ability to capture the broad spectrum of mycotoxin metabolites using both accurate-mass of full-scan MS and/or fragmentation mass spectral data (MS/MS or MS^n). For instance, Yang et al. used HRMS to study T-2 and HT-2 metabolism in different species and identify main metabolic pathways and novel metabolites [32,33]. However, such single-analyte metabolism studies relied on a variety of analytical platforms and methods, thus hindering the creation of a comprehensive metabolite LC-MS library using a single analytical method and its further application in human biomonitoring. As such, it is of utmost importance to include mycotoxin metabolites in ongoing biomonitoring efforts and to use this information to prioritize the most commonly observed mycotoxin metabolites that may contribute to under-estimation of exposure. To achieve this goal, the first step is to fully characterize and build a comprehensive LC-MS library of mycotoxin metabolites using a single well-characterized LC-MS method.

In this work we present in vitro metabolism studies of 17 mycotoxins detected in the Canadian food supply: aflatoxins B1, B2, G1 and G2 (AFB1, AFB2, AFG1, AFG2), zearalenone (ZEN), 7-α-hydroxy-zearalenol (α-ZOL), 7-β-hydroxy-zearalenol (β-ZOL), zearalanone (ZAN), 7-α-hydroxy-zearalanol (α-ZAL), 7-β-hydroxy-zearalanol (β-ZAL), T-2 toxin (T-2), HT-2 toxin (HT-2), DON, nivalenol (NIV), 15-acetyldeoxynivalenol (15-AcDON), 3-acetyldeoxynivalenol (3-AcDON) and fusarenon X (FUS-X) in order to characterize phase I and glucuronide phase II mycotoxin metabolites. Mycotoxin metabolites were generated in vitro using pooled human liver microsomes to build an extensive in-house library of these species, for which standard compounds are often not commercially available. The final in-house LC-MS library was built using a previously published validated method for sensitive quantitation of 17 mycotoxins in plasma using liquid chromatography–high-resolution mass spectrometry (LC-HRMS) [29]. The use of this method allowed excellent chromatographic separation of many isomers and the optimized highly sensitive HRMS detection allowed detailed characterization of both known and novel metabolites.

2. Results

Human liver microsomes are important and common tool for in vitro investigations of toxin metabolism because they express a variety of enzymes which are involved in phase I metabolism such as microsomal cytochrome P450 (P450) and flavin-containing monooxygenases (FMO). These enzymes are responsible for the most common phase I reactions, such as oxidation. Usually, toxins are converted to more polar compounds due to phase I reactions. Phase II metabolism provides an additional mechanism to clear toxins from the body by adding water-soluble groups, such as glucuronic, methyl, sulfate and acetyl groups [20,37]. In this work, phase II glucuronidation reaction was chosen as a major human metabolic pathway of toxins in addition to phase I metabolism. In total, 17 mycotoxins, such as trichothecene type A (T-2 and HT-2), trichothecene type B (NIV, FUS-X, DON, 3-AcDON and 15-AcDON), aflatoxins (AFB1, AFB2, AFG1, and AFG2) and group of zearalenone (ZEN, α-ZOL, β-ZOL, ZAN, α-ZAL, and β-ZAL) were incubated individually in the presence of human microsomes and 188 different fungal metabolites were characterized and detected. The analysis of all microsomal incubation samples was performed with high-resolution mass spectrometer (LTQ Orbitrap Velos) coupled with liquid chromatography in order to detect and identify mycotoxin metabolites. Structural elucidation of metabolites was performed using data-dependent MS/MS acquisition and collision-induced dissociation (CID) fragmentation technique. Metabolite characterization and data analysis was performed using Compound Discoverer software 2.1, which contains extensive libraries of common metabolic pathways and mass spectral libraries.

To confirm enzymatic origin of metabolites, besides test samples for the phase I and II, several controls were used as shown in Figure 1 and Supplementary Table S1: standard that contains toxin dissolved in PBS buffer, control without any co-factors, control without NADPH, but with UDPGA, and controls with previously heated microsomes (45 °C) for both phase I and II samples. The mycotoxin standard control and the controls without cofactors were used to highlight and eliminate the metabolites that are not enzymatically produced from the final LC-MS library. Finally, the controls without toxin ensured that any endogenous species present in microsomes would not be misidentified as mycotoxin metabolites. The controls with pre-heated microsomes were included in the experiment to test the stability of microsomal enzymes. In heat-inactivated samples, the metabolic activity was changed and the generation of metabolites was reduced during the phase I metabolic reactions indicating that the responsible enzymes were sensitive to heat (Figure 2a). However, an opposite effect was observed in the phase II reactions, whereby an increased rate of glucuronidation was observed in all mycotoxin samples (Figure 2b). The deactivation of phase I metabolism observed in our study matches the previously published work about enzyme stability [38,39]. In contrast, uridine 5′-diphospho-glucuronosyltransferases (UGTs), key enzymes used in our phase II glucuronidation reactions, appear to be thermally stable enzymes [40], and the heat-inactivation step was beneficial to generating additional glucuronide metabolites in sufficient quantities for detailed characterization.

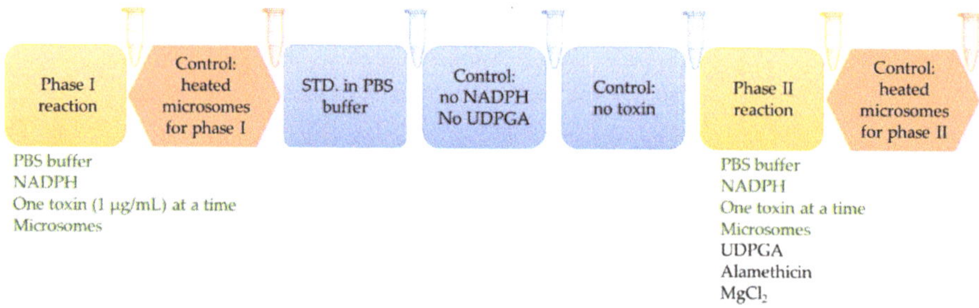

Figure 1. Scheme of microsomal incubation experiment to generate phase I and glucuronide phase II metabolites.

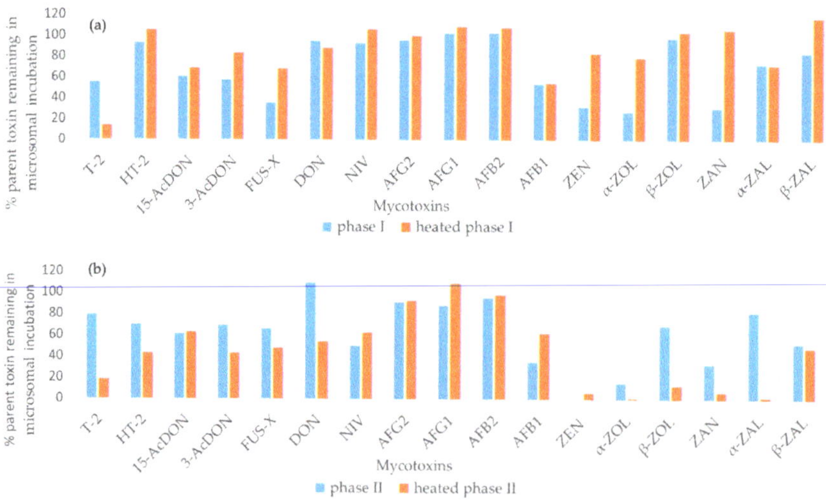

Figure 2. Comparison of the amount of parent toxin remaining after incubation in phase I and heated phase I (**a**) and phase II and heated phase II (**b**) microsomal incubation samples.

2.1. Trichothecene Type A and B

2.1.1. Trichothecene Type A

The list of T-2 generated metabolites is shown in the Supplementary Table S2. There were two main pathways for T-2 metabolites, hydrolysis and oxidation in phase I (Figure 3). The chromatographic separation of T-2 and its metabolites is shown in Supplementary Figures S1 and S2. The identification of metabolites was performed by comparing $[M+Na]^+$ product ion mass spectra of T-2 and its metabolites. The fragmentation pattern of T-2 showed some characteristic fragments, 387.2 m/z, 327.2 m/z and 267.2 m/z due to the loss of isovaleric acid ($C_5H_{10}O_2$, 102.1 Da) at position 8, and acetic acid (CH_3COOH) at position 15 or 4, respectively, (Supplementary Figure S3d). The extracted ion chromatogram of $[M+Na]^+$ ion at 447.1989 m/z revealed two peaks at 7.93 min (447.1988, 0.22 ppm) and 8.22 min (447.1986, 0.67 ppm), indicating the presence of two metabolites that were 42.0 Da less then T-2 (Supplementary Figures S1 and S4a,b). The peak observed at 8.22 min was identified as HT-2, since it had the same RT and MS^2 as the authentic standard of HT2. The second peak could be putatively identified as 15-deacetyl-T-2 (15-de-Ac-T-2). 15-de-Ac-T-2 had been previously observed as a metabolite of T-2 in Wistar rats [41]. Based on the structure of T-2, the possible loss of 42.0 Da can be due to the loss of the second acetyl group at position 15. Also, both metabolites had identical MS^2 spectra with the typical losses of isovaleric side chain (102.1 Da) and acetic acid (60.0 Da) at fragments of 345.2 Da and 285.2 Da, respectively (Supplementary Table S2 and Figure S4a,b). There was also another ion at 405.1881 (0.74 ppm) m/z corresponding to a loss of two acetyl groups from T-2, but it was very low intensity ion, so further identification was not possible (Supplementary Table S2 and Figure S1). The literature reports two possible compounds with this mass, neosolaniol and T-2 triol [41,42].

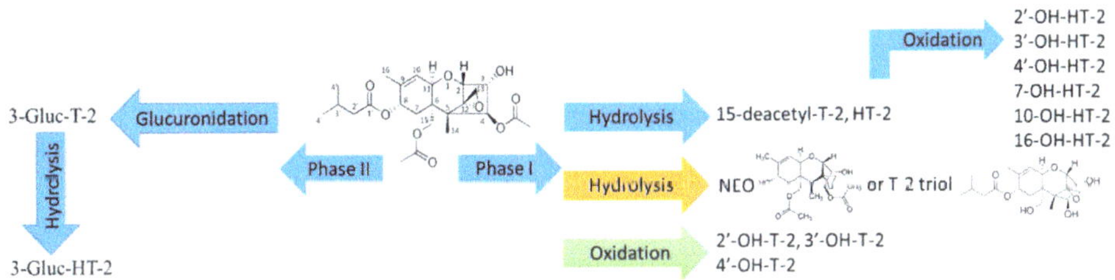

Figure 3. Microsomal biotransformation of T-2 toxin in phase I and phase II reactions.

The second pathway observed in phase I reactions was oxidation of both T-2 and HT-2. The theoretical masses of $[M+Na]^+$ ions of T-2 (505.2044) and HT-2 (463.1939) hydroxyl-metabolites were 16.0 Da higher than T-2 and HT-2, which confirmed the presence of additional oxygen in those compounds (Supplementary Table S2). There were three T-2 hydroxy metabolites observed at 6.54 min (505.2042, 0.40 ppm), 6.60 min 505.2041, 0.59 ppm), and 6.81 min (505.2043, 0.2 ppm), Supplementary Table S2 and Figure S2. All three hydroxyl metabolites had similar MS^2 spectra, Supplementary Table S2 and Figure S3a–c). The position of hydroxyl group was identified by comparing $[M+Na]^+$ product ion spectra of the hydroxyl-metabolites and T-2. In MS^2 spectrum of hydroxyl metabolites there was a fragment with 387.2 m/z that could be generated as the loss of isovaleric acid side chain plus oxygen atom ($C_5H_{10}O_3$, 118.1 Da). This fragment showed a 16.0 Da shift that indicated the position of hydroxyl group on the isovaleric side chain. Therefore, all three hydroxyl metabolites have OH group on isovaleric side at position 3′, 4′ or 2′, (Supplementary Table S2 and Figure S3a–c). Hydroxy metabolites of HT-T-2 also had a 16.0 Da shift. The numbering of HT-2 metabolites was chosen to match numbering of the peaks described for HT-2 where incubation and detailed characterization was performed with HT-2 toxin. T-2 microsomal incubation samples had 5 out of 6 metabolites of these hydroxyl metabolites, but their intensity was ~4 × less than in HT-2 incubations, Supplementary Table S2 and Figure S2. Overall, only 26% of T-2 metabolized in phase I reactions, with HT-2 as the predominant metabolite. The metabolism of T-2 has already been investigated by Yang et al. in farm animals and humans [33]. Our data are in agreement with their results, HT-2 is predominant metabolite of T-2. However, our study also generated two additional new hydroxyl metabolites of T-2. LC chromatographic separation of isomers using our pentafluorophenyl stationary phase and/or excellent limits of detection of the method may have facilitated the detection of these additional metabolites versus previous work. Furthermore, these newly detected metabolites are consistent with the available sites on T-2 for hydroxyl modifications.

In phase II reaction samples of T-2, there were two glucuronide forms, glucuronide of T-2 and HT-2 (Figure 3, Supplementary Table S2, and Figure S5). The different metabolic activity was observed in the phase II sample and heated control, about 79% and 18% of T-2 did not metabolize, respectively (Figure 2b). The most predominant glucuronide form was glucuronide of HT-2 (51%) and only 8% composed T-2 glucuronide in the heated control. Their product ion mass spectra of $[M+Na]^+$, 665.2413 m/z (−0.43 ppm) and 623.2304 m/z (−0.96 ppm) showed the indicative loss for glucuronides, 176.0 Da and typical fragments, 489.2 m/z and 447.5 m/z of T-2 and HT-2 respectively, confirming the T-2 and HT-2 origin of glucuronides, Supplementary Table S2 and Figure S6a,b. Further comparison of product ion spectra of $[M+NH_4]^+$ (Supplementary Figure S6c) to literature spectra confirmed this glucuronide as 3-glucuronide-HT-2 by the presence of fragment ions of 425.2 and 499.0 and their relative intensities to each other [33,43]. Ion with 499.0 m/z should be less intense than 425.2 m/z according to published data. According to the structure of T-2 and literature data there was only one possible glucuronide of T-2 [43].

The generated metabolites of HT-2 are presented in Supplementary Table S3. HT-2 is a main metabolite of T-2, and has hydroxyl group at position 4 instead of acetyl group. Two pathways were observed in phase I reactions, hydrolysis and oxidation. The chromatographic separation of HT-2 and its metabolites is shown in Supplementary Figures S7 and S8. Two hydrolysis products were observed as shown in Supplementary Table S3 and Figure S7. The extracted ion of $[M+Na]^+$ at m/z 363.1413 (0.28 ppm) shows the 84.1 Da mass difference from HT-2 $[M+Na]^+$ ion, which can be attributed to the loss of isovaleric group at position 8 and the addition of OH group. The first peak at 3.56 min can be putatively identified as 4-deacetyl-neosolaniol (4-de-Ac-NEO) which has OH group instead of isovaleric group. The product ion spectrum of the first peak has fragments with mass of 345.1 Da and 303.1 Da that confirm the water loss and further loss of acetic acid which were also found in the product ion spectra of HT-2, Supplementary Table S3. The $[M+Na]^+$ ion at m/z 463.1939 was 16.0 Da higher than HT-2 $[M+Na]^+$ ion, m/z 447.1989, and confirmed the hydroxylation pathway, Supplementary Table S3. The extracted ion chromatogram displayed 6 peaks with the same m/z 463.1936 (0.65 ppm) at 5.67 min, 5.77 min, 5.95 min, 6.11 min, 6.21 min and 8.23 min, Supplementary Figure S8. The first two peaks at RTs of 5.67 min and 5.77 min have similar product ion spectra, containing an indicative fragment ion at m/z 345.5 and 345.2, respectively, Supplementary Table S3 and Figure S9a,b. These ions were generated as the loss of isovaleric side chain ($C_5H_{10}O_2$) plus oxygen atom resulting in the neutral loss of 118.1 Da. The presence of these fragments confirmed the position of hydroxyl group at isovaleric side chain either at position 3′ or 4′. 3′ and 4′-hydroxy-HT-2 metabolites were observed in human and animals, respectively, by Yang et al. [33]. The third peak at 5.95 min was very low intensity, and its product ion spectra were similar to the previous peaks (Supplementary Figure S9c), assuming that OH is present at isovaleric group at position 2′. For the next three peaks, the loss of 102.1 Da results in an ion fragment with m/z 361.2 Da, so it indicates that the isovaleric side chain is not changed, Supplementary Table S3 and Figure S9d–f. Therefore, the position of OH group can be found at the position 7, 10 or 16 carbon atoms. However, the product mass spectra are similar, so further identification is not possible. Overall, six hydroxyl metabolites were also detected by Yang et al., but only four of their metabolites were observed in human liver microsomes [32]. Additionally, two peaks at RT of 5.44 min and 6.55 min were observed with the mass of 405.1880 (0.99 ppm) which corresponds to 42.0 Da difference from the parent compound (HT-2) which could indicate the loss of acyl group at position 15, Supplementary Table S3. However, their product ion mass spectra showed similar losses to HT-2. Based on the fragments at 303.2 m/z and 345.2 m/z which were generated as the loss of isovaleric acid (102 Da) and acetic acid (60 Da) respectively it was concluded that the main structure is not changed, and from the known metabolites it was not possible to propose putative structures. Overall, our data are similar to the previous metabolism studies done by Yang et al. [32], confirming hydroxylation as the major pathway of HT-2.

In phase II reaction samples of HT-2, the 3-glucuronide of HT-2 was generated as described when discussing the observed T-2 metabolites.

2.1.2. Trichothecene Type B

The common phase I pathways of type B trichothecenes are de-acetylation for 3/15-AcDON and FUS-X and de-epoxidation for DON (Figure 4) and NIV. Microsomal biotransformation of DON is summarized in Figure 4 as an example representative for this family. Chromatographic separation and MS^2 spectra of 3/15-AcDON, FUS-X, DON, NIV and their metabolites are shown in Supplementary Tables S4–S8 and Figures S10–S21. In phase I, all metabolites were generated non-enzymatically—all of these metabolites were observed not only in the phase I sample, but also in controls without NADPH and in heat-inactivated controls. The examples of non-enzymatic reactions included the removal of acetyl group in 3-AcDON converting into DON and de-epoxy-deoxynivalenol DOM-1, 15-AcDON into DON, FUS-X into NIV, DON into DOM-1, and NIV converted into de-epoxy-nivalenol (DNIV), Supplementary Tables S4–S8. 1% of 15-AcDON and 50% of 3-AcDON was converted into DON, whereas 54% of FUS-X was converted into NIV. Only 5% of NIV was converted into DNIV and less than 1% of DON to DNIV. Higher deacetylation rate of 3-AcDON than 15-AcDON had already been demonstrated in literature [44]. In our studies, these metabolites were clearly of non-enzymatic origin; however, other studies have also shown that 3-AcDON can be metabolized to DON (78%) during incubation with human feces [45].

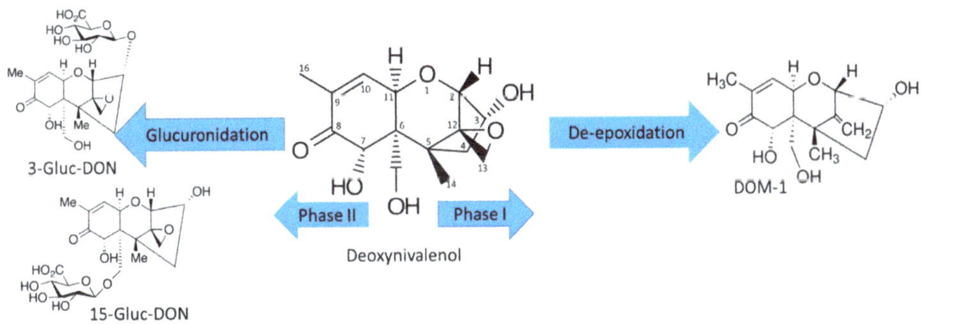

Figure 4. Microsomal biotransformation of DON in phase I and II reactions.

During phase II incubations, type B trichothecenes generated 3- and 15-Gluc-DON (1%), shown in Figure 4, Gluc-3-AcDON (11%), and Gluc-15-AcDON (1%) (Supplementary Table S4–S8 and Figures S10–S15, S17–S21), whereas heated samples generated 3- and 15-Gluc-DON (2%), 41% of Gluc-3-AcDON and 2% of Gluc-15-AcDON. However, to observe the glucuronidation of NIV (<1%) and FUS-X (<1%), it was necessary to increase the mycotoxin concentration x10 and incubation time (20 h), and they were only observed in heated samples (Supplementary Figures S18 and S20). The identifications of glucuronides were based on the product ion spectra of $[M-H]^-$ for Gluc-FUS-X (529.1561, 0.32 ppm), Gluc-NIV (487.1457, 0.23 ppm), 3- and 15-Gluc-DON (471.1508, 0 ppm), and Gluc-3-AcDON (513.1614, 0.19 ppm) and $[M+Na]^+$ for Gluc-15-AcDON (537.1575, 0.74 ppm), Supplementary Tables S4–S8. It is interesting to note that all glucuronides in ESI(-) generated only $[M-H]^-$ and not $[M+HAc-H]^-$ as their parent mycotoxins. Some mycotoxins, like DON (2 forms), FUS-X (2 forms), and NIV (3 forms), could have more than one glucuronic form based on their structures. In our experiment, we possibly observed two glucuronides of DON based on two distinctive product mass spectra, 3 and 15-Gluc-DON. However, the peaks were not fully resolved and MS^2 spectra could be a mixture of the two, Supplementary Figures S14 and S17. According to the literature, the first peak can be assigned as 3-Gluc-DON and the second as 15-Gluc-DON [19]. MS^2 spectrum of the first peak has intense fragment of 441.1 m/z that can happen due to the loss of CH_2O at position 15 when it is not glucuronidated, Supplementary Figure S17. The partial chromatographic separation of 3-Gluc-DON and 15-Gluc-DON shows that the predominant form is 15-Gluc-DON. FUS-X glucuronide showed only one chromatographic peak as shown in Supplementary Figure S18. Previously, FUS-X glucuronides were not reported either in animal nor human samples (Supplementary Figure S19). NIV glucuronides showed two not fully resolved peaks, assuming that there are at least two glucuronic forms present, Supplementary Figure S20. Only one MS^2 spectrum was obtained for the second peak, Supplementary Figure S21b. However, previous studies of nivalenol metabolism in rats exhibited only one 3-glucuronide-NIV and DNIV [46]. De-epoxidation of DON also was observed in both human and animals [17,47,48]. In contrast to rat metabolism studies, NIV incubation with human feces showed no de-epoxydated metabolites [45]. To the best of our knowledge, NIV glucuronides have not been previously observed in human samples, possibly due to the low extent of glucuronidation and/or poor limits of detection for the polar nivalenol and its metabolites using most LC-MS methods. The human exposure studies to DON revealed that the predominant species were 15-Gluc-DON (49%), then free DON (27%), and 3-Gluc-DON (14%) in urine and proposed to use them as biomarkers of DON exposure [17]. Despite trichothecene type B mycotoxins, including DON, being extensively studied, we found new metabolites, showing the importance of these detailed incubation studies and the need to build more systematic libraries of mycotoxin metabolites.

2.2. *Aflatoxins*

AFB1 microsomal biotransformations included the following three types of reactions: oxidative (hydroxylation, epoxidation), reductive (keto-reduction), and hydrolytic (hydrolysis) in phase I, as summarized in Figure 5 and Supplementary Table S9. AFB1 generated various metabolites, including aflatoxin M1 (AFM1, 3%), AFB1 8,9 endo/exo-epoxide (AFBO, <1%), aflatoxin B1 di-hydrodiol (AFB1-diol, <1%), and minor metabolites, AFP1 (<1%), ((H2)+(O))-AFB1 (<1%), and AFL (<1%). The chromatographic separation of AFB1 and its metabolites is shown in Supplementary Figure S22. All metabolites were identified based on their MS1 and comparison of their MS^2 spectra to the literature data as described in the Supplementary Table S9. Two hydroxy-metabolites at m/z 329.0651 (1.8 ppm) showed shift of 16.0 Da versus $[M+H]^+$ ion of AFB1 at 313.0707 (0 ppm), thus confirming the presence of additional oxygen in those compounds, Supplementary Table S9 and Figure S23a,b. The first peak at RT of 5.03 min was identified as AFBO based on the fact of in-source AFB1-diol formation, Supplementary Figures S22 and S23e. AFBO was previously described as a non-stable compound that reacts with water to form AFB1-diol [49,50]. The second peak at RT 6.32 min was identified as hydroxy-metabolite, AFM1. The identification of this hydroxy metabolite was performed by comparing its product ion mass spectra (Supplementary Figure S23b) to the published one [51]. The main distinctive fragment ion of AFM1 is 273.0757 m/z, which can be present only in AFM1 and not in its isomer aflatoxin Q1 (AFQ1) based on the previously published work by Walton et al. [52]. Also, fragment ions of 273.1 Da and 259.0 Da observed in our product ion spectra were chosen as quantifier and qualifier ion for AFM1 in other published papers [51,53,54]. Finally, MS^2 of AFM1 is similar to product mass spectra obtained by Everley et al. [51]. $[M+H]^+$ ion at m/z 347.0760 (0.23 ppm) was 34.0 mass units greater than AFB1, Supplementary Table S9 and Figure S23c,d. This difference indicated the presence of two hydroxyl groups, whereas the presence of two chromatographic peaks indicates the presence of two isomers as shown in Supplementary Figure S22. Their MS^2 spectra exhibited the intense water loss fragment, 329.0650 (3.3 ppm) and 329.0653 (2.1 ppm) for the first and the second peaks, which confirms the presence of hydroxyl groups, Supplementary Figure S23c,d. Additionally, both peaks showed the loss of two water molecules that yielded fragments, 311.0545 (3.5 ppm) and

311.0549 (2.3 ppm), Supplementary Figure S23c,d. The first peak can be identified as AFB1-diol with hydroxyl groups at positions 8 and 9. Its product mass spectra fragments, 283.0597 (38%) and 329.0650 (100%), have similar intensity as shown by Walton et al., namely, 329.1 (100%) and 283.0 (32%) [52]. The identification of O-demethylated products with theoretical mass of $[M+H]^+$ ion at 299.0550 m/z resulted in two chromatographic peaks at 7.05 min (299.0549, 0.33 ppm) and 7.31 min (299.0549, 0.33 ppm), Supplementary Table S9 and Figure S22. Both peaks had similar product ion mass spectra, Supplementary Figure S24. Based on the comparison of fragment ions 271.0602 Da and 299.0554 Da (Supplementary Figure S24) observed in their product ion mass spectra to the literature data, it was possible to determine these peaks as aflatoxin P1 (AFP1) and its isomer [52]. Two metabolites of keto-reduction pathway with measured m/z 337.0682 (0 ppm) of $[M+Na]^+$ ion at 7.66 min and 8.60 min were putatively identified as an isomer of aflatoxicol and aflatoxicol (AFL), respectively, since they were found at trace level, Supplementary Table S9 and Figure S22. The conversion of AFB1 to AFL was previously confirmed using in vitro studies of placental human microsomal proteins [55]. One more type of reduction reaction with the further oxidation resulted in metabolites (+(H2)+(O)-AFB1) with m/z 331.0812 (0 ppm) of $[M+H]^+$ ion at and RT at 5.45 min, Supplementary Table S9 and Figures S22 and S23f. This metabolite definitely has AFB1 origin, since its product mass spectrum has the same fragments, 285.0757 m/z and 313.0705 m/z as AFB1 (Supplementary Figure S23). Also, the loss of H_2O (18.0106, 0 ppm) in product ion mass spectra confirmed the OH group in this molecule. Dohnal et al. reviewed aflatoxin metabolism and concluded that besides the interspecies differences there were also regional, inter-individual differences [56]. The main urinary metabolite of AFB1 was AFM1, which was observed in Brazilian volunteers [57]. Also, AFM1 was found in Italian adult urine samples [58] and Italian children urine and serum samples [24]. However, AFQ1 was found as the most predominant form of aflatoxins in Chinese urinary and fecal samples [59]. Also, previously it was shown that different enzymes are responsible for the conversion of AFB1 to AFQ1 and AFM1 [60,61].

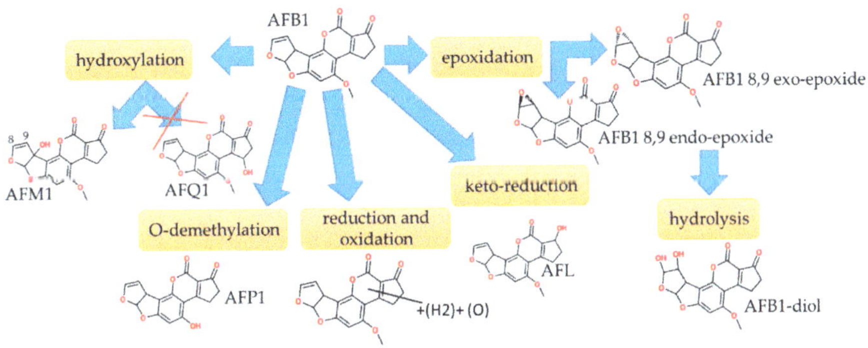

Figure 5. Microsomal biotransformations of aflatoxin B1 in phase I reactions.

The remaining aflatoxins, AFG1, AFB2, AFG2 were unstable during experiment and produced non-enzymatic hydroxyl metabolites. AFG1, AFB2, AFG2 and their metabolites are summarized in Supplementary Tables S10–S12. The stability of aflatoxins in plasma at room temperature was evaluated and it was shown that AFG1 and AFG2 were not stable in plasma for more than 3 h [29]. Another stability study demonstrated the dependence of aflatoxin stability on the temperature and the composition of the solvent [62]. One hydroxy metabolite with enzymatic origin was observed corresponding to hydroxy-metabolite of AFG1, $[M+H]^+$ ion at m/z 345.0604 (0.3 ppm), Supplementary Table S11. MS^2 spectra of AFG1 and its hydroxyl metabolite are shown in Supplementary Figure S25a,b. This AFG1 hydroxy metabolite can be putatively identified as aflatoxin GM1 (AFGM1) metabolite [63]; however, this metabolite, to our knowledge, had not been previously found in human samples. Studies of the prevalence of different aflatoxins in Egyptian infant blood and urine samples performed by Hatem et al. did not confirm its presence [64]. To our knowledge, there were no in vitro metabolism studies performed for AFG1 or AFG2. In our experiment, four non-enzymatic hydroxy metabolites of AFG2 (Supplementary Table S12) were observed with 347.0761 m/z, at least two of them could be aflatoxin GM2 (AFGM2) and aflatoxin G2A (AFG2A) as mentioned in the previous review paper [61]. Product mass spectrum of AFG2 is shown in Supplementary Figure S26. AFB2 was converted non-enzymatically to three hydroxy metabolites with 331.0813 m/z (Supplementary Table S10). Product mass spectra of AFB2 and its hydroxyl metabolites are shown in Supplementary Figure S27a–c. Putatively, they can be identified as previously mentioned aflatoxin M2 (AFM2), aflatoxin Q2 (AFQ2), and aflatoxin B2A (AFB2A) [61,65]. Roebuck et al. performed in vitro metabolism studies of AFB2 which showed the presence of trace levels of AFQ2, aflatoxin P2 (AFP2) and either AFM1 or AFM2 in human samples [66]. In phase II, no glucuronides for any aflatoxins were generated.

2.3. Group of Zearalenone

Microsomal biotransformation of ZEN is summarized in Figure 6 as an example representative for this family. The group of zearalenone, ZEN, α-ZOL, β-ZOL, ZAN, α-ZAL, and β-ZAL, was metabolized most extensively out of all chosen mycotoxin groups, resulting in total of 133 metabolites (Figure 7, Supplementary Tables S13–S18). The most predominant phase I metabolic pathway for this class of mycotoxins is oxidation. There were seven types of oxidation reactions, desaturation with oxidation (-(H4) +(O)), desaturation with oxidation (-(H2) +(O)), oxidation (+(O)), reduction with oxidation (+(H2)+ (O)), oxidation (+(O2)), and desaturation with oxidation (-(H2) +(O2) and (-(H4) +(O2)). Among these oxidation (+(O)) reactions resulted in the formation of the highest number of metabolites for ZEN, 9 metabolites, α-ZAL (4), β-ZAL (8), α-ZOL (8), β-ZOL (7), except ZAN for which the reduction with oxidation (+(H2)+ (O)) resulted in the highest number of metabolites (8), as shown in Table 1. Also, the total pattern number of oxidized metabolites of ZEN and its two metabolites, α-ZOL and β-ZOL differed from ZAN and its two metabolites, α-ZAL and β-ZAL. ZAN metabolized the most extensively and resulted in 22 metabolites, but ZEN had only 12 metabolites. α-ZOL (15) had more oxidized metabolites than β-ZOL (8), but α-ZAL (10) had less than β-ZAL (15). According to the percentage of metabolized parent toxin in phase I reactions, ZEN (27%), ZAN (66%), α-ZAL (29%), β-ZAL (23%), α-ZOL (70%), and β-ZOL (7%), this metabolic pathway is not predominant, except for ZEN and α-ZOL. The metabolism of ZEN has already been investigated by Yang et al. [31], and they reported a variety of ZEN oxidized metabolites. Also, ZEN, α-ZAL and ZAN oxidized metabolites were reported in other studies [67,68], but there were no metabolism studies performed for α-ZOL, β-ZOL and β-ZAL.

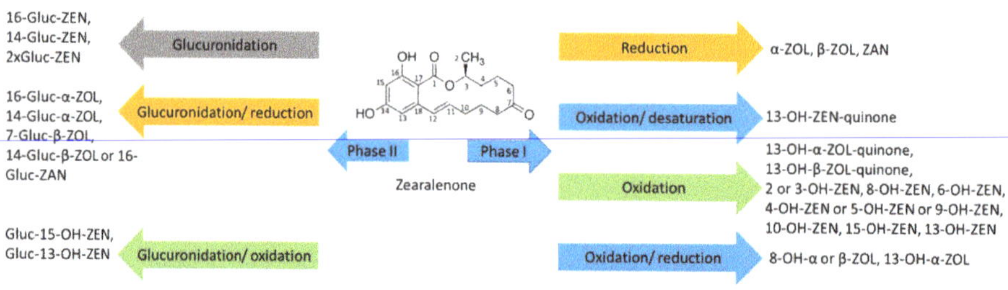

Figure 6. Microsomal biotransformation of ZEN in phase I and II reactions.

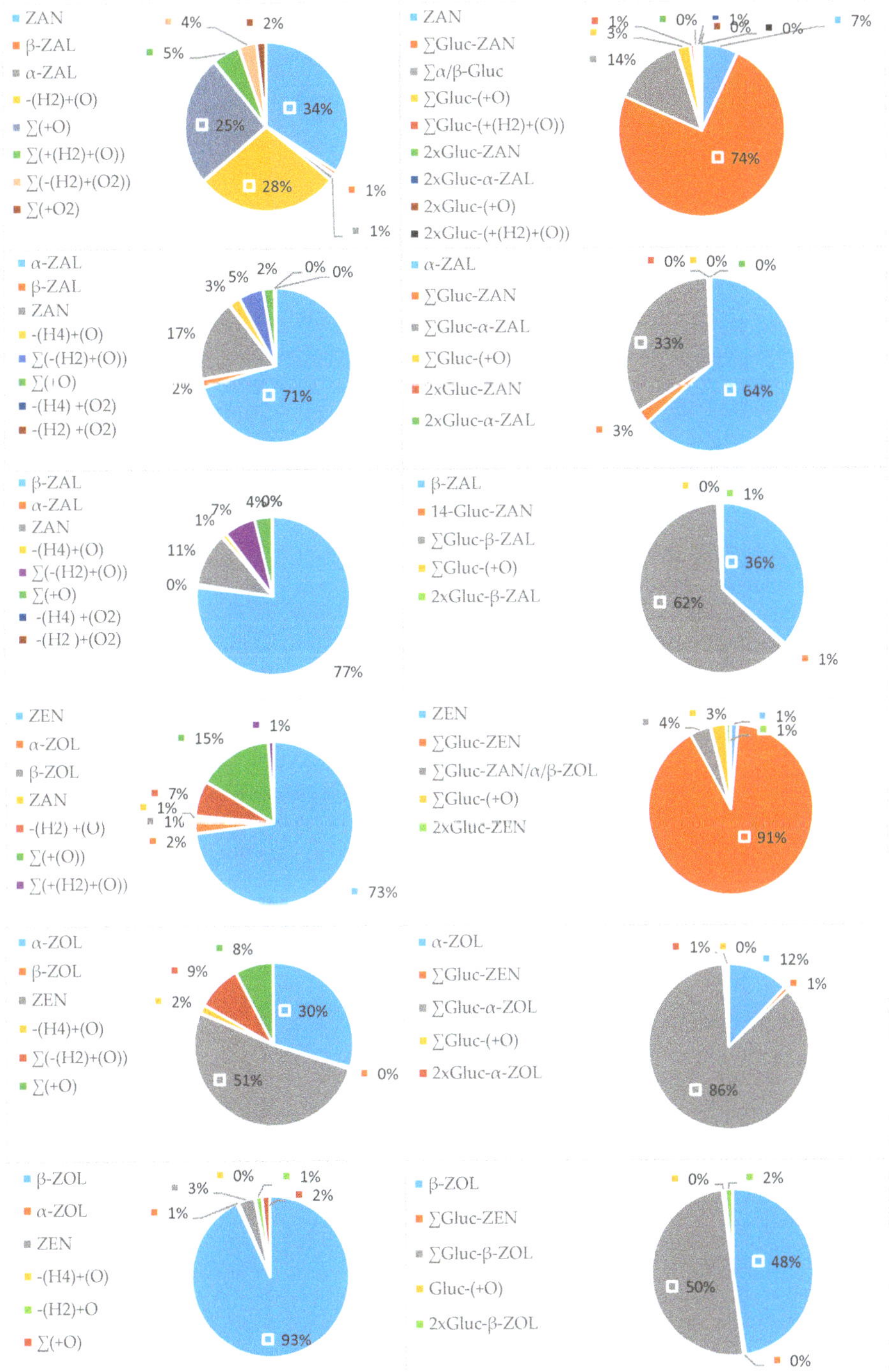

Figure 7. Summary of phase I and phase II metabolism of zearalenone group.

Table 1. Summary of Phase I oxidation metabolites observed for zearalenone group

Mycotoxin	Oxidation Reactions and Number of Metabolites, (*n*)							$\sum$ n
	Desaturation, oxidation, −(H4) +(O)	Desaturation, oxidation, −(H2) +(O)	Oxidation +(O)	Reduction, oxidation, +(H2) +(O)	Oxidation, +(O2)	Desaturation oxidation, −(H2) +(O2)	Oxidation, −(H4) +(O2)	
ZEN	0	1	9(75%)	2	0	0	0	12
α-ZOL	1	6	8(53%)	0	0	0	0	15
β-ZOL	1	1	7(88%)	0	0	0	0	8
ZAN	0	1	5(23%)	8(36%)	6(27%)	2	0	22
α-ZAL	1	3	4(40%)	0	0	1	1	10
β-ZAL	1	4	8(53%)	0	0	1	1	15

Based on the present results, the glucuronidation pathway is the predominant metabolic pathway for almost all the group of zearalenones. 93% of ZAN, 99% of ZEN, 64% of β-ZAL, 51% of β-ZOL, and 88% of α-ZOL were converted into glucoronides, except α-ZAL where conversion was 36% (Figure 7). The identifications of both phase I and II reaction products were based on the comparison MS^2 spectra to literature data and/or analysis of MS^2 spectra. Phase II reactions resulted in various glucuronide forms of parent toxin, its metabolites of phase I reactions, and double glucuronide forms (denoted as 2 × Gluc). The numbers of observed glucuronide forms for each mycotoxin from zearalenone group are summarized in Table 2. Common glucuronide forms of parent toxins were glucuronides at position C-14 or C-16 for ZEN and ZAN and at additional position C-7 for α-ZOL, β-ZOL, α-ZAL and β-ZAL. These glucuronides were previously generated by Stevenson et al. [35], and their studies are in accordance with ours. Among the most predominant glucuronides were glucuronides of parent toxins at position C-14, ZEN (86%), ZAN (73%), α-ZOL (72%), β-ZOL (27%), α-ZAL (23%) and β-ZAL (45%). The sum of parent glucuronides of group of zearalenone composed 91% for ZEN, 74% for ZAN, 86% for α-ZOL, 49% for β-ZOL, 33% for α-ZAL and 62% β-ZAL as shown in Figure 7. The glucuronides of oxidized metabolites and double glucuronides were only minor products, 19% for ZAN, 3% for α-ZAL, 2% for β-ZAL, 8% for ZEN, 2% for α-ZOL, 2% for β-ZOL. Yang et al. have already reported ZEN glucuronides of oxidized metabolites and di-glucuronide forms [69]. Overall, ZAN was the most metabolized toxin in phase II and resulted in 24 glucuronides. β-ZOL and β-ZAL were least metabolized toxins and each generated only 7 glucoronic forms. Comparing phase II reaction samples to heated controls, it was noticed that the glucururonidation process was more efficient in heated samples (45 °C) vs. phase II reaction samples, except ZAN.

Table 2. Number of glucuronides observed in zearalenone group.

Mycotoxin	Total Number of Glucuronides
ZEN	10
α-ZOL	10
β-ZOL	7
ZAN	24
α-ZAL	10
β-ZAL	7

3. Conclusions

In conclusion, the newly generated LC-MS library containing 188 metabolites represents the most comprehensive resource of mycotoxin metabolites that can be analyzed using a single LC-MS method. The in vitro microsomal incubation workflow used in this work was able to successfully generate metabolites from hydrolysis, oxidation, de-epoxidation, epoxidation, demethylation, reduction and glucuronidation pathways as summarized in Table 3. The excellent limits-of-detection and isomer separation capability of our LC-MS method allowed us to characterize for the first time 100 metabolites that had not been previously reported in the literature, to the best of our knowledge. Among the known phase I and phase II metabolites of 17 mycotoxins that were the focus of this study, only four metabolites—aflatoxin Q1 (AFQ1), aflatoxin P2 (AFP2), Gluc-4-HT-2, Gluc-3-4-de-acetyl-neosolaniol—could not be generated using our microsomal incubation workflow. The remaining 88 known metabolites were successfully generated, thus showing the power of our workflow and high-confidence identification capability. Table 4 summarizes the main subclasses of the newly characterized metabolites in this study. To ensure the high confidence of our library identifications we used three key strategies: (i) incubation with one mycotoxin at a time to properly assign the origin of metabolites to a given parent mycotoxin, (ii) extensive controls to eliminate endogenous biomolecules present in microsomes, impurities in standards and metabolites that could be generated non-enzymatically, and (iii) MS/MS comparison to the published literature spectra when available and to the parent compounds since the generated metabolites share many of the same structural features as the parent compounds. In the absence of authentic standards for all of these metabolites, our identifications are putative. In future, this new LC-MS library will be used during biomonitoring studies to characterize which of these metabolites may be observed in various biological samples in vivo and to provide semi-quantitative information on their concentrations using parent calibration curves. This relevant subset of metabolites then can be synthesized for further confirmation of identity and full quantification. Additionally, the clarification of some metabolite structures that remain ambiguous in our library (e.g., exact position of hydroxyl groups in several ZEN metabolites) can be improved in future work by the application of isotopically labeled standards, as was previously demonstrated in the literature [68,70], or through synthesis of authentic standards.

Table 3. Summary of metabolic pathways of 17 mycotoxins.

Mycotoxins	Hydrolysis	Oxidation	De-Epoxidation	Epoxidation	Demethylation	Reduction	Glucuronidation
T-2	✓	✓					✓
HT-2	✓	✓					✓
3-AcDON	✓						✓
15-AcDON	✓						✓
FUS-X	✓						✓
DON			✓				✓
NIV			✓				✓
AFB1	✓	✓		✓	✓	✓	
AFB2		✓					
AFG1		✓					
AFG2		✓					
ZEN		✓					✓
α-ZOL		✓					✓
β-ZOL		✓					✓
ZAN		✓					✓
α-ZAL		✓					✓
β-ZAL		✓					✓

Table 4. Comparison of literature expected metabolites and generated LC-MS library metabolites.

Mycotoxin	Expected Metabolites	Missing Metabolites	LC-MS Library	New Metabolites
T-2	Phase I metabolites: HT-2, 15-deacetyl-T-2 (15-de-Ac-T-2), 3′-OH-T-2, neosolaniol (NEO), T-2 triol, 3′-OH-HT-2, T-2 triol, Glucuronides: Gluc-3-T-2	NO	Phase I metabolites: HT-2, 15-de-Ac-T-2, 3′-OH-T-2 and its two isomers, NEO or T-2 triol, 3′-OH-HT-2, Glucuronides: Gluc-3-T-2	Two isomers of 3′-OH-T-2, 4 isomers of 3′-OH-HT-2

Table 4. *Cont.*

Mycotoxin	Expected Metabolites	Missing Metabolites	LC-MS Library	New Metabolites
HT-2	Phase I metabolites: 4-de-Ac-NEO, 3′-OH-HT-2, 4′-OH-HT-2 7-OH-HT-2 and its isomer, 10-OH-HT-2, Glucuronides: Gluc-3-HT-2, Gluc-4-HT-2, Gluc-3-4-de-Ac-NEO	Gluc-4-HT-2, Gluc-3-4-de-Ac-NEO	Phase I metabolites: 4-de-Ac-NEO and its isomer, 3′-OH-HT-2, 4′-OH-HT-2 and its isomer Three OH-T-2 metabolites at 7 or 10 or 16-OH-HT-2 Two unknown metabolites Glucuronides: Gluc-3-HT-2	4-de-Ac-NEO isomer
3-AcDON	Phase I metabolites: DON Glucuronides: Gluc-3-AcDON	NO	Phase I metabolites: DON Glucuronides: Gluc-3-AcDON	NO
15-AcDON	Phase I metabolites: DON Glucuronides: Gluc-15-AcDON	NO	Phase I metabolites: DON Glucuronides: Gluc-15-AcDON	NO
FUS-X	Phase I metabolites: NIV	NO	Phase I metabolites: NIV Glucuronides: Gluc-FUS-X	Gluc-FUS-X
DON	Phase I metabolites: De-epoxy-DON (DOM-1) Glucuronides: 15-Gluc-DON, 3-Gluc-DON	NO	Phase I metabolites: NIV, DOM-1 and its two isomers, Glucuronides: 15-Gluc-DON, 3-Gluc-DON	NIV, isomers of DOM-1
NIV	Phase I metabolites: De-epoxy-NIV (DENIV), Glucuronides: Gluc-3-NIV	NO	Phase I metabolites: DENIV and its two isomer Glucuronides: Two Gluc-NIVs	Gluc-NIV
AFB1	Phase I metabolites: AFM1, AFQ1, AFBO, AFP1, AFL, AFB1-diol Glucuronides: NO	AFQ1	Phase I metabolites: AFM1, AFBO, AFP1 and its isomer, AFL and its isomer, AFB1-diol and its isomer, ((H2)+(O)-AFB1 Glucuronides: NO	((H2)+(O)-AFB1
AFB2	Phase I metabolites: AFM2, AFQ2, AB2A, AFP2 Glucuronides: NO	AFP2	Phase I metabolites: AFM2, AFQ2 and AFB2A Glucuronides: NO	NO
AFG1	Phase I metabolites: AFGM1 Glucuronides: NO	NO	Phase I metabolites: AFGM1 Glucuronides: NO	NO
AFG2	Phase I metabolites: AFGM2, AFG2A Glucuronides: NO	NO	Phase I metabolites: AFGM2, AFG2A Glucuronides: NO	NO
ZEN	Phase I metabolites: (-(H2) +(O))-ZEN, (+(O))-ZEN, (+(H2)+ (O))-ZEN Glucuronides: Gluc-16-ZEN, Gluc-14-ZEN, Gluc-(+O)-ZEN, 2xGluc-ZEN	NO	Phase I metabolites: (-(H2) +(O))-ZEN, (+(O))-ZEN, (+(H2)+ (O))-ZEN Glucuronides: Gluc-16-ZEN, Gluc-14-ZEN, Gluc-(+O)-ZEN, 2xGluc-ZEN	NO

Table 4. *Cont.*

Mycotoxin	Expected Metabolites	Missing Metabolites	LC-MS Library	New Metabolites
α-ZOL	Phase I metabolites: (-(H4)+(O))- α-ZOL (-(H2)+(O))- α-ZOL (+O)- α-ZOL Glucuronides: Gluc-16- α-ZOL, Gluc-14- α-ZOL, Gluc-7- α-ZOL	NO	Phase I metabolites: (-(H4)+(O))- α-ZOL (-(H2)+(O))- α-ZOL (+O)- α-ZOL Glucuronides: Gluc-16- α-ZOL, Gluc-14- α-ZOL, Gluc-7- α-ZOL, Gluc-(+O)- α-ZOL, 2 × Gluc- α-ZOL	Gluc-(+O)- α-ZOL, (2 × Gluc)- α-ZOL
β-ZOL	Phase I metabolites: NO Glucuronides: Gluc-16- β-ZOL, Gluc-14- β-ZOL, Gluc-7- β-ZOL,	NO	Phase I metabolites: (-(H4)+(O))- β-ZOL (-(H2)+(O))- β-ZOL (+O)- β-ZOL Glucuronides: Gluc-16- β-ZOL, Gluc-14- β-ZOL, Gluc-7- β-ZOL, Gluc-(+O)- β-ZOL, 2 × Gluc- β-ZOL	(-(H4)+(O))- β-ZOL (-(H2)+(O))- β-ZOL (+O)- β-ZOL Gluc-(+O)- β-ZOL, (2 × Gluc)- β-ZOL
ZAN	Phase I metabolites: NO Glucuronides: Gluc-16- ZAN, Gluc-14-ZAN	NO	Phase I metabolites: (-(H2) +(O))-ZAN, (+(O))-ZAN, (+(H2)+ (O))-ZAN, (+(O2))-ZAN, (-(H2) +(O2))-ZAN Glucuronides: Gluc-16-ZAN, Gluc-14-ZAN, Gluc-(+O)- ZAN, 2xGluc- ZAN, Gluc-(+(H2)+(O))-ZAN, 2 × Gluc-(+O)-ZAN, 2 × Gluc-(+(H2)+(O))-ZAN	Gluc-(+O)- ZAN, (2 × Gluc)- ZAN, Gluc-(+(H2)+(O))-ZAN, (2 × Gluc)-(+O)-ZAN, (2 × Gluc)-(+(H2)+(O))-ZAN
α-ZAL	Phase I metabolites: (-(H4)+(O))- α-ZAL (-(H2)+(O))- α-ZAL (+O)- α-ZAL (-(H4) +(O2))- α-ZAL (-(H2)+(O2))- α-ZAL Glucuronides: Gluc-16- α-ZAL, Gluc-14- α-ZAL, Gluc-7- α-ZAL,	NO	Phase I metabolites: (-(H4)+(O))- α-ZAL (-(H2)+(O))- α-ZAL (+O)- α-ZAL (-(H4) +(O2))- α-ZAL (-(H2)+(O2))- α-ZAL Glucuronides: Gluc-16- α-ZAL, Gluc-14- α-ZAL, Gluc-7- α-ZAL, Gluc-(+O)- α-ZAL 2 × Gluc-α-ZAL	Gluc-(+O)- α-ZAL (2 × Gluc)-α-ZAL
β-ZAL	Phase I metabolites: NO Glucuronides: Gluc-16-β-ZAL, Gluc-14-β-ZAL, Gluc-7-β-ZAL	NO	Phase I metabolites: (-(H4)+(O))- β-ZAL (-(H2)+(O))- β-ZAL (+O)- β-ZAL (-(H4) +(O2))- β-ZAL (-(H2)+(O2))- β-ZAL Glucuronides: Gluc-16-β-ZAL, Gluc-14-β-ZAL, Gluc-7-β-ZAL, Gluc-(+O)- β-ZAL, 2xGluc-β-ZAL	(−(H4)+(O))- β-ZAL (−(H2)+(O))- β-ZAL (+O)- β-ZAL (−(H4) +(O2))- β-ZAL (−(H2)+(O2))- β-ZALGluc-(+O)- β-ZAL, (2xGluc)-β-ZAL

4. Materials and Methods

4.1. Chemicals

Water ((H_2O, LC-MS grade), methanol ((MeOH), LC-MS grade), acetonitrile ((MeCN), LC-MS grade), and acetic acid ((AA, LC-MS grade) were purchased from Fisher Scientific (Ottawa, Ontario, Canada). Sodium chloride ((NaCl), meets specifications of American Chemical Society grade (ACS), ≥99.0%), sodium phosphate dibasic ((Na_2HPO_4), ACS, ≥99.0%), potassium phosphate monobasic

(($KH2PO_4$), ACS, ≥99.0%), and magnesium chloride (($MgCl_2$),anhydrous, ≥98%), β-nicotinamide adenine dinucleotide 2′-phosphate reduced tetrasodium salt hydrate ((NADPH), ≥97%), uridine 5′-diphosphoglucuronic acid trisodium salt ((UDPGA), 98–100%), alamethicin from *Trichoderma viride* (≥98%, HPLC grade), and human microsomes from liver (pooled, CMV-negative, 20 mg/mL) were purchased from Sigma-Aldrich Canada (Oakville, Ontario, Canada). Potassium chloride ((KCl), reagent grade, 99.0%) was purchased from BioShop Canada (Burlington, Ontario, Canada).

4.2. Mycotoxin Standards

All mycotoxins were purchased from Sigma-Aldrich Canada, unless otherwise indicated. AFG1, T-2, HT-2, α-ZAL, β-ZAL were purchased from Toronto Research Chemicals Inc. (Toronto, ON, Canada). Zearalenone (ZEN) was purchased from Cayman Chemicals (Ann Arbor, MI, USA). Individual standard stock solutions of all mycotoxins at 1 mg/ml concentration were prepared in methanol and kept at −80 °C.

4.3. Experimental Design and Microsomal Incubations

The purpose of this work was to generate phase I and phase II (glucuronidation) metabolites of 17 mycotoxins using standard in vitro microsomal incubation protocol. Each toxin was incubated individually with microsomes in the presence of NADPH for phase I reactions. For phase II glucuronidation reactions, UDPGA, alamethicin and $MgCl_2$ were also added. In all cases, the following controls were used in order to confirm product formation during enzymatic reaction: (i) microsomal incubation without toxin added, (ii) microsomal incubation without co-factors added (iii), microsomal incubation without NADPH, but containing UDPGA, alamethicin and $MgCl_2$, (iv) incubation with heated microsomes, and (v) standard solution of each toxin dissolved in PBS buffer. This experimental design is summarized in Figure 1.

100 mM PBS buffer (pH 7.4), 20 mM NADPH dissolved in 100 mM phosphate buffer, 100 mM UDPGA in water, 5 mg/mL alamethicin in methanol, 100 mM $MgCl_2$ in water and 200 μg/mL standard solution of each mycotoxin in acetonitrile were prepared before the start of microsomal incubations. Microsomes were thawed on ice. In an Eppendorf tube, 182 μL of PBS buffer, 2 μL of NADPH and 5 μL of microsomes were transferred for phase I reactions. For phase II reactions, all of the reagents for phase I reactions plus alamethicin, 10 μl of UDPGA and $MgCl_2$ were transferred. Microsomes were then pre-incubated for 5 min, followed by the addition of mycotoxin (final concentration of 1 μg/mL) and then the remaining amount of NADPH (10 μL). All samples were incubated for 1 h at 37 °C, reactions were stopped by adding 200 μL of acetonitrile. Detailed description of test samples and controls is shown in Supplementary Table S1.

4.4. LC-HRMS Analysis

All LC-MS measurements were performed according to the validated multi-mycotoxin method for 17 parent mycotoxins [29]. Briefly, the method combined HPLC 1100 (Agilent Technologies, Santa Clara, CA, USA) and reversed-phase chromatographic separation on pentafluorophenyl stationary phase and gradient elution using water and methanol containing 0.1% AA (v/v) for ESI(+), and 0.02% for ESI(-) [29]. The flow rate of 0.3 mL/min, the column temperature of 30°C, and 10 μL injection volume were used for all analyses. MS analysis was performed on LTQ Orbitrap Velos at 60,000 resolving power using the mass range of 200–700 m/z. In addition, MS^n analysis with data-dependent acquisition (DDA) mode was used for the identification and elucidation of metabolite structures. In DDA mode, the three most intense ions from the full MS scan were selected for MS^2 fragmentation. MS^2 analysis used collision-induced dissociation (CID) and signal threshold: 5,000; normalized collision energy: 35; isolation width: 2 Da; activation time: 30 ms. MS^3 used targeted parent and product mass lists to trigger MS^3 for the selected ions of interest. MS^3 was performed with CID as activation type; minimal signal threshold: 5000; isolation width: 2 Da; activation time: 30 ms; normalized collision energy: 45. For AFB1 and its metabolites, MS^2 analysis used higher energy collisional dissociation (HCD) with signal threshold: 5000; normalized collision energy: 35; isolation width: 2 Da; activation time: 0.1 ms, lock mass was used for ESI(−) and ESI(+).

Data was processed using Compound Discoverer 2.1 (ThermoFisher Scientific). Raw data files were uploaded to Compound Discoverer and analyzed using generic metabolism workflow. General settings in the workflow were mass tolerance, 5 ppm; signal threshold, 3; minimum peak intensity 10000. Parameters used to generate expected compounds were parent toxin structure, metabolic transformations for phase I and II reactions, and preferred ions.

Supplementary Materials: The following are available online at http://www.mdpi.com/2072-6651/11/8/433/s1, Table S1: Microsomal incubation protocol for phase I and II reactions, Table S2: T-2 and its metabolites, detected in ESI(+), as CID product ion spectra of $[M+Na]^+$, Table S3: HT-2 and its metabolites, detected in ESI(+), as $[M+Na]^+$ ions, Table S4: Metabolites of 3-AcDON generated in phase I and phase II, detected in ESI(−), as $[M+CH_3COO-H]^-$ ions, except Gluc-3AcDON which was detected as $[M-H]^-$ ion, Table S5: Metabolites of 15-AcDON generated in phase I and phase II, detected in ESI(+), as $[M+Na]^+$ ions of 15-AcDON and Gluc-15-AcDON, except DON which was detected as $[M+H]^+$ ion, Table S6: Metabolites of DON generated in phase I and phase II, detected in ESI(−), as $[M+CH_3COO-H]^-$ ions, except Gluc-DON which was detected as $[M-H]^-$ ion, Table S7: Metabolites of FUS-X generated in phase I and phase II, detected in ESI(−), as $[M+CH_3COO-H]^-$ ions, except Gluc-FUS-X which was detected as $[M-H]^-$ ion, Table S8: Metabolites of NIV generated in phase I and phase II, detected in ESI(−), as $[M+CH_3COO-H]^-$ ions, except Gluc-NIV which was detected as $[M-H]^-$ ion, Table S9: AFB1 and its metabolites of phase I reactions, detected in ESI(+), as $[M+H]^+$ ions and * $[M+Na]^+$ ions for AFL, Table S10: AFB2 and its metabolites of phase I reactions, detected in ESI(+), as $[M+H]^+$ ions, Table S11: AFG1 and its metabolite, OH-AFG1, detected in ESI(+), as $[M+H]^+$ ions, Table S12: AFG2 and its metabolites of phase I reactions, detected in ESI(+), as $[M+H]^+$ ions, Table S13: Metabolites of ZEN generated in phase I and phase II, detected in ESI(−), as $[M-H]^-$ ions, Table S14: Metabolites of α-ZOL generated in phase I and phase II, detected in ESI(−), as $[M-H]^-$ ions, Table S15: Metabolites of β-ZOL generated in phase I and phase II, detected in ESI(−), as $[M-H]^-$ ions, Table S16: Metabolites of ZAN generated in phase I and phase II, detected in ESI(−), as $[M-H]^-$ ions, Table S17: Metabolites of α-ZAL generated in phase I and phase II, detected in ESI(−), as $[M-H]^-$ ions, Table S18: Metabolites of β-ZAL generated in phase I and phase II, detected in ESI(−), as $[M-H]^-$ ions, Figure S1: Extracted ion chromatograms of T-2 and its metabolites (505.2044 m/z, peak 1-505 to peak 3-505, 463.1939 m/z, peak 1-463 to peak 5-463, 405.1884 m/z, peak 1-405, 447.1989 m/z, peak 1-447 and peak 2-448) in ESI(+), detected as $[M+Na]^+$ ions, Figure S2: Zooming into extracted ion chromatograms of T-2 (505.2044 m/z) and HT-2 (463.1938 m/z) hydroxy metabolites in ESI(+), detected $[M+Na]^+$ ions. Panel (a) shows T-2 hydroxyl metabolites, panel (b) shows HT-2 hydroxyl metabolites, Figure S3: Product ion spectra of T-2 hydroxy metabolites at 505.2044 m/z, peak 1-505 (a), peak 2-505 (b), peak 3-505 (c) and at 489.2095, T-2, detected in ESI(+), as $[M+Na]^+$ ions, Figure S4: Product ion mass spectra of the peak 1-447 (a) which was tentatively identified as 15-deacetyl-T-2 and the peak 2-447 (b) which was identified as HT-2, detected in ESI(+), as $[M+Na]^+$ ions, Figure S5: Extracted ion chromatograms of T-2 and HT-2 glucuronides (665.2416 m/z, Gluc-T-2, and 623.2310 m/z, Gluc-HT-2) in ESI(+), detected as $[M+Na]^+$ ions, Figure S6: Product ion spectra of HT-2 glucuronide (a) at 623.2310 m/z; T-2 glucuronide (b) at 665.2416 m/z and HT-2 glucuronide (c) at 618.2756 m/z, detected in ESI(+), as [M+Na]+ ions for (a) and (b) and [M+NH4]+ions for (c), Figure S7: Extracted ion chromatograms of HT-2 and its metabolites (463.1939 m/z, peak 1-463 to peak 5-463, 405.1884 m/z, peak 1-405 and peak 2-405, 363.1414 m/z, peak 1-363 and peak 2-363) in ESI(+), detected as $[M+Na]^+$ ions, Figure S8: Zooming into extracted ion chromatogram of HT-2 hydroxyl-metabolites (463.1939 m/z, peak 1-463 to 5-463), Figure S9: Product ion mass spectra of HT-2 hydroxy metabolites at 463. 1939 m/z: peak 1-463(a), peak 2-463 (b), peak 3-463 (c), peak 4-463 (d), peak 5-463 (e), peak 6-463 (f), detected in ESI(+), as $[M+Na]^+$ ions, Figure S10: Extracted ion chromatograms of 3-AcDON (397.1505 m/z) and its metabolites (355.1399 m/z, DON, 339.1448 m/z, DOM-1, and 513.1613 m/z, Gluc-3-AcDON) in ESI(-), detected as $[M+CH_3COO-H]^-$ ions for all except Gluc-3AcDON ($[M-H]^-$), Figure S11: Product ion mass spectra of 3-AcDON at 397.1505 m/z and its glucuronide at 513.1613 m/z, detected in ESI(-), as $[M+CH_3COO-H]^-$ and Gluc-3AcDON ($[M-H]^-$ ions, respectively, Figure S12: Extracted ion chromatogram of 15-AcDON (361.1258 m/z) and its metabolites (297.1333 m/z, DON, and 537.1579 m/z, Gluc-15-AcDON) in ESI(+), detected as $[M+Na]^+$ ions, Figure S13: Product mass spectra of 15-AcDON (361.1258 m/z) and its glucuronide (537.1579 m/z), detected in ESI(+) as $[M+Na]^+$ ions, Figure S14: Extracted ion chromatograms of DON and its metabolites (371.1348 m/z, NIV, 339.1449 m/z, peak 1-339 to peak 4-339, and 471.1508 m/z, Gluc-DON) in ESI(-), detected as $[M+CH3COO-H]^-$ ions for all except Gluc-DON ($[M-H]^-$), Figure S15: Product mass spectra of DON (355.1399 m/z) and NIV (371.1348 m/z) detected in ESI(-), as $[M+CH_3COO-H]^-$ ions, Figure S16: Product ion mass spectra of de-epoxy-deoxynivalenol at 339.1348 m/z, detected in ESI(-), as $[M+CH_3COO-H]^-$ ions. Peak 1-339 (a) was observed as phase I metabolite of DON and 3-AcDON, peak 2-339 (b) and peak 3-339 (c) were observed as phase I metabolite of DON only, Figure S17: Product mass spectra of DON glucuronides, 471.1508 m/z, detected in ESI(-), as $[M-H]^-$ ions, Figure S18: Extracted ion chromatogram of FUS-X (413.1454 m/z) and its metabolites (371.1348 m/z, NIV, and 529.1563 m/z, Gluc-FUS-X) in ESI(-), detected as $[M+CH3COO-H]^-$ ions for all except Gluc-FUS-X ($[M-H]^-$), Figure S19: Product ion mass spectra of FUS-X at 413.1454 m/z and its glucuronide at 529.1563 m/z, detected in ESI(-), as $[M+CH_3COO-H]^-$ and Gluc-3AcDON ($[M-H]^-$ ions, respectively. Figure S20: Extracted ion chromatograms of NIV, de-epoxy-metabolite and its isomers (355.1398 m/z, peak 1-355, peak 2-355, peak 3-355) and its glucuronides (487.1457 m/z) in ESI(-), NIV and de-epoxy-metabolite were detected as $[M+CH3COO-H]^-$ ions and glucuronides as $[M-H]^-$ ion, Figure S21: Product ion mass spectra of NIV metabolites: de-epoxy-nivalenol, peak 1-355 (a) at 355.1398 m/z and NIV glucuronide (b) at 487.1458 m/z, detected in ESI(-), as $[M-H]^-$ and $[M+CH_3COO-H]^-$ ions, respectively., Figure S22: Chromatographic separation of AFB1 metabolites generated in phase I reactions. Extracted ion chromatograms of AFB1 (313.0707 m/z), 337.0682 m/z (peak 1-337 and peak 2-337), 299.0550 m/z (peak 1-299 and peak 2-299), 331.0812 m/z (peak 1-331), 329.0661 m/z (peak 1-329, AFBO, and peak 2-329, AFM1), 347.0761 m/z (peak 1-347 and peak 2-347) detected in ESI(+), Figure S23: Product mass spectra of AFB1 (a); AFM1, peak 2-329 (b); AFB-diol, peak 1 (c); AFB-diol, peak 2 (d); AFB-8,9-endo/exo-epoxide (AFBO), peak 1-329 (e); peak 1-331 (f), detected in ESI(+), as $[M+H]^+$ ions, Figure S24: Product ion mass spectra of the peak 1-447 (a) which was tentatively identified as 15-deacetyl-T-2 and the peak 2-447 which was identified as HT-2, detected in ESI(+), as $[M+Na]^+$ ions, Figure S25: Product mass spectra of AFG1 (a) at 329.0656 m/z and its hydroxyl metabolite peak 1-345 (b), detected in ESI(+), as $[M+H]^+$ ions, Figure S26: Product mass spectra of AFG2 at 331.0813 m/z, detected in ESI(+), as $[M+H]^+$ ion:. Figure S27 Product mass spectra of AFB2 (a) at 315.0863 m/z and its hydroxyl metabolites at 331.0813, peak 2-331 (b) and peak 1-331 (c), detected in ESI(+), as $[M+H]^+$ ions,

Author Contributions: Conceptualization, I.S. and D.V.; Data curation, I.S.; Formal analysis, I.S. and D.V.; Funding acquisition, D.V.; Investigation, I.S., R.S. and M.S.R.; Methodology, I.S. and D.V.; Project administration, D.V.; Resources, D.V.; Supervision, D.V.; Visualization, I.S. and D.V.; Writing—original draft, I.S.; Writing—review & editing, I.S., R.S., M.S.R. and D.V.

Funding: This research was funded by Fonds du Recherche du Quebec [FRQNT grant number 2014-NC-172924 and FRQS grant number 33273]. IS was funded by scholarship from the Centre of Biological Applications of Mass Spectrometry at Concordia University.

Acknowledgments: We gratefully acknowledge technical support and the use of LC-HRMS instrumentation from the Centre of Biological Applications of Mass Spectrometry at Concordia University.

Conflicts of Interest: The authors declare no conflict of interest.

References

1. Alshannaq, A.; Yu, J. Occurrence, Toxicity, and Analysis of Major Mycotoxins in Food. *Int. J. Environ. Res. Public Health* **2017**, *14*, 632. [CrossRef] [PubMed]
2. Roscoe, V.; Lombaert, G.; Huzel, V.; Neumann, G.; Melietio, J.; Kitchen, D.; Kotello, S.; Krakalovich, T.; Trelka, R.; Scott, P. Mycotoxins in breakfast cereals from the Canadian retail market: A 3-year survey. *Food Addit. Contam. Part A Chem. Anal. Control. Expo. Risk Assess.* **2008**, *25*, 347–355. [CrossRef] [PubMed]
3. Turcotte, A.-M.; Scott, P. Ochratoxin A in cocoa and chocolate sampled in Canada. *Food Addit. Contam. Part A* **2011**, *28*, 762–766. [CrossRef] [PubMed]
4. Kolakowski, B.; O'Rourke, S.M.; Bietlot, H.P.; Kurz, K.; Aweryn, B. Ochratoxin A Concentrations in a Variety of Grain-Based and Non – Grain-Based Foods on the Canadian Retail Market from 2009 to 2014. *J. Food Prot.* **2016**, *79*, 2143–2159. [CrossRef] [PubMed]
5. Ng, L.; Mankotia, M.; Pantazopoulos, P.; Neil, R.J.; Scott, P.M. Ochratoxin A in wine and grape juice sold in Canada. *Food Addit. Contam.* **2004**, *21*, 971–981. [CrossRef] [PubMed]
6. Lombaert, G.A.; Pellaers, P.; Roscoe, V.; Mankotia, M.; Neil, R.; Scott, P.M. Mycotoxins in infant cereal foods from the Canadian retail market. *Food Addit.Contam* **2003**, *20*, 494–504. [CrossRef] [PubMed]
7. Hooker, D.C.; Schaafsma, A.W. Agronomic and environmental impacts on concentrations of deoxynivalenol and fumonisin B1 in corn across Ontario. *Can. J. Plant Pathol.* **2005**, *27*, 347–356. [CrossRef]
8. Schatzmayr, G.; Streit, E. Global occurrence of mycotoxins in the food and feed chain: Facts and figures. *World Mycotoxin J.* **2013**, *6*, 213–222. [CrossRef]
9. Canadian Food Inspection Agency. *2013–2015 Multi-Mycotoxin Analysis in Selected Foods;* Canadian Food Inspection Agency: Ottawa, ON, Canada, 2016.
10. Smith, M.; Madec, S.; Coton, E.; Hymery, N. Natural Co-Occurrence of Mycotoxins in Foods and Feeds and Their in vitro Combined Toxicological Effects. *Toxins* **2016**, *8*, 94. [CrossRef]
11. Serrano, A.B.; Font, G.; Ruiz, M.J.; Ferrer, E. Co-occurrence and risk assessment of mycotoxins in food and diet from Mediterranean area. *Food Chem.* **2012**, *135*, 423–429. [CrossRef]
12. Wells, L.; Hardie, L.; Williams, C.; White, K.; Liu, Y.; De Santis, B.; Debegnach, F.; Moretti, G.; Greetham, S.; Brera, C.; et al. Deoxynivalenol Biomarkers in the Urine of UK Vegetarians. *Toxins* **2017**, *9*, 196. [CrossRef] [PubMed]
13. Rodríguez-carrasco, Y.; Moltó, J.C.; Mañes, J.; Berrada, H. Exposure assessment approach through mycotoxin/creatinine ratio evaluation in urine by GC—MS/MS. *Food Chem. Toxicol.* **2014**, *72*, 69–75.

14. Solfrizzo, M.; Gambacorta, L.; Lattanzio, V.M.T.; Powers, S.; Visconti, A. Simultaneous LC-MS/MS determination of aflatoxin M 1, ochratoxin A, deoxynivalenol, de-epoxydeoxynivalenol, α and β-zearalenols and fumonisin B 1 in urine as a multi-biomarker method to assess exposure to mycotoxins. *Anal. Bioanal. Chem.* **2011**, *401*, 2831–2841. [CrossRef] [PubMed]
15. Heyndrickx, E.; Sioen, I.; Huybrechts, B.; Callebaut, A.; De Henauw, S.; De Saeger, S. Human biomonitoring of multiple mycotoxins in the Belgian population: Results of the BIOMYCO study. *Environ. Int.* **2015**, *84*, 82–89. [CrossRef] [PubMed]
16. Turner, P.C.; Burley, V.J.; Rothwell, J.A.; White, K.L.M.; Cade, J.E.; Wild, C.P. Deoxynivalenol: Rationale for development and application of a urinary biomarker. *Food Addit. Contam. Part A Chem. Anal. Control. Expo. Risk Assess.* **2008**, *25*, 864–871. [CrossRef] [PubMed]
17. Vidal, A.; Claeys, L.; Mengelers, M.; Vanhoorne, V.; Vervaet, C.; Huybrechts, B.; De Saeger, S.; De Boevre, M. Humans significantly metabolize and excrete the mycotoxin deoxynivalenol and its modified form deoxynivalenol-3-glucoside within 24 hours. *Sci. Rep.* **2018**, *8*, 1–11. [CrossRef] [PubMed]
18. Maul, R.; Warth, B.; Kant, J.S.; Schebb, N.H.; Krska, R.; Koch, M.; Sulyok, M. Investigation of the hepatic glucuronidation pattern of the Fusarium mycotoxin deoxynivalenol in various species. *Chem. Res. Toxicol.* **2012**, *25*, 2715–2717. [CrossRef] [PubMed]
19. Warth, B.; Sulyok, M.; Fruhmann, P.; Berthiller, F.; Schuhmacher, R.; Hametner, C.; Adam, G.; Fröhlich, J.; Krska, R. Assessment of human deoxynivalenol exposure using an LC-MS/MS based biomarker method. *Toxicol. Lett.* **2012**, *211*, 85–90. [CrossRef] [PubMed]
20. Vidal, A.; Mengelers, M.; Yang, S.; De Saeger, S.; De Boevre, M. Mycotoxin Biomarkers of Exposure: A Comprehensive Review. *Compr. Rev. Food Sci. Food Saf.* **2018**, *17*, 1127–1155. [CrossRef]
21. Rodríguez-Carrasco, Y.; Moltó, J.C.; Mañes, J.; Berrada, H. Development of microextraction techniques in combination with GC–MS/MS for the determination of mycotoxins and metabolites in human urine. *J. Sep. Sci.* **2017**, *40*, 1572–1582. [CrossRef]
22. Föllmann, W.; Ali, N.; Blaszkewicz, M.; Degen, G.H. Biomonitoring of Mycotoxins in Urine: Pilot Study in Mill Workers. *J. Toxicol. Environ. Health. A* **2016**, *79*, 1015–1025. [CrossRef] [PubMed]
23. Cao, X.; Li, X.; Li, J.; Niu, Y.; Shi, L.; Fang, Z.; Zhang, T.; Ding, H. Quantitative determination of carcinogenic mycotoxins in human and animal biological matrices and animal-derived foods using multi-mycotoxin and analyte-specific high performance liquid chromatography-tandem mass spectrometric methods. *J. Chromatogr. B Anal. Technol. Biomed. Life Sci.* **2018**, *1073*, 191–200. [CrossRef] [PubMed]
24. De Santis, B.; Raggi, M.E.; Moretti, G.; Facchiano, F.; Mezzelani, A.; Villa, L.; Bonfanti, A.; Campioni, A.; Rossi, S.; Camposeo, S.; et al. Study on the association among mycotoxins and other variables in children with autism. *Toxins* **2017**, *9*, 203. [CrossRef] [PubMed]
25. Escrivá, L.; Manyes, L.; Font, G.; Berrada, H. Mycotoxin analysis of human urine by LC-MS/MS: A comparative extraction study. *Toxins* **2017**, *9*, 330. [CrossRef] [PubMed]
26. Warth, B.; Sulyok, M.; Fruhmann, P.; Mikula, H.; Berthiller, F.; Schuhmacher, R.; Hametner, C.; Abia, W.A.; Adam, G.; Fröhlich, J.; et al. Development and validation of a rapid multi-biomarker liquid chromatography/tandem mass spectrometry method to assess human exposure to mycotoxins. *Rapid Commun. Mass Spectrom.* **2012**, *26*, 1533–1540. [CrossRef] [PubMed]
27. Ediage, E.N.; Di Mavungu, J.D.; Song, S.; Wu, A.; Van Peteghem, C.; De Saeger, S. A direct assessment of mycotoxin biomarkers in human urine samples by liquid chromatography tandem mass spectrometry. *Anal. Chim. Acta* **2012**, *741*, 58–69. [CrossRef] [PubMed]
28. Huybrechts, B.; Martins, J.C.; Debongnie, P.; Uhlig, S.; Callebaut, A. Fast and sensitive LC–MS/MS method measuring human mycotoxin exposure using biomarkers in urine. *Arch. Toxicol.* **2015**, *89*, 1993–2005. [CrossRef] [PubMed]
29. Slobodchikova, I.; Vuckovic, D. Liquid chromatography – high resolution mass spectrometry method for monitoring of 17 mycotoxins in human plasma for exposure studies. *J. Chromatogr. A* **2018**, *1548*, 51–63. [CrossRef]
30. Osteresch, B.; Viegas, S.; Cramer, B.; Humpf, H.U. Multi-mycotoxin analysis using dried blood spots and dried serum spots. *Anal. Bioanal. Chem.* **2017**, *409*, 3369–3382. [CrossRef] [PubMed]

31. Yang, S.; Zhang, H.; Sun, F.; De Ruyck, K.; Zhang, J.; Jin, Y.; Li, Y.; Wang, Z.; Zhang, S.; De Saeger, S.; et al. Metabolic Profile of Zearalenone in Liver Microsomes from Different Species and Its in Vivo Metabolism in Rats and Chickens Using Ultra High-Pressure Liquid Chromatography-Quadrupole/Time-of-Flight Mass Spectrometry. *J. Agric. Food Chem.* **2017**, *65*, 11292–11303. [CrossRef]
32. Yang, S.; DE Boevre, M.; Li, Y.; De Saeger, S. The Toxicokinetics of HT-2 Toxin in Rats and Its Metabolic Profile in Livestock and Human Liver Microsomes. *J. Agric. Food Chem.* **2018**, *66*, 8160–8168. [CrossRef] [PubMed]
33. Yang, S.; De Boevre, M.; Zhang, H.; De Ruyck, K.; Sun, F.; Zhang, J.; Jin, Y.; Li, Y.; Wang, Z.; Zhang, S.; et al. Metabolism of T-2 Toxin in Farm Animals and Human In Vitro and in Chickens In Vivo Using Ultra High-Performance Liquid Chromatography- Quadrupole/Time-of-Flight Hybrid Mass Spectrometry Along with Online Hydrogen/Deuterium Exchange Technique. *J. Agric. Food Chem.* **2017**, *65*, 7217–7227. [CrossRef] [PubMed]
34. Pfeiffer, E.; Hildebrand, A.; Mikula, H.; Metzler, M. Glucuronidation of zearalenone, zeranol and four metabolites in vitro: Formation of glucuronides by various microsomes and human UDP-glucuronosyltransferase isoforms. *Mol. Nutr. Food Res.* **2010**, *54*, 1468–1476. [CrossRef] [PubMed]
35. Stevenson, D.E.; Hansen, R.P.; Loader, J.I.; Jensen, D.J.; Cooney, J.M.; Wilkins, A.L.; Miles, C.O. Preparative Enzymatic Synthesis of Glucuronides of Zearalenone and Five of Its Metabolites. *J. Agric. Food Chem.* **2008**, *56*, 4032–4038. [CrossRef] [PubMed]
36. Frizzell, C.; Uhlig, S.; Miles, C.O.; Verhaegen, S.; Elliott, C.T.; Eriksen, G.S.; Sørlie, M.; Ropstad, E.; Connolly, L. Biotransformation of zearalenone and zearalenols to their major glucuronide metabolites reduces estrogenic activity. *Toxicol. Vitr.* **2015**, *29*, 575–581. [CrossRef] [PubMed]
37. Wen, J.; Mu, P.; Deng, Y. Mycotoxins: Cytotoxicity and biotransformation in animal cells. *Toxicol. Res.* **2016**, *5*, 377–387. [CrossRef] [PubMed]
38. Yanni, S.; Annaert, P.P.; Augustijns, P.; Bridges, A.; Gao, Y.; Daniel, K.; Thakker, D.R. Role of Flavin-Containing Monooxygenase in Oxidative Metabolism of Voriconazole by Human Liver Microsomes. *Drug Metab. Dispos.* **2009**, *36*, 1119–1125. [CrossRef] [PubMed]
39. Grothusen, A.; Hardt, J.; Bräutigam, L.; Lang, D.; Böcker, R. A convenient method to discriminate between cytochrome P450 enzymes and flavin containing monooxygenases in human liver microsomes. *Arch. Toxicol.* **1996**, *71*, 64–71. [CrossRef] [PubMed]
40. Fujiwara, R.; Nakajima, M.; Yamamoto, T.; Nagao, H.; Yokoi, T. In silico and in vitro Approaches to Elucidate the Thermal Stability of Human UDP-glucuronosyltransferase UGT 1A9. *Drug Metab. Pharmacokinet.* **2009**, *24*, 235–244. [CrossRef] [PubMed]
41. Yang, S.; Li, Y.; Cao, X.; Hu, D.; Wang, Z.; Wang, Y.; Shen, J.; Zhang, S. Metabolic pathways of T-2 toxin in in vivo and in vitro systems of Wistar rats. *J. Agric. Food Chem.* **2013**, *61*, 9734–9743. [CrossRef]
42. Sun, Y.; Zhang, G.; Zhao, H.; Zheng, J.; Hu, F.; Fang, B. Liquid chromatography-tandem mass spectrometry method for toxicokinetics, tissue distribution, and excretion studies of T-2 toxin and its major metabolites in pigs. *J. Chromatogr. B Anal. Technol. Biomed. Life Sci.* **2014**, *958*, 75–82. [CrossRef] [PubMed]
43. Welsch, T.; Humpf, H.U. HT-2 toxin 4-glucuronide as new T-2 toxin metabolite: Enzymatic synthesis, analysis, and species specific formation of T-2 and HT-2 toxin glucuronides by rat, mouse, pig, and human liver microsomes. *J. Agric. Food Chem.* **2012**, *60*, 10170–10178. [CrossRef] [PubMed]
44. Ajandouz, E.H.; Berdah, S.; Moutardier, V.; Bege, T.; Birnbaum, D.J.; Perrier, J.; Di Pasquale, E.; Maresca, M. Hydrolytic Fate of 3/15-Acetyldeoxynivalenol in Humans: Specific Deacetylation by the Small Intestine and Liver Revealed Using in Vitro and ex Vivo Approaches. *Toxins* **2016**, *8*, 232. [CrossRef] [PubMed]
45. Eriksen, G.S.; Pettersson, H. Lack of de-epoxidation of type B trichothecenes in incubates with human faeces. *Food Addit. Contam.* **2003**, *20*, 579–582. [CrossRef] [PubMed]
46. Schwartz-Zimmermann, H.E.; Binder, S.B.; Hametner, C.; Miró-Abella, E.; Schwarz, C.; Michlmayr, H.; Reiterer, N.; Labudova, S.; Adam, G.; Berthiller, F. Metabolism of nivalenol and nivalenol-3-glucoside in rats. *Toxicol. Lett.* **2019**, *306*, 43–52. [CrossRef] [PubMed]
47. Gratz, S.W.; Duncan, G.; Richardson, A.J. The Human Fecal Microbiota Metabolizes Deoxynivalenol and Deoxynivalenol-3-Glucoside and May Be Responsible for Urinary. *Appl. Environ. Microbiol.* **2013**, *79*, 1821–1825. [CrossRef]
48. Brezina, U.; Rempe, I.; Kersten, S.; Valenta, H.; Humpf, H.U.; Dänicke, S. Diagnosis of intoxications of piglets fed with Fusarium toxin-contaminated maize by the analysis of mycotoxin residues in serum, liquor and urine with LC-MS/MS. *Arch. Anim. Nutr.* **2014**, *68*, 425–447. [CrossRef] [PubMed]

49. Guengerich, F.P.; Arneson, K.O.; Williams, K.M.; Deng, Z.; Harris, T.M. Reaction of aflatoxin B1 oxidation products with lysine. *Chem. Res. Toxicol.* **2002**, *15*, 780–792. [CrossRef]
50. Johnson, W.W.; Guengerich, F.P. Reaction of aflatoxin B1 exo-8,9-epoxide with DNA: Kinetic analysis of covalent binding and DNA-induced hydrolysis. *Proc. Natl. Acad. Sci. USA* **1997**, *94*, 6121–6125. [CrossRef]
51. Everley, R.A.; Ciner, F.L.; Zhang, D.; Scholl, P.F.; Groopman, J.D.; Croley, T.R. Measurement of aflatoxin and aflatoxin metabolites in urine by liquid chromatography-tandem mass spectrometry. *J. Anal. Toxicol.* **2007**, *31*, 150–156. [CrossRef]
52. Walton, M.; Egner, P.; Scholl, P.F.; Walker, J.; Kensler, T.W.; Groopman, J.D. Liquid chromatography electrospray-mass spectrometry of urinary aflatoxin biomarkers: Characterization and application to dosimetry and chemoprevention in rats. *Chem. Res. Toxicol.* **2001**, *14*, 919–926. [CrossRef] [PubMed]
53. Andrade, P.D.; Gomas da Silva, J.L.; Caldas, E.D. Simultaneous analysis of aflatoxins B1, B2, G1, G2, M1 and ochratoxin A in breast milk by high-performance liquid chromatography/fluorescence after liquid-liquid extraction with low temperature purification (LLE-LTP). *J. Chromatogr. A* **2013**, *1304*, 61–68. [CrossRef] [PubMed]
54. Fan, S.; Li, Q.; Sun, L.; Du, Y.; Xia, J.; Zhang, Y. Simultaneous determination of aflatoxin B 1 and M 1 in milk, fresh milk and milk powder by LC-MS/MS utilising online turbulent flow chromatography. *Food Addit. Contam. Part A* **2015**, *32*, 1175–1184. [CrossRef] [PubMed]
55. Leppänen, J.M.; Partanen, H.A.; Vähäkangas, K.H.; Woodhouse, H.J.; Myllynen, P.K.; El-Nezami, H.S. Aflatoxin B1 Transfer and Metabolism in Human Placenta. *Toxicol. Sci.* **2009**, *113*, 216–225.
56. Dohnal, V.; Wu, Q.; Kuc, K. Metabolism of aflatoxins: Key enzymes and interindividual as well as interspecies differences. *Arch. Toxicol.* **2014**, *88*, 1635–1644. [CrossRef] [PubMed]
57. Jager, A.V.; Tonin, F.G.; Souto, P.C.M.C.; Privatti, R.T.; Oliveira, C.A.F. Determination of urinary biomarkers for assessment of short-term human exposure to aflatoxins in São Paulo, Brazil. *Toxins* **2014**, *6*, 1996–2007. [CrossRef] [PubMed]
58. Solfrizzo, M.; Gambacorta, L.; Visconti, A. Assessment of multi-mycotoxin exposure in southern Italy by urinary multi-biomarker determination. *Toxins* **2014**, *6*, 523–538. [CrossRef] [PubMed]
59. Mykkänen, H.; Zhu, H.; Salminen, E.; Juvonen, R.O.; Ling, W.; Ma, J.; Polychronaki, N.; Kemiläinen, H.; Mykkänen, O.; Salminen, S.; et al. Fecal and urinary excretion of aflatoxin B1 metabolites (AFQ1, AFM1 and AFB-N7-guanine) in young Chinese males. *Int. J. Cancer* **2005**, *115*, 879–884. [CrossRef] [PubMed]
60. Gallagher, E.P.; Wienkers, L.C.; Stapleton, P.L.; Kunze, K.L.; Eaton, D.L. Role of Human Microsomal and Human Complementary DNA-expressed Cytochromes P4501A2 and P4503A4 in the Bioactivation of Aflatoxin B11. *Cancer Res.* **1994**, *54*, 101–108.
61. Wild, C.P.; Turner, P.C. The toxicology of aflatoxins as a basis for public health decisions. *Mutagenesis* **2002**, *17*, 471–481. [CrossRef]
62. Diaz, G.J.; Cepeda, S.M.; Martos, P.A. Stability of aflatoxins in solution. *J. AOAC Int.* **2012**, *95*, 1084–1088. [CrossRef] [PubMed]
63. Bbosa, G.S.; Kitya, D.; Lubega, A.; Ogwal-Okeng, J.; Anokbonggo, W.W.; Kyegombe, D.B. Review of the Biological and Health Effects of Aflatoxins on Body Organs and Body Systems. *Aflatoxins Recent Adv. Futur. Prospect.* **2013**, *12*, 239–265.
64. Hatem, N.L.; Hassab, H.M.A.; Abd Al-Rahman, E.M.; El-Deeb, S.A.; El-Sayed Ahmed, R.L. Prevalence of afl atoxins in blood and urine of Egyptian infants with protein – energy malnutrition. *Food Nutr. Bull.* **2005**, *26*, 49–56. [CrossRef] [PubMed]
65. Leong, Y.H.; Latiff, A.A.; Ahmad, N.I.; Rosma, A. Exposure measurement of aflatoxins and aflatoxin metabolites in human body fluids. A short review. *Mycotoxin Res.* **2012**, *28*, 79–87. [CrossRef] [PubMed]
66. Roebuck, B.D.; Siegel, W.G.; Wogan, G.N. In vitro metabolism of aflatoxin B2 by animal and human liver. *Cancer Res.* **1978**, *38*, 999–1002. [PubMed]
67. Bravin, F.; Duca, R.C.; Balaguer, P.; Delaforge, M. In Vitro Cytochrome P450 Formation of a Mono-Hydroxylated Metabolite of Zearalenone Exhibiting Estrogenic Activities: Possible Occurrence of This Metabolite in Vivo. *Int. J. Mol. Sci.* **2009**, *10*, 1824–1837. [CrossRef] [PubMed]
68. Hildebrand, A.; Pfeiffer, E.; Metzler, M. Aromatic hydroxylation and catechol formation: A novel metabolic pathway of the growth promotor zeranol. *Toxicol. Lett.* **2010**, *192*, 379–386. [CrossRef]

69. Yang, S.; Zhang, H.; Zhang, J.; Li, Y.; Jin, Y.; Zhang, S.; De Saeger, S.; Li, Y.; Sun, F.; De Boevre, M. Deglucosylation of zearalenone-14-glucoside in animals and human liver leads to underestimation of exposure to zearalenone in humans. *Arch. Toxicol.* **2018**, *92*, 2779–2791. [CrossRef]
70. Kluger, B.; Bueschl, C.; Neumann, N.; Stückler, R.; Doppler, M.; Chassy, A.W.; Waterhouse, A.L.; Rechthaler, J.; Kampleitner, N.; Thallinger, G.G.; et al. Untargeted profiling of tracer-derived metabolites using stable isotopic labeling and fast polarity-switching LC-ESI-HRMS. *Anal. Chem.* **2014**, *86*, 11533–11537. [CrossRef]

toxins

MDPI

Article

Multi LC-MS/MS and LC-HRMS Methods for Determination of 24 Mycotoxins including Major Phase I and II Biomarker Metabolites in Biological Matrices from Pigs and Broiler Chickens

Marianne Lauwers [1,2], Siegrid De Baere [1], Ben Letor [2], Michael Rychlik [3], Siska Croubels [1] and Mathias Devreese [1,*]

1 Department of Pharmacology, Toxicology and Biochemistry, Faculty of Veterinary Medicine, Ghent University, 9820 Merelbeke, Belgium; marianne.lauwers@ugent.be (M.L.); siegrid.debaere@ugent.be (S.D.B.); siska.croubels@ugent.be (S.C.)

2 Innovad, Postbaan 69, 2910 Essen, Belgium; b.letor@innovad-global.com

3 Chair of Analytical Food Chemistry, Technische Universität München, Maximus-von-Imhof-Forum 2, 85354 Freising, Germany; michael.rychlik@tum.de

* Correspondence: mathias.devreese@ugent.be; Tel.: +32-9-264-7527

Received: 15 January 2019; Accepted: 12 March 2019; Published: 19 March 2019

check for updates

Abstract: A reliable and practical multi-method was developed for the quantification of mycotoxins in plasma, urine, and feces of pigs, and plasma and excreta of broiler chickens using liquid chromatography–tandem mass spectrometry. The targeted mycotoxins belong to the regulated groups, i.e., aflatoxins, ochratoxin A and *Fusarium* mycotoxins, and to two groups of emerging mycotoxins, i.e., *Alternaria* mycotoxins and enniatins. In addition, the developed method was transferred to a LC-high resolution mass spectrometry instrument to qualitatively determine phase I and II metabolites, for which analytical standards are not always commercially available. Sample preparation of plasma was simple and generic and was accomplished by precipitation of proteins alone (pig) or in combination with removal of phospholipids (chicken). A more intensive sample clean-up of the other matrices was needed and consisted of a pH-dependent liquid–liquid extraction (LLE) using ethyl acetate (pig urine), methanol/ethyl acetate/formic acid (75/24/1, *v*/*v*/*v*) (pig feces) or acetonitrile (chicken excreta). For the extraction of pig feces, additionally a combination of LLE using acetone and filtration of the supernatant on a HybridSPE-phospholipid cartridge was applied. The LC-MS/MS method was in-house validated according to guidelines defined by the European and international community. Finally, the multi-methods were successfully applied in a specific toxicokinetic study and a screening study to monitor the exposure of individual animals.

Keywords: Biomarkers; exposure; LC-MS/MS; LC-HRMS; pig; broiler chicken; multi-mycotoxin

Key Contribution: An LC-MS/MS multi-method was developed, validated and transferred to the LC-HRMS to determine mycotoxins and their metabolites in plasma, urine and feces of pigs and plasma and urine of broiler chickens. This method enables the determination of biomarkers for exposure to mycotoxins in biological samples.

1. Introduction

The worldwide contamination of feed with mycotoxins is of major agro-economic importance. In addition to crop and feed loss and damage, these mycotoxins can have a large impact on animal health. Surveys show that mycotoxins occur in more than 70% of the tested feed samples and 38% of these samples contain multiple mycotoxins. Co-contamination of several mycotoxins can result in additive or synergistic effects. Consequently, multi-methods are an asset in mycotoxin analysis [1].

This study focused on the mycotoxins regulated by the European Union [2,3], as well as the mycotoxins for which legislation is currently lacking but which reveal evident toxicity and high prevalence in feed [4]. More specifically, aflatoxin B1 (AFB1), ochratoxin A (OTA), fumonisin B1 (FB1), T2-toxin (T2), zearalenone (ZEN) and deoxynivalenol (DON), as well as enniatins (ENN), beauvericin (BEA), alternariol (AOH) and tenuazonic acid (TeA) were included.

Traditionally, mycotoxins are determined and regulated at the level of the feed. However, feed analysis has some major disadvantages. First, the possible presence of mycotoxin hotspots, i.e., local areas in the feed with a higher concentration, can cause an unequal distribution of mycotoxins in the feed and make it difficult to obtain a representative sample [5]. Second, analyzing feed gives no information about the individual exposure. Fluctuations can arise from differences in food consumption or in absorption, distribution, biotransformation and excretion (ADME) processes between the animals. Third, the risk associated with exposure can be underestimated because feed analysis does not include alternative routes of exposure such as dermal and respiratory exposure [6,7]. Fourth, modified or conjugated forms, previously called masked mycotoxins, can convert back to their free forms and hence contribute to the adverse effects related to mycotoxin exposure. This has been demonstrated for 3- and 15-acetyldeoxynivalenol (3ADON and 15ADON, respectively) and DON-3-glucoside (DON3G) in pigs and broiler chickens [8,9]. Therefore, both ADONs and DON3G can be considered as toxic as DON itself. Detection of these modified forms in feed can be difficult and is not always possible with conventional methods where the non-modified mycotoxin is determined. This creates a possible mismatch between the feed contamination level and the exposure of the animals [10]. Finally, clinical signs of mycotoxin exposure can appear when the contaminated feed has already been consumed, thus complicating or preventing diagnosis of herd problems associated with mycotoxins [11].

These issues can be resolved by biomonitoring of the animals and determining the exposure to mycotoxins in biological matrices with the use of so-called biomarkers of exposure. Biomarkers are molecules related to the exposure and are often the mycotoxin itself, the in vivo formed metabolites or interaction products with macromolecules such as nucleic acids or proteins [12,13].

The selected mycotoxins and their phase I and II metabolites or interaction products (e.g., aflatoxin-guanine) can be measured in several biological matrices. In this study, plasma, urine and feces of pigs and plasma and excreta of broiler chickens were chosen as biological matrix since combining these enables studying all the in vivo toxicokinetic parameters and the complete metabolic profile. Moreover, these matrices can be used to determine the efficacy of mycotoxin detoxifiers according to the guidelines of the European Food Safety Authority (EFSA) [14]. In addition, they can be easily applied for the detection of mycotoxins in these animals under field conditions. Especially, the non-invasive character of urine and feces collection can be of added value when sampling on farm.

Nowadays, state-of-the-art equipment such as liquid chromatography (LC) coupled to a tandem mass spectrometer (MS/MS) or a high-resolution mass spectrometer (HRMS) have become the standard in determining mycotoxins in biological matrices. Indeed, several LC-MS/MS multi-methods ($\geq$2 mycotoxins) have already been developed for the determination of mycotoxins in biological matrices of pigs and broiler chickens. However, most of these multi-methods are limited to one group of mycotoxins. Such methods are available for (the metabolites of) DON [9,15–18], ZEN [19,20], ENNs [21,22], T2 [23–25], AFB1 [26] and TeA [27]. Only few multi-methods combine mycotoxins from different families. To the best of our knowledge, this is the case for urine [28–33] and plasma [33–35] of pigs, and plasma [35] of broiler chickens. These multiclass methods are also available for other animal species such as for fish plasma [36], human plasma and urine [5,37–40], and rat plasma and urine [39]. The methods for urine clean-up often apply immunoaffinity columns or solid phase columns [30–33]. This approach increases the sensitivity but also the cost of analyzing a large number of samples and it limits the number of analytes that can be detected. To avoid these limitations, the dilute and shoot approach without further pre-treatment is frequently used [5,37,39,40]. However, this method demands careful optimization of the dilution factor, and often matrix effects and interfering matrix peaks are observed, which might decrease the sensitivity. Therefore, liquid–liquid extractions (LLE) are preferred as urine sample preparation because they are easy to perform in large quantity and enough sensitivity can be obtained. The developed plasma methods by Brezina et al. [41] and De Baere et al. [35] use OASIS HLB columns that require conditioning before use, limiting the number of samples that can be processed in a period of time. To increase the possible number of samples processed, LLE and protein precipitation are preferred. This study therefore aimed to develop a practical, fast and cost-efficient multi-method for analysis of plasma and urine of pigs and plasma of chickens, using LLE and protein precipitation, avoiding the use of affinity/solid phase columns.

Multi-methods are to the best of our knowledge currently not available for pig feces and excreta of broiler chickens. This might be due to high matrix complexity and the diversity of the physicochemical characteristics of the mycotoxins. Hence, the aim of this study was also to address these difficulties and to develop a multi-method in these highly complex matrices.

LC-HRMS is especially interesting to elucidate and determine phase I and II metabolites and interaction products for which analytical standards are not easily available. Especially, since phase II metabolites might be the most appropriate biomarkers for some mycotoxins due to the extensive biotransformation by these pathways. The glucuronidated and sulfated metabolites of DON and ZEN, as well as the metabolites formed after hepatic biotransformation of ENNB and B1 have already been determined using LC-HRMS [9,20,22,42]. However, to our knowledge, these metabolites of several toxins belonging to multiple classes have never been determined in a single chromatographic run using LC-HRMS.

Thus, the combination of LC-MS/MS and LC-HRMS enables not only determining mycotoxins with an appropriate sensitivity in the lower ng/mL or ng/g range but also detecting a broad range of mycotoxins, i.e., the mycotoxins and metabolites for which standards are readily available as well as other phase I and II metabolites and interaction products. This is especially interesting when assessing the most appropriate biomarker for exposure.

Therefore, the aim of this study was to develop and validate multi-methods using LC-MS/MS to determine relevant mycotoxins in biological matrices of pigs and broiler chickens. In addition, the LC-MS/MS method was transferred to LC-HRMS to determine mycotoxins for which analytical standards are not always commercially available. It is important to consider the main challenges including the high complexity of the matrix, the large range of different physicochemical characteristics of the mycotoxins and the need for a method with an appropriate sensitivity (in the lower ng/mL or ng/g range), so that they can be used not only for specific toxicokinetic studies but also for screening studies to monitor the exposure of individual animals.

2. Results and Discussion

2.1. Method Development

2.1.1. Sample Preparation and Extraction

Three requirements were prioritized in the development of adequate sample pre-treatments for the different matrices. First, the sample preparation should be simple and practical, thereby enabling analysis of a large quantity of samples in a time and cost efficient way. Second, the sample preparation should be as generic as possible, to allow the extraction of the 24 mycotoxins including some relevant metabolites, which was a real challenge taking into account the various physicochemical properties of the different classes of mycotoxins. Third, the method should be sensitive and, therefore, an LOQ of 1 ng/mL of ng/g was aimed for all analytes in all matrices.

Initially, all methods started from the most generic and simple sample preparation: deproteinization with an organic solvent. However, due to high matrix complexity and the large variety in physicochemical characteristics of the different mycotoxins more complex methods were needed for urine, feces and excreta. A summary of the final protocols can be found in the flowchart below (Figure 1).

Pig and Chicken Plasma

Method development was started with the optimization of the extraction of the analytes of interest from pig and chicken plasma. Proteins and phospholipids are often removed from plasma samples before analysis on LC-MS/MS equipment to prevent clotting and contamination of the equipment. Deproteinization can be accomplished using organic solvents, such as MeOH and ACN. The elimination of phospholipids could be obtained using an Oasis Ostro®-plate or HybridSPE®-phospholipid 30 mg/1 mL solid-phase extraction (SPE) tubes. The use of the Ostro®-plate required the addition of 0.1% formic acid. Therefore, 0.1% formic acid was added to ACN and MeOH.

For pig plasma, deproteinization with 0.1% formic acid in MeOH and 0.1% formic acid in ACN were tested in triplicate on spiked pig plasma samples (analyte concentration: 10 ng/mL). In this study, ACN was preferred as deproteinization solvent for pig plasma, compared to MeOH, since it gave a clearer supernatant after centrifugation. This is in accordance with previous studies where deproteinization with ACN was successfully applied to detect mycotoxins in pig plasma [20,34].

Plasma of broiler chickens contains more phospholipids compared to pigs [43]. Therefore, an additional removal of the phospholipids was necessary to obtain clear samples and to prevent clogging of tubings of the LC-MS/MS and LC-HRMS instruments during routine sample analysis. Spiked broiler chicken plasma samples (analyte concentration: 10 ng/mL) were extracted using the Oasis Ostro®-plate or the hybrideSPE® phospholipid SPE tubes (n = 3 per protocol). As can be seen in Figure 2, the use of SPE tubes resulted in a lower peak area, especially for ENNs, BEA, TEA, AME and AOH. Consequently, for broiler chicken plasma, deproteinization with ACN and 0.1% formic acid was combined with the Oasis Ostro®-plate to obtain the best results. This combination was already successfully applied to detect mycotoxins in chicken plasma by our group [44].

Pig Urine

Since methanol and ACN are mixable with urine, it was not possible to use the same method for urine as for plasma. The use of immunoaffinity columns is avoided due to the high cost. Dilute and shoot methods were eliminated to avoid matrix effects. Therefore, LLE was used to extract the mycotoxins from urine. The main parameters affecting the extraction of all components were optimized: type of extraction solvent, pH, solvent volume and extraction time. During initial experiments, ethyl acetate was evaluated as extraction solvent at neutral pH as in literature extraction of mycotoxins from human urine with this solvent can be found [5,28]. Next, the pH of extraction was optimized using urine spiked at analyte concentrations of 10 ng/mL (n = 2 per protocol): acidic (pH 2), neutral (pH 8) and basic (pH 10) extraction conditions were evaluated. Most components showed good extraction at neutral pH (see Figure 3); only for TeA, HT2 and OTA neutral pH showed insufficient results. For these components, significantly higher peak areas were observed at pH 2. This can be explained by the pKa-values of these components. The pKa of TeA (4.5 ± 1) and OTA (3.2 ± 0.1) [45,46] indicate that these components are weak acids and at pH 2 all these components will be neutral and can thus easily be extracted. The pKa values and chemical formulas of the other toxins can be found in Supplementary Table S12. As a result, it was decided to perform two extraction protocols for pig urine, i.e., one in acidic medium (pH 2, tube 1) and another in neutral medium (pH 7, tube 2). Since the same internal standard was used for TeA as for AME and AOH and these components were also detected in acidic medium, they were also added to the acidic protocol. Finally, the volume of ethyl acetate and the extraction time were also optimized. Different volumes (1.5, 3, 5, and 7.5 mL) and different extraction times (15, 30, 45, and 60 min) were tested. It was found that extraction was optimal using a solvent volume of 3 mL and an extraction time of 15 min (results not shown). The combination of both extracts into one vial was not possible since DOM1 and ADON were no longer recovered.

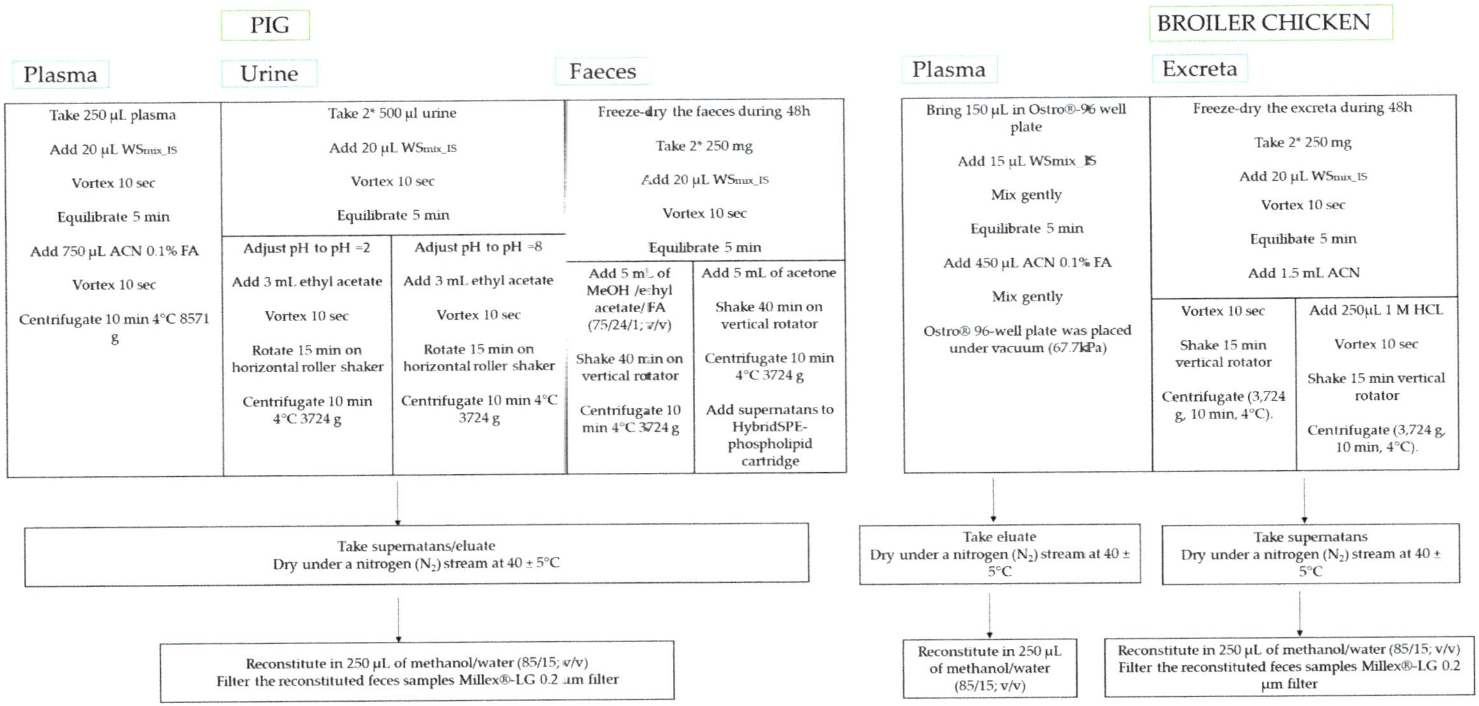

Figure 1. Flowchart of the final methods to determine different mycotoxins in plasma, urine and feces of pigs, and plasma and excreta of broiler chickens.

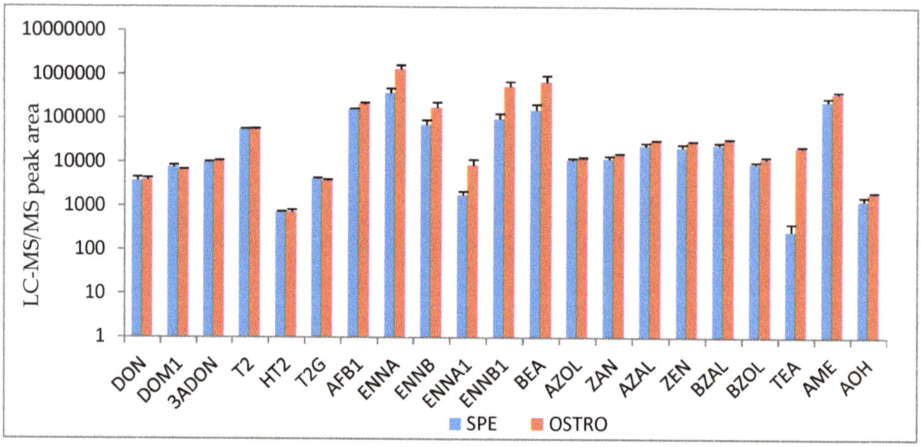

Figure 2. Comparison of the chromatographic peak areas of the different mycotoxins (mean + SD) after extraction from broiler chicken plasma (spiked at 10 ng/mL) using Oasis Ostro® plate or hybrideSPE® phospholipid SPE tubes ($n = 3$ per protocol).

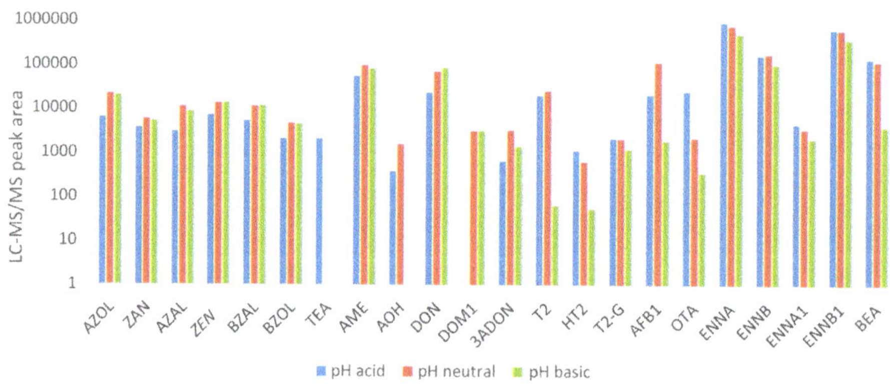

Figure 3. Comparison of the chromatographic peak areas of the different mycotoxins after extraction with ethyl acetate (spiked at 10 ng/mL) from pig urine at three different pH levels: acid (pH 2), neutral (pH 8), and basic (pH 10) ($n = 2$ per protocol).

Chicken Excreta

For the excreta of broiler chickens, the same solvents as used for extraction of urine and deproteinization of plasma were evaluated: ACN, MeOH and ethyl acetate. The excreta samples were spiked at an analyte concentration of 10 ng/g ($n = 3$ per protocol). MeOH extraction of broiler chicken excreta did not contain all the metabolites of ZEN. In addition, the peak areas of the measured metabolites of ZEN and DON and AOH/AME were much lower. Next, the chromatograms after extraction with ethyl acetate and ACN were compared. The ethyl acetate extract showed a lower S/N ratio for ZEN and its metabolites. Moreover, the peak areas for the ENNs, AFB1 and DON family were much higher after extraction with ACN compared to ethyl acetate (Figure 4). Therefore, ACN was chosen as optimal extraction solvent. However, to improve the recovery, the influence of the addition of salts ($MgSO_4$ and Na_2SO_4), acid (HCl), base (NaOH) and water on the extraction recovery was also evaluated. The extraction recovery of TeA and OTA increased by a factor 100 by adding HCl, whereas for all the other components extraction with ACN alone showed the highest recovery. Therefore, it was decided to perform the extraction of chicken excreta twice: with and without HCl (only for TeA and OTA). The final protocol for broiler chicken excreta used 1.5 mL of ACN as an extraction solvent.

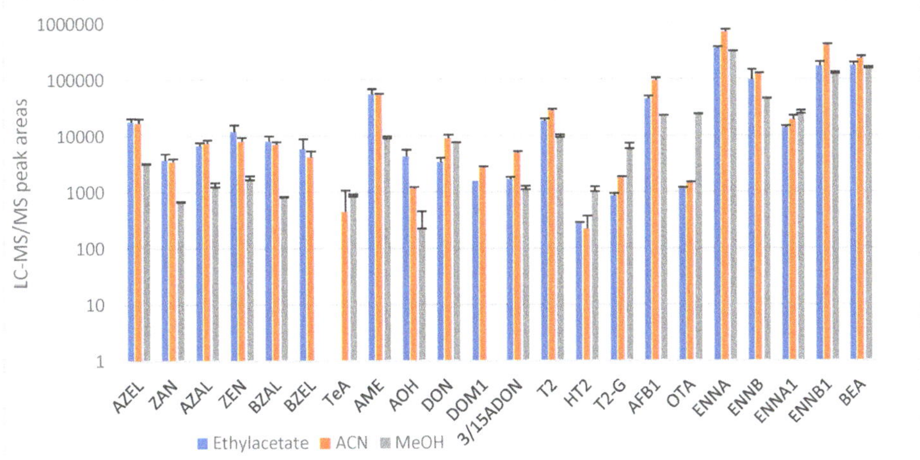

Figure 4. Comparison of the chromatographic peak areas of the different mycotoxins after extraction from broiler chicken excreta (spiked at 10 ng/g) with ethyl acetate, acetonitrile and methanol (n = 3 per protocol).

Pig Feces

The extraction of mycotoxins from pig feces was initially evaluated using the same solvents as for the optimization of chicken excreta: MeOH, ACN, and ethyl acetate. However, these solvents did not give satisfactory results in recovery and the extracts were not sufficiently clean to inject into the instrument. Therefore, other extraction solvents (acetone, diethyl ether) were also evaluated, but the recovery of the mycotoxins was still insufficient. This could be explained by the complexity of the matrix. This challenge was solved by testing different combination of solvents and the combination of the different solvents with SPE columns. This led to a double extraction procedure. The first extraction was a liquid extraction using MeOH/ethyl acetate/formic acid (75/24/1; v/v) to extract OTA, TeA, AME and AOH. The second protocol to extract the other mycotoxins was a combination of a liquid extraction with acetone and a solid phase extraction with a HybridSPE-phospholipid column. All extraction procedures were tested in triplicate at analyte concentrations of 10 ng/g.

Both feces and excreta extraction required the use of a filtration step using the Millex®-LG filter unit (0.2 μm) to obtain samples that were sufficiently clean to inject on the equipment.

For all matrices, the dried extract was reconstituted in 250 μL (or 150 μL for chicken plasma) of MeOH/water (85/15; v/v). The combination of water and MeOH was crucial to redissolve all mycotoxins with their various physicochemical characteristics.

2.1.2. Optimization of LC-MS/MS and HRMS Parameters

Four different reversed phase columns (Hypersil Gold 50 mm × 2.1 mm, dp: 1.9, Thermo Scientific, Breda, The Netherlands; Zorbax Eclipse C18 50 mm × 2.1 mm, dp: 1.8, Agilent, Sint-Katelijne-Waver, Belgium; Acquity BEH-C18 50 mm × 2.1 mm, dp: 1.7, Waters, Milford, MA, USA; and Acquity HSS-T3 100 mm × 2.1 mm, dp: 1.8, Waters, Milford, MA, USA) were tested to achieve chromatographic separation of the selected mycotoxins. The best separation of all components was obtained on the HSS-T3 column.

The multi-methods were developed with two subsequent analytical runs, i.e., ESI+ and ESI− mode respectively. This was necessary to be able to detect all the mycotoxins with sufficient sensitivity without increasing the run time. Therefore, the mobile phases for each ionization mode were optimized separately. In the literature, the most common mobile phases for mycotoxin detection consist of water and an organic solvent (such as ACN or MeOH). These solvents are often combined with mobile phase modifiers such as volatile acids (formic acid and acetic acid) and ammonium formate or ammonium acetate [34,35]. Different combinations of these solvents and modifiers were evaluated to identify the optimal combination for each ionization mode.

In ESI− mode, ZEN, AZEL, BZEL, AZAL, BZAL, ZAN, TeA, AOH and AME were determined. Baseline separation among ZAN, AZEL, and BZEL as well as between AZAL and BZAL was achieved using water (MP A) and ACN (MP B) as mobile phases [20]. The use of mobile phase with a pH close to neutral led to an impaired peak shape for TeA [47]. The peak shape was optimal when using 1% acetic acid. Therefore, 1% acetic acid in water and 1% acetic acid in ACN were chosen as final mobile phases, since this combination gave satisfactory results for all analytes. Figure 5a shows the chromatographic separation of the mycotoxins in ESI− mode with the optimized parameters as described here.

DON, DOM-1, 3-ADON, 15-ADON, T2, HT2, T2G, OTA, AFB1, AFM1, FB1, ENNA, ENNA1, ENNB, ENNB1 and BEA were determined in ESI+ mode. Taking into account all analytes, the combination of water (MP A) and MeOH (MP B) was most suitable [34]. These mobile phases were further optimized using ammonium formate and formic acid to evaluate the formation of ammonium adducts (M + [NH_4^+]). These adducts are generally easier to fragment than sodium adducts, thus enhancing the sensitivity of the method. The final combination of mobile phases was 10 mM ammonium formate and 0.3% formic acid in water (MP A) and in methanol (MP B). Figure 5b,c shows the chromatographic separation of the mycotoxins in ESI+ mode with the optimized parameters as described here.

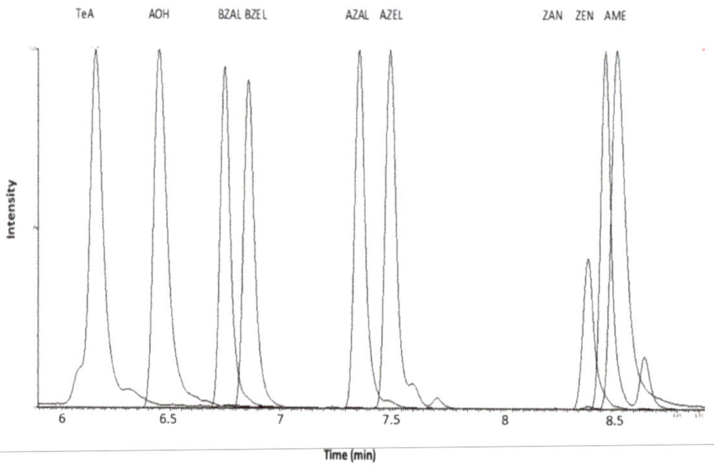

(a)

Figure 5. *Cont.*

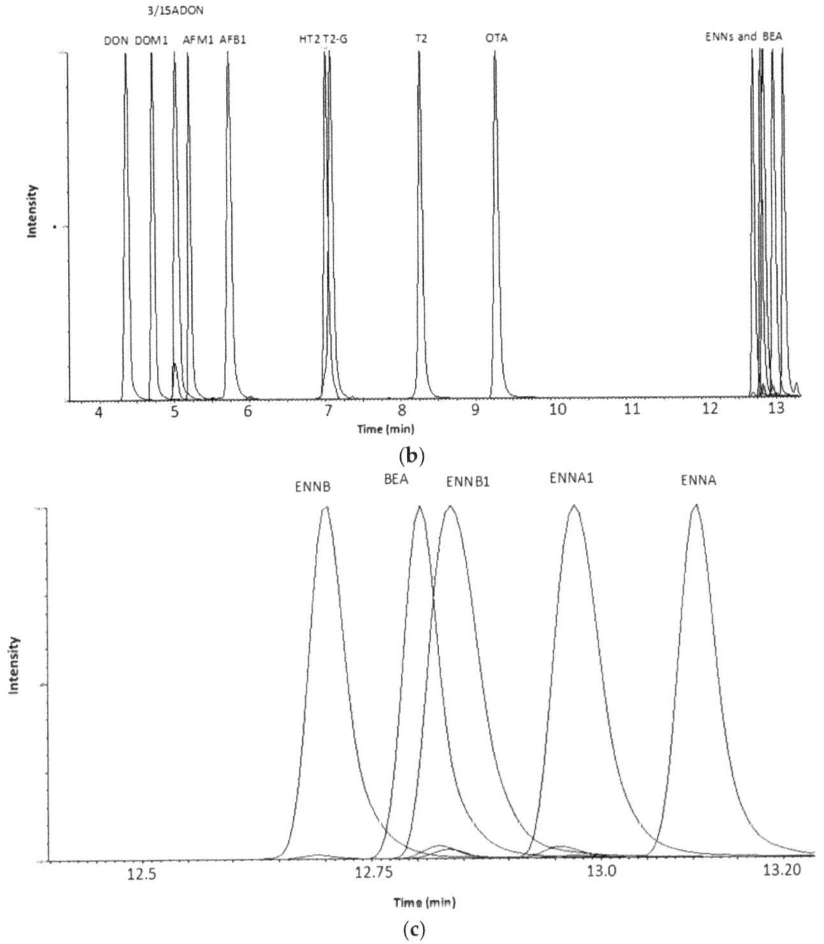

Figure 5. (**a**) LC-MS/MS chromatogram showing the separation of tenuazonic acid (TeA, 6.16 min), alternariol (AOH, 6.45 min), α-zearalanol (AZAL, 7.34 min), α-zearalenol (AZEL, 7.47 min), β-zearalanol (BZAL, 6.73 min), β-zearalenol (BZEL, 6.84 min), zearalanone (ZAN, 8.38 min), zearalenone (ZEN, 8.47 min) and alternariol-monomethyl ether (AME, 8.52 min) at a concentration of 10 ng/mL in broiler chicken plasma; (**b**) LC-MS/MS chromatogram showing the separation of deoxynivalenol (DON, 4.37 min), de-epoxy-deoxynivalenol (DOM1, 4.71 min), 3/15-acetyl-deoxynivalenol (3/15-ADON, 5.02 min), aflatoxin M1 (AFM1, 5.20 min), aflatoxin B1 (AFB1, 5.73), HT2-toxin (HT2, 7.01 min), T2-glucoside (T2G, 7.08 min), T2 toxin (T2, 8.27 min), ochratoxin A(OTA , 9.28 min), beauvericin (BEA) and the enniatins (ENNA, A1, B and B1) in broiler chicken plasma at a concentration of 10 ng/mL; and (**c**) enlargement of (**b**): LC-MS/MS chromatogram showing the separation of BEA and the enniatins (ENNA, A1, B and B1) in broiler chicken plasma at a concentration of 10 ng/mL.

Since an isotopically labeled IS for each single mycotoxin is too expensive and not commercially available, an IS labeled with [^{13}C] or [^{15}N] was used for each group of mycotoxins. [$^{13}C_{15}$]-deoxynivalenol was used as IS for DON, DOM-1 and 3/15ADON; [$^{13}C_{17}$]-Aflatoxin B1 for AFB1 and AFM1; [$^{13}C_{20}$]-Ochratoxin A for OTA; [$^{13}C_{34}$]-Fumonisin B1 for FB1; [$^{13}C_6{}^{15}N$]-Tenuazonic acid for TeA, AME and AOH; [$^{13}C_{18}$]-Zearalenone for ZEN, AZAL, BZAL, AZEL, BZEL and ZAN;

and [$^{15}N_3$]-Enniatin B for ENNA, ENNA1, ENNB, ENNB1 and BEA. Hence, an optimal correction for matrix effects and losses during sample preparation was obtained, which was confirmed during method validation (see Table 2, Tables S6–S9, and Results Section for accuracy and precision).

Data acquisition on the high-resolution mass spectrometer (HRMS) was done in the positive or negative ESI resolution mode, using the MS^E continuum scan function. The results were processed using the Unify version 1.8 software (Milford, MA, USA) to determine the phase I and II metabolite, for which no commercial analytical standards were available.

Metabolites known in the literature were added to the accurate mass—MSe screening method and additionally a pathway profiling MS^E processing method (the chemical formulas and theoretical accurate masses were added) with additional adducts and transformations was made. Peaks were identified based on the found accurate mass in the low energy spectrum and the product ions generated in the high energy spectrum. An additional confirmation criterion was the observed profile of peak areas versus time, seen in the samples obtained during the toxicokinetic study. An example of an extracted ion chromatogram of a glucuronidated metabolite of ZEN and the corresponding low energy and high energy spectra, which was detected in a plasma sample of a pig administered an intra-gastric bolus of ZEN (3 mg/kg bw), is shown in Figure 6.

The Unify version 1.8. software (Waters, Milford, MA, USA) detected two peaks with the exact mass of ZEN-glucuronide [m/z 494.1788]. However, only the MS/MS spectrum of the second peak (4.68 min) showed the product ions [m/z 317 and 175] of ZEN-glucuronide as defined in the literature [19]. The first fragment ion (m/z = 317), corresponds with the loss of glucuronic acid (176 amu) and the second at m/z = 175 results from the loss of the aglycone from the quasimolecular ion [48]. The data of the first peak can be found in Figure S1.

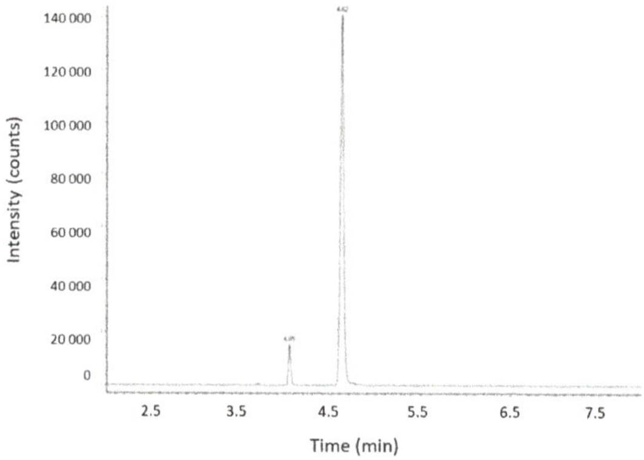

(**A**) Extracted Ion Chromatogram

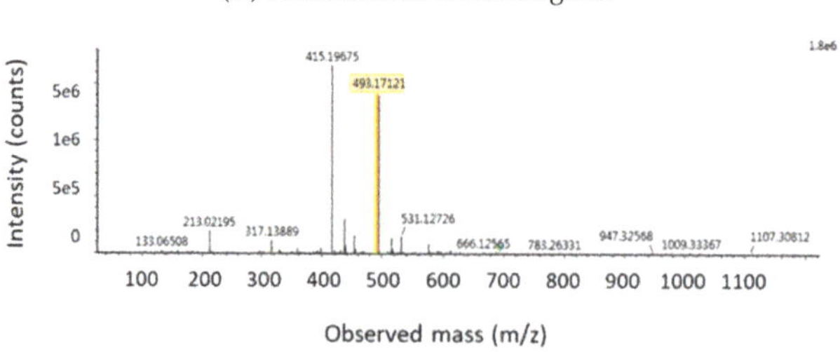

(**B**) Low Energy spectrum

Figure 6. *Cont.*

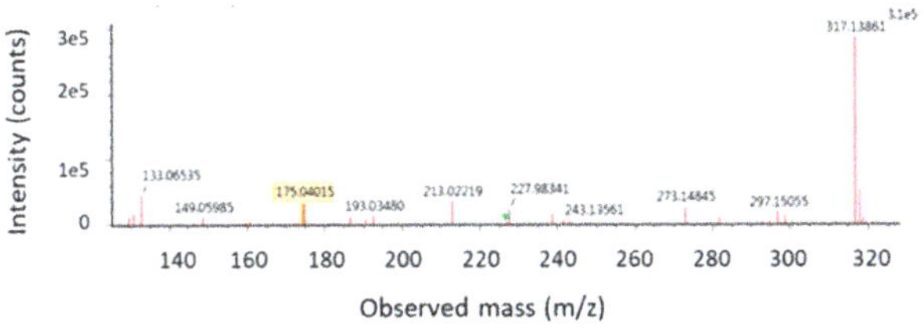

(C) High Energy spectrum

Figure 6. (**A**) LC high-resolution extracted mass chromatogram of a plasma sample that was taken from a pig that received an intra-gastric bolus of zearalenone (ZEN) (3 mg/kg bw); the following mass-to-charge (m/z) values, corresponding to the theoretical exact mass of the deprotonated molecular ions [M − H] , were extracted from the total ion chromatogram using Unify 1.8 software: ZEN-glucuronide: [m/z 493.1788]. (**B**) In the low energy MS/MS spectrum, this mass was confirmed as m/z-value 493.17121 (target mass error = 10 ppm). (**C**) In the high energy MS/MS spectrum, the corresponding product ions at m/z 317.13877 and 175.04015 (target mass error = 10 ppm) are shown.

2.2. Method Validation

The most optimal extraction protocol for plasma, urine and feces of pigs and plasma of broiler chickens was not validated for fumonisins due to low recovery. Therefore, it was decided to only validate fumonisin B2 in broiler chicken excreta, hence in this matrix 25 mycotoxins were validated.

The correlation coefficient (r) and the goodness-of-fit (g) are shown in Table 1 as an average ± standard deviation of three curves made across three different analysis days. The linearity results for the other matrices can be found in Tables S2–S5. They ranged for pig plasma from 0.993 to 0.998 (r) and 9.0% to 17.5% (g); for pig urine from 0.995 to 0.999 (r) and 3.5% to 17.0% (g); for pig feces from 0.993 to 0.999 (r) and 7.1% to 18.7% (g); for chicken plasma from 0.994 to 0.999 (r) and 5.9% to 17.7% (g); and for chicken excreta from 0.995 to 0.999 (r) and 5.1% to 16.80% (g). Most of the calibration curves matched a linear calibration model with a 1/x weighing factor, except for the ENNs and BEA. These components show a quadratic 1/x model. Linearity results of each component separately for pig plasma are shown in Table 1.

Table 1. Validation results for linearity (linear range, correlation coefficient (r) and goodness-of-fit coefficient (g)) and limit of quantification (LOQ) of 24 mycotoxins in pig plasma.

Analyte	Linearity (n = 3 Different Days)			LOQ (ng/mL)
	Linear Range (ng/mL)	r ± SD	g (%) ± SD	
ZEN	1–200	0.996 ± 0.003	12.9 ± 3.8	1.0
AZEL	1–200	0.995 ± 0.002	15.0 ± 1.2	1.0
AZAL	1–200	0.995± 0.003	15.2 ± 3.6	1.0
BZAL	1–200	0.996 ± 0.001	10.7 ± 2.0	1.0
BZEL	1–200	0.996 ± 0.002	14.9 ± 3.7	1.0
ZAN	1–200	0.997 ± 0.001	16.3 ± 3.1	1.0
TEA	1–200	0.998 ± 0.001	12.0 ± 4.9	1.0
AOH	1–200	0.997 ± 0.002	12.9 ± 5.2	1.0
AME	1–200	0.996± 0.004	13.1± 5.1	1.0
DON	1–200	0.998 ± 0.002	13.9 ± 4.4	1.0
DOM-1	1–200	0.997 ± 0.003	16.9 ± 2.5	1.0
3/15 ADON	1–200	0.998 ± 0.001	9.3 ± 3.1	1.0
T2	1–200	0.998 ± 0.001	9.9 ± 1.6	1.0
HT2	1–100	0.993 ± 0.002	17.5 ± 2.8	1.0
T2G	2–200	0.995 ± 0.003	14.4 ± 1.2	2.0
AFB1	1–200	0.996 ± 0.002	12.7 ± 2.8	1.0
AFM1	1–200	0.997 ± 0.002	13.8 ± 5.6	1.0
OTA	1–200	0.993 ±0.004	9.3 ± 2.9	1.0

Table 1. *Cont.*

Analyte	Linearity (*n* = 3 Different Days)			LOQ (ng/mL)
	Linear Range (ng/mL)	r ± SD	g (%) ± SD	
ENNA1	1–200	0.998 ± 0.001	9.0 ± 0.9	1.0
ENNA	1–50	0.995 ± 0.003	13.9 ± 3.2	1.0
ENNB	1–100	0.993 ± 0.002	9.3 ± 1.2	1.0
ENNB1	1–100	0.998 ± 0.001	15.9 ± 2.3	1.0
BEA	1–100	0.998 ± 0.000	16.6 ± 4.5	1.0

Note: SD, standard deviation; acceptance criteria: r ≥ 0.990 and g ≤ 20.

The LOQ that was aimed for during method development was 1 ng/mL or ng/g. This could be obtained for the majority of the components in the different matrices, with the following exceptions: DOM1 (4 ng/mL) and T2G (2 ng/mL) in pig urine; T2G (2 and 5 ng/mL) in pig and broiler chicken plasma; and T2G (2 ng/g), HT2 (4 ng/g) and FB2 (10 ng/g) in broiler chicken excreta. In pig feces, the LOQ for ZEN, AZAL, AOH, DON, DOM-1, HT2 and T2G was established at 5 ng/g.

No peaks were detected at the retention time zone of the analytes of interest in the solvent sample that was injected after the highest calibrator sample, thus demonstrating the absence of carry over. Moreover, for none of the components a signal was observed at the elution zone of the analytes of interest in a blank matrix sample. This indicates a good specificity of the method.

The results of the within-day and between-day precision and accuracy met the specifications for all mycotoxins and matrices. The results can be found in Table 2 for pig plasma and Tables S6–S9 for the other matrices.

The results for matrix effects (signal enhancement and suppression) and extraction recovery are shown in the Tables S10 and S11. Most components gave acceptable results (range 60–140%). However, for some components, matrix effects were more pronounced and recovery was rather low. However, for all mycotoxins, an adequate internal standard and matrix-matched calibration curves were used, resulting in validation results for accuracy and precision matching the acceptance criteria.

The validation results for the other matrices can be found in Tables S6–S9.

This resulted in a fully validated quantitative targeted LC-MS/MS method and additionally a qualitative untargeted LC-HRMS method. Both methods together enable not only determining mycotoxins with good sensitivity but also targeting a broad range of mycotoxins and their metabolites, not limited by the commercial availability of standards. This approach makes is possible to determine 24 mycotoxins and their relevant metabolites in easily obtainable biological matrices (plasma, urine and feces) of pigs and broiler chickens. This is the first time that a paper covers such a broad range of matrices and mycotoxins with a simple and practical sample preparation. This leads to a general applicable method that can be applied in, among others, in vivo toxicokinetic studies and screening studies to investigate the exposure of individual animals to mycotoxins, as shown in Section 2.3.

2.3. Biological Samples: Toxicokinetic Study

2.3.1. Pigs

In the pig plasma samples, low concentrations (1–15 ng/mL) of DON and ZEN were found using LC-MS/MS analysis. The plasma concentration–time curves for these components are shown in Figure S2. However, analysis of the samples using the LC-HRMS instrument showed that DON-GlcA and ZEN-GlcA are better biomarkers for exposure as their observed peak areas are much higher than those of the respective parent components DON and ZEN [9,20]. Since no DON-GlcA and ZEN-GlcA standards were available at our laboratory, these components were tentatively identified using the LC-HRMS multi-method. DON-GlcA and ZEN-GlcA plasma response–time curves are shown in Figure 7. The highest response for ZEN-GlcA was achieved at 30 min and for DON-GlcA at 4 h post-administration.

Table 2. Results of the within-day and between-day precision and accuracy experiments for 24 mycotoxins in pig plasma.

Analyte	Within-Day Precision and Accuracy (n = 6)						Between-Day Precision and Accuracy (n = 3 × 3)					
	Theoretical Concentration LOQ		Theoretical Concentration 10 ng/mL		Theoretical Concentration 100 ng/mL		Theoretical Concentration LOQ		Theoretical Concentration 10 ng/mL		Theoretical Concentration 100 ng/mL	
	Precision (RSD %)	Accuracy (%)	Precision (RSD %)	Accuracy (%)	Precision (RSD %)	Accuracy (%)	Precision (RSD %)	Accuracy (%)	Precision (RSD %)	Accuracy (%)	Precision (RSD %)	Accuracy (%)
ZEN	12.1	6.5	7.3	1	5.4	−1.5	13.4	2.7	7.5	2.4	5.7	−2.2
AZEL	6.5	19.7	13	2.3	7.4	6.6	35.4	−5.3	10.9	−0.7	11.6	1.5
AZAL	17.4	−14.5	7.7	−1.4	4.4	−0.4	20.9	−3.3	8.2	−5.2	8.1	−4.9
BZAL	5.2	7	3.5	8.2	2.9	8.9	9.3	6.2	3.9	6.6	10.7	4.2
BZEL	13.0	−9.2	5.1	−1.8	6.1	−2.4	15.9	−1.8	8.1	−4.9	6.6	−5.3
ZAN	11.7	−40.5	3.9	8.8	4.6	3.0	21.5	−32.4	3.6	9.8	7.1	0.4
TEA	2.8	19.0	3.2	8.2	3.6	4.5	10.7	18.2	3.4	8.7	5.8	4.8
AOH	17.6	−32.2	3.5	8.6	3.8	−0.4	26.9	−20.8	4.5	9.5	5.3	−3.1
AME	14.1	10.1	7.7	2.4	4.8	−12.3	18.7	−5.3	7.0	4.8	5.9	−11.9
DON	24.9	−8.0	6.9	1.0	5.6	−5.7	22.3	4.4	10.0	2.3	8	−4.6
DOM-1	17.5	−0.4	15.0	−3.3	5.9	−6.9	14.7	−2.0	14	−2.7	7.9	−7.3
3/15 ADON	15.6	7.5	5.2	5.2	6.4	−3.5	16.4	10.9	5.4	6.9	8.6	−3.5
T2	15.4	3.3	1.6	7.7	2.7	8.4	12.7	3.7	1.5	8.1	6.2	5.0
HT2	21.1	−29.8	6.5	−14.0	9.9	5.8	30.5	−21.4	10.0	−11.5	5.7	2.1
T2G	10.8	−3.8	7.8	6.9	7.0	1.3	23.8	−2.6	9.5	4.1	13.8	6.6
AFB1	13.1	−14.3	3.0	3.9	4.9	−2.0	16.4	−16.9	3.6	4.5	6.3	−2.8
AFM1	11.2	−38.8	10.6	−19.5	8.8	−15.6	28.0	−28.5	18.8	−7.1	20.2	−5.4
OTA	7.5	13.4	8.7	−13.3	4.5	−12.3	14.4	2.5	7.2	−12.5	9.2	−8.0
ENN A1	15.7	−11.3	12.6	−3.7	6.7	1.8	14.6	−2.3	10.8	−0.8	9.6	−0.9
ENNA	19.4	−1.0	9.4	−14.1	11.7	−14.7	41.7	−11.2	13.8	−6.8	13.5	−5.6
ENNB	16.7	−0.1	11.8	9.6	2.6	9.4	16.8	−1.4	13.9	−1.3	6.6	4.1
ENNB1	7.6	16.1	3.9	−0.1	3.5	−3.9	31.8	1.6	8.7	5.5	3.2	−3.2
BEA	13.3	−2.9	3.2	6.9	3.2	7.4	29.7	−6.3	2.3	7.2	11.9	8.5

Note: The acceptance criteria: Accuracy, ≤1 ng/mL: −50% to +20%; 1–10 ng/mL: −30% to +10%; ≥10ng/mL: −20% to +10%. Within-day precision: RSD% < RSDmax with RSDmax for ≥1 to <10 ng/mL: <25% and ≥10 to <100 ng/mL: <15%. Between-day precision: the RSD% < RSDmax with RSDmax 22.6%, 32% and 45% for the respective concentrations of 100 ng/mL, 10 ng/mL and 1 ng/mL, respectively.

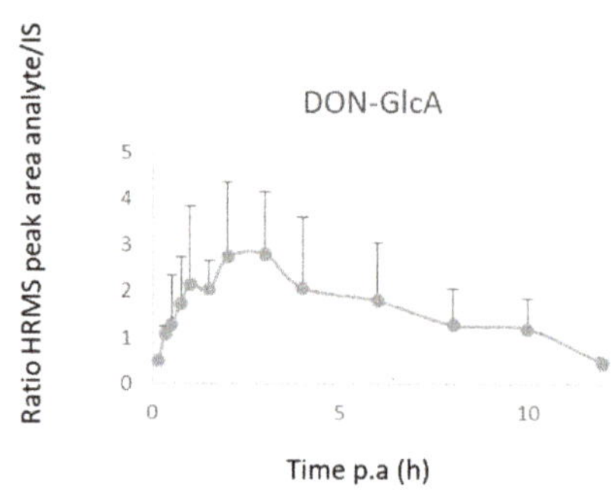

(A)

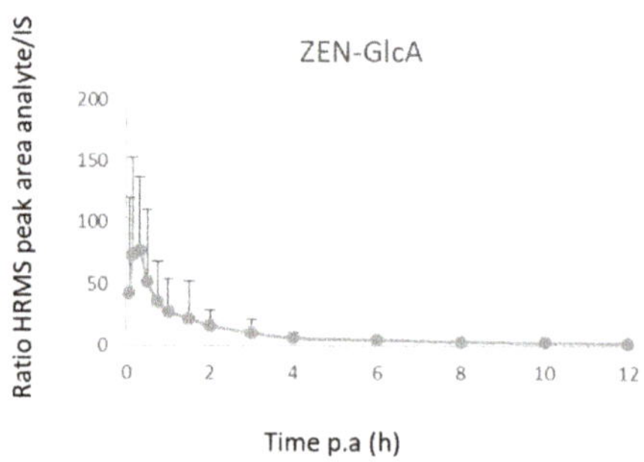

(B)

Figure 7. (**A**) HRMS response–time curve of deoxynivalenol-glucuronide (DON-GlcA) in plasma after intra-gastric administration of deoxynivalenol (DON, 36 μg/kg bw) to pigs (n = 8). The mean ratio of the HRMS peak areas of DON-GlcA/$^{13}C_{15}$-DON + SD is shown. (**B**) HRMS response–time curve of zearalenone-glucuronide (ZEN-GlcA) in plasma after intra-gastric administration of zearalenone (ZEN, 3 mg/kg bw) to pigs (n = 8). The mean ratio of the HRMS peak areas of ZEN-GlcA/$^{13}C_{18}$-ZEN + SD is shown.

In pig feces, no traces of DON were observed. This can be explained by the complete absorption and fast elimination of DON in urine, while only 1–3% of the administered dose is reported to be excreted via feces [16]. The concentration–time profiles of ZEN and its phase I metabolites in feces showed maximum levels from the first 12 h after exposure onwards (Figure 8). The highest amounts were excreted during 12–24 h. This is in line with the observations of Binder et al., who found the highest amounts of ZEN and metabolites were excreted during 24–48 h [19].

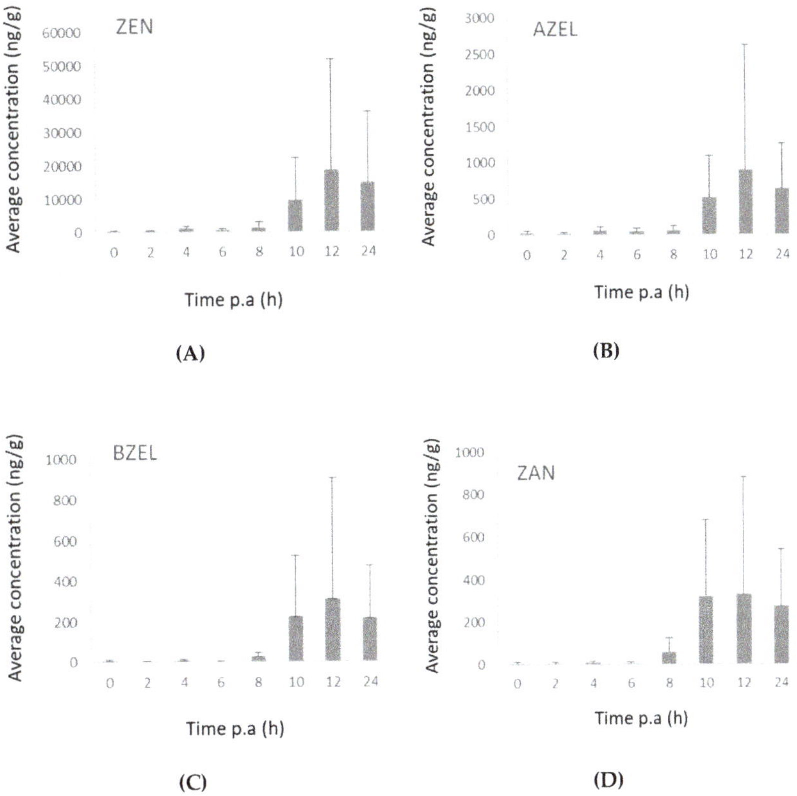

Figure 8. (**A**) Concentration–time curve of zearalenone (ZEN) in feces after intra-gastric administration of ZEN (3 mg/kg bw) to pigs (n = 8). The mean + SD is shown. (**B**) Concentration–time curves of α-zearalenol (AZEL) in feces after intra-gastric administration of ZEN (3 mg/kg bw) to pigs (n = 8). The mean + SD is shown. (**C**) Concentration–time curve of β-zearalenol (BZEL) in feces after intra-gastric administration of ZEN (3 mg/kg bw) to pigs (n = 8). The mean + SD is shown. (**D**) Concentration–time curve zearalanone (ZAN) in feces after intra-gastric administration of ZEN (3 mg/kg bw) to pigs (n = 8). The mean + SD is shown.

In pig urine, DON, ZEN and ZEN-GlcA were detected. The maximum concentrations were achieved after 4–8 h for ZEN, ZEN-GlcA and DON. Nagl et al. also demonstrated a fast elimination of DON in urine with a maximum concentration in the first 4 h. DON showed to be the most important urinary metabolite [16]. Binder et al. also detected ZEN and ZEN-GlcA in pig urine after oral administration of ZEN, with ZEN-GlcA as the major metabolite. In this study, the response ratio of ZEN-GlcA/IS was also much higher than the area ratio of ZEN/IS, indicating ZEN-GlcA as a major metabolite [19]. The concentration (or HRMS response)–time curves of these molecules are depicted in Figure 9.

In all feces and urine samples, low concentrations (or HRMS response areas) were observed at the time of administration due to the presence of low levels DON and ZEN in the feed. In urine, this concentration was negligible, especially when compared to the concentration after administration. In feces, the effect of the administration was only seen after 10 h. Twelve hours fasting before administration was not enough to eliminate the concentration of mycotoxins in feces after long-term exposure in this study. However, blank samples were obtained in previous studies, as shown in Figure S3.

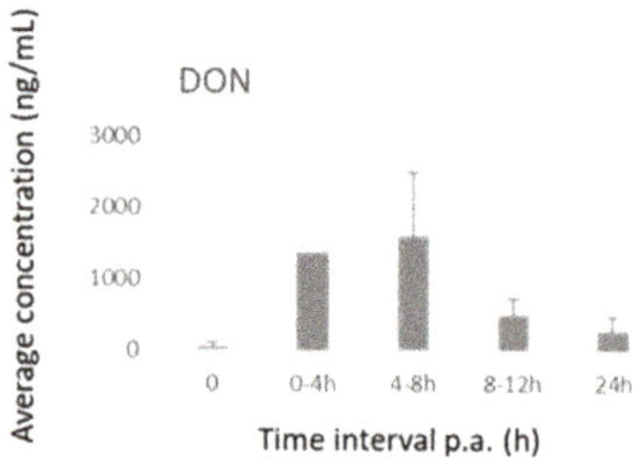

(A)

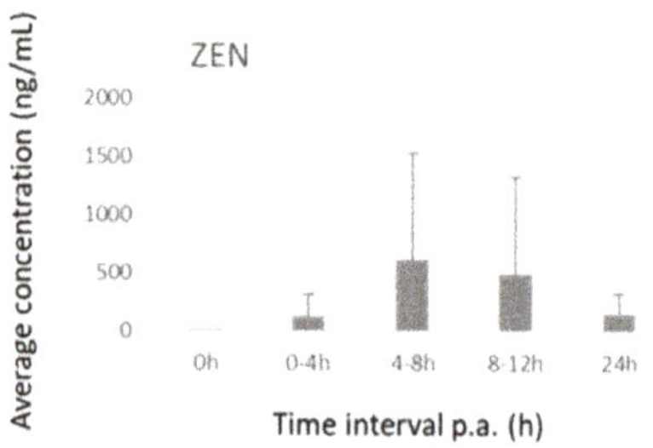

(B)

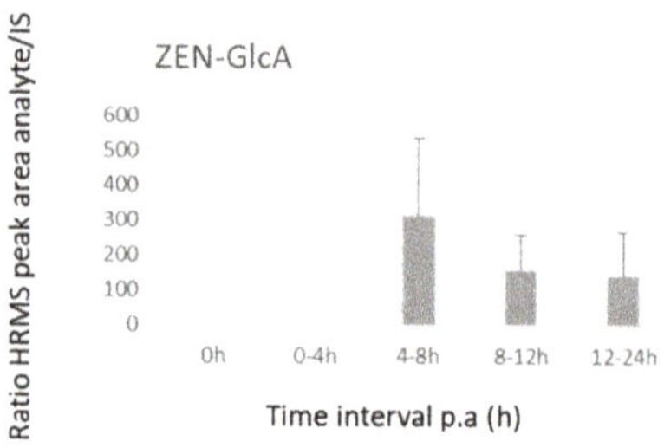

(C)

Figure 9. (**A**) Concentration–time curves of deoxynivalenol (DON) in urine after intra-gastric administration of DON (36 μg/kg bw) to pigs (n = 8). The mean + SD is shown (**B**) Concentration–time curves of zearalenone (ZEN) in urine after intra-gastric administration of ZEN (3 mg/kg bw) to pigs (n = 8). The mean + SD is shown. (**C**) The HRMS response–time curves of zearalenone-glucuronide (ZEN-GlcA) in urine after intra-gastric administration of ZEN (3 mg/kg bw) to pigs (n = 8). The mean ratio of the HRMS peak areas of ZEN-GlcA/$^{13}C_{18}$-ZEN + SD is shown.

2.3.2. Broiler Chickens

AFB1 and OTA were detected in plasma, as well as in excreta samples. No other relevant metabolites were found by LC-HRMS. The concentration (response)–time curves are shown in Figure 10. OTA and AFB1 showed a second peak in the plasma concentration–time curve around 4 h p.a. This can be attributed to enterohepatic recirculation, which has previously been described for OTA [49,50]. The maximum concentration measured for AFB1 was 8.4 ng/mL and for OTA was 50 ng/mL, for both toxins observed after 15 min. After administration of DON, only DON-sulfate was found in plasma and excreta due to the high conversion rate of DON to DON-sulfate in broiler chickens, confirming previous literature reports [9,16]. The maximum response was achieved after 30 min.

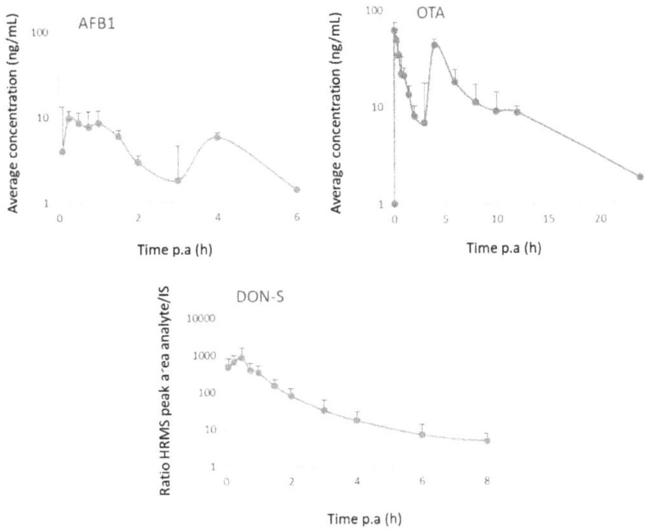

(**A**) Broiler chicken plasma

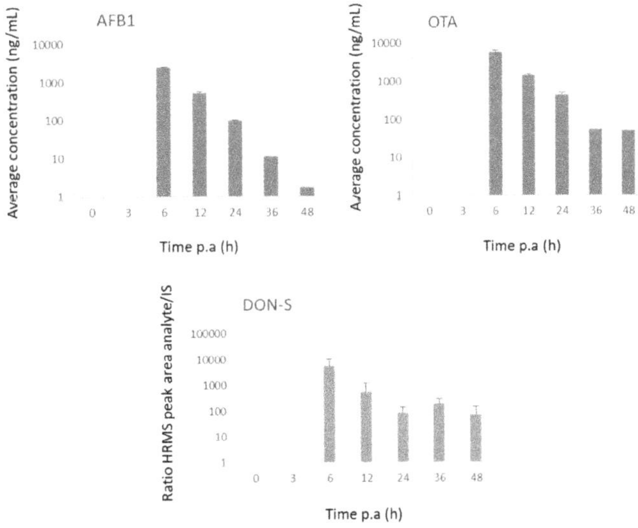

(B) Broiler chicken excreta

Figure 10. (**A**) Concentration–time curves of aflatoxin B1 (AFB1) and ochratoxin A (OTA), and HRMS response–time curves of deoxynivalenol-sulfate (DON-S) in plasma after PO administration of deoxynivalenol (DON, 0.5 mg/kg bw), AFB1 (2 mg/kg bw) and OTA (0.25 mg/kg bw) to broiler chickens (n = 16). The mean + SD is shown for AFB1 and OTA and the mean ratio of the HRMS peak areas of DON-S/$^{13}C_{15}$-DON + SD is shown. (**B**) Concentration–time curves of AFB1 and OTA and HRMS response–time curves of DON-sulfate (DON-S) in excreta after PO administration of DON (0.5 mg/kg bw), AFB1 (2 mg/kg bw) and OTA (0.25 mg/kg bw) to broiler chickens (n = 16). The mean + SD is shown for AFB1 and OTA and the mean ratio of the HRMS peak areas of DON-S/$^{13}C_{15}$-DON + SD is shown.

2.4. Screening Study

Plasma samples from pigs and broiler chickens were obtained from the field and analyzed with the presented method for the presence of mycotoxins in plasma. Two interesting samples are highlighted. The first sample is a pig plasma sample from Belgium. This sample contained ZEN, TeA and DON with respective concentrations of 1, 1.9 and 8.6 ng/mL. The chromatograms are shown in Figure 11.

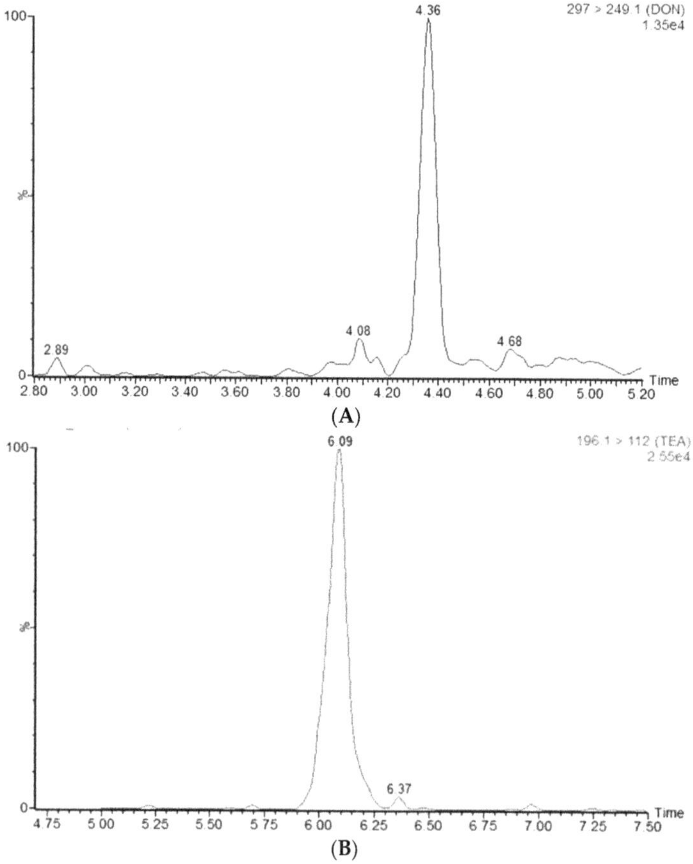

Figure 11. *Cont.*

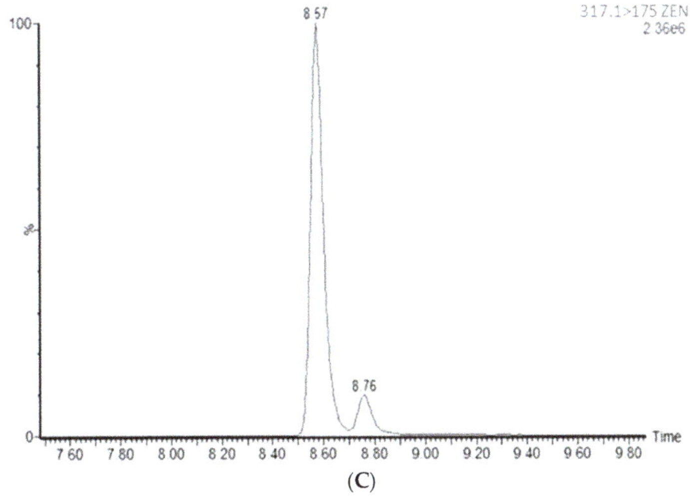

Figure 11. The extracted ion chromatograms showing the presence of: deoxynivalenol (DON) (**A**); tenuazonic acid (TeA) (**B**); and zearalenone (ZEN) (**C**) in pig plasma samples obtained from a farm in Belgium.

The second sample is a broiler chicken plasma sample from Lithuania. This sample contained TeA and DON with respective concentrations of 1,016 and 70,617 ng/mL. The chromatograms are shown in Figure 12.

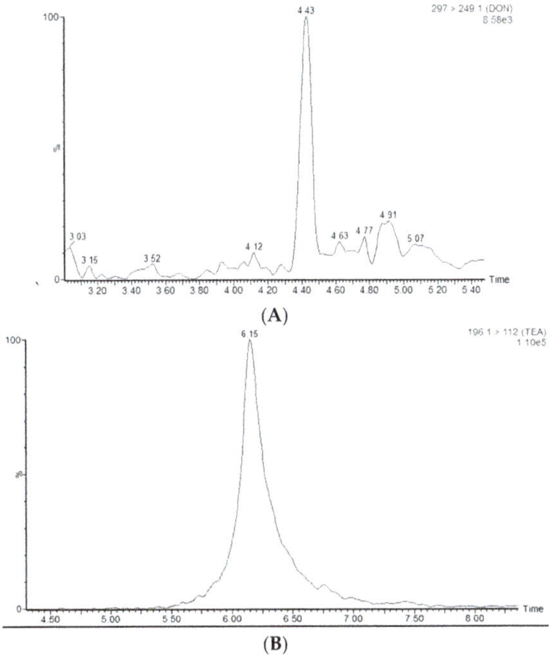

Figure 12. The extraction ion chromatograms showing the presence of: deoxynivalenol (DON) (**A**); and tenuazonic acid (TeA) (**B**) in broiler chicken plasma samples obtained from a farm in Lithuania.

3. Conclusions

This paper describes a fully validated quantitative targeted LC/MS-MS method, and a qualitative untargeted LC-HRMS approach to determine mycotoxins and their relevant metabolites in easily obtainable biological matrices of pigs and broiler chickens. The methods were applied to plasma, urine, feces and/or excreta samples that were obtained during in vivo toxicokinetic studies with DON and ZEN in pigs, and with DON, AFB1 and OTA in broiler chickens and during a pilot field screening study to monitor exposure to mycotoxins. These results show the successful applicability of the multi-method to pig and broiler chicken samples, providing a proof-of-concept of the developed methods.

4. Materials and Methods

4.1. Chemicals, Products and Reagents

The analytical standards of ZEN, T2, HT2 toxin (HT2), OTA, AFB1, aflatoxin M1 (AFM1), FB2, AOH, alternariol monomethyl ether (AME), TeA, DON, 3ADON, 15ADON, enniatin A (ENNA), enniatin A1 (ENNA1), enniatin B1 (ENNB1), enniatin B (ENNB) and BEA were obtained from Fermentek (Jerusalem, Israel). Zearalanone (ZAN), α-zearalenol (AZEL), β-zearalenol (BZEL), α-zearalanol (AZAL) and β-zearalanol (BZAL) were purchased from Sigma-Aldrich (Bornem, Belgium). De-epoxy-deoxynivalenol (DOM-1) was obtained from Biopure (Tulln, Austria). T-2 toxin-3α-glucoside (T2G) was synthesized by the U.S. Department of Agriculture (USDA) as described [51,52]. Internal standards (IS) of $^{13}C_{15}$-DON, $^{13}C_{24}$-T2, $^{13}C_{18}$-ZEN, $^{13}C_{20}$-OTA, $^{13}C_{34}$-FB1 and $^{13}C_{17}$-AFB1 were purchased from Biopure. The internal standard $^{13}C_6{}^{15}N$-TeA was synthesized according to the method of Asam et al. [53], and $^{15}N_3$-ENN B was synthesized according to the method of Hu and Rychlik [54]. All standards were stored according to the recommendations of the supplier. Water, methanol (MeOH), acetonitrile (ACN), ammonium formate, glacial acetic acid and formic acid for the preparation of mobile phases were of LC-MS grade and were obtained from Biosolve (Valkenswaard, The Netherlands). Acetone, ammonium formate, formic acid and ethyl acetate were of analytical grade and were purchased from VWR (Leuven, Belgium). Millex®-LG filter units (0.2 μm), sodium hydroxide (NaOH) pellets and hydrochloric acid (HCl) 37% fuming solution were obtained from Merck (Overijse, Belgium). Ostro®-96 well plates were obtained from Waters (Milford, MA, USA). HybridSPE®-phospholipid 30 mg/1 mL SPE tubes were purchased from Sigma-Aldrich. Merck Alcalit pH indicator paper pH 0–14 was obtained from Novolab (Geraardsbergen, Belgium).

4.2. Preparation of Standard Solutions

Standard stock solutions (SS) for ZEN, AZAL, BZAL, AZEL, BZEL, ZAN, DON, T2, T2-G, HT2, AFB1, AFM1, AOH, AME, ENNs, BEA and FB1 were prepared in ACN at 100 μg/mL. Standard SS for OTA was prepared in ACN at 10 μg/mL. The standard SS of TeA was prepared in methanol at 100 μg/mL. Following standards were purchased as solutions: 3ADON (100 μg/mL in ACN), 15ADON (100 μg/mL in ACN) and DOM-1 (50 μg/mL in ACN). A standard SS of 10 μg/mL in ACN was prepared for DOM-1. All internal standards (IS) were obtained as solutions: $^{13}C_{15}$-DON (25 μg/mL in ACN), $^{13}C_{24}$-T2 (25 μg/mL in ACN), $^{13}C_{18}$-ZEN (25 μg/mL in ACN), $^{13}C_{20}$-OTA (10 μg/mL in ACN), $^{13}C_{34}$-FB1 (25 μg/mL in ACN/water) and $^{13}C_{17}$-AFB1 (0.5 μg/mL in ACN). Standard SS of the synthesized internal standards were prepared at a concentration of 100 μg/mL in MeOH for $^{13}C_6{}^{15}N$-TeA and 10 μg/mL in ACN for $^{15}N_3$-ENN B. The SS were stored at ≤ -15 °C.

A combined working solution of all analytical standards (WS_{mix}, without IS) at a concentration of 1 μg/mL was prepared by transferring 10 μL of the stock solutions with a concentration of 100 μg/mL and 100 μL of a 10 μg/mL solution of DOM-1 and OTA, followed by further dilution with ACN up to a total volume of 1 mL. Serial dilutions of the WS_{mix} were prepared, yielding concentrations of 100 ng/mL and 10 ng/mL. Individual working solutions of 1 μg/mL were made for all IS, except for $^{13}C_{17}$-AFB1 (100 ng/mL) and $^{15}N_3$-ENN B (100 ng/mL). Next, a combined working solution of all IS (WS_{mix_IS}) was prepared with a final concentration of 100 ng/mL for all components, except $^{13}C_{17}$-AFB1 (10 ng/mL) and $^{15}N_3$-ENN B (10 ng/mL). All working solutions were stored at ≤−15 °C.

4.3. Biological Samples

A toxicokinetic study was performed to demonstrate the applicability of the developed method. Incurred plasma samples were obtained from 8 hybrid pigs (6 weeks of age, 9.94 ± 1.24 kg, ♂/♀4/4), dosed with a single oral (intragastric) bolus of DON (36 µg/kg bodyweight (bw)) and ZEN (3 mg/kg bw) after a fasting period of 12 h, and from 16 12 h-fasted broiler chickens (Ross 308, 3 weeks of age, 1.05 ± 0.11 kg, ♂/♀5/11) administered AFB1 (2 mg/kg bw), DON (0.5 mg/kg bw) and OTA (0.25 mg/kg bw). All mycotoxin doses were administered as a single oral bolus (acute exposure, by gavage), and corresponded with doses previously used in toxicokinetic studies and studies to test the efficacy of mycotoxin detoxifiers. For DON, this dose was in agreement with the EU legislation in feed. The maximum guidance level in pig feed is set at 0.9 mg/kg DON [3]. Pigs of this age category consume on average 40 g feed/kg bw/day. This resulted in the administration of 36 µg DON/kg bw as described in [8]. For broiler chickens, the EU regulations set the maximum guidance level at 5 mg/kg feed [3]. Broilers (±1 kg bw) consume on average 100 g feed/kg bw/day. This resulted in the administration of 0.5 mg DON/kg bw as described in [8].

For ZEN in pigs and AFB1 and OTA in broiler chickens, the administered doses were higher than set by the EU legislation. These doses corresponded to the doses previously administered in toxicokinetic studies and studies to determine the efficacy of mycotoxin detoxifiers [50]. These higher doses were necessary to obtain sufficiently high plasma concentrations to evaluate the toxicokinetic parameters of the toxins, as well as to demonstrate the efficacy of mycotoxin detoxifiers. Although the doses were higher than the guidance levels, they did not evoke clinical toxicity after this single administration. This was the case in the previously mentioned studies [50], as well as in the present study. The most important read-out of this single oral bolus dosing in animals is the area under the plasma concentration–time curve (AUC), which has to be high enough to be able to demonstrate a statistical significant reduction in AUC when combined with the detoxifier. As the goal here was to evaluate the use of the analytical methods in this type of studies, the same doses were used.

Blood was sampled before administration (0 min) and at 5, 10, 20, 30, 45, 60, 90, and 120 min, and 3, 4, 6, 8, 10, 12 and 24 h after administration of mycotoxins. Blood was collected via the vena jugularis in EDTA tubes using a Venoject® system and centrifuged (2851× *g*, 10 min, 4 °C) to obtain plasma, which was stored at ≤−15 °C until analysis. Urine was collected from male pigs using pediatric urine collection bags as described by Gasthuys et al.[55]. Sampling was done at 4 intervals: 0–4 h, 4–8 h, 8–12 h and 12–24 h. Feces were collected every 2 h by rectal stimulation of the pigs. Excreta were collected every 2 h by placing the broiler chickens in separate boxes for sampling. The animal trial was approved by the ethical committee of the Faculty of Veterinary Medicine and the Faculty of Bioscience Engineering of Ghent University (EC2017/05) on 30 March 2017.

A multi-mycotoxin LC-MS/MS analysis of the pig and chicken feed (Primoris, Zwijnaarde, Belgium) showed only low amounts of DON (respectively, 139 and 140 µg/kg) and ZEN (respectively, 12 and 20 µg/kg). The feed was conform the EU legislation since these amounts are below the guidance values of the European commission [3]. Blank plasma, urine and feces samples were obtained from pigs and broiler chickens on the mycotoxin control diet. The blank samples were used for the preparation of matrix-matched calibration curves and quality control samples.

Besides the toxicokinetic studies, a preliminary screening study (n = 1 farm per animal species) was performed to monitor the exposure of pigs and broiler chickens to mycotoxins and to demonstrate the applicability of the developed method in the field. Therefore, farms with problems that might be related to mycotoxins (e.g., postpartum problems, tail necrosis and refused feed intake) and where mycotoxins were found in feed were selected. Blood of 10 pigs (5 sows and 5 piglets) and 10 broiler chickens (2 weeks of age) was collected around 30 min after feeding. The blood was collected in EDTA tubes and centrifuged (2851× *g*, 10 min, 4 °C) to obtain plasma. The animal trial was also approved by the ethical committee of the Faculty of Veterinary Medicine and the Faculty of Bioscience Engineering of Ghent University (EC2017/115) on 30 March 2017.

4.4. *Sample Pre-Treatment*

4.4.1. Pig Plasma

To 250 μL of plasma, 20 μL of a 100 ng/mL WS_{mix_IS} were added, vortex mixed and set for equilibration at room temperature for 5 min. Next, 750 μL of ACN with 0.1% formic acid of analytical grade were added, followed by vortex mixing (10 s) and centrifugation (8,517× *g*, 10 min, 4 °C). The total supernatant was collected and dried under a nitrogen (N_2) stream at 40 ± 5 °C. The dried supernatant was reconstituted in 250 μL of methanol/water (85/15; *v*/*v*), followed by vortex mixing. The reconstituted sample was transferred into an autosampler vial and an aliquot (5 μL) was injected onto the LC-MS/MS and LC-HRMS instrument.

4.4.2. Pig Urine

To two tubes (tube 1 and 2), each containing 500 μL of urine, were added 20 μL of a 100 ng/mL WS_{mix_IS}, followed by vortex mixing and equilibration at room temperature for 5 min. The pH was determined by means of pH test strips and was adjusted to pH 8 in tube 1 and to pH 2 in tube 2, using a 0.1 M NaOH solution and a 1 M HCl solution, respectively. This pH adjustment was necessary to allow extraction of the different mycotoxins. At pH 8 most mycotoxins were extracted except for TeA, AOH, AME and OTA that extracted best at pH 2.

Thereafter, 3 mL of ethyl acetate were added to each tube, followed by vortex mixing for 10 s and rotating during 15 min on a horizontal roller shaker (Staffordshire, UK). Finally, the tubes were centrifuged for 10 min at 3724× *g* and 4 °C. The organic phase of each tube was transferred to separate polypropylene tube and evaporated to dryness using a gentle N_2-stream at 40 ± 5 °C. The dried extracts of tube 1 and 2 were reconstituted in 250 μL of MeOH/water (85/15; *v*/*v*) and vortex mixed. The redissolved extracts of tube 1 and 2 were transferred to a separate autosampler vial and an aliquot (5 μL) was injected onto the analytical instruments.

4.4.3. Pig Feces

Feces samples were first freeze dried for 48 h to eliminate variation due to different moisture contents. Then, 20 μL of a 100 ng/mL WS_{mix_IS} were added to two tubes (tube 1 and 2), each containing 250 mg of freeze-dried feces. The samples were vortex mixed during 10 s and equilibrated at room temperature for 5 min. The extraction of TeA, OTA, AME and AOH was performed using tube 2, with 5 mL of methanol/ethyl acetate/formic acid of analytical grade (75/24/1; *v*/*v*), whereas the other mycotoxins were extracted using tube 1 with 5 mL of acetone. Both tubes were shaken for 40 min on an in-house made vertical rotator (75 rpm), followed by centrifugation (3724× *g*, 10 min, 4 °C). The supernatant of tube 1 was transferred onto a HybridSPE-phospholipid cartridge. The eluate of tube 1 and the supernatant of tube 2 were evaporated until dryness using a gentle N_2-stream at 40 ± 5 °C. The dried extracts were reconstituted in 250 μL of MeOH/water (85/15; *v*/*v*) and filtered through a Millex®-LG 0.2 μm filter. The redissoved extracts of tube 1 and 2 were transferred to a separate vial and an aliquot (5 μL) was injected onto the LC-MS/MS and the LC-HRMS instrument.

4.4.4. Broiler Chicken Plasma

First, 150 μL of chicken plasma was brought in a well of an Ostro® 96-well plate. Next, 15 μL of a 100 ng/mL WS_{mix_IS} were added, followed by a gentle up and down pipetting for mixing and equilibration for 5 min at room temperature. Next, 450 μL of ACN containing 0.1% formic acid of analytical were added. After gentle mixing, the Ostro® 96-well plate was placed under vacuum (67.7 kPa) to allow the sample to pass through the plate. The eluate of each sample was transferred to a polypropylene tube, dried under a gentle N_2-stream at 40 ± 5 °C and reconstituted in 150 µL of MeOH/water (85/15; *v*/*v*). An aliquot of 5 µL was injected onto the analytical instruments.

4.4.5. Broiler Chicken Excreta

Excreta samples were first freeze dried for 48 h to eliminate variation due to different moisture contents. The extraction of TEA, AME, OTA and AOH was performed in acidic medium (tube 1), whereas the other mycotoxins were extracted without adjustment of the pH (tube 2). Hence, 250 mg of freeze-dried chicken excreta were added to tube 1 and 2, followed by the addition of 20 µL of a 100 ng/mL WS_{mix_IS} and equilibration for 5 min at room temperature. Next 1.5 mL of ACN was added to both tubes, followed by adding 250 µL of a 1 M HCl solution to tube 1. The pH of the excreta in tube 2 was not adjusted. The two tubes were vortex mixed and shaken for 15 min on a vertical rotator, followed by centrifugation (3724× *g*, 10 min, 4 °C). The supernatants were transferred to separate polypropylene tubes and dried under a N_2-stream at 40 ± 5 °C. The dried extracts were reconstituted in 250 µL of MeoH/water (85/15; *v*/*v*) and filtered through a Millex®-LG 0.2 µm filter. Each sample was transferred to a separate vial. An aliquot of 5 µL was injected onto the LC-MS/MS and LC-HRMS instruments.

4.5. Chromatography

The chromatographic systems consisted of an Acquity H-Class ultra-performance liquid chromatograph (UPLC) coupled to a Xevo® TQ-S mass spectrometer and an Acquity I-Class UPLC coupled to a Synapt® G2-Si high definition mass spectrometer (HDMS), all from Waters. Chromatographic separation was achieved on an Acquity HSS T3 column (100 mm × 2.1 mm i.d., dp: 1.8 µm) and a VanGuard pre-column of the same type (5 mm × 2.1 mm i.d., dp: 1.8 µm), both from Waters. The temperatures of the column oven and autosampler tray were set at 45 °C and 8 °C, respectively.

The chromatographic conditions were optimized for the different ionization modes, i.e., positive and negative electro-spray ionization (ESI). The optimal chromatographic conditions for the ESI positive mode were obtained with the mobile phases (MP) containing 10 mM ammonium formate, 0.3% formic acid in water (MP A), and 10 mM ammonium formate, 0.3% formic acid in methanol (MP B), all of LC-MS grade. In ESI negative mode, the most suitable mobile phases consisted of 1% acetic acid in water (MP C) and 1% acetic acid in acetonitrile (MP D), all of LC-MS grade. A gradient elution program was run for each ionization mode separately. For ESI positive: 0–1.5 min, 95% A, 5% B; 1.5–3 min, linear gradient to 40% A; 3–5 min, 40% A, 60% B; 5.0–10 min, linear gradient to 20% A; 10–10.50 min, linear gradient to 1% A; 10.50–13.0 min, 1% A, 99% B; 13–14 min, linear gradient 95% A; 14.0–16.0 min, 95% A, 5% B. For ESI negative: 0–1.5 min, 95% C, 5% D; 1.5–3 min, linear gradient to 60% C; 3.0–4.0 min, 60% C, 40% D; 4.0–7.0 min, linear gradient to 40% C; 7.0–9.0 min, 40% C, 60% D; 9.0–9.5 min, linear gradient 95% C; 9.5–12.0 min, 95% C, 5% D. The flow rate was set at 300 µL/min. The chromatographic parameters were the same for both the LC-MS/MS and LC-HRMS instruments.

4.6. Mass Spectrometry

Instrument parameters were optimized by syringe infusion of working solutions of 10 µg/mL of each compound (flow rate 10 µL/min).

4.6.1. LC-MS/MS

The settings on the Xevo® TQ-S mass spectrometer were as follows: desolvation gas flow rate: 800 L/h; desolvation temperature: 600 °C; cone gas flow rate: 150 L/h; source temperature: 150 °C. The capillary voltage was optimized at 3.2 kV for ESI positive and 3.0 kV for ESI negative mode, respectively. Dwell times of 25 and 10 ms/transition were selected for each component separately. The Xevo® TQ-S mass spectrometer was operated in the selected reaction monitoring (SRM) mode. For every compound, the two most intense product ions were selected for quantification and qualification, respectively. In Table 3; Table 4, an overview is given of the compound specific MS/MS parameters at the selected ionization mode (ESI negative and ESI positive, respectively).

4.6.2. LC-HRMS

Following instrument parameters were selected: desolvation gas flow: 800 L/h; desolvation temperature: 600 °C; cone gas flow: 50 L/h; source temperature: 150 °C. The capillary voltage was 3.2 kV for ESI positive and and 3.0 kV for ESI negative. The HRMS acquisition was performed in resolution mode using the MS^E continuum scan function. The MS^E data acquisition was optimal for use in the Unify 1.8. software (Waters, Milford, MA, USA). The settings were as follows: low mass, 50 dalton (Da); high mass, 1200 Da; scan time, 0.1 s; data format, continuum. The lock mass solution consisted of leucine encephalin (200 pg/μL). The lock spray capillary voltage was 2.8 kV for positive and 2.15 kV for the negative ionisation mode. The additional lock spray settings were as follows: scan time, 0.1 s; interval, 30 s; scans to average, 3; mass window, 0.5 Da. The lock spray was acquired during HRMS acquisition, but not corrected. The lock spray correction (m/z 556.276575; m/z 554.26202) and data processing was performed using Unify 1.8 software (Waters, Milford, MA, USA). Identification of analytes for which analytical standards were available was based on retention time (target T_R tolerance: 0.1 min) and mass (target mass tolerance: 10 ppm). Identification of phase I and II metabolites for which no analytical standards existed was based on the found exact mass (target mass tolerance: 10 ppm).

For every mycotoxin and some of their phase I and II metabolites, the accurate masses were defined in the Unify 1.8 processing method, as shown in Table S1. The search for phase I and II metabolites was performed using a pathway profiling approach. Additional adducts (e.g., Na^+, NH_4^+, CH_3COO^-, $HCOO^-$) and the following transformations were added to the method: glucuronidation, sulfation, oxidation, glutathione conjugations, glucosylations. After detection of a peak based on the accurate mass by Unify 1.8, the given MS/MS spectrum was inspected to confirm the proposed structure based on the present product ions.

4.7. Method Validation

The LC-MS/MS method was validated according to a protocol previously described by De Baere et al. [35], using spiked blank plasma, urine and feces samples obtained from healthy, untreated animals. The validation requirements are in compliance with the recommendations and guidelines defined by the European and international Community [56–58]. Following parameters were evaluated: linearity, within-day and between-day precision and accuracy, limit of quantification (LOQ), carry over, specificity, extraction recovery and matrix effects.

Table 3. Overview of the compound specific MS/MS parameters for mycotoxins, measured in the ESI negative mode.

Name	Measured Form/Adduct	Precursor Ion (*m/z*)	Quantifier Ion (*m/z*)	Qualifier Ion (*m/z*)	Cone Voltage (V)	Collision Energy (eV) (a-b)	Retention Time (min)
ZEN	$[M-H]^-$	317.1	175.0	130.8	15	25-30	8.50
ZAN	$[M-H]^-$	319.1	275.0	205.0	20	20-22	8.41
BZEL	$[M-H]^-$	319.2	275.0	301.0	20	20-22	6.87
BZAL	$[M-H]^-$	321.2	277.2	303.3	30	23-20	6.75
AZEL	$[M-H]^-$	319.2	275.0	301.0	20	20-22	7.51
AZAL	$[M-H]^-$	321.2	277.2	303.3	30	23-20	7.36
TeA	$[M-H]^-$	196.1	112.0	139.0	55	23-23	6.06
AOH	$[M-H]^-$	256.3	213.0	185.2	20	28-28	6.33
AME	$[M-H]^-$	271.1	256.0	228.0	48	24-30	8.38
$[^{13}C_{18}]$-zearalenone	$[M-H]^-$	335.3	185.1	169.1	15	28-40	8.50
$[^{13}C_6{}^{15}N]$-tenuazonic acid	$[M-H]^-$	202.3	113.1	141.9	40	23-20	6.06

Note: m/z = mass-to-charge ratio; (a-b): collision energy for the quantifier (a) and qualifier ion (b), respectively.

Table 4. Overview of the compound specific MS/MS parameters for mycotoxins, measured in the ESI positive mode.

Name	Measured Form/Adduct	Precursor Ion (*m/z*)	Quantifier Ion (*m/z*)	Qualifier Ion (*m/z*)	Cone Voltage (V)	Collision Energy (eV) (a-b)	Retention Time (min)
DON	$[M+H]^+$	297.0	249.1	203.4	20	9-14	4.36
DOM1	$[M+H]^+$	281.0	215.0	233.0	30	12-12	4.70
3/15-ADON	$[M+H]^+$	339.2	213.1	230.9	25	12-8	5.02
T2	$[M+NH_4]^+$	484.0	215.2	304.8	26	18-15	8.24
HT2	$[M+NH_4]^+$	442.0	263.0	215.1	20	10-10	7.01
T2-G	$[M+NH_4]^+$	646.0	245.0	215.3	35	20-22	7.20
AFB1	$[M+H]^+$	313.0	285.1	241.1	35	23-34	5.63
AFM1	$[M+H]^+$	328.9	272.9	229.0	30	20-35	5.18
OTA	$[M+H]^+$	404.0	238.9	220.8	35	20-32	9.23
ENN A1	$[M+H]^+$	668.2	210.1	227.9	80	20-20	12.96
ENN A	$[M+Na]^+$	704.5	350.1	232.2	35	48-48	13.12
ENN B	$[M+H]^+$	640.1	213.8	527.2	80	22-21	12.69
ENN B1	$[M+NH_4]^+$	671.5	196.2	214.3	30	28-28	12.84
BEA	$[M+H]^+$	784.1	244.0	262.1	80	25-30	12.81
FB2	$[M+H]^+$	706.10	318.2	336.2	60	35-42	10.07
$[^{13}C_{15}]$-Deoxynivalenol	$[M+H]^+$	312.0	263.0	245.0	20	10-10	4.37
$[^{13}C_{17}]$-Aflatoxin B1	$[M+H]^+$	330.10	255.1	301.0	20	35-28	5.63
$[^{13}C_{20}]$-Ochratoxin A	$[M+H]^+$	424.0	250.0	377.1	20	25-15	9.23
$[^{13}C_{24}]$-T2-toxin	$[M+NH_4]^+$	508.40	229.1	198.2	25	20-20	8.24
$[^{13}C_{34}]$-Fumonisin B1	$[M+H]^+$	756.50	356.2	374.3	15	40-35	9.67
$[^{15}N_3]$-Enniatin B	$[M+H]^+$	643.30	197.1	215.30	80	18-18	12.69

Note: m/z = mass-to-charge ratio; (a-b): collision energy for the quantifier (a) and qualifier ion (b), respectively.

4.7.1. Linearity

Linearity was assessed by preparing three matrix-matched calibration curves over a concentration range of 1–200 ng/mL or 1–200 ng/g. Ten concentrations were included: 0, 1, 2, 4, 5, 10, 30, 50, 100 and 200 ng/mL or ng/g. The correlation coefficients (r) and the goodness-of-fit coefficients (g) were calculated and acceptance criteria were set, respectively, at $\geq$0.99 and $\leq$20% [59].

4.7.2. Precision and Accuracy

Within-day precision and accuracy were evaluated by analyzing six blank samples spiked at low (LOQ), medium (10 ng/mL or ng/g) and high (100 ng/mL or ng/g) concentration levels. Between-day precision and accuracy were determined by analyzing in threefold three quality control samples spiked at low (LOQ), medium (10 ng/mL or ng/g) and high (100 ng/mL or ng/g) concentration level on three different days. The acceptance criteria for within-day and between-day accuracy were as follows: −50% to +20%, −30% to +10% and −20 to +10% for concentrations of $\leq$1 ng/mL or ng/g, 1–10 ng/mL or ng/g and $\geq$10 ng/mL or ng/g, respectively. For the within-day precision, the relative standard deviation (RSD%) had to be lower than the maximum relative standard deviation (RSDmax), which was <25%, and <15% for concentrations $\geq$1 to <10 ng/mL or ng/g and $\geq$10 to <100 ng/mL or ng/g, respectively [58]. For the between-day precision, the RSD% had to be lower than the RSDmax, which was defined by the Horwitz equation: $RSDmax = 2^{(1-0.5 \log Concentration(g/mL))}$. The RSDmax was 22.6%, 32% and 45% for the respective concentrations of 100 ng/mL, 10 ng/mL and 1 ng/mL [56,57].

4.7.3. Limit of Quantification (LOQ)

The LOQ was the lowest concentration of the analyte for which the method was validated with an acceptable accuracy and precision according to the guidelines described above. The LOQ was also the lowest concentration of the calibration curves. The LOQ was determined by analyzing different concentrations spiked in six-fold on the same day.

4.7.4. Carry Over

Carry over was assessed by analyzing a mixture of MeOH/water (85/15; *v*/*v*) directly after the highest calibrator (200 ng/mL or ng/g). The concentration of mycotoxins in this mixture had to be below a signal-to-noise ratio of 3/1.

4.7.5. Specificity

The specificity of the method was evaluated with respect to interference of endogenous components. Hence, a blank sample was analyzed. The signal of the eventual interference at the elution zone of the analytes of interest should be below the signal-to-noise ratio of 3/1.

4.7.6. Extraction Recovery and Matrix Effects

The extraction recovery and matrix effects of the method were calculated according to the method of Matuszewski et al. [60] Therefore, three types of samples were prepared. The first samples were matrix-matched and prepared by spiking blank samples before extraction (=Spiked). The second samples consisted of matrix-matched blank samples that were spiked after extraction (=SpikedExtract). The third samples were prepared using standard solutions (=Standard). All samples were spiked at 10 ng/mL or 10 ng/g and were made in triplicate. The peak areas of these samples were compared to calculate the recovery of the extraction step (RE) and the signal suppression/enhancement (SSE) due to matrix effects.

$$SSE = 100 \times (\text{Area SpikedExtract}/\text{Area Standard})$$
$$RE = 100 \times (\text{Area Spiked}/\text{Area SpikedExtract})$$

Supplementary Materials: The following are available online at http://www.mdpi.com/2072-6651/11/3/171/s1, Table S1: Overview of the theoretical accurate mass of the 24 mycotoxins and possible phase I and/or phase II metabolites, Figure S1: Extracted ion chromatogram of the first peak (4.09 min) in the trace of ZEN-glucuconide [*m/z* 493.1788] with HRMS spectrum, Table S2: Validation results for linearity (linear range, correlation coefficient (r) and goodness-of-fit coefficient (g)) and limit of quantification (LOQ) of 24 mycotoxins in pig feces, Table S3: Validation results for linearity (linear range, correlation coefficient (r) and goodness-of-fit coefficient (g)) and limit of quantification (LOQ) of 24 mycotoxins in pig urine, Table S4: Validation results for linearity (linear range, correlation coefficient (r) and goodness-of-fit coefficient (g)) and limit of quantification (LOQ) of 24 mycotoxins in broiler chicken plasma, Table S5: Validation results for linearity (linear range, correlation coefficient (r) and goodness-of-fit coefficient (g)) and limit of quantification (LOQ) of 25 mycotoxins in broiler chicken excreta, Table S6: Results of the within-day and between-day precision and accuracy experiments for 24 mycotoxins in pig feces, Table S7: Results of the within-day and between-day precision and accuracy experiments for 24 mycotoxins in pig urine, Table S8: Results of the within-day and between-day precision and accuracy experiments for 24 mycotoxins in broiler chicken plasma, Table S9: Results of the within-day and between-day precision and accuracy experiments for 25 mycotoxins in broiler chicken excreta, Figure S2: Concentration–time curves of zearalenone (ZEN) and deoxynivalenol (DON) in plasma after intra-gastric administration of DON (36 μg/kg bw) and ZEN (3 mg/kg bw) to pigs (n = 8). The mean + SD is shown, Figure S3: ZEN metabolites extracted ion chromatograms for a blank pig feces sample, Table S10. Results of signal suppression enhancement (SSE) end extraction recovery (RE) in pig plasma, urine and feces spiked at 10 ng/mL or 10 ng/g, Table S11: Results of signal suppression enhancement (SSE) end extraction recovery (RE) in broiler chicken plasma and excreta spiked at 10 ng/mL or 10 ng/g, Table S12: Chemical structure and pKa-values of the 26 mycotoxins.

Author Contributions: Conceptualization, S.C. and M.D.; Methodology, M.L. and S.D.B.; Validation, M.L.; Investigation, M.L.; Resources, M.R. Writing—Original Draft Preparation, M.L.; Writing—Review and Editing, S.C., M.D., S.D.B., B.L., and M.R.; and Supervision, S.C., M.D. and S.D.B.

Funding: The Synapt High Definition Mass Spectrometer (Synapt G2-Si HDMS, Waters) was funded by the Hercules project (No. AUGE 13/13).

Acknowledgments: The authors are grateful towards Laura Dhondt, Elke Gasthuys, Wim Schelstraete, Amelie Catteuw, Femke Vandael, Gunther Antonissen, Sophie Fraeyman, Joske Millecam, Joren De Smet, Nathan Broekaert, and Sandra Debevere for the experienced assistance during the animal trial.

Conflicts of Interest: The authors declare no conflict of interest.

References

1. Kovalsky, P.; Kos, G.; Nährer, K.; Schwab, C.; Jenkins, T.; Schatzmayr, G.; Sulyok, M.; Krska, R. Co-occurrence of regulated, masked and emerging mycotoxins and secondary metabolites in finished feed and maize–An extensive survey. *Toxins* **2016**, *8*, 363. [CrossRef] [PubMed]
2. European Parliament and the Council of the EU Directive of The European Parliament and of the Council of 7 May 2002 on undesirable substances in animal feed 2002/32. *Off. J. Eur. Communities* **2002**, *L110*, 1–15.
3. European Commission Commission Recommendation 2006/576/EC of 17 August 2006 on the presence of deoxynivalenol, zearalenone, ochratoxin A, T-2 and HT-2 and fumonisins in products intended for animal feeding. *Off. J. Eur. Union* **2006**, *L229*, 7–9.
4. Fraeyman, S.; Croubels, S.; Devreese, M.; Antonissen, G. Emerging fusarium and alternaria mycotoxins: Occurrence, toxicity and toxicokinetics. *Toxins* **2017**, *9*, 228. [CrossRef] [PubMed]
5. Heyndrickx, E.; Sioen, I.; Huybrechts, B.; Callebaut, A.; De Henauw, S.; De Saeger, S. Human biomonitoring of multiple mycotoxins in the Belgian population: Results of the BIOMYCO study. *Environ. Int.* **2015**, *84*, 82–89. [CrossRef] [PubMed]
6. Boonen, J.; Malysheva, S.V.; Taevernier, L.; Diana Di Mavungu, J.; De Saeger, S.; De Spiegeleer, B. Human skin penetration of selected model mycotoxins. *Toxicology* **2012**, *301*, 21–32. [CrossRef]
7. Niculita-Hirzel, H.; Hantier, G.; Storti, F.; Plateel, G.; Roger, T. Frequent occupational exposure to Fusarium mycotoxins of workers in the swiss grain industry. *Toxins* **2016**, *8*, 370. [CrossRef]
8. Broekaert, N.; Devreese, M.; De Mil, T.; Fraeyman, S.; Antonissen, G.; De Baere, S.; De Backer, P.; Vermeulen, A.; Croubels, S. Oral Bioavailability, Hydrolysis, and Comparative Toxicokinetics of 3-Acetyldeoxynivalenol and 15-Acetyldeoxynivalenol in Broiler Chickens and Pigs. *J. Agric. Food Chem.* **2015**, *63*, 8734–8742. [CrossRef] [PubMed]

9. Broekaert, N.; Devreese, M.; van Bergen, T.; Schauvliege, S.; De Boevre, M.; De Saeger, S.; Vanhaecke, L.; Berthiller, F.; Michlmayr, H.; Malachová, A.; et al. In vivo contribution of deoxynivalenol-3-β-d-glucoside to deoxynivalenol exposure in broiler chickens and pigs: Oral bioavailability, hydrolysis and toxicokinetics. *Arch. Toxicol.* **2017**, *91*, 699–712. [CrossRef] [PubMed]
10. Freire, L.; Sant'Ana, A.S. Modified mycotoxins: An updated review on their formation, detection, occurrence, and toxic effects. *Food Chem. Toxicol.* **2018**, *111*, 189–205. [CrossRef] [PubMed]
11. Sherif, S.O.; Salama, E.E.; Abdel-Wahhab, M.A. Mycotoxins and child health: The need for health risk assessment. *Int. J. Hyg. Environ. Health* **2009**, *212*, 347–368. [CrossRef] [PubMed]
12. Crews, H.; Alink, G.; Andersen, R.; Braesco, V.; Holst, B.; Maiani, G.; Ovesen, L.; Scotter, M.; Solfrizzo, M.; van den Berg, R.; et al. A critical assessment of some biomarker approaches linked with dietary intake. *Br. J. Nutr.* **2001**, *86* (Suppl. 1), S5–S35. [CrossRef] [PubMed]
13. Baldwin, T.; Riley, R.; Zitomer, N.; Voss, K.; Coulombe, R., Jr.; Pestka, J.; Williams, D.; Glenn, A. The current state of mycotoxin biomarker development in humans and animals and the potential for application to plant systems. *World Mycotoxin J.* **2011**, *4*, 257–270. [CrossRef]
14. Efsa Statement on the establishment of guidelines for the assessment of additives from the functional group 'substances for reduction of the contamination of feed by mycotoxins' 1 EFSA Panel on Additives and Products or Substances used in Animal Feed (FEEDA. *EFSA J.* **2010**, *8*, 1–8. [CrossRef]
15. Broekaert, N.; Devreese, M.; De Mil, T.; Fraeyman, S.; De Baere, S.; De Saeger, S.; De Backer, P.; Croubels, S. Development and validation of an LC-MS/MS method for the toxicokinetic study of deoxynivalenol and its acetylated derivatives in chicken and pig plasma. *J. Chromatogr. B Anal. Technol. Biomed. Life Sci.* **2014**, *971*, 43–51. [CrossRef]
16. Nagl, V.; Woechtl, B.; Schwartz-Zimmermann, H.E.; Hennig-Pauka, I.; Moll, W.D.; Adam, G.; Berthiller, F. Metabolism of the masked mycotoxin deoxynivalenol-3-glucoside in pigs. *Toxicol. Lett.* **2014**, *229*, 190–197. [CrossRef]
17. Schwartz-Zimmermann, H.E.; Fruhmann, P.; Dänicke, S.; Wiesenberger, G.; Caha, S.; Weber, J.; Berthiller, F. Metabolism of deoxynivalenol and deepoxy-deoxynivalenol in broiler chickens, pullets, roosters and turkeys. *Toxins* **2015**, *7*, 4706–4729. [CrossRef]
18. Dänicke, S.; Valenta, H.; Döll, S. On the toxicokinetics and the metabolism of deoxynivalenol (don) in the pig. *Arch. Anim. Nutr.* **2007**, *58*, 169–180. [CrossRef]
19. Binder, S.B.; Schwartz-Zimmermann, H.E.; Varga, E.; Bichl, G.; Michlmayr, H.; Adam, G.; Berthiller, F. Metabolism of zearalenone and its major modified forms in pigs. *Toxins* **2017**, *9*, 56. [CrossRef]
20. De Baere, S.; Osselaere, A.; Devreese, M.; Vanhaecke, L.; De Backer, P.; Croubels, S. Development of a liquid-chromatography tandem mass spectrometry and ultra-high-performance liquid chromatography high-resolution mass spectrometry method for the quantitative determination of zearalenone and its major metabolites in chicken and pig plasma. *Anal. Chim. Acta* **2012**, *756*, 37–48. [CrossRef]
21. Devreese, M.; De Baere, S.; De Backer, P.; Croubels, S. Quantitative determination of the Fusarium mycotoxins beauvericin, enniatin A, A1, B and B1 in pig plasma using high performance liquid chromatography-tandem mass spectrometry. *Talanta* **2013**, *106*, 212–219. [CrossRef] [PubMed]
22. Fraeyman, S.; Devreese, M.; Antonissen, G.; De Baere, S.; Rychlik, M.; Croubels, S. Comparative Oral Bioavailability, Toxicokinetics, and Biotransformation of Enniatin B1 and Enniatin B in Broiler Chickens. *J. Agric. Food Chem.* **2016**, *64*, 7259–7264. [CrossRef]
23. Broekaert, N.; Devreese, M.; De Boevre, M.; De Saeger, S.; Croubels, S. T-2 Toxin-3α-glucoside in Broiler Chickens: Toxicokinetics, Absolute Oral Bioavailability, and in Vivo Hydrolysis. *J. Agric. Food Chem.* **2017**, *65*, 4797–4803. [CrossRef] [PubMed]
24. Yoshizawa, T.; Swanson, S.P.; Mirocha, C.J. T-2 metabolites in the excreta of broiler chickens administered 3H-labeled T-2 toxin. *Appl. Environ. Microbiol.* **1980**, *39*, 1172–1177. [PubMed]
25. Sun, Y.-X.; Yao, X.; Shi, S.N.; Zhang, G.J.; Xu, L.X.; Liu, Y.J.; Fang, B.H. Toxicokinetics of T-2 toxin and its major metabolites in broiler chickens after intravenous and oral administration. *J. Vet. Pharmacol. Ther.* **2015**, *38*, 80–85. [CrossRef] [PubMed]
26. Di Gregorio, M.C.; Jager, A.V.; Costa, A.A.; Bordin, K.; Rottinhghaus, G.E.; Petta, T.; Souto, P.C.M.C.; Budiño, F.E.L.; Oliveira, C.A.F. Determination of Aflatoxin B_1 -Lysine in Pig Serum and Plasma by Liquid Chromatography–Tandem Mass Spectrometry. *J. Anal. Toxicol.* **2016**, 236–241. [CrossRef] [PubMed]

27. Fraeyman, S.; Devreese, M.; Broekaert, N.; De Mil, T.; Antonissen, G.; De Baere, S.; De Backer, P.; Rychlik, M.; Croubels, S. Quantitative Determination of Tenuazonic Acid in Pig and Broiler Chicken Plasma by LC-MS/MS and Its Comparative Toxicokinetics. *J. Agric. Food Chem.* **2015**, *63*, 8560–8567. [CrossRef]
28. Song, S.; Ediage, E.N.; Wu, A.; De Saeger, S. Development and application of salting-out assisted liquid/liquid extraction for multi-mycotoxin biomarkers analysis in pig urine with high performance liquid chromatography/tandem mass spectrometry. *J. Chromatogr. A* **2013**, *1292*, 111–120. [CrossRef] [PubMed]
29. Thanner, S.; Czeglédi, L.; Schwartz-Zimmermann, H.E.; Berthiller, F.; Gutzwiller, A. Urinary deoxynivalenol (DON) and zearalenone (ZEA) as biomarkers of DON and ZEA exposure of pigs. *Mycotoxin Res.* **2016**, *32*, 69–75. [CrossRef]
30. Razzazi-Fazeli, E.; Böhm, J.; Jarukamjorn, K.; Zentek, J. Simultaneous determination of major B-trichothecenes and the de-epoxy-metabolite of deoxynivalenol in pig urine and maize using high-performance liquid chromatography-mass spectrometry. *J. Chromatogr. B Anal. Technol. Biomed. Life Sci.* **2003**, *796*, 21–33. [CrossRef]
31. Thieu, N.Q.; Pettersson, H. Zearalenone, deoxynivalenol and aflatoxin B1 and their metabolites in pig urine as biomarkers for mycotoxin exposure. *Mycotoxin Res.* **2009**, *25*, 59–66. [CrossRef] [PubMed]
32. Solfrizzo, M.; Gambacorta, L.; Lattanzio, V.M.T.; Powers, S.; Visconti, A. Simultaneous LC-MS/MS determination of aflatoxin M 1, ochratoxin A, deoxynivalenol, de-epoxydeoxynivalenol, α and β-zearalenols and fumonisin B 1 in urine as a multi-biomarker method to assess exposure to mycotoxins. *Anal. Bioanal. Chem.* **2011**, *401*, 2831–2841. [CrossRef] [PubMed]
33. Gambacorta, S.; Solfrizzo, H.; Visconti, A.; Powers, S.; Cossalter, A.M.; Pinton, P.; Oswald, I.P. Validation study on urinary biomarkers of exposure for aflatoxin B_1, ochratoxin A, fumonisin B_1, deoxynivalenol and zearalenone in piglets. *World Mycotoxin J.* **2013**, *6*, 299–308. [CrossRef]
34. Devreese, M.; De Baere, S.; De Backer, P.; Croubels, S. Quantitative determination of several toxicological important mycotoxins in pig plasma using multi-mycotoxin and analyte-specific high performance liquid chromatography-tandem mass spectrometric methods. *J. Chromatogr. A* **2012**, *1257*, 74–80. [CrossRef] [PubMed]
35. De Baere, S.; Goossens, J.; Osselaere, A.; Devreese, M.; Vandenbroucke, V.; De Backer, P.; Croubels, S. Quantitative determination of T-2 toxin, HT-2 toxin, deoxynivalenol and deepoxy-deoxynivalenol in animal body fluids using LC-MS/MS detection. *J. Chromatogr. B Anal. Technol. Biomed. Life Sci.* **2011**, *879*, 2403–2415. [CrossRef]
36. Tolosa, J.; Font, G.; Mañes, J.; Ferrer, E. Multimycotoxin analysis in water and fish plasma by liquid chromatography-tandem mass spectrometry. *Chemosphere* **2016**, *145*, 402–408. [CrossRef] [PubMed]
37. Brezina, U.; Valenta, H.; Rempe, I.; Kersten, S.; Humpf, H.U.; Dänicke, S. Development of a liquid chromatography tandem mass spectrometry method for the simultaneous determination of zearalenone, deoxynivalenol and their metabolites in pig serum. *Mycotoxin Res.* **2014**, *30*, 171–186. [CrossRef]
38. Warth, B.; Petchkongkaew, A.; Sulyok, M.; Krska, R. Utilising an LC-MS/MS-based multi-biomarker approach to assess mycotoxin exposure in the Bangkok metropolitan area and surrounding provinces. *Food Addit. Contam. Part A Chem. Anal. Control. Expo. Risk Assess.* **2014**, *31*, 2040–2046. [CrossRef]
39. Slobodchikova, I.; Vuckovic, D. Liquid chromatography—High resolution mass spectrometry method for monitoring of 17 mycotoxins in human plasma for exposure studies. *J. Chromatogr. A* **2018**, *1548*, 51–63. [CrossRef]
40. Cao, X.; Li, X.; Li, J.; Niu, Y.; Shi, L.; Fang, Z.; Zhang, T.; Ding, H. Quantitative determination of carcinogenic mycotoxins in human and animal biological matrices and animal-derived foods using multi-mycotoxin and analyte-specific high performance liquid chromatography-tandem mass spectrometric methods. *J. Chromatogr. B Anal. Technol. Biomed. Life Sci.* **2018**, *1073*, 191–200. [CrossRef]
41. Warth, B.; Sulyok, M.; Fruhmann, P.; Mikula, H.; Berthiller, F.; Schuhmacher, R.; Hametner, C.; Abia, W.A.; Adam, G.; Fröhlich, J.; et al. Development and validation of a rapid multi-biomarker liquid chromatography/tandem mass spectrometry method to assess human exposure to mycotoxins. *Rapid Commun. Mass Spectrom.* **2012**, *26*, 1533–1540. [CrossRef]
42. Ivanova, L.; Uhlig, S.; Devreese, M.; Croubels, S.; Fæste, C.K. Biotransformation of the mycotoxin enniatin B1 in pigs: A comparative in vitro and in vivo approach. *Food Chem. Toxicol.* **2017**, *105*, 506–517. [CrossRef] [PubMed]

43. Ferlazzo, A.M.; Bruschetta, G.; Di Pietro, P.; Medica, P.; Notti, A.; Rotondo, E. Phospholipid composition of plasma and erythrocyte membranes in animal species by 31P NMR. *Vet. Res. Commun.* **2011**, *35*, 521–530. [CrossRef] [PubMed]
44. De Baere, S.; Croubels, S.; Novak, B.; Bichl, G.; Antonissen, G. Development and validation of a UPLC-MS/MS and UPLC-HR-MS method for the determination of fumonisin B1 and its hydrolysed metabolites and fumonisin b2 in broiler chicken plasma. *Toxins* **2018**, *10*, 62. [CrossRef] [PubMed]
45. Wishart, D.; Arndt, D.; Pon, A.; Sajed, T.; Guo, A.C.; Djoumbou, Y.; Knox, C.; Wilson, M.; Liang, Y.; Grant, J.; et al. T3DB: The toxic exposome database. *Nucleic Acids Res.* **2015**, *43*, D928–D934. [CrossRef] [PubMed]
46. el Khoury, A.E.; Atoui, A. Ochratoxin a: General overview and actual molecular status. *Toxins* **2010**, *2*, 461–493. [CrossRef] [PubMed]
47. Walravens, J.; Mikula, H.; Rychlik, M.; Asam, S.; Ediage, E.N.; Di Mavungu, J.D.; Van Landschoot, A.; Vanhaecke, L.; De Saeger, S. Development and validation of an ultra-high-performance liquid chromatography tandem mass spectrometric method for the simultaneous determination of free and conjugated Alternaria toxins in cereal-based foodstuffs. *J. Chromatogr. A* **2014**, *1372*, 91–101. [CrossRef]
48. Pfeiffer, E.; Hildebrand, A.; Mikula, H.; Metzler, M. Glucuronidation of zearalenone, zeranol and four metabolites in vitro: Formation of glucuronides by various microsomes and human UDP-glucuronosyltransferase isoforms. *Mol. Nutr. Food Res.* **2010**, *54*, 1468–1476. [CrossRef] [PubMed]
49. Ringot, D.; Chango, A.; Schneider, Y.J.; Larondelle, Y. Toxicokinetics and toxicodynamics of ochratoxin A, an update. *Chem. Biol. Interact.* **2006**, *159*, 18–46. [CrossRef]
50. Devreese, M.; Croubels, S.; De Baere, S.; Gehring, R.; Antonissen, G. Comparative Toxicokinetics and Plasma Protein Binding of Ochratoxin A in Four Avian Species. *Agric Food Chem.* **2018**, *66*, 2129–2135. [CrossRef] [PubMed]
51. McCormick, S.P.; Price, N.P.J.; Kurtzman, C.P. Glucosylation and other biotransformations of T-2 toxin by yeasts of the trichomonascus clade. *Appl. Environ. Microbiol.* **2012**, *78*, 8694–8702. [CrossRef] [PubMed]
52. McCormick, S.P.; Kato, T.; Maragos, C.M.; Busman, M.; Lattanzio, V.M.T.; Galaverna, G.; Dall-Asta, C.; Crich, D.; Price, N.P.J.; Kurtzman, C.P. Anomericity of T-2 toxin-glucoside: Masked mycotoxin in cereal crops. *J. Agric. Food Chem.* **2015**, *63*, 731–738. [CrossRef] [PubMed]
53. Asam, S.; Liu, Y.; Konitzer, K.; Rychlik, M. Development of a stable isotope dilution assay for tenuazonic acid. *J. Agric. Food Chem.* **2011**, *59*, 2980–2987. [CrossRef] [PubMed]
54. Hu, L.; Rychlik, M. Biosynthesis of 15N3-labeled enniatins and beauvericin and their application to stable isotope dilution assays. *J. Agric. Food Chem.* **2012**, *60*, 7129–7136. [CrossRef] [PubMed]
55. Gasthuys, E.; Schauvliege, S.; van Bergen, T.; Millecam, J.; Cerasoli, I.; Martens, A.; Gasthuys, F.; Vandecasteele, T.; Cornillie, P.; Van den Broeck, W.; et al. Repetitive urine and blood sampling in neonatal and weaned piglets for pharmacokinetic and pharmacodynamic modelling in drug discovery: A pilot study. *Lab. Anim.* **2017**, *51*, 498–508. [CrossRef]
56. Heitzman, R. *Veterinary Drug Residues Residues in Food Producing Animals and Their Products: Reference Materials and Methods*; Blackwell Scientific Publications: Oxford, UK, 1996; Volume 636, ISBN 0-8412-3419-1.
57. European Commission Commission Decision 2002/657/Implementing Council Directive 96/23/EC concerning the performance of analytical methods and the interpretation of results. *Off. J. Eur. Communities* **2002**, *L221*, 29. [CrossRef]
58. *EMA/CMVP/VICH Steering Committee VICH Topic GL49: Studies to Evaluate the Metabolism and Residues Kinetics of Veterinary Drugs in Human Food-Producing Animals: Validation of Analytical Methods Used in Residue Depletion Studies*; European Medicine Agency: Brussels, Belgium, 2016; Volume 44.
59. Knecht, J.; Stork, G. Percentage and logarithmic procedures for calculation of calibration curves. *Fresenius Z. Anal. Chem.* **1974**, *270*, 97–99. [CrossRef]
60. Matuszewski, B.K.; Constanzer, M.L.; Chavez-Eng, C.M. Strategies for the assessment of matrix effect in quantitative bioanalytical methods based on HPLC-MS/MS. *Anal. Chem.* **2003**, *75*, 3019–3030. [CrossRef]

toxins

MDPI

Article

Dispersive Solid-Phase Extraction Using Magnetic Carbon Nanotube Composite for the Determination of Emergent Mycotoxins in Urine Samples

Natalia Arroyo-Manzanares, Rosa Peñalver-Soler, Natalia Campillo and Pilar Viñas *

Department of Analytical Chemistry, Faculty of Chemistry, Regional Campus of International Excellence "Campus Mare Nostrum", University of Murcia, E-30100 Murcia, Spain; natalia.arroyo@um.es (N.A.-M.); rosamaria.penalver@um.es (R.P.-S.); ncampi@um.es (N.C.)

* Correspondence: pilarvi@um.es; Tel.: +34-8-6888-7415

Received: 9 December 2019; Accepted: 13 January 2020; Published: 15 January 2020

check for updates

Abstract: Dispersive magnetic solid-phase extraction (DMSPE) has received growing attention for sample treatment preconcentration prior to the separation of analytes due to its many advantages. In the present work, the potential of DMSPE for the determination of emergent mycotoxins (enniatins A, A1, B and B1, and beauvericin) is investigated for the first time. Different magnetic nanoparticles were tested and a magnetic multiwalled carbon nanotube (Fe_3O_4@MWCNT) composite was selected for the extraction and preconcentration of the five target mycotoxins in human urine samples before their analysis by ultrahigh performance liquid chromatography coupled to high resolution mass spectrometry (UHPLC-HRMS). The nanocomposite was characterized by energy dispersive X-ray spectrometry, scanning electron microscopy, Fourier transform infrared spectrophotometry, and X-ray diffraction. Several parameters affecting the adsorption and desorption of DMSPE steps were optimized and the method was fully validated. Due to a matrix effect, matrix-matched calibration curves were necessary to carry out quantification. In this way, limits of quantification of between 0.04 and 0.1 μg/L, relative standard deviation values lower than 12% and recoveries between 89.3% and 98.9% were obtained. Finally, a study of the reuse of the Fe_3O_4@MWCNT composite was carried out, confirming that it can be reused at least four times.

Keywords: emergent mycotoxins; urine; dispersive solid-phase extraction; magnetic carbon nanotube composite

Key Contribution: The first application of dispersive magnetic solid-phase extraction using a magnetic multiwalled carbon nanotube nanocomposite is described for the determination of the emergent mycotoxins enniatins and beauvericin in urine.

1. Introduction

Mycotoxins are low molecular weight noxious secondary metabolites produced by toxicogenic strains of some mould species that mainly belong to the genera *Aspergillus*, *Fusarium*, *Penicillium*, *Alternaria*, and *Claviceps*. To date, approximately 400 different mycotoxins have been described, although the most common are aflatoxins (B_1, B_2, G_1, and G_2), ochratoxin A, fumonisins, trichothecenes, zearalenone, and patulin. Mycotoxins can contaminate animal feed, food, or the raw materials used for their production, originating diseases and disorders, both in humans and animals. Mycotoxins can bioaccumulate in fluids, tissues and organs, especially the liver and kidney, as well as affect the nervous, endocrine, and immune systems. Due to their toxicity, the European Union has established or recommended maximum permissible contents for some of these contaminants in various foods [1–3].

Among mycotoxins, enniatins and beauvericins have aroused great interest in recent years. These compounds, also known as emerging mycotoxins, due to their recent discovery, are produced by *Fusarium* species and, although no specific legislation has dealt with them yet, their toxicity-including genotoxicity, cytotoxicity, and effects on the reproductive system, has been demonstrated in in vitro studies [4]. The European Food Safety Authority (EFSA) suggests that chronic exposure to these mycotoxins might be a concern and should be confirmed with more acute toxicological data [5].

Enniatins and beauvericins are cyclic hexadepsipeptide compounds. Enniatins are made of six residues that vary between N-methyl amino acids and hydroxylated carboxylic acids, whereas beauvericins have a core structure made up of three N-methyl-L-phenylalanine blocks connected alternately with three 2-hydroxy-D-isovaleric acid units [6]. About 44 enniatins and beauvericins have been isolated and determined, the most commonly found in food and animal feed being enniatins A (ENNA), A1 (ENNA1), B (ENNB), B1 (ENNB1), and beauvericin (BEA) (Figure 1), all of which have been described on a worldwide scale [5,7–12]. Some studies have demonstrated the presence of these toxins in 100% of the samples analyzed [11,12]. However, it is difficult to measure real exposure to mycotoxins based on their occurrence described in animal feed and food—an effective alternative might be monitoring them in biological liquids such as urine or blood (serum and plasma). The mycotoxins present in biological fluids could be considered as biomarkers for the consumption of contaminated foodstuff. Among biological fluids, urine is the most often used for measuring toxin exposure [13] since big amounts can be collected simply and non-invasively.

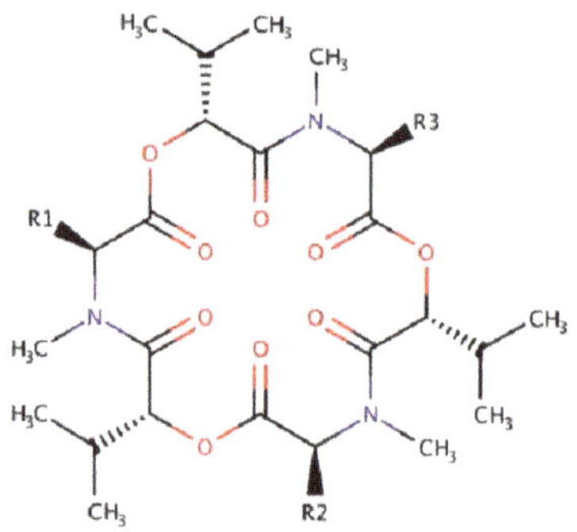

	R1	R2	R3
ENNA (M=681.9 Da)	sec-butyl	sec-butyl	sec-butyl
ENNA1 (M=667.9 Da)	sec-butyl	sec-butyl	iso-propyl
ENNB (M=639.8 Da)	iso-propyl	iso-propyl	iso-propyl
ENNB1 (M=653.9 Da)	iso-propyl	iso-propyl	sec-butyl
BEA (M=784.0 Da)	phenylmethyl	phenylmethyl	phenylmethyl

Figure 1. Chemical structures of enniatins and beauvericin.

The determination of mycotoxins in biological samples requires sensitive, effective, and accurate methods, as these compounds appear at very low concentration levels. According to EFSA recommendations, the application of liquid chromatography with tandem mass spectrometry (LC-MS/MS) is recommended rather than LC with UV detection, since matrix effects can be better handled, and concentrations below 1 µg/kg can be quantified [5].

The determination of ENNA, ENNA1, ENNB, ENNB1, and BEA has been carried out in urine [14–19], and serum [14–16,18,20–24], from both humans and animals, but also in feces samples [15,16,18] and in different organs and tissues from animals [25–28]. The vast majority of these analytical methods are based on LC-MS/MS, although LC coupled to high resolution mass spectrometry (HRMS) has also been proposed [17–19].

Regarding sample treatment, solid-phase extraction (SPE) [14,26,27] and liquid–liquid extraction (LLE) using ethyl acetate or acetonitrile [15,16,18,21–23,25–28] are the most commonly used. Miniaturized techniques like dispersive liquid–liquid microextraction (DLLME) have also been proposed [24] and, recently, Rodríguez-Carrasco et al. [19] compared and evaluated three different sample preparation approaches (dilute and shoot, DLLME, and salting-out liquid–liquid extraction (SALLE)) for the determination of ENNB and its phase I metabolites in human urine samples, concluding that SALLE showed satisfactory validation results. Moreover, Escrivá et al. [17] also compared three extraction methods, SALLE, DLLME, and miniQuEChERS (quick, easy, cheap, effective, rugged, and safe) for the determination of ENNA, ENNA1, ENNB, ENNB1, and BEA together with another six mycotoxins in human urine and, in this case, DLLME was selected as the most suitable methodology.

In recent years, the use of magnetic nanoparticles (MNPs) has increased for sample treatment. More specifically, MNPs can be used as magnetic sorbents in SPE (MSPE) for the separation and preconcentration of different analytes [29]. The magnetic sorbent can also be dispersed in the sample directly (dispersive magnetic solid-phase extraction, DMSPE), instead of using a solid-phase packed cartridge, enhancing the mass transfer and improving the extraction efficiency [30], while reducing the volume of organic solvent needed in the desorption step. In DMSPE, MNPs can be separated simply and rapidly by applying an external magnetic field, avoiding filtration or centrifugation steps. However, despite its multiple advantages, to the best of our knowledge, DMSPE has not been applied for the determination of emerging mycotoxins.

In this study, the potential of DMSPE has been investigated for the determination of ENNA, ENNA1, ENNB, ENNB1, and BEA in human urine samples, and a magnetic multiwalled carbon nanotube (Fe_3O_4@MWCNT) composite is applied for the first time as sorbent material for the preconcentration of these mycotoxins before determination by UHPLC-HRMS.

2. Results and Discussion

2.1. Optimization of Sample Treatment

In order to get the best extraction conditions, different materials were evaluated for the preparation of the magnetic nanoparticles: β-cyclodextrin (β-CD) [31], polydopamine (PDA) [32], chitosan [33], oleic acid [34], polystyrene (PS) [35], Fe_3O_4@MWCNTs [36], MWCNTS/Fe_3O_4/polypyrrole (PPy) [36], PPy-nanotubes (NTs) [37] and magnetic cellulose particles [38], and desorption solvents, methanol (MeOH), and acetonitrile (MeCN). For this study, a mass of 20 mg of each nanoparticle type was added to 5 mL of sample spiked at 100 ng/mL before submitting the mixture to orbital shaking for 20 min. Then, 3 mL of extraction solvent (MeOH or MeCN) was used to desorb the mycotoxins from the nanomaterial. A total of 18 experiments were carried out and the best results were obtained with the combination of Fe_3O_4@MWCNTs and MeCN. The results are shown in Supplementary Materials (Figure S1).

Once the type of nanoparticle and the composition of the desorption solvent had been selected, the other factors influencing the extraction efficiency in DMSPE were optimized: sample volume, nanoparticle mass, absorption and desorption times, volume of desorption solvent, and percentage of NaCl in the sample. Initially, sample volume was investigated between 5 and 40 mL. The best preconcentration factor was obtained using 30 mL, which was therefore selected as the optimum volume. Then, the mass of nanoparticles, and absorption and desorption times were optimized by means of a multivariate method, to consider potential interactions between the variables. A central composite design (2^3 + star, face centered), with three spaced central points, including 17 runs, was used to create the response surface, using the peak area as analytical response. The different factors were evaluated in the following ranges: mass of nanoparticles (5–35 mg), absorption time (5–40 min), and desorption time (1–10 min).

The nanoparticle mass and its quadratic term had a significant effect on the recovery of the target mycotoxins. Subsequently, multiple response optimization was carried out using the desirability function; the response surface plot is shown in Figure 2. The determination coefficient (R^2) obtained for the target analytes ranged between 83.7% and 92.1%, confirming the suitability of the design. The optimum conditions were as follows: nanoparticles mass, 30 mg; absorption time, 35 min; and desorption time, 5 min.

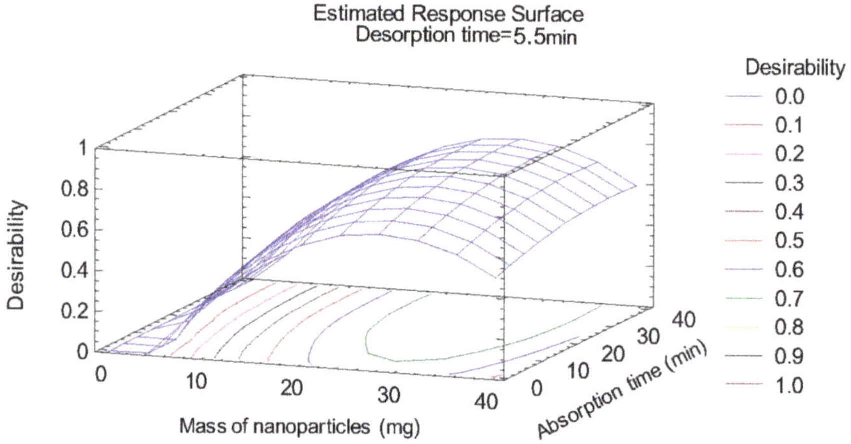

Figure 2. Estimated response surface for the study of significant variables in the dispersive magnetic solid-phase extraction (DMSPE) step obtained by multiple response optimization.

After the multivariate study, the desorption solvent volume between (0.5 and 5 mL) and percentage of NaCl in the sample between (0 and 10% wt.) were optimized using a univariate method. No significant differences were found for the different salt percentages, and so NaCl was not added to the sample for the DMSPE procedure. In the case of desorption solvent, the best preconcentration and recovery results were obtained using 1.5 mL of MeCN.

Under these optimum conditions, the obtained preconcentration factors were 95, 35, 44, 19, and 34 for ENNB, ENNB1, ENNA1, ENNA, and BEA, respectively. The preconcentration factors were calculated as the ratio of the slope of the calibration curves applying the DMSPE method and the slope of the calibration curves obtained by direct injection of the standards in urine into the UHPLC-HRMS system.

2.2. Characterization of Nanocomposite

The characterization of the multiwalled carbon nanotube (MWCNT) and magnetic multiwalled carbon nanotube (Fe_3O_4@MWCNT) nanocomposite was carried out by scanning electron microscopy (SEM), energy dispersive X-ray spectrometry (EDS), X-ray diffraction (XRD), and Fourier transform infrared spectrophotometry (FT-IR).

The pictures obtained by SEM show the morphology of the MWCNT (Figure 3a) and the synthetized Fe_3O_4@MWCNT (Figure 3b) material, where it can be appreciated how the iron oxide nanoparticles were fixed on the surface of the MWCNTs to produce Fe_3O_4@MWCNTs.

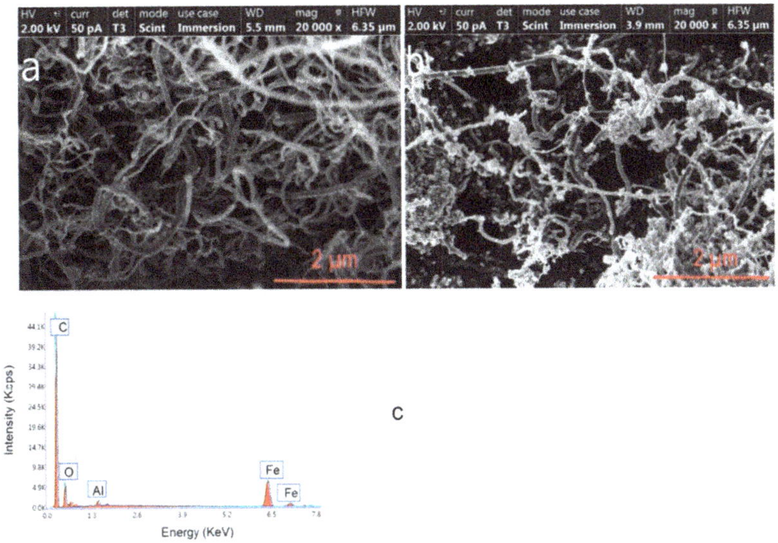

Figure 3. Scanning electron microscopy (SEM) images of multiwalled carbon nanotube (MWCNT) (**a**) and Fe_3O_4@MWCNT (**b**) nanocomposite. Energy dispersive X-ray spectrometry (EDS) spectrum of Fe_3O_4@MWCNT (**c**).

Peaks related to Fe, C, and O atoms were observed in the EDS spectrum of the Fe_3O_4@MWCNT adsorbent material (Figure 3c). The quantitative analysis gave weight ratios of 14.95%, 68.06%, and 15.49% for Fe, C, and O, respectively. Aluminum was used as support for the measurements.

Analysis of the XRD data using XPOWDER software [39] showed the presence of two types of magnetic iron oxide; magnetite (Fe_3O_4, around 10% wt.) and maghemite (gamma-Fe_3O_4, around 61% wt.). The standard X-ray diffraction peaks (2θ = 30.014°, 35.578°, 43.180°, 53.652°, 56.788°, and 62.757°) which can be assigned to maghemite or magnetite match with those observed in the spectrum depicted in Figure 4. The other signals are related to the oxidized and non-magnetized carbon nanotubes (Carbon and Graphite).

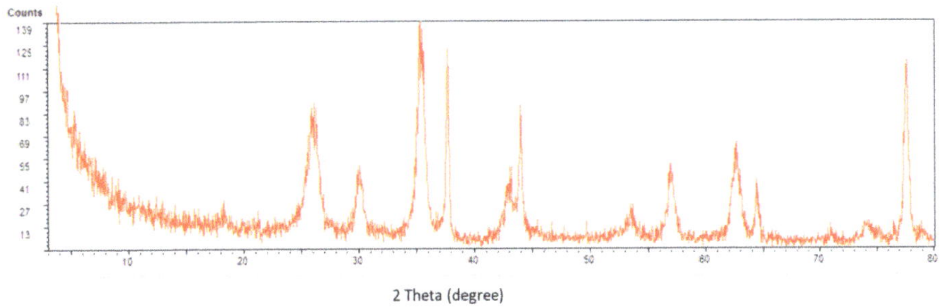

Figure 4. X-ray diffraction (XRD) spectrum of Fe_3O_4@MWCNT nanocomposite.

The Fe_3O_4@MWCNT material was also analyzed by FTIR to determine the functional groups present on the surface of the nanocomposite particles. Figure 5 shows several bands related to the carboxylic acid groups formed during the oxidation step of the nanoparticle synthesis process. Note the COO^- asymmetric stretching band at 1633 cm^{-1}, the 3435 cm^{-1} signal due to the stretching band for O-H, and the 578 cm^{-1} signal for characteristic Fe-O band.

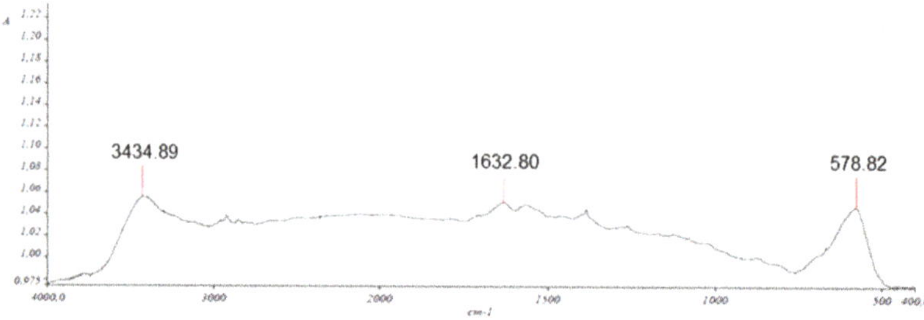

Figure 5. Fourier transform infrared spectrophotometry (FTIR) spectrum of MWCNT nanocomposite.

2.3. Method Validation

In order to evaluate the suitability of the proposed method for the determination of emergent mycotoxins in urine, an in-depth validation was carried out by evaluating the linear dynamic ranges, limits of detection (LOD) and quantification (LOQ), matrix effect (ME), trueness, and precision.

The ME (Table 1) was calculated as 100 × [(signal of analyte in sample extract−signal of analyte in neat solvent)/signal of analyte in neat solvent] and was evaluated at three concentration levels for each emergent mycotoxin (0.1, 5, and 25 μg/L). A significant signal suppression effect was found, and matrix-matched calibration curves were therefore necessary for quantification purposes.

Table 1. Method validation data for the determination of emergent mycotoxins in urine.

Mycotoxin	Equation	Linear Range (μg/L)	Linearity R^2	LOD (μg/L)	LOQ (μg/L)
ENNA	y = 629312 x + 871432	0.04–50	0.996	0.01	0.04
ENNA1	y = 1755799 x + 1381024	0.10–50	0.995	0.03	0.10
ENNB	y = 833213 x + 669371	0.04–50	0.995	0.01	0.04
ENNB1	y = 1883668 x + 3420346	0.04–50	0.992	0.01	0.04
BEA	y = 1468318 x − 1256332	0.04–50	0.993	0.01	0.04
	Matrix Effect (%)			Trueness (%)		
	0.1 μg/L	5 μg/L	25 μg/L	0.1 μg/L	5 μg/L	25 μg/L
ENNA	−12.5	−18.6	−17.0	97.0	92.2	89.3
ENNA1	−33.6	−34.3	−22.3	96.9	92.4	93.9
ENNB	−33.7	−37.8	−37.2	97.6	95.7	98.9
ENNB1	−5.1	−8.7	−12.1	98.6	98.3	98.0
BEA	−35.5	−20.3	−35.1	96.7	90.5	94.4
	Repeatability, %RSD (n = 9)			Intermediate Precision, %RSD (n = 12)		
	0.1 μg/L	5 μg/L	25 μg/L	0.1 μg/L	5 μg/L	25 μg/L
ENNA	6.0	6.7	6.9	10.1	8.2	7.4
ENNA1	6.8	8.0	8.6	7.5	8.5	10.2
ENNB	8.9	9.4	8.7	10.2	9.3	9.9
ENNB1	8.5	6.9	5.9	8.6	9.1	11.7
BEA	8.7	8.8	9.0	8.9	9.0	9.1

Matrix-matched calibration curves were set up using urine samples spiked at six concentrations levels between 0.1 and 50 μg/L of emergent mycotoxins. Each concentration level was processed in triplicate and also injected in triplicate, considering peak area as the analytical signal. Determination coefficients above 0.99 (Table 1) were obtained in all cases, confirming satisfactory linearity over the whole studied range.

The LODs and LOQs were calculated using the criteria of 3 and 10 times the signal-to-noise ratio (S/N), respectively. Low LOQs were obtained for all the target mycotoxins, which were similar to [14,17] or lower than [15,16,18] those obtained with other analytical methods described in the literature.

The precision of the method was studied in terms of repeatability (intraday precision) and intermediate precision (interday precision). Repeatability was evaluated by applying the complete procedure to three samples (experimental replicates) spiked at three different concentration levels of each emergent mycotoxin (0.1, 5, and 25 μg/L). All the samples were measured on the same day, and each extract was injected in triplicate (instrumental replicates). Intermediate precision was assessed with a similar procedure, spiking and analyzing four different samples on four different days. The results of the precision study, expressed as the relative standard deviation (RSD, %) of analyte peak area are shown in Table 1 and, in all cases, RSD values lower than 12% were obtained, conforming with the legislation concerning other mycotoxins [40].

In order to evaluate the trueness of the proposed method, recovery experiments were performed with urine samples previously analyzed to establish the lack of detectable mycotoxins. None of them provided a positive result above the LODs of the method. These samples were spiked at three different concentration levels (0.1, 5, and 25 μg/L), treated as described previously and injected in triplicate into the UPLC-HRMS equipment. Recoveries ranged between 89.3% and 98.9%, fulfilling the requirements of current legislation for other mycotoxins [40].

2.4. Application to Urine Samples

The suitability of the method was finally evaluated by analyzing ten samples of human urine according to the optimized method. ENNA, ENNA1, ENNB, ENNB1, and BEA were not detected in any of the samples.

2.5. Nanomaterial Reuse

One of the disadvantages of DMSPE is that it requires a stage of magnetic material synthesis, which must be carried out very carefully to obtain reproducible results, thereby slowing down the analytical procedure. One way to avoid this disadvantage would be the possibility of reusing the nanomaterial. For this reason, a study of the reuse of the Fe_3O_4@MWCNT composite synthetized for the determination of emergent mycotoxins in urine was carried out. For the assay, 30 mg of Fe_3O_4@MWCNTs was used to extract the five mycotoxins at 10 μg/L from 30 mL of urine and, after a desorption step, the nanocomposite was used to consecutively extract another three urine samples fortified at the same concentration level. Although a small loss of nanoparticle mass was assumed during the sample treatment, the results, shown in Figure 6, confirmed that it can be reused at least 4 times. The only significant difference in concentration was observed for ENNA1, where a decrease in the signal of approximately 6% was observed.

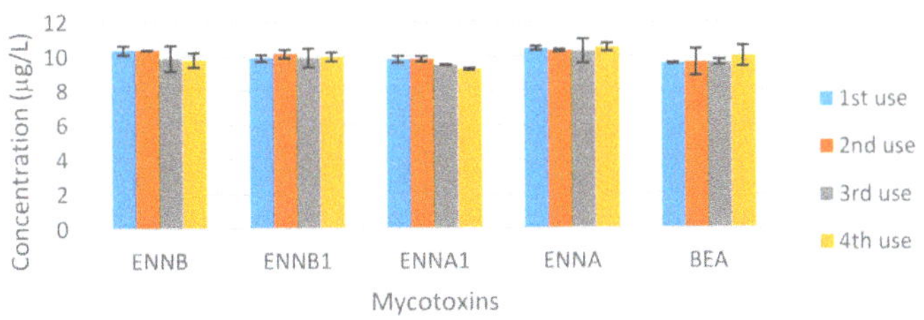

Figure 6. Fe_3O_4@MWCNTs reuse study.

3. Conclusions

In this work, the applicability of MNPs in sample treatment for the determination of emergent mycotoxins (ENNA, ENNA1, ENNB, ENNB1, and BEA) before their determination by UHPLC-HRMS has been demonstrated for the first time. The proposed method based on DMSPE has multiple advantages over other sample treatments, such as the enhancement of the mass transfer, the improvement of the extraction efficiency, and the reduction of organic volume solvent in the desorption step, avoiding filtration or centrifugation steps. A wide variety of the most used nanoparticles for other types of contaminants were explored (β-CD, PDA, chitosan, oleic acid, PS, Fe_3O_4@MWCNTs, MWCNTS/Fe_3O_4/PPy, PPy-NTs, and magnetic cellulose particles) and Fe_3O_4@MWCNTs showed the best result in term of recovery. The proposed methodology was applied for the determination of the target mycotoxins in urine samples since the determination of mycotoxins in biological samples requires sensitive, effective, and accurate methods, as these compounds appear at very low concentration levels. The DMSPE-UHPLC-HRMS method allowed to achieve low LOD and LOQ, enabling the detection of target mycotoxins at levels normally found in urine. In addition, the precision was lower than 12% and recoveries ranged between 89.3% and 98.9%, demonstrating that the Fe_3O_4@MWCNT composite can be reused at least four times.

4. Materials and Methods

4.1. Chemicals, Reagents, and Standards

Individual standards of ENNA, ENNA1, ENNB, ENNB1, and BEA were obtained from Sigma Aldrich (St. Louis, MO, USA). Mycotoxin stock solutions were prepared at 1 mg/L in acetonitrile (MeCN) and stored at −20 °C. Ethanol and MeCN were provided by ChemLab (Zedelgem, Belgium). Nitric acid (65%) was purchased from Panreac (Barcelona, Spain). Iron chloride (III) (>99%) ($FeCl_3$), ammonium iron (II) sulfate hexahydrate ($(NH_4)_2Fe(SO_4)_2·6H_2O$), ammonia solution, and sodium chloride were also purchased from Sigma-Aldrich. MeOH and formic acid (HCOOH) used as mobile phase were LC-MS grade and were also obtained from Sigma-Aldrich.

MWCNTs (average diameters between 40 and 60 nm, average length of >5 μm, specific surface area of 40–70 m^2/g) were provided by Shenzhen Nanotech Port Co., Ltd. (Shenzhen, China). For the filtration of samples before the chromatographic analysis, nylon syringe filters, 0.22 μm × 25 mm (Agela Technologies, New York, NY, USA) were used.

4.2. Instrumentation and Software

The UHPLC system consisted of an Agilent 1290 Infinity II Series HPLC (Agilent Technologies, Santa Clara, CA, USA) provided with an automated multisampler unit and a high speed binary pump, coupled to an Agilent 6550 Q-TOF mass spectrometer (Agilent Technologies, Santa Clara, CA, USA) equipped with an ESI source (Agilent Jet Stream Dual electrospray, AJS-Dual ESI, Santa Clara, CA, USA). For data processing, MassHunter Workstation Data Acquisition software (Agilent Technologies, Rev. B.08.00, Santa Clara, CA, USA) was used. The statistic software Statgraphics Centurion XV.II was used for data treatment.

A Unicen-21 centrifuge (TQTech, Shenzhen, China), an IKA-KS-130-Basic orbital agitator (IKA Werke GmbH & Co KG, Staufen, Germany), a vortex stirrer LLG-uniTEXER (Serviquimia, Constantí, Tarragona), a rotavapor (BUCHI, Labortechnink:AG: Flawil, Switzerland), and an Xcelvap air-drying system (Horizon Technology Inc., Salem, NH, USA) were also used for sample treatment.

Permanent magnets were purchased from Supermagnete (Gottmadingen, Germany). The magnets were blocks composed of Nd-Fe-B (50 × 15 × 15 mm and 86 g weight) with a strength of 33 kg.

4.3. Synthesis of Fe3O4@MWCNTs Composite

The magnetic Fe_3O_4@MWCNT composite material was synthetized following the methodology described by Asgharinezhad and Ebrahimzadeh [36]. In a first step, MWCNTs were purified with 1 M nitric acid solution for 6 h at room temperature and then washed many times with distilled water and dried in an oven at 100 °C. Then, about 0.5 g of the purified nanotubes were added into 250 mL of a solution containing 0.85 g $(NH_4)_2Fe(SO_4)_2·6H_2O$ and 0.4222 g $FeCl_3$ at 50 °C. After that, the suspension was sonicated for 20 min and 20 mL of 8 M ammonia solution was added dropwise to precipitate the magnetized particles while the solution was still under sonication. The pH was controlled to ensure that it remained in the 10–11 range by adding ammonia solution 25% (*w/w*). In order to enhance the whole growth of the nanoparticle crystals, the reaction was allowed to proceed at 50 °C for 30 min. The dispersion was cooled to room temperature, the Fe_3O_4@MWCNT were collected by a strong permanent magnet and washed three times with deionized water followed by ethanol. The magnetic composite substance was dried at 60 °C overnight and lastly ground in a mortar and kept at room temperature in an amber glass vial.

4.4. Sample Preparation

Urine samples were obtained from healthy volunteers and were collected in sterile plastic containers and refrigerated until analysis. For sample treatment, 30 mg of Fe_3O_4@MWCNTs and 30 mL of urine were placed in a test tube, which was orbitally shaken for 35 min at room temperature. The nanocomposite was then attracted with the external neodymium magnet and the supernatant solution was discarded.

In order to desorb the emergent mycotoxins, 1.5 mL of MeCN was added and the mixture was shaken using the orbital shaker for 5 min at room temperature. After that, the nanomaterial was again attracted with the magnet and in this case, the supernatant solution was recovered. A 250 µL volume of the recovered organic phase was mixed with 250 µL of water and filtered through a 0.2 mm filter before injection in the UHPLC-Q-TOF system.

4.5. UHPLC-HRMS Analysis

The separation and determination of emergent mycotoxins was performed in a ZORBAX RRHD Eclipse Plus C18 (1.8 µm, 2.1 × 100 mm) column equipped with a 0.3 µm inline filter from Agilent Technologies and using a mobile phase consisting of H_2O:MeOH (95:5, *v/v*) containing 0.1% HCOOH (solvent A) and MeOH:H_2O (95:5, *v/v*) containing 0.1% HCOOH (solvent B) at a flow rate of 0.4 mL/min. The elution program was applied as follows: 0–1 min: 70% B; 1–3 min: 70–100% B; 3–5 min: 100% B; 5–5.2 min: 100–70% B; 5.2–9 min: 70% B. The column temperature was set at 35 °C and the autosampler temperature was 5 °C. A volume of 20 µL of the sample was injected.

The mass spectrometer worked in the positive mode. The nebulizer gas pressure was established to 30 psi, while the drying gas flow was set to 16 L/min at a temperature of 130 °C, and the sheath gas flow was set to 11 L/min at a temperature of 300 °C. The capillary spray, nozzle, fragmentor, and 1 RF Vpp octopole voltages were 4000, 500, 360, and 750 V, respectively. Data acquisition was carried out using "All-ion mode" and three collision energies (0, 10, and 40 V) were measured. Profile data in the 50–1500 *m/z* range were acquired for MS scans in 2 GHz extended dynamic range mode with 3 spectra/s, 333.3 ms/spectrum, and 2675 transients/spectrum. Reference masses of 121.0509 and 922.0098 *m/z* were used for mass correction during the analysis. For quantification purposes, extracted ion chromatograms (EIC) from the full-scan MS data were obtained for the protonated molecule of each analyte with a 5 ppm window. The monitored ions of the target analytes are shown in Table 2, as well as their retention times, the theoretical and experimental *m/z* values, and the corresponding instrumental error, which was calculated (in terms of percentage) as the difference between the experimental and theoretical *m/z* values divided by the theoretical *m/z* value multiplied by 10^6.

Table 2. Monitored ions of the target analytes.

Compound	t_R (min)	Formula	*m/z* Theoretical	*m/z* Experimental	Error (ppm)	Q1, *m/z*	Q2, *m/z*
ENNB	4.00	$C_{33}H_{58}N_3O_9^+$	640.4168	640.4173	0.8	196.1341	214.1441
ENNB1	4.21	$C_{34}H_{60}N_3O_9^+$	654.4324	654.4326	0.3	196.1333	210.1489
BEA	4.21	$C_{45}H_{58}N_3O_9^+$	784.4168	784.4163	−0.6	244.1334	262.1438
ENNA1	4.43	$C_{35}H_{62}N_3O_9^+$	668.4481	668.4485	0.6	210.1491	228.1592
ENNA	4.67	$C_{36}H_{64}N_3O_9^+$	682.4637	682.4636	−0.1	210.1491	228.1593

Figure 7 shows a chromatogram obtained with the method proposed for the determination of emergent mycotoxins in urine.

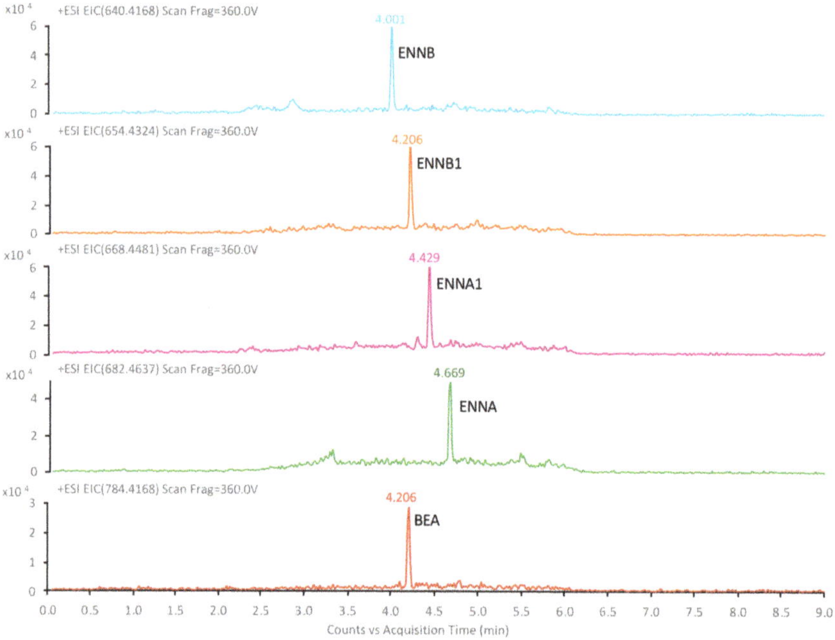

Figure 7. UHPLC–HRMS extracted ion chromatograms of a spiked urine sample at 0.1 µg/L analyzed with the proposed method.

Supplementary Materials: The following are available online at http://www.mdpi.com/2072-6651/12/1/51/s1, Figure S1: Study of different materials for the preparation of the magnetic nanoparticles and desorption solvents (MeCN (**a**) and MeOH (**b**)) for the extraction of emergent mycotoxins (n = 3).

Author Contributions: Conceptualization, N.A.-M. and P.V.; methodology, N.A.-M. and R.P.-S.; software, N.A.-M.; validation, N.A.-M., R.P.-S., N.C. and P.V.; formal analysis, N.A.-M., R.P.-S.; writing—original draft preparation, N.A.-M., R.P-S.; writing—review and editing, N.A.-M., R.P.-S., N.C. and P.V.; supervision, N.C. and P.V.; project administration, P.V.; funding acquisition, N.A.-M., R.P.-S., N.C. and P.V. All authors have read and agreed to the published version of the manuscript.

Funding: This research was funded by the Comunidad Autónoma de la Región de Murcia (CARM, Fundación Séneca, Project 19888/GERM/15), the Spanish MICINN (PGC2018-098363-B-I00), and the European Commission (FEDER/ERDF).

Acknowledgments: The authors acknowledge Alejandro Torrecillas Sánchez for the technical support given by the Research and Development Support Centre (CAID) of the University of Murcia.

Conflicts of Interest: The authors declare no conflict of interest.

References

1. Commission of the European Communities. Regulation (EC) No. 1881/2006 setting maximum levels for certain contaminants in foodstuffs. *Off. J. Eur. Commun.* **2006**, *L364*, 5–24.
2. European Commission. Commission recommendation No. 2012/154/UE on the monitoring of the presence of ergot alkaloids in feed and food. *Off. J. Eur. Commun.* **2012**, *L77*, 20–21.
3. Commission of the European Communities. Commission recommendation No. 2013/165/EU on the presence of T-2 and HT-2 toxin in cereals and cereal products. *Off. J. Eur. Commun.* **2013**, *L91*, 12–15.

4. Mamur, S.; Yuzbasioglu, D.; Yılmaz, S.; Erikel, E.; Unal, F. Assessment of cytotoxic and genotoxic effects of enniatin-A *in vitro*. *Food Addit. Contam. Part A* **2018**, *35*, 1633–1644. [CrossRef]
5. EFSA Panel on Contaminants in the Food Chain (CONTAM). Scientific opinion on the risks to human and animal health related to the presence of beauvericin and enniatins in food and feed. *EFSA J.* **2014**, *12*, 3802. [CrossRef]
6. Wang, X.; Gong, X.; Li, P.; Lai, D.; Zhou, L. Structural diversity and biological activities of cyclic depsipeptides from fungi. *Molecules* **2018**, *23*, 169. [CrossRef]
7. Juan, C.; Mañes, J.; Raiola, A.; Ritieni, A. Evaluation of beauvericin and enniatins in Italian cereal products and multicereal food by liquid chromatography coupled to triple quadrupole mass spectrometry. *Food Chem.* **2013**, *140*, 755–762. [CrossRef] [PubMed]
8. Juan, C.; Ritieni, A.; Mañes, J. Occurrence of Fusarium mycotoxins in Italian cereal and cereal products from organic farming. *Food Chem* **2013**, *141*, 1747–1755. [CrossRef] [PubMed]
9. Kim, D.-B.; Song, N.-E.; Nam, T.G.; Lee, S.; Seo, D.; Yoo, M. Occurrence of emerging mycotoxins in cereals and cereal-based products from the Korean market using LC-MS/MS. *Food Addit. Contam. Part A* **2019**, *36*, 289–295. [CrossRef] [PubMed]
10. Han, X.; Xu, W.; Zhang, J.; Xu, J.; Li, F. Natural occurrence of beauvericin and enniatins in corn-and wheat-based samples harvested in 2017 collected from Shandong province, China. *Toxins* **2019**, *11*, 9. [CrossRef] [PubMed]
11. Arroyo-Manzanares, N.; Rodríguez-Estévez, V.; Arenas-Fernández, P.; García-Campaña, A.M.; Gámiz-Gracia, L. Occurrence of mycotoxins in swine feeding from Spain. *Toxins* **2019**, *11*, 342. [CrossRef] [PubMed]
12. Uhlig, S.; Torp, M.; Heier, B.T. Beauvericin and enniatins A, A1, B and B1 in Norwegian grain: A survey. *Food Chem.* **2006**, *94*, 193–201. [CrossRef]
13. Huybrechts, B.; Martins, J.C.; Debongnie, P.; Uhlig, S.; Callebau, A. Fast and sensitive LC–MS/MS method measuring human mycotoxin exposure using biomarkers in urine. *Arch. Toxicol.* **2015**, *89*, 1993–2005. [CrossRef] [PubMed]
14. Serrano, A.B.; Capriotti, A.L.; Cavaliere, C.; Piovesana, S.; Samperi, R.; Ventura, S.; Laganà, A. Development of a rapid LC-MS/MS method for the determination of emerging fusarium mycotoxins enniatins and beauvericin in human biological fluids. *Toxins* **2015**, *7*, 3554–3571. [CrossRef]
15. Juan, C.; Manyes, L.; Font, G.; Juan-Garcia, A. Evaluation of immunologic effect of Enniatin A and quantitative determination in feces, urine and serum on treated Wistar rats. *Toxicon* **2014**, *87*, 45–53. [CrossRef]
16. Escrivá, L.; Font, G.; Manyes, L. Quantitation of enniatins in biological samples of Wistar rats after oral administration by LC-MS/MS. *Toxicol Mech. Methods* **2015**, *25*, 552–558. [CrossRef]
17. Escrivá, L.; Manyes, L.; Font, G.; Berrada, H. Mycotoxin analysis of human urine by LC-MS/MS: A comparative extraction study. *Toxins* **2017**, *9*, 330. [CrossRef]
18. Lauwers, M.; De Baere, S.; Letor, B.; Rychlik, M.; Croubels, S.; Devreese, M. Multi LC-MS/MS and LC-HRMS methods for determination of 24 mycotoxins including major phase I and II biomarker metabolites in biological matrices from pigs and broiler chickens. *Toxins* **2019**, *11*, 171. [CrossRef]
19. Rodríguez-Carrasco, Y.; Izzo, L.; Gaspari, A.; Graziani, G.; Mañesa, J.; Ritieni, A. Urinary levels of enniatin B and its phase I metabolites: First human pilot T biomonitoring study. *Food Chem. Toxicol.* **2018**, *118*, 454–459. [CrossRef]
20. Taevernier, L.; Bracke, N.; Veryser, L.; Wynendaele, E.; Gevaert, B.; Peremans, K.; De Spiegeleer, B. Blood-brain barrier transport kinetics of the cyclic depsipeptide mycotoxins beauvericin and enniatins. *Toxicol. Lett.* **2016**, *258*, 175–184. [CrossRef]
21. Fraeyman, S.; Devreese, M.; Antonissen, G.; De Baere, S.; Rychlik, M.; Croubels, S. Comparative Oral Bioavailability, Toxicokinetics, and Biotransformation of Enniatin B1 and Enniatin B in Broiler Chickens. *J. Agric. Food Chem.* **2016**, *64*, 7259–7264. [CrossRef] [PubMed]
22. Devreese, M.; De Baere, S.; De Backer, P.; Croubels, S. Quantitative determination of the Fusarium mycotoxins beauvericin, enniatin A, A1, B and B1 in pig plasma using high performance liquid chromatography–tandem mass spectrometry. *Talanta* **2013**, *106*, 212–219. [CrossRef] [PubMed]
23. Devreese, M.; Broekaert, N.; De Mil, T.; Fraeyman, S.; De Backer, P.; Croubels, S. Pilot toxicokinetic study and absolute oral bioavailability of the Fusarium mycotoxin enniatin B1 in pigs. *Food Chem. Toxicol.* **2014**, *63*, 161–165. [CrossRef] [PubMed]

24. Tolosa, J.; Font, G.; Mañes, J.; Ferrer, E. Multimycotoxin analysis in water and fish plasma by liquid chromatography-tandem mass spectrometry. *Chemosphere* **2016**, *145*, 402–408. [CrossRef] [PubMed]
25. Manyes, L.; Escriva, L.; BelenSerrano, A.; Rodriguez-Carrasco, Y.; Tolosa, J.; Meca, G.; Font, G. A preliminary study in Wistar rats with enniatin A contaminated feed. *Toxicol. Mech. Methods* **2014**, *24*, 179–190. [CrossRef]
26. Kongkapan, J.; Giorgi, M.; Poapolathep, S.; Isariyodom, S.; Poapolathep, A. Toxic kinetics and tissue distribution of nivalenol in broiler chickens. *Toxicon* **2016**, *111*, 31–36. [CrossRef]
27. Tolosa, J.; Font, G.; Mañes, J.; Ferrer, E. Natural occurrence of emerging Fusarium mycotoxins in feed and fish from aquaculture. *J. Agric. Food Chem.* **2014**, *62*, 12462–12470. [CrossRef]
28. Rodríguez-Carrasco, Y.; Heilos, D.; Richter, L.; Süssmuth, R.D.; Heffeter, P.; Sulyok, M.; Kenner, L.; Berger, W.; Dornetshuber-Fleiss, R. Mouse tissue distribution and persistence of the food-born fusariotoxins Enniatin B and Beauvericin. *Toxicol. Lett.* **2016**, *247*, 35–44. [CrossRef]
29. Herrero-Latorre, C.; Barciela-García, J.; García-Martín, S.; Peña-Crecente, R.M.; Otárola-Jiménez, J. Magnetic solid-phase extraction using carbon nanotubes as sorbents: A review. *Anal. Chim. Acta* **2015**, *892*, 10–26. [CrossRef]
30. Jing, W.; Zhou, Y.; Wang, J.; Ni, M.; Bi, W.; Chen, D.D.Y. Dispersive Magnetic Solid-Phase Extraction Coupled to Direct Analysis in Real Time Mass Spectrometry for High-Throughput analysis of trace environmental contaminants. *Anal. Chem.* **2019**, *9*, 11240–11246. [CrossRef]
31. Zhang, X.; Wang, Y.; Yang, S. Simultaneous removal of Co (II) and 1-naphthol by core-shell structured Fe_3O_4@cyclodextrin magnetic nanoparticles. *Carbohydr. Polym.* **2014**, *114*, 521–529. [CrossRef] [PubMed]
32. Huang, Z.; Lee, H.K. Study and comparison of polydopamine and its derived carbon decorated nanoparticles in the magnetic solid-phase extraction of estrogens. *J. Chromatogr. A* **2015**, *1414*, 41–50. [CrossRef] [PubMed]
33. Gopal, J.; Abdelhamid, H.N.; Hua, P.Y.; Wu, H.F. Chitosan nanomagnets for effective extraction and sensitive mass spectrometric detection of pathogenic bacterial endotoxin from human urine. *J. Mater. Chem.* **2013**, *1*, 2463–2475. [CrossRef]
34. Benedé, J.L.; Chisvert, A.; Giokas, D.L.; Salvador, A. Development of stir bar sorptive-dispersive microextraction mediated by magnetic nanoparticles and its analytical application to the determination of hydrophobic organic compounds in aqueous media. *J. Chromatogr. A* **2014**, *1362*, 25–33. [CrossRef] [PubMed]
35. Yu, X.; Sun, Y.; Jiang, C.; Sun, X.; Gao, Y.; Wang, Y.; Zhang, H.; Son, D. Magnetic solid-phase extraction of five pyrethroids from environmental water samples followed by ultrafast liquid chromatography analysis. *Talanta* **2012**, *98*, 257–264. [CrossRef] [PubMed]
36. Asgharinezhad, A.A.; Ebrahimzadeh, H. Coextraction of acidic, basic and amphiprotic pollutants using multiwalled carbon nanotubes/magnetite nanoparticles@polypyrrole composite. *J. Chromatogr. A* **2015**, *1412*, 1–11. [CrossRef]
37. Upadhyay, J.; Kumar, A.; Gogoi, B.; Buragohain, A.K. Antibacterial and hemolysis activity of polypyrrole nanotubes decorated with silver nanoparticles by an in-situ reduction process. *Mater. Sci. Eng. C* **2015**, *54*, 8–13. [CrossRef]
38. Periyasamy, S.; Gopalakannan, V.; Viswanathan, N. Fabrication of magnetic particles imprinted cellulose based biocomposites for chromium (VI) removal. *Carbohyd. Polym.* **2017**, *174*, 352–359. [CrossRef]
39. Martin, J.D. XPowder. 2006. Available online: www.xpowder.com (accessed on 6 December 2019).
40. Commission of the European Communities. Regulation (EC) No. 401/2006 of laying down the methods of sampling and analysis for the official control of the levels of mycotoxins in foodstuffs. *Off. J. Eur. Commun.* **2006**, *L70*, 12–34.

toxins

MDPI

Article

Multiple Mycotoxins Determination in Food by LC-MS/MS: An International Collaborative Study

Thomas Bessaire *, Claudia Mujahid, Pascal Mottier and Aurélien Desmarchelier

Nestlé Research, Route du Jorat 57, Vers-chez-les-Blanc, 1000 Lausanne 26, Switzerland; claudia.mujahid@rdls.nestle.com (C.M.); pascal.mottier@rdls.nestle.com (P.M.); aurelien.desmarchelier@rdls.nestle.com (A.D.)

* Correspondence: thomas.bessaire@rdls.nestle.com; Tel.: +41-21-785-8323

Received: 4 October 2019; Accepted: 5 November 2019; Published: 12 November 2019

Abstract: An intercollaborative study was organized to evaluate the performance characteristics of a liquid chromatography tandem mass spectrometry procedure for the simultaneous determination of 12 mycotoxins in food, which were ochratoxin A, aflatoxins B1, B2, G1, G2, and M1, deoxynivalenol, zearalenone, fumonisins B1 and B2, and T-2 and HT-2 toxins. The method combined the simplicity of the QuEChERS (Quick, Easy, Cheap, Efficient, Rugged and Safe) approach with the efficiency of immunoaffinity column cleanup (the step used to enhance sensitivity and sample cleanup for some matrices only). Twenty-three entities were enrolled and were European reference laboratories for mycotoxin analysis, U.S. and European service laboratories, and Nestlé laboratories. Each participant analyzed 28 incurred and/or spiked blind samples composed of spices, nuts, milk powder, dried fruits, cereals, and baby food using the protocol given. Method performances were assessed according to ISO 5725-2. Relative standard deviations of repeatability and reproducibility and trueness values for each of the 115 mycotoxin/sample combinations ranged from 5% to 23%, 7% to 26%, and 85% to 129%, respectively, in line with requirements defined in EC 401/2006. The overall set of data gathered demonstrated that the method offered a unique platform to ensure compliance with EC 1881/2006 and EC 165/2013 regulations setting maximum limits for mycotoxins in food samples, even at low regulated levels for foods intended for infants and young children. The method was applicable regardless of the food, the regulated mycotoxin, and the concentration level, and thus is an excellent candidate for future standardization.

Keywords: LC-MS/MS; mycotoxins; collaborative study; isotopic dilution; compliance; infant food

Key Contribution: Validation data gathered through this international collaborative study demonstrated that the analytical method proposed was suitable to ensure full compliance with regards to EC 1881/2006 and EC 165/2013 regulations, even at low regulated levels for foods intended for infants and young children. The method was applicable regardless of the food commodity, the regulated mycotoxin, and the concentration level, and thus is an excellent candidate for future standardization.

1. Introduction

Mycotoxins are a group of toxic chemical substances produced by filamentous fungi (molds) that commonly grow on a number of food commodities such as cereals, nuts, spices, fruits, oil seeds, or coffee. These toxins can be produced before harvest in the crop and even after harvest if climate conditions are favorable for further fungal growth. Mycotoxins are undoubtedly one of the most severe food safety hazards amplified by global climate change. Indeed, the ability of fungi to produce mycotoxins is largely influenced by temperature, precipitation, relative humidity, and stress conditions in the plants. For instance, aflatoxin-producing fungi (mainly *Aspergillus flavus* and *Aspergillus parasiticus*), usually more prevalent in tropical and sub-tropical regions, are expected to be found in areas such as southern and eastern Europe or the USA, where temperatures >30 °C (close to the optimum for aflatoxin production) may become usual [1,2].

Among several hundreds of mycotoxins identified so far, a few are of concern from a food safety perspective. To protect consumer health, maximum levels (MLs) for mycotoxins in foodstuffs have been established worldwide. In particular, the European Union legislation (often considered as the most stringent one) has established MLs for aflatoxins (AFLAs), ochratoxin A (OTA), zearalenone (ZEN), fumonisins (FBs), and deoxynivalenol (DON) (EC 1881:2006 [3]) and recently indicative levels for T-2 and HT-2 toxins (EC 165/2013 [4]) in a broad range of food commodities. Lower MLs have been also established for food intended for infants and young children.

International organizations such as ISO (International Organization for Standardization), CEN (European Committee for Standardization), or AOAC International have continued gathering experts over the years to develop internationally recognized analytical standards. The goal is to avoid discrepancies in results that may arise from the use of different analytical approaches, with the risk to distorted global food trade. Today, there are 72 official methods available from these organizations for the monitoring of mycotoxins in food (Table 1). These methodologies are often limited to a single compound or to certain family of mycotoxins only and generally validated for a single food category. Indeed, current regulations were established on toxicological data from studies taking into account only one mycotoxin exposure at a time and did not consider the combined effects of mycotoxins [5]. Noteworthy, the co-occurrence of mycotoxins has been extensively described over the last decade. In 2016, data from 107 publications were compiled to summarize the findings on mycotoxins and their co-occurrence in various foods and feeds from all over the world [5]. AFLAs + FBs, DON + ZEN, AFLAs + OTA, and FBs + ZEN were the most observed combinations. Another study on barley and wheat (n = 72 and n = 83, respectively) evidenced that among the mycotoxin-positive samples, 70% of barley samples and 54% of wheat samples were co-contaminated with at least two mycotoxins [6]. The need to develop methods able to screen several mycotoxins at once was justified in a large-scale global survey in feed where mycotoxin co-contamination was common [7]. Concentrations of aflatoxin B1 (AFB1), ZEN, FBs, OTA, DON, and T-2 toxin were analyzed in 74,821 samples of feed and feed raw materials (e.g., maize, wheat, soybean) collected from 100 countries from 2008 to 2017. In total, a large fraction of samples (64%) was co-contaminated with more than two mycotoxins, whilst 88% of the samples were contaminated with at least one mycotoxin. In that regard, moving from single residue analysis toward "multi-analyte and multi-matrix" ones is of interest to speed up efficiently and rationalize mycotoxin analysis in high-throughput routine environments.

Table 1. Official methods (n = 72) for the determination of mycotoxins in food (AOAC, CEN, ISO). ELISA, enzyme-linked immunosorbent assay; HPLC-FLD, high performance liquid chromatography with fluorescence detection; IAC, immunoaffinity column; TLC, thin layer chromatography; HPLC-UV, high performance liquid chromatography with ultra violet detection; GC, gas chromatography.

Mycotoxin(s)	Document	Scope	Technique
AFB1	AOAC 990.32	Corn, roasted peanuts	ELISA
AFB1	AOAC 2000.16	Infant formula	HPLC-FLD (IAC)
AFB1	EN 15851:2010	Cereals, cereal based foods [1]	HPLC-FLD (IAC)
AFB1	AOAC 978.15	Egg	TLC
AFB1, AFB2, AFG1, AFG2	AOAC 990.34	Corn, cotton seed, peanuts, peanut butter	ELISA
AFB1, AFB2, AFG1, AFG2	AOAC 991.31	Corn, raw peanuts, peanut butter	HPLC-FLD (IAC)
AFB1, AFB2, AFG1, AFG2	AOAC 2005.08	Corn, raw peanuts, peanut butter	HPLC-FLD (IAC)
AFB1, AFB2, AFG1, AFG2	AOAC 994.08	Corn, almonds, nuts, peanuts, pistachio nuts	HPLC-FLD
AFB1, AFB2, AFG1, AFG2	AOAC 968.22	Peanuts and peanut products	TLC
AFB1, AFB2, AFG1, AFG2	AOAC 970.45	Peanuts and peanut products	TLC
AFB1, AFB2, AFG1, AFG2	AOAC 972.27	Soybeans	TLC
AFB1, AFB2, AFG1, AFG2	AOAC 971.23	Cocoa beans	TLC
AFB1, AFB2, AFG1, AFG2	AOAC 971.24	Coconut, copra	TLC
AFB1, AFB2, AFG1, AFG2	AOAC 972.26	Corn	TLC
AFB1, AFB2, AFG1, AFG2	AOAC 993.17	Corn, peanuts	TLC
AFB1, AFB2, AFG1, AFG2	AOAC 970.46	Green coffee	TLC
AFB1, AFB2, AFG1, AFG2	AOAC 974.16	Pistachio nuts	TLC
AFB1, AFB2, AFG1, AFG2	AOAC 998.03	Shelled peanuts	TLC

Table 1. *Cont.*

Mycotoxin(s)	Document	Scope	Technique
AFB1, Total AFLAs	AOAC 990.33	Corn, peanut butter	HPLC-FLD
AFB1, Total AFLAs	ISO 16050:2003	Cereals, nuts, oilseed products, dried fruits	HPLC-FLD
AFB1, Total AFLAs	AOAC 999.07	Peanut butter, pistachio paste, fig paste, paprika	HPLC-FLD (IAC)
AFB1, Total AFLAs	EN 14123:2007	Hazelnuts, peanuts, pistachios, figs, paprika	HPLC-FLD (IAC)
Total AFLAs	AOAC 991.45	Peanut butter	ELISA
Total AFLAs	AOAC 993.16	Corn	ELISA
Total AFLAs	AOAC 2013.05	Olive oil, peanut oil, sesame oil	HPLC-FLD (IAC)
Total AFLAs	AOAC 975.36	Corn, peanut, peanut butter, pistachio nuts	UV Lamp
Total AFLAs	AOAC 979.18	Corn, raw and shelled peanuts	Visual fluorescence
AFM1	ISO 14675:2003	Milk products	ELISA
AFM1	ISO 14501:2007	Milk products	HPLC-FLD
AFM1	AOAC 2000.08	Liquid milk	HPLC-FLD (IAC)
AFM1	ISO 14674:2005	Milk products	TLC
AFM1	AOAC 974.17	Dairy products	TLC
AFM1	AOAC 980.21	Milk, cheese	TLC
AFM1	AOAC 982.26	Liver	TLC
AFM1, AFB1	AOAC 982.24	Liver	TLC
AFM1, AFB1	AOAC 982.25	Liver	TLC
AFM1, AFM2 [2]	AOAC 986.16	Liquid milk	HPLC-FLD
CIT [3]	EN 17203:2018	Cereals, red yeast rice	LC-MS/MS
DON	AOAC 986.18	Wheat	GC
DON	EN 15891:2010	Cereals, cereal based foods [1]	HPLC-UV
DON	AOAC 986.17	Wheat	TLC
FB1, FB2	AOAC 2001.04	Corn, corn flakes	HPLC-FLD (IAC)
FB1, FB2	EN 16187:2015	Maize based food [1]	HPLC-FLD (IAC)
FB1, FB2	EN 14352:2004	Maize based food	HPLC-FLD (IAC)
FB1, FB2, FB3 [4]	AOAC 995.15	Corn	HPLC-FLD
Total FBs	AOAC 2001.06	Corn	ELISA
OTA	AOAC 991.44	Barley, wheat, rye, corn	HPLC-FLD
OTA	ISO 15141:2018	Cereals	HPLC-FLD
OTA	AOAC 2000.03	Barley	HPLC-FLD (IAC)
OTA	AOAC 2000.09	Roasted coffee	HPLC-FLD (IAC)
OTA	AOAC 2001.01	Wines, beer	HPLC-FLD (IAC)
OTA	AOAC 2004.10	Green coffee	HPLC-FLD (IAC)
OTA	EN 15835:2010	Cereal based food	HPLC-FLD (IAC)
OTA	EN 15829:2010	Dried fruits (currants, raisins, sultanas, figs)	HPLC-FLD (IAC)
OTA	EN 14133:2009	Wine, beer	HPLC-FLD (IAC)
OTA	EN 14132:2009	Barley, roasted coffee	HPLC-FLD (IAC)
OTA	ISO 15141:2018	Cereals, cereal based products	HPLC-FLD (IAC)
OTA	AOAC 975.38	Green coffee	TLC
OTA, OTB [5]	AOAC 973.37	Barley	TLC
Patulin	ISO 8128-2:1993	Apple juice and apple juice based products	HPLC-UV
Patulin	EN 15890:2010	Fruit juice and fruit based purée [1]	HPLC-UV
Patulin	EN 14177:2003	Clear and cloudy apple juice and puree	HPLC-UV
Patulin	AOAC 995.10	Apple juice	HPLC-UV
Patulin	AOAC 2000.02	Apple juice, apple puree	HPLC-UV
Patulin	ISO 8128-1:1993	Apple juice and apple juice based products	TLC
Patulin	AOAC 974.18	Apple juice	TLC
T-2, HT-2	EN 16923:2017	Cereals, cereal based foods [1]	LC-MS/MS
ZEN	AOAC 994.01	Corn, wheat, feed	ELISA
ZEN	EN 15850:2010	Cereals, cereal based foods [1]	HPLC-FLD (IAC)
ZEN	EN 16924:2017	Edible vegetable oils	HPLC-FLD or LC-MS/MS
ZEN	AOAC 976.22	Corn	TLC
ZEN, α-ZEL [6]	AOAC 985.18	Corn	HPLC-FLD

[1] Including foods intended for infants and young children; [2] aflatoxin M2; [3] citrinin; [4] fumonisin B3; [5] ochratoxin B; [6] α-zearalenol.

Most of the current reference methods still make use of non-confirmatory approaches for quantitative analysis of mycotoxins, e.g., TLC, HPLC-UV, and HPLC-FLD. Use of IAC cleanup often compensates for the lack of specificity of these detection techniques, as authorized in EC 2002/657 [8]. However, LC-MS/MS is currently considered as the state-of-the-art technique to analyze hundreds of contaminants (pesticides, veterinary drugs, etc.) in various food commodities, including mycotoxins, as emphasized in a recent review [9]. The first official LC-MS based methodologies for the determination of mycotoxins in foods were published in 2017 by the CEN and are limited to ZEN [10], and T-2/HT-2 toxins' [11] determination. Indeed, the need for standardized LC-MS methods for mycotoxins' determination was highlighted only recently by regulatory authorities and scientific advisory bodies, triggering the CEN organization to establish European standards within the M/520 standardization mandate [12]. Among the several LC-MS methods still under development or at the final approval stage at CEN, only two are multi-mycotoxin methods for food analysis [13]. The first one is devoted to the screening of OTA, AFB1, DON, ZEN, FB1, FB2, T-2, and HT-2 toxins and excludes foods for infants and young children from the scope. The second one does not include regulated aflatoxins, zearalenone, and fumonisins and is limited to cereals and cereal based products. Such limited analyte and/or matrix scopes are constraints for food business operators willing to use efficient standard protocols to ensure the safety and compliance of a broad range of food commodities.

To fill this gap on our side, an analytical method was developed and internally validated in 2013 [14] before its deployment in 10 Nestlé Quality Assurance Centers (NQACs, control laboratories) in Brazil, Chile, the USA, France, Italy, Poland, Russia, China, Singapore, and India. The procedure enables the quantitative LC-MS/MS determination of regulated mycotoxins (OTA, AFB1, AFB2, AFG1, AFG2, AFM1, DON, ZEN, FB1, FB2, and T-2 and HT-2 toxins) in a broad range of food items including cereals and cereal based baby foods (infant cereals, biscuits), spices, nuts, coffee, tea, cocoa, vegetable oils, dried fruits, infant formula, dairy products, feed/pet food, etc. This routine method has the advantage of offering a unique platform to ensure full compliance with the EC 1881/2006 regulation [3], even at low regulated levels for foods intended for infants and young children. Its sample preparation combines the simplicity of the QuEChERS approach (widely used for pesticide residues analysis; European Norm EN 15662:2018 [15]) with the efficiency of IAC cleanup (this step being used for some matrices only). Quantification is performed by the isotopic dilution approach using ^{13}C-labeled mycotoxins as internal standards (ISTD). The collection of thousands of validation data over time [16], the high number of analyses conducted each year (15,000 samples analyzed in 2017 [17]), and the regular enrollment of NQACs to internal (under ISO 17043:2010 accreditation [18]) and external proficiency tests have demonstrated the high robustness of this method.

In this context, we set up an international collaborative study to support its standardization. Twenty-three (23) laboratories from authorities and private sectors were involved in this study, representing entities from 14 countries. This collaborative study was organized according to the AOAC International guideline [19], and statistical evaluation was performed following ISO 5725-2 document [20]. The present paper summarizes the overall set of data gathered and demonstrates the method's applicability over a broad range of concentrations across different food categories, including foods for infants and young children.

2. Results and Discussion

2.1. Samples and Homogeneity Testing

Providing blank matrices to be further spiked by each participant as done in previous studies [21] was not considered to avoid potential analytical bias introduced by operators. Alternatively, the preference was to use blind duplicates of former proficiency test samples still available as quality control (QC) materials (paprika, black pepper, almonds, hazelnuts, dried raisins, dried figs, wheat, maize (2), milk powder (2), maize (2), rice, and wheat based infant cereals samples). The homogeneity of such materials was extensively tested by their providers. However, since the study samples were repacked from 200-g aluminum bags to 25-g units before shipping, a small-scale testing was performed (six replicates from six different units for each sample) to ensure sample homogeneity. The resulting coefficients of relative standard deviation of repeatability RSD_r ranged from 0.5% to 12%, in line with internal validation data, evidencing the satisfactory homogeneity of the set of samples. The exception was a maize sample with unacceptable RSD_r for ZEN and HT-2, at 40% and 21%, respectively. The particle size distribution of this material was heterogeneous (evidence by visual inspection), which led us to re-open and merge all individual units. The resulting bulk material (ca. 2 kg) was ground in our laboratory by means of a cryogenic grinder (SPEX 6875D Freezer/Mill, Stanmore, U.K.) and further dispatched again into 25-g units. Subsequent homogeneity testing (duplicate analysis of six different units) successfully validated this new preparation with a global decrease of RSD_r for the twelve mycotoxins (ZEN and HT-2 at 6.7 and 10.6%, respectively).

2.2. Participants Instrumental Method Setup

Knowing the diversity of laboratory equipment available worldwide, participants were free to select their instrumental setup, meaning choice of LC columns, mobile phases, MS parameters, etc. Typical LC-MS/MS conditions were provided by the study director for information purposes only. The use of HR-MS instruments was accepted, but none of the twenty-three participants reported its usage. As shown in Figure 1, 13 different LC columns and 11 models of MS instruments were engaged. Mobile phases and LC gradient provided in the guidance document were chosen by 16 and 18 out the 23 participants, respectively (sometimes with a reduction of the final equilibration time). Various column temperatures (from 30 °C to 50 °C) and injection volumes (from 4 µL to 20 µL) were considered. All instruments operated in MS/MS mode using polarity switching, except three labs for which ESI was only used in positive mode. In this last case, ZEN was analyzed as its $[M + H]^+$ adduct as already reported [22,23]. The diversity of the instrumentation (i.e., different generations, varying degrees of performance) used by participants was an additional proof of the applicability of the method.

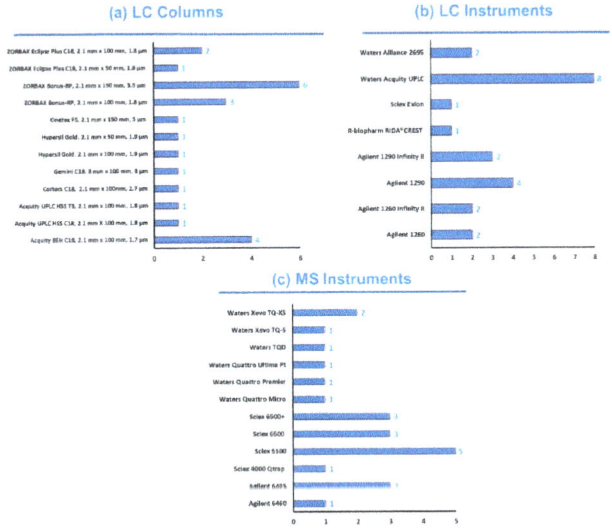

Figure 1. Instrumental setup used by the participating laboratories (n = 23).

To facilitate lab work, calibrant solutions (nine levels) were provided as ready-to-use to each participant. The extent of mycotoxin contaminations is known to be unpredictable and variable; thus, the calibration range of each analyte was set broad enough to avoid reinjection or re-analysis of highly contaminated samples. Participants were asked to consider the two highest levels (calibration point 7 (CAL 7) and CAL 8) only when facing such highly contaminated samples with concentrations out of the classical calibration range (CAL 0 to CAL 6). Use of a weighing factor (1/x or $1/x^2$) for drawing calibration curves was strongly recommended, or alternatively to force regression lines through the origin (i.e., intercept = 0), as done elsewhere [14], this to maintain good precision of data at low contamination levels. Typically, such an approach enabled the direct quantification of either AFB1 from 0.025 µg/kg to 32 µg/kg or OTA from 0.125 µg/kg to 32 µg/kg in cereals within one single analysis, avoiding a tedious re-extraction of the sample using a reduced test portion.

2.3. Laboratory Qualification

Participants were first asked to analyze one single sample (practice sample) to get familiar with the protocol and to communicate generated results to the study director. This was to ensure that the method was correctly set up before engaging laboratories in the second part of the study, consisting of the analysis of 28 samples. This practice sample being a maize based infant cereal, an IAC cleanup was required to get extra sensitivity for AFLAs and OTA. Other mycotoxins were extracted using the "QuEChERS" procedure (Figure 2). The 11 assigned values derived from the proficiency test were 0.26 μg/kg AFB1, 0.28 μg/kg AFB2, 0.15 μg/kg AFG1, 0.15 μg/kg AFG2, 0.81 μg/kg AFTOT, 0.71 μg/kg OTA, 138 μg/kg DON, 31 μg/kg ZEN, 31 μg/kg T-2, 27 μg/kg HT-2, 56 μg/kg T-2 + HT-2, 61 μg/kg FB1, 79 μg/kg FB2, and 140 μg/kg FBTOT, thus very close to the low regulated levels for foods intended for infants and young children [3]. Twenty (20) out of 23 participants successfully reported data with z-scores (Z) and recoveries (Rec) within $-2 < Z < +2$ and $70\% < Rec < 130\%$ for these 11 mycotoxins and were thus qualified for the second part of the study. At this stage, assistance provided by the study director was limited to a few participants facing issues in using the correct ISTD concentrations or eliminating FBs' carry-over on their MS instrument. One laboratory did not see any peak for ZEN, OTA, FBs, HT-2, T-2, and DON, which are compounds to be detected by the QuEChERS stream. Investigations to identify the root cause failed. As the method could not be properly implemented in this lab, all data further generated were not considered for method performance evaluation. Another laboratory missed reporting practice sample data before starting the full study. Unfortunately, the z-scores for ZEN and AFG2 and recoveries for AFB1, AFB2 and AFG1 were unsatisfactory, which highlighted a lack of control of the procedure together with a possible instrumental issue. A third laboratory reported both technical and organizational problems and did not provide data for the practice sample. For consistency with other participants, the decision was made to remove these three laboratories for the final statistical evaluation. Consequently, 20 laboratories were eventually approved in the full collaborative study.

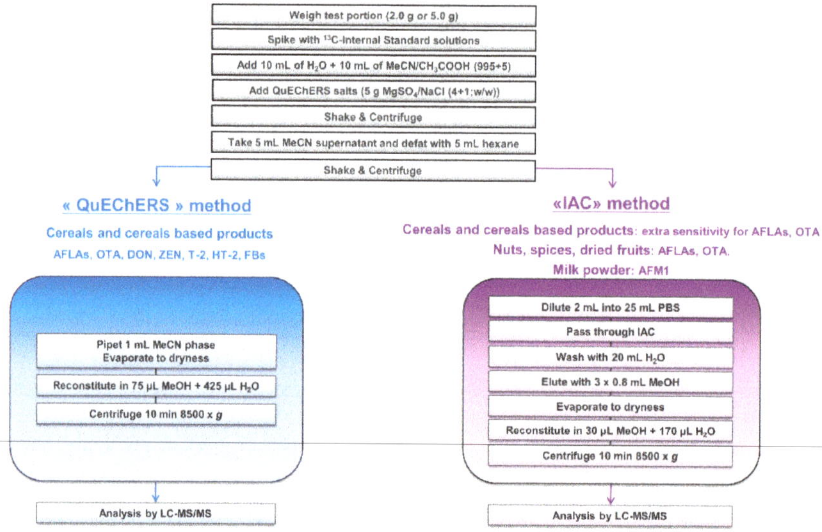

Figure 2. Overview of the sample preparation.

2.4. Full Collaborative Study

Among the 3700 individual data tentatively collected, only a few of them were missing or excluded ($n = 84$, i.e., 2.3% of the data) before statistical evaluation for the following reasons: (a) one lab did not report DON and HT-2 toxin concentration data in the wheat based baby food due to ion ratios out of tolerance; (b) OTA level in the black pepper sample was discarded for one laboratory facing a low signal-to-noise-ratio (<3) along with a very bad peak shape; (c) one lab did not report data for ZEN in maize and wheat samples due to important baseline interference for ZEN ISTD; (d) all data for black pepper were removed for one lab because the sample preparation was not followed; (e) data for T-2 and HT-2 toxins were not considered for another laboratory reporting major instrumental issues for these two toxins; (f) an error during the ISTD spiking step (DON, ZEN, T-2, HT-2) of the wheat based infant cereal was mentioned by one laboratory.

The precision of the method was characterized by the repeatability RSD_r and the reproducibility RSD_R, after removal of outliers as recommended in ISO 5725-2 [20]. As shown in Tables 2–4 and Figure 3, RSD_r ranged from 3% to 22% (average 7%) and RSD_R from 5% to 28% (average 12%), most values being largely below 20%. All these data fulfilled Commission Regulation (EC) No. 401/2006 [24], which has established method performance criteria for the official control of the levels of mycotoxins in foodstuffs. As expected, reproducibility values were higher than the repeatability ones for all combinations except three.

Table 2. Method performance evaluation for AFB1, AFB2, AFG1, AFG2, Total AFLAs and OTA in twelve samples.

		S1	**S2**	**S3**	**S4**	**S5**	**S6**	**S7**	**S8**	**S9**	**S10**	**S11**	**S12**
	Assigned Value (µg/kg)	5.4	11.4	11.3	6.54	8.2	2.48	4.86	2	4.58	0.26	0.51	0.0857
	No. of Laboratories	20	20	20	19	20	20	20	20	20	20	20	19
	No. of Outliers	1	2	2	0	2	1	2	0	0	0	2	0
	No. of Accepted Results	19	18	18	19	18	19	18	20	20	20	18	19
	Mean (µg/kg)	6.46	12.7	12.7	8.42	8.44	2.89	4.86	2.33	4.97	0.237	0.491	0.0848
AFB1	SDr (µg/kg)	0.32	0.53	0.47	0.71	0.3	0.14	0.29	0.45	0.39	0.013	0.028	0.0101
	RSDr (%)	5	4	4	8	4	5	6	19	8	5	6	12
	SDR (µg/kg)	0.67	1.15	1.31	1.22	0.6	0.3	0.69	0.54	0.57	0.031	0.036	0.0155
	RSDR (%)	10	9	10	15	7	10	14	23	12	13	7	18
	Rec (%)	120	111	113	129	103	116	100	116	109	91	96	99
	HorRat Values	0.5	0.4	0.5	0.7	0.3	0.5	0.6	1.1	0.5	0.6	0.3	0.8
	Assigned Value (µg/kg)	3.3	12.5	0.9	5.59	4.4	0.96	n.d.	1.9	2.01	0.25	0.5	0.0792
	No. of Laboratories	20	20	20	19	20	20	-	18	20	20	20	19
	No. of Outliers	2	2	0	2	1	1	-	1	0	0	1	2
	No. of Accepted Results	18	18	20	17	19	19	-	17	20	20	19	17
	Mean (µg/kg)	4.06	13.7	0.991	6.52	4.22	1.09	-	2.07	2.1	0.269	0.483	0.0778
AFB2	SDr (µg/kg)	0.12	0.71	0.062	0.46	0.17	0.05	-	0.15	0.26	0.022	0.022	0.0077
	RSDr (%)	3	5	6	7	4	4	-	7	12	8	5	10
	SDR (µg/kg)	0.22	0.91	0.115	0.84	0.36	0.09	-	0.25	0.24	0.027	0.031	0.0121
	RSDR (%)	5	7	12	13	9	8	-	12	11	10	6	16
	Rec (%)	123	109	110	117	96	113	-	109	105	108	97	98
	HorRat Values	0.2	0.3	0.5	0.6	0.4	0.4	-	0.5	0.5	0.5	0.3	0.7
	Assigned Value (µg/kg)	2.1	20.9	9.5	3.03	1.8	2.65	n.d.	1.7	7.77	0.2	0.5	0.0628
	No. of Laboratories	20	20	20	19	20	20	-	18	20	20	20	19
	No. of Outliers	1	1	1	2	0	0	-	1	0	1	0	3
	No. of Accepted Results	19	19	19	17	20	20	-	17	20	19	20	16
	Mean (µg/kg)	2.06	21.6	10.2	3.58	1.79	3.02	-	1.71	8.42	0.199	0.504	0.0613
AFG1	SDr (µg/kg)	0.094	0.75	0.49	0.26	0.093	0.24	-	0.17	0.65	0.027	0.018	0.0049
	RSDr (%)	5	3	5	7	5	8	-	10	8	13	4	8
	SDR (µg/kg)	0.15	1.55	0.79	0.35	0.15	0.27	-	0.24	1.1	0.023	0.034	0.0083
	RSDR (%)	7	7	8	10	8	9	-	14	13	11	7	14
	Rec (%)	98	103	107	118	99	114	-	101	108	100	101	98
	HorRat Values	0.3	0.3	0.4	0.4	0.4	0.4	-	0.6	0.6	0.5	0.3	0.6
	Assigned Value (µg/kg)	2.2	15	11.9	2.84	0.9	0.83	n.d.	3.7	5.79	0.31	0.52	0.052
	No. of Laboratories	20	20	20	19	19	19	-	19	20	20	20	15
	No. of Outliers	2	1	0	2	1	3	-	0	0	1	2	1
	No. of Accepted Results	18	19	20	17	18	16	-	19	20	19	18	14
	Mean (µg/kg)	2.32	16.4	11.3	3.46	0.814	0.885	-	3.13	6.14	0.304	0.488	0.052
AFG2	SDr (µg/kg)	0.1	1.01	0.65	0.33	0.059	0.055	-	0.48	0.56	0.041	0.045	0.0077
	RSDr (%)	4	6	6	9	7	6	-	15	9	14	9	15
	SDR (µg/kg)	0.19	1.85	0.97	0.37	0.133	0.072	-	0.66	1.05	0.048	0.043	0.0077
	RSDR (%)	8	11	9	11	16	8	-	21	17	16	9	15
	Rec (%)	105	109	95	122	90	107	-	84	106	98	94	100
	HorRat Values	0.4	0.5	0.4	0.5	0.7	0.4	-	1	0.8	0.7	0.4	0.7

Table 2. *Cont.*

		S1	S2	S3	S4	S5	S6	S7	S8	S9	S10	S11	S12
Total AFLAs	Assigned Value (μg/kg)	12.5	58.4	33.8	19.2	15.2	6.81	4.86	9.2	20.2	1.05	2.07	0.28
	No. of Laboratories	20	20	20	19	20	19	20	20	20	20	20	19
	No. of Outliers	0	0	0	0	0	1	2	0	0	1	1	1
	No. of Accepted Results	20	20	20	19	20	18	18	20	20	19	19	18
	Mean (μg/kg)	15	65.4	35.6	21.7	15.3	7.83	4.86	8.9	21.6	1	1.95	0.268
	SDr (μg/kg)	0.67	3.04	2.09	1.3	0.76	0.35	0.29	0.65	1.36	0.051	0.058	0.0305
	RSDr (%)	4	5	6	6	5	4	6	7	6	5	3	11
	SDR (μg/kg)	1.22	5.22	3.16	2.36	1.15	0.56	0.69	1.36	2.16	0.073	0.097	0.0351
	RSDR (%)	8	8	9	11	8	7	14	15	10	7	5	13
	Rec (%)	120	112	105	113	101	115	100	97	107	95	94	96
	HorRat Values	0.4	0.4	0.4	0.5	0.3	0.3	0.6	0.7	0.5	0.3	0.2	0.6
OTA	Assigned Value (μg/kg)	16.9	7	12	17.2	1	9.05	3.46	4.1	2.2	0.52	0.51	0.448
	No. of Laboratories	20	20	20	18	20	20	20	20	20	20	20	20
	No. of Outliers	1	0	0	1	0	1	2	0	2	1	0	0
	No. of Accepted Results	19	20	20	17	20	19	18	20	18	19	20	20
	Mean (μg/kg)	17.7	8.3	14.5	20.9	1.05	9.81	3.37	4.51	2.72	0.6	0.535	0.444
	SDr (μg/kg)	0.64	0.21	0.77	1.26	0.06	0.31	0.37	0.54	0.22	0.03	0.0345	0.0341
	RSDr (%)	4	3	5	6	5	3	11	12	8	5	6	8
	SDR (μg/kg)	2.73	1.27	1.91	4.16	0.16	1.33	0.43	1.27	0.44	0.069	0.06	0.056
	RSDR (%)	15	15	13	20	15	14	13	28	16	11	11	13
	Rec (%)	105	119	121	122	105	108	97	110	123	116	105	99
	HorRat Values	0.7	0.7	0.6	0.9	0.7	0.6	0.6	1.3	0.7	0.5	0.5	0.6

S1: paprika; S2: hazelnuts, S3: dried raisins; S4: black pepper; S5: almond powder; S6: Figs; S7: maize; S8: maize; S9: wheat; S10: wheat based infant cereal; S11: maize based infant cereal; S12: rice based infant cereal; SDr: standard deviations of repeatability; RSDr: relative standard deviations of repeatability; SDR: standard deviations of reproducibility; RSDR: relative standard deviations of reproducibility; Rec: recovery; n.d.: not detected.

Table 3. Method performance evaluation for ZEN, DON, T-2, HT-2, T-2+HT-2, FB1, FB2, Total FBs in six samples.

		S7	S8	S9	S10	S11	S12
ZEN	Assigned Value (μg/kg)	5.4	11.4	11.3	6.54	8.2	2.48
	No. of Laboratories	20	20	20	19	20	20
	No. of Outliers	1	2	2	0	2	1
	No. of Accepted Results	19	18	18	19	18	19
	Mean (μg/kg)	6.46	12.7	12.7	8.42	8.44	2.89
	SDr (μg/kg)	0.32	0.53	0.47	0.71	0.3	0.14
	RSDr (%)	5	4	4	8	4	5
	SDR (μg/kg)	0.67	1.15	1.31	1.22	0.6	0.3
	RSDR (%)	10	9	10	15	7	10
	Rec (%)	120	111	113	129	103	116
	HorRat Values	0.5	0.4	0.5	0.7	0.3	0.5
DON	Assigned Value (μg/kg)	743	506	176	220	295	45.2
	No. of Laboratories	20	20	20	18	20	20
	No. of Outliers	1	0	1	1	0	0
	No. of Accepted Results	19	20	19	17	20	20
	Mean (μg/kg)	739	527	195	220	269	43.7
	SDr (μg/kg)	22	41.5	12.3	10.4	18.9	3.3
	RSDr (%)	3	8	6	5	7	8
	SDR (μg/kg)	48.5	52.5	13.6	15.5	25.3	3.86
	RSDR (%)	7	10	7	7	9	9
	Rec (%)	99	104	111	100	91	97
	HorRat Values	0.4	0.6	0.3	0.3	0.5	0.4
T-2	Assigned Value (μg/kg)	57.9	43	10.3	15	23.3	11.3
	No. of Laboratories	19	19	19	18	19	19
	No. of Outliers	1	0	0	0	0	2
	No. of Accepted Results	18	19	19	18	19	17
	Mean (μg/kg)	56.9	44.4	9.66	16.1	22.2	10.6
	SDr (μg/kg)	3.5	3.29	0.78	0.89	1.52	0.7
	RSDr (%)	6	7	8	6	7	7
	SDR (μg/kg)	3.78	5.98	2.15	1.54	2.2	0.87
	RSDR (%)	7	13	22	10	10	8
	Rec (%)	98	103	94	107	95	94
	HorRat Values	0.3	0.6	1	0.4	0.5	0.4

Table 3. *Cont.*

		S7	S8	S9	S10	S11	S12
HT-2	Assigned Value (μg/kg)	81.8	27	28.6	19	14.5	9.5
	No. of Laboratories	19	19	19	17	19	19
	No. of Outliers	0	0	0	1	0	0
	No. of Accepted Results	19	19	19	16	19	19
	Mean (μg/kg)	81.1	30.3	32.3	15.9	16.1	8.1
	SDr (μg/kg)	5.19	2.34	2.65	1.24	1.79	0.82
	RSDr (%)	6	8	8	8	11	10
	SDR (μg/kg)	10.2	4.67	4.65	1.8	2.88	1.03
	RSDR (%)	13	15	14	11	18	13
	Rec (%)	99	112	113	83	111	85
	HorRat Values	0.6	0.7	0.7	0.5	0.8	0.6
T-2 + HT-2	Assigned Value (μg/kg)	132	68	38.9	36	38.2	20.8
	No. of Laboratories	19	19	19	17	19	19
	No. of Outliers	0	0	0	1	0	0
	No. of Accepted Results	19	19	19	16	19	19
	Mean (μg/kg)	137	74.7	42	31.8	38.4	18.7
	SDr (μg/kg)	6.49	4.43	2.89	1.98	2.83	1.17
	RSDr (%)	5	6	7	6	7	6
	SDR (μg/kg)	12.6	6.72	5.45	2.49	3.97	1.92
	RSDR (%)	9	9	13	8	10	10
	Rec (%)	104	110	108	88	100	90
	HorRat Values	132	68	38.9	36	38.2	20.8
FB1	Assigned Value (μg/kg)	275	4262	n.d	72	121	31.1
	No. of Laboratories	20	18	-	20	20	20
	No. of Outliers	0	0	-	1	0	0
	No. of Accepted Results	20	18	-	19	20	20
	Mean (μg/kg)	291	4735	-	76.5	125	33.1
	SDr (μg/kg)	14.9	258	-	4.32	8.15	3.14
	RSDr (%)	5	5	-	6	7	9
	SDR (μg/kg)	35.2	541	-	11.9	18.6	5.53
	RSDR (%)	12	11	-	16	15	17
	Rec (%)	106	111	-	106	103	106
	HorRat Values	0.6	0.9	-	0.7	0.7	0.8
FB2	Assigned Value (μg/kg)	223	1299	n.d	72	130	44.2
	No. of Laboratories	20	19	-	20	20	20
	No. of Outlier	0	0	-	2	2	2
	No. of Accepted Results	20	19	-	18	18	18
	Mean (μg/kg)	245	1500	-	74.8	134	41.1
	SDr (μg/kg)	10	72.2	-	5.46	7.45	4.17
	RSDr (%)	4	5	-	7	6	10
	SDR (μg/kg)	25	180	-	7.25	10	4.63
	RSDR (%)	10	12	-	10	7	11
	Rec (%)	110	115	-	104	103	93
	HorRat Values	0.5	0.8	-	0.4	0.3	0.5
Total FBs	Assigned Value (μg/kg)	485	5528	n.d.	150	245	75.3
	No. of Laboratories	20	18	-	20	20	20
	No. of Outliers	0	0	-	2	1	0
	No. of Accepted Results	20	18	-	18	19	20
	Mean (μg/kg)	536	6233	-	152	263	76
	SDr (μg/kg)	18.6	291	-	8.9	12.3	5.6
	RSDr (%)	3	5	-	6	5	7
	SDR (μg/kg)	51.5	588	-	16.1	26.7	11.3
	RSDR (%)	10	9	-	11	10	15
	Rec (%)	110	113	-	101	107	101
	HorRat Values	0.5	0.8	-	0.5	0.5	0.7

Table 4. Method performance evaluation for AFM1 in two milk powder samples (S13 and S14).

		S13	S14
	Assigned Value (μg/kg)	0.1121	0.0342
	No. of Laboratories	20	16
	No. of Outliers	2	2
	No. of Accepted Results	18	14
	Mean (μg/kg)	0.127	0.0333
AFM1	SDr (μg/kg)	0.013	0.0073
	RSDr (%)	10	22
	SDR (μg/kg)	0.017	0.0087
	RSDR (%)	13	26
	Rec (%)	113	97
	HorRat Values	0.6	1.2

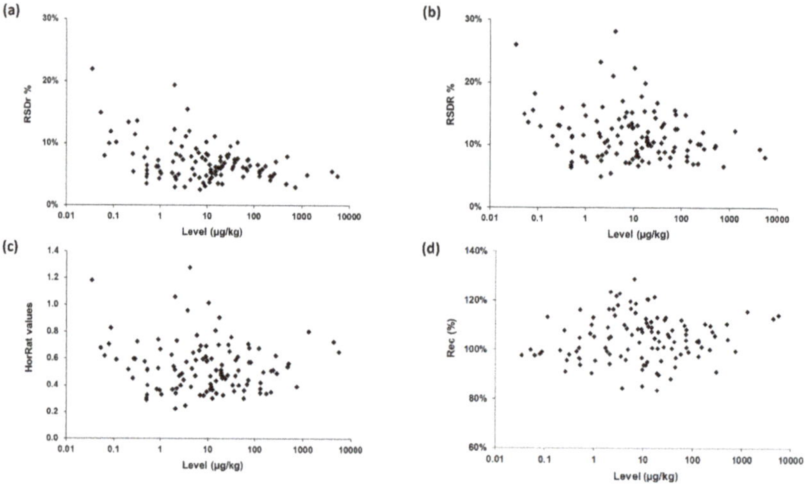

Figure 3. Method performances on 115 mycotoxin/sample combinations: (**a**) repeatability; (**b**) reproducibility; (**c**) Horwitz ratio (HorRat) values; (**d**) trueness. Rec, recovery.

The Horwitz ratio (HorRat) is a useful index of method performance with respect to precision and is calculated as the ratio between the RSD_R as obtained during the collaborative study and the RSD_R as predicted by the modified Horwitz equation [25,26]. Method reproducibility is considered as normal when the HorRat value is between 0.5 and 1.5 [19]. In this study, HorRat values ranged from 0.2 to 1.3 for the 115 mycotoxin/sample combinations (Tables 2–4, Figure 3). Approximately 50% of HorRat values were even lower than 0.5. This might be explained by: (a) the high robustness of the method (already known since it has been heavily used for >5 years in routine environments at Nestlé), (b) by the use of isotopically labelled internal standards for quantification purpose, and (c) by the analytical skills of participating laboratories.

The trueness of the method was then assessed by calculating the recovery (Rec, %) for each of the 115 mycotoxin/matrix combinations. Rec figures ranged from 84% to 129% with eleven data slightly above the requirements defined in EC 401/2006 [24]. We mention that three of them concerned FB1, FB2, and FBs in the maize sample for which levels were out of the calibration range (data reported for information purposes only). The overall set of data unambiguously demonstrated the high confidence level of the method regardless of the sample/concentration/mycotoxin combinations.

Focusing on individual participant data, all mycotoxins were correctly detected of the sample except in two cases. While all laboratories detected AFG2 at 0.31 μg/kg in the wheat based infant cereal, five of them could not detect AFG2 at a lower level in the rice based infant cereal (0.052 μg/kg), which could jeopardize the conclusion with respect to method sensitivity. However, neither AFG2 nor AFTOT (sum of the four AFLAs) are regulated for infant cereals. AFM1 in each of the two milk powders was detected by 16 and 20 out of the 20 laboratories, respectively. According to the related supplier recommendation, respective concentration levels in reconstituted milk correspond to 0.0034 µg/kg and 0.0117 µg/kg. Consequently, the highest AFM1 validated level that was detected by all participants was still half of the ML set for AFM1 in milk-based products (0.025 µg/kg [3]).

As shown in Table 5, the lowest level of the sum of T-2 and HT-2 validated in this study was 18.6 µg/kg (i.e., 10.6 µg/kg and 8.0 µg/kg for T-2 and HT-2 as individual toxin, respectively). This is slightly higher than the 15 µg/kg level set for cereal based foods for infants and young children as an indicative level for which subsequent investigations should be performed according to EC 165/2013 [4]. However, internal validation data at 10 µg/kg for the sum of T-2 and HT-2 (i.e., 5 µg/kg for each T-2 and HT-2 toxins) [14,16] demonstrated the good performance of this method and its compliance with EC 165/2013 [4]. For all other mycotoxins considered, the lowest validated level in this study was lower than the lowest EU MLs [3], evidencing the method as fit for purpose to ensure compliance of food materials with regards to this regulation.

Table 5. Fitness-for-purpose of the method with regards to EU regulations for foods intended for infants and young children.

Mycotoxins	ML (µg/kg) [a]	Lowest Validated Level (µg/kg)	RSDr (%)	RSDR (%)	Rec%
AFB1	0.1	0.084	12	18	99
OTA	0.5	0.44	8	13	99
DON	200	43.6	8	9	97
ZEN	20	8.8	8	13	92
T-2 + HT-2	15 [b]	18.6	6	10	90
FBTOT	200	76.2	7	15	101
AFM1	0.025 [c]	0.0117	10	13	113

[a] Commission Regulation (EC) No. 1881/2006; [b] Commission Regulation (EC) No. 165/2013; [c] refers to the products ready to use (marketed as such or after reconstitution as instructed by the manufacturer).

2.5. Output: Future Perspective

Whilst LC-MS/MS based methods are recognized as state-of-the-art approaches for chemical contaminants' analysis in food, current standards for mycotoxin monitoring are still mainly based on TLC or HPLC-UV/FLD and/or often limited to a single compound, or family at best [13]. This study proved that the proposed methodology is an excellent candidate for future standardization since it offers a highly trustful and efficient approach for multi-residue analysis of regulated mycotoxins. Its applicability whatever food type, mycotoxin, and concentration level is well adapted for food control environments, today facing shorter and shorter turn-around times.

The cost effectiveness of any routine method is not to be neglected, and the price of ^{13}C-labeled compounds used as ISTD might impede their usage. However, our own estimation showed that the sample throughput by using this isotopic dilution approach was increased by a factor of three, which was balanced with the additional cost of the ISs. The matrix matched calibration curve (which is restricted to one or two sample types analyzed per batch with the additional requirement to have blank matrices available for spiking purposes) or standard addition (requiring several extractions for one single analysis) quantifications are indeed time consuming, not user friendly, and not adapted for laboratories dealing with a broad range of food matrices on a daily basis. The use of labelled ISTD also improves method precision and accuracy, as recently highlighted in a study evaluating different approaches (extraction, clean-up, quantification) for mycotoxins determination in cereals [27]. The best performances were achieved by a method making use of ^{13}C-labeled internal standards for quantification.

Finally, the proposed extraction procedure (QuEChERS) is generic and already known to be efficient for a wide range of compounds with different polarity. Our trials already demonstrated that such an extraction approach is valid as well for deoxynivalenol acetylated or modified forms (e.g., 3-acetyl-DON, 15-acetyl-DON, DON-3-glucoside). These compounds are not regulated yet, but the European Commission initiated discussions to review the existing maximum levels for DON, taking into account DON acetylated and modified forms [28]. Related isotopically labelled compounds are now commercially available, making these compounds good candidates for future scope extension.

3. Conclusions

Standardized and internationally accepted methods for food compliance testing are of utmost importance to avoid discrepancies in results, as this may generate unsubstantiated disputes and ultimately distort global food trade. The current standardized methodologies devoted to the analysis of mycotoxins in food are not adapted anymore to high-throughput routine environments, facing today a broader range of items to monitor, an increasing pressure to shorten turn-around time, and cost constraints. This paper proposes an efficient analytical approach for multi-residue analysis of regulated mycotoxins, even at low regulated levels for foods intended for infants and young children. The overall set of data derived from this study proved that the proposed methodology is an excellent candidate for future standardization.

4. Materials and Methods

4.1. Study Organization

4.1.1. Study Materials

Each participant was supplied with AFLAOCHRA PREP® IACs (R-Biopharm, Darmstadt, Germany, $n = 26$), ready-to-use QuEChERS salt mixtures (Agilent, Geneva, Switzerland, $n = 35$), and ready-to-use PBS tablets (Oxoid, Basingstoke, U.K., $n = 10$).

4.1.2. Study Samples

Each participant was given 30 bottles containing ca. 25.0 g of foods. Samples were composed of 28 blind duplicate samples randomly coded from A to β and of one "practice sample" in duplicate. Samples A to L ($n = 12$) belonged to the group "nuts, spices, and dried fruits" and were paprika, black pepper, almonds, hazelnuts, dried raisins, and dried figs. Samples M to R were "raw cereals" composed of wheat or maize ($n = 6$). Samples S to X were either rice, wheat, or maize based "infant cereals" ($n = 6$). Samples Y to β were "milk powder" ($n = 4$). The "practice sample" was a maize based infant cereal. All samples were in powdered form except hazelnuts and dried raisins, which were provided as slurry paste.

These samples were prepared by Fapas (York, U.K.) or BIPEA (Paris, France). All items but one were former proficiency test materials from the last 2 years, still commercially available as QC materials. They were initially packaged in 200-g aluminum bags, but further re-dispatched in smaller 25-g units after extensive mixing. Mycotoxins were either naturally present in the material and/or spiked by the supplier to reach the level of interest. Only the rice based infant cereal sample was specifically prepared by BIPEA using a mycotoxin-free product available from a supermarket in France. The item was further spiked at levels close to the first calibration point (i.e., at the limit of quantification).

4.1.3. Study Analytical Standards Solutions for Spiking Purposes

Five ready-to-use ^{13}C-labelled mycotoxin mixtures to be used as internal standards were prepared by Romer Labs (Tulln, Austria): (a) the ^{13}C-AFLA mixture was composed of ($^{13}C_{17}$)-AFB1, ($^{13}C_{17}$)-AFB2, ($^{13}C_{17}$)-AFG1, and ($^{13}C_{17}$)-AFG2, each at 0.5 µg/mL in acetonitrile; (b) the ^{13}C-AFM1 solution was composed of ($^{13}C_{17}$)-AFM1 at 0.1 µg/mL in acetonitrile; (c) the ^{13}C-[DON, T-2, HT-2, ZEN] mixture was composed of ($^{13}C_{15}$)-DON, ($^{13}C_{24}$)-T-2 toxin, ($^{13}C_{22}$)-HT-2 toxin, and ($^{13}C_{18}$)-ZEN, at 5, 2.5, 2.5, and 2 µg/mL in acetonitrile, respectively; (d) the ^{13}C-FBs solution was composed of ($^{13}C_{34}$)-FB1 and

($^{13}C_{34}$)-FB2, each at 10 μg/mL in acetonitrile/water (50 + 50); (e) ^{13}C-OTA was composed of ($^{13}C_{20}$)-OTA at 10 μg/mL in acetonitrile.

These prepared solutions were dispatched as such to all participants with the exception of ^{13}C-OTA, which was subsequently diluted at 0.1 μg/mL in methanol-water (85 + 15) by the study organizer.

4.1.4. Study Analytical Standard Solutions for External Calibration Curves

Unlabeled mycotoxin working standard solutions were provided by Romer Labs and were: (a) the AFLA mixture composed of AFB1, AFB2, AFG1, and AFG2, each at 1 μg/mL in acetonitrile; (b) AFM1 at 0.1 μg/mL in acetonitrile; (c) the DON, T-2, HT-2, and ZEN mixture at 5, 2.5, 2.5, and 2 μg/mL in acetonitrile, respectively; (d) the FB mixture composed of FB1 and FB2, each at 5 μg/mL in acetonitrile-water (50 + 50); (e) OTA at 10 μg/mL in acetonitrile. The AFLA mixture was then diluted at 0.1 μg/mL and 0.01 μg/mL in acetonitrile and OTA at 0.1 μg/mL in methanol-water (15 + 85).

To ease lab work, a set of nine individual "ready-to-be injected" calibration points (named CAL 0 to CAL 8) was also provided. Related concentrations are given in the Supplementary Data. The same batch of each labelled mycotoxin mixture was intended to be used for both making calibration solutions and spiking test portions. These ready-to-use calibration solutions were all prepared at Nestlé Research Lausanne (Lausanne, Switzerland). A total of 30 mL of each of individual calibration solution was prepared and aliquoted in crimp vials before being stored at −20 °C. A previous internal stability study evidenced that such calibration solutions are stable for at least 4 months when stored at −20 °C.

Before use, all calibration solutions were brought to room temperature, extensively vortexed (minimum 1 min), and sonicated (approximately 10 min) to ensure efficient (re)solubilization of all analytes.

4.1.5. Shipping Study Materials

Initially, 26 laboratories enrolled in this study, but two labs could not receive materials due to customs issues (Romania, Russia). One lab withdrew its participation due to a lack of available time. Thus, 23 packages were finally dispatched to laboratories in Austria, France, Germany (2), Hungary, India, Ireland (2), Italia (5), Poland, Serbia, Singapore, Switzerland, The Netherlands (2), The United Kingdom (2), and the USA (2) in June 2019.

Samples and mycotoxin standards' solutions were all sent in a frozen state (dry ice), whereas other materials were sent at room temperature.

4.2. Sample Preparation

Participants were asked to follow strictly the provided protocol. Samples were extracted as previously reported [14] with minor modifications.

Test portions (5 g for cereals and cereal based products and 2 g for milk powders, nuts, spices, and dried fruits) were weighed in 50 mL polypropylene tubes and each subsequently spiked with 50 μL of ^{13}C-labelled working standard solutions of interest. Water (10 mL) was added, and the tube was vigorously shaken by hand until complete dissolution. Acetonitrile containing 1% acetic acid (10 mL) was added, and the tube was then mechanically shaken for 10 min at approximately 300 rpm. Ready-to-use QuEChERS salt mixture containing 4.0 ± 0.1 g of $MgSO_4$ and 1.0 ± 0.1 g of NaCl was supplemented to initiate phase separation. The tube was immediately hand shaken to prevent any lump formation and then placed onto a mechanical shaker for 10 min. After centrifugation (4000× *g*, 10 min, room temperature (RT)), 5 mL of the supernatant acetonitrile phase were mixed with 5 mL of n-hexane and shaken for approximately 10 min on a mechanical shaker. After centrifugation (4000× *g*, 1 min, RT), the upper n-hexane phase was discarded.

The sample extract was then divided into two portions and submitted to two different clean-up protocols, named "QuEChERS" and "IAC", depending on the mycotoxin/matrix combination and the sensitivity required for AFLAs and OTA. A general scheme is presented in Figure 1:

- "QuEChERS": Generic cleanup for all mycotoxins potentially present in cereals when an improved sensitivity for AFLAs and OTA is not required. An aliquot of the defatted acetonitrile layer (1 mL) was evaporated to dryness under a stream of nitrogen at about 40 °C and reconstituted in 75 μL methanol and 425 μL water. The resulting extract was mixed for about 5 s using a vortex mixer and ultracentrifuged at 8500× *g* at room temperature for 10 min.
- "IAC": Specific cleanup for AFLAs and OTA for sensitivity purposes when dealing with infant foods (e.g., infant cereals) and "difficult" matrices (e.g., spices, nuts, dried fruits). An aliquot of the acetonitrile layer (2 mL) was diluted in a PBS solution (25 mL), and the whole extract was applied onto IAC containing antibodies specific to AFLAs and OTA. The IAC was then washed with 20 mL water and the toxins finally eluted with methanol (3 × 800 μL). The eluate was evaporated to dryness under a stream of nitrogen at about 40 °C and reconstituted in 30 μL of methanol and 170 μL of water. The resulting extract was mixed for about 5 s using a vortex mixer and ultracentrifuged at 8500× *g* at room temperature for 10 min.

4.3. LC-MS/MS Analysis

Instrumental conditions described hereafter were given to the participants, who were free to adapt them for their own instrument.

LC analysis was performed with an Agilent 1290 binary pump system (Agilent, Geneva, Switzerland). Optimal LC conditions were obtained using a Waters Acquity BEH C18 column (2.1 × 100 mm, 1.7 μm) equipped with a BEH C18 VanGuard precolumn (2.1 × 5 mm, 1.7 μm), both thermostated at 50 °C. The mobile phases were constituted of formic acid (0.15%) in water containing 10 mM of ammonium formate (Solvent A) and formic acid (0.05%) in methanol (Solvent B). A gradient program was set up as follows: 0–0.3 min with 85% A; 0.3–4.0 min linear gradient down to 0% A; hold at 0% A for 3 min; return to 85% A in 0.05 min and hold at 85% A for 3.95 min (total run time 10 min). The flow rate was 0.4 mL/min, and the injection volume was 10 μL. The LC flow was directed into the MS detector between 1.0 and 6.0 min.

MS detection was performed using a Sciex TRIPLE QUAD 6500+ instrument (Foster City, USA) equipped with a Turbo V™ ion source. MS parameters were first obtained by syringe-infusing (each individual standard solution in electrospray ionization (ESI) mode at concentrations of ca. 0.1 μg/mL to 1.0 μg/mL along with the LC flow (0.4 mL/min, constituted of 50% Aqueous Mobile Phase A and 50% Organic Mobile Phase B) using a T connector. Analyses were conducted using tandem MS in scheduled selected reaction monitoring (scheduled MRM™) mode alternating two transition reactions for each compound with an acquisition window of 40 s and a target scan time of 250 ms. The block source temperature was maintained at 550 °C, and gas values were set as follows: curtain gas 35 psi, GS1 40 psi, GS2 40 psi, and collision activated dissociation (CAD) gas at 10 psi. Other parameters were ion spray voltage (5.0 kV or −4.0 kV), entrance potential (±10 eV), and collision exit potential (±15 eV). MS/MS operated in positive/negative ionization switching mode. Individual MS parameters for the 24 compounds are provided in the Supplementary Data. Data acquisition was carried out using Analyst software 1.7 and subsequent data processing done using Multiquant software 3.0 (both from Sciex).

4.4. Identification of Mycotoxins

Mycotoxins were considered as positively identified in the sample when all confirmation criteria defined in the SANTE/12089/2016 document [29] were fulfilled: (a) a signal visible at least at two diagnostic transition reactions selected for each mycotoxin and each corresponding IS; (b) the retention time of the analyte in the sample extract corresponds to that of the average of the calibration standards measured in the same sequence with a tolerance of ±0.2 min; (c) the retention time of the analyte corresponds to that of its labelled internal standard with a tolerance of ±0.05 min; (d) the peak area ratio from the different transition reactions recorded for each analyte is ±30%.

4.5. Quantification

Quantification was performed by the isotopic dilution approach using ^{13}C-labeled mycotoxins as internal standard. For each mycotoxin, the calibration curve was built by plotting the peak area ratio of each mycotoxin and its ISTD using the transition reaction for quantitation (= y axis) against the concentration ratio of each mycotoxin and its ISTD (= x axis) using calibration solutions from CAL 0 to CAL 6. CAL 7 and CAL 8 were only considered in the case of a highly contaminated sample when mycotoxin levels were out of the classical calibration range.

To improve the precision of the results at the low calibration points, participants were encouraged to use a 1/x or 1/x^2 weighing factor for drawing calibration curves. Alternatively, the regression line was forced through the origin (i.e., intercept = 0) as initially validated [14]. Deviations of back calculated concentrations of calibrant standards from the true concentration were checked to be below ±20% [30].

The mass fraction of each analyte in the sample (w_a) in µg/kg was calculated using the following equation:

$$W_a = \frac{\left(\frac{A_a}{A_{is}}\right) - I}{S} \times \frac{m_{is}}{m_a} \quad (1)$$

where A_a is the peak area of a given analyte; A_{is} is the peak area of the corresponding ISTD; I is the intercept of the regression line; S is the slope of the regression line; ma is the mass of the test portion, in g (either 2.0 g or 5.0 g); m_{is} is the mass of *ISTD* added to the test portion, in ng.

4.6. Statistical Evaluation

Each participating laboratory was first asked to analyze a "practice sample" by strictly following the protocol given and to report results to the study director. Such a preliminary trial was intended to familiarize the participant with the procedure and to ensure that the method was correctly set up before starting the full study. The practice sample was an old proficiency test sample. Therefore, for each laboratory, z-scores and recoveries were calculated based on the p-test assigned values (BIPEA, Babyfood, round 12-3931, December 2018). When z-scores (Z) and recoveries (Rec) were within $-2 < Z < +2$ and $70\% < \text{Rec} < 130\%$ ranges, the laboratory was qualified for the second part of the study. When a laboratory did not return satisfactory results, technical support was provided to fix issues.

Method performances were then assessed based on ISO 5725-2 [20] and AOAC International [19] procedures using the data from the 28 blind duplicate samples that were analyzed over a maximum of four different days by the qualified labs. After data collection, outliers and stragglers were detected using Cochran and Grubbs tests (with critical values set at 1% and 5%, respectively). Outliers were removed prior to calculations, but stragglers were retained. Statistics for AFTOT, FBTOT, and T2 + HT-2 toxins were derived from the sum of individual mycotoxin concentrations. Averaged concentrations, standard deviations of repeatability (SDr), and relative standard deviations of repeatability (RSD_r) were estimated from blind duplicates for each sample and each mycotoxin. Standard deviations of reproducibility (SDR), relative standard deviations of reproducibility (RSD_R), and HorRat values were also calculated. Trueness was evaluated based on the assigned values given by proficiency test manufacturers for each individual material, except for Sample S12 (Tables 2 and 3) for which reference values were obtained using our internal method (six replicates).

Supplementary Materials: The following are available online at http://www.mdpi.com/2072-6651/11/11/658/s1: Document 1: Determination of mycotoxins by liquid chromatography-tandem mass spectrometry (LC-MS/MS), standard operation procedure for the method validation study, 3 July 2019.

Author Contributions: Conceptualization, T.B.; Methodology, A.D., T.B. and C.M.; Validation, C.M. and T.B.; Formal analysis, T.B.; Investigation, T.B. and C.M.; Supervision, P.M.; Writing—original draft, T.B.; Writing—review & editing, T.B., C.M., P.M. and A.D.; Project administration, T.B.

Funding: This research received no external funding.

Acknowledgments: The authors would like to thank R-Biopharm AG (Germany), who kindly provided the immunoaffinity columns for this study, Romer Labs Division Holding GmbH (Austria) for their help in preparing ready-to-use analytical standard mixtures, and FAPAS (United Kingdom) and BIPEA (France) for their support to prepare the set of samples. The authors wish to warmly thank the following people for their strong involvement in this collaborative study: Sonia Colicchia, Helmut Rost, Ervin Tanyi, and Daniel Piller (Romer Labs Diagnostic GmbH, Austria); Nina Tomcic, Dragana Marinkovic, and Biljana Marosanovic (SP LABORATORIJA, Serbia); Daniel Tan and Quentin Baslé (Nestlé Quality Assurance Center, Singapore); Marta Ferro and Silvia Anselmo (Arpal Dipartimento Laboratorio Regionale, Italy); Jozsef Kereszturi and Gabor Kiraly (WESSLING, Hungary); Aengus O'Briain, Audrey Nugent, and Julie Tierney (The State Laboratory, Ireland); Ernest Capraro and Kimberly Bowman (Nestlé Quality Assurance Center Dublin, USA); Phillip D Anderson, Dan Hengst, and John Zulkoski (Eurofins Food Integrity and Innovation, USA); Wim Broer and Manuska Ganpat (NOFALAB, The Netherlands); Joanna Oleniacz, Justyna Kłeczek-Panek, and Magdalena Migut (Nestlé Quality Assurance Center Rzeszow, Poland); Elisabetta Caprai, Giuseppe Luca Polonini, and Simonetta Menotta (Istituto Zooprofilattico Sperimentale della Lombardia e dell'Emilia Romagna, Italy); Manfred Schneider and Marco Feyer (CVUA Rheinland, Germany); Amrit Kaur, Gurpreet Raj, and Mohit Mogla (Nestlé Quality Assurance Center Moga, India); Ed Boers, and Ruud van Dam (Wageningen Food Safety Research, The Netherlands); Marc Zollinger, Daniela Matera, and Thomas Gude (SQTS, Swiss Quality Testing Services, Switzerland); Alessandro Scantamburlo (Nestlé Quality Assurance Center Padova, Italy); Cristina Poligani, Gaia Alice Romeo and Marco Meschiari (Neotron S.p.A, Italy); Elaine Marley, Carol Donnelly and Dave Leeman (R-Biopharm Rhone LTD, United Kingdom); Nicolas O'Sullivan, Alison Brazil, and Patrick English (Public Analyst's Laboratory, Ireland); Florence Bouveret, Martine Briand, Fiona Gasparini, Marie-Thérèse Renon, and Vincent Dufailly (Nestlé Quality Assurance Center Cergy, France); Barbara De Santis, Carlo Brera, and Gabriele Moracci (Italian National Institute of Health, Italy); Simone Staiger, Christina Pille, and Niels Eick (Eurofins WEJ Contaminants GmbH, Germany); Susan MacDonald and Joanna Stratton (Fera Science Ltd., U.K.).

Conflicts of Interest: The authors declare no conflict of interest.

References

1. Paterson, R.R.M.; Lima, N. Toxicology of mycotoxins. In *Molecular, Clinical and Environmental Toxicology: Volume 2: Clinical Toxicology*; Luch, A., Ed.; Birkhäuser Basel: Basel, Switerland, 2010; pp. 31–63. [CrossRef]
2. Battilani, P.; Toscano, P.; Van der Fels-Klerx, H.J.; Moretti, A.; Camardo Leggieri, M.; Brera, C.; Rortais, A.; Goumperis, T.; Robinson, T. Aflatoxin B1 contamination in maize in Europe increases due to climate change. *Sci. Rep.* **2016**, *6*, 24328. [CrossRef] [PubMed]
3. European Union. COMMISSION REGULATION (EC) No 1881/2006 of 19 December 2006 setting maximum levels for certain contaminants in foodstuffs. *Off. J. Eur. Union* **2006**, *L 364*, 5–24.
4. European Union. 2013/165/EU Commission Recommendation of 27 March 2013 on the presence of T-2 and HT-2 toxin in cereals and cereal products Text with EEA relevance. *Off. J. Eur. Union* **2013**, *L 91*, 12–15.
5. Smith, M.-C.; Madec, S.; Coton, E.; Hymery, N. Natural Co-Occurrence of Mycotoxins in Foods and Feeds and Their in vitro Combined Toxicological Effects. *Toxins* **2016**, *8*, 94. [CrossRef] [PubMed]
6. Shi, H.; Schwab, W.; Yu, P. Natural Occurrence and Co-Contamination of Twelve Mycotoxins in Industry-Submitted Cool-Season Cereal Grains Grown under a Low Heat Unit Climate Condition. *Toxins* **2019**, *11*, 160. [CrossRef] [PubMed]
7. Gruber-Dorninger, C.; Jenkins, T.; Schatzmayr, G. Global Mycotoxin Occurrence in Feed: A Ten-Year Survey. *Toxins* **2019**, *11*, 375. [CrossRef] [PubMed]
8. European Union. Commission Decision (EC) No 657/2002 of 12 August 2002 implementing Council Directive 96/23/EC concerning the performance of analytical methods and the interpretation of results. *Off. J. Eur. Union* **2002**, *L 221*, 8–36.
9. Tittlemier, S.A.; Cramer, B.; Dall'Asta, C.; Iha, M.H.; Lattanzio, V.M.T.; Malone, R.J.; Maragos, C.; Solfrizzo, M.; Stranska-Zachariasova, M.; Stroka, J. Developments in mycotoxin analysis: an update for 2017–2018. *World Mycotoxin J.* **2019**, *12*, 3–29. [CrossRef]
10. EN 16924:2017. *Foodstuffs—Determination of Zearalenone in Edible Vegetable Oils by LC-FLD or LC-MS/MS*; European Committee for Standardization: Brussels, Belgium, 2017.
11. EN 16923:2017. *Foodstuffs—Determination of T-2 Toxin and HT-2 Toxin in Cereals and Cereal Products for Infants and Young Children by LC-MS/MS after SPE Cleanup*; European Committee for Standardization: Brussels, Belgium, 2017.
12. European Union. MANDATE FOR STANDARDISATION ADDRESSED TO CEN FOR METHODS OF ANALYSIS FOR MYCOTOXINS IN FOOD. Available online: https://law.resource.org/pub/eu/mandates/m520.pdf (accessed on 19 September 2019).

13. Pascale, M.; De Girolamo, A.; Lippolis, V.; Stroka, J.; Mol, H.G.J.; Lattanzio, V.M.T. Performance Evaluation of LC-MS Methods for Multimycotoxin Determination. *J. AOAC Int.* **2019**. [CrossRef] [PubMed]
14. Desmarchelier, A.; Tessiot, S.; Bessaire, T.; Racault, L.; Fiorese, E.; Urbani, A.; Chan, W.C.; Cheng, P.; Mottier, P. Combining the quick, easy, cheap, effective, rugged and safe approach and clean-up by immunoaffinity column for the analysis of 15 mycotoxins by isotope dilution liquid chromatography tandem mass spectrometry. *J. Chromatogr. A* **2014**, *1337*, 75–84. [CrossRef] [PubMed]
15. EN 15662:2018. *Foods of Plant Origin. Multimethod for the Determination of Pesticide Residues Using GC- and LC-Based Analysis Following Acetonitrile Extraction/Partitioning and Clean-up by Dispersive SPE. Modular QuEChERS-Method*; European Committee for Standardization: Brussels, Belgium, 2018.
16. Chan, W.-C.; Cheng, P.; Desmarchelier, A.; Fiorese, E.; Stephan, M. Mycotoxins in Foodstuffs by Isotope Dilution LC-MS/MS. In Proceedings of the World Mycotoxin Forum 8th Conference, Vienna, Austria, 10–12 November 2014.
17. Bessaire, T. Prevention & Control of Mycotoxins at Nestlé—The Pivotal Role of Analytics. In Proceedings of the 12th International Fresenius Conference—Contaminants and Residues in Food, Cologne, Germany, 10 April 2019.
18. ISO 17043:2010. *Conformity Assessment—General Requirements for Proficiency Testing*; International Organization for Standardization: Geneva, Switzerland, 2010.
19. AOAC International. Appendix D: Guidelines for Collaborative Study Procedures To Validate Characteristics of *a* Method of Analysis. *J. AOAC Int.* **1995**, *78*, 143A–160A.
20. ISO 5725-2. *Accuracy (Trueness and Precision) of Measurement Methods and Results—Part 2: Basic Method for the Determination of Repeatability and Reproducibility of a Standard Measurement Method*; International Organization for Standardization: Geneva, Switzerland, 1994.
21. Zhang, K.; Schaab, M.R.; Southwood, G.; Tor, E.R.; Aston, L.S.; Song, W.; Eitzer, B.; Majumdar, S.; Lapainis, T.; Mai, H.; et al. A Collaborative Study: Determination of Mycotoxins in Corn, Peanut Butter, and Wheat Flour Using Stable Isotope Dilution Assay (SIDA) and Liquid Chromatography–Tandem Mass Spectrometry (LC-MS/MS). *J. Agric. Food Chem.* **2017**, *65*, 7138–7152. [CrossRef] [PubMed]
22. De Boevre, M.; Di Mavungu, J.D.; Maene, P.; Audenaert, K.; Deforce, D.; Haesaert, G.; Eeckhout, M.; Callebaut, A.; Berthiller, F.; Van Peteghem, C.; et al. Development and validation of an LC-MS/MS method for the simultaneous determination of deoxynivalenol, zearalenone, T-2-toxin and some masked metabolites in different cereals and cereal-derived food. *Food Addit. Contam. Part A Chem. Anal. Control. Expo. Risk Assess.* **2012**, *29*, 819–835. [CrossRef] [PubMed]
23. De Santis, B.; Debegnach, F.; Gregori, E.; Russo, S.; Marchegiani, F.; Moracci, G.; Brera, C. Development of a LC-MS/MS Method for the Multi-Mycotoxin Determination in Composite Cereal-Based Samples. *Toxins* **2017**, *9*, 169. [CrossRef] [PubMed]
24. European Union. Commission Regulation (EC) No 401/2006 of 23 February 2006 laying down the methods of sampling and analysis for the official control of the levels of mycotoxins in foodstuffs. *Off. J. Eur. Union* **2006**, *L 70*, 12–34.
25. Horwitz, W.; Kamps, L.R.; Boyer, K.W. Quality assurance in the analysis of foods and trace constituents. *J. Assoc. Off. Anal. Chem.* **1980**, *63*, 1344–1354. [PubMed]
26. Thompson, M. Recent trends in inter-laboratory precision at ppb and sub-ppb concentrations in relation to fitness for purpose criteria in proficiency testing. *Analyst* **2000**, *125*, 385–386. [CrossRef]
27. Solfrizzo, M.; Gambacorta, L.; Bibi, R.; Ciriaci, M.; Paoloni, A.; Pecorelli, I. Multimycotoxin Analysis by LC-MS/MS in Cereal Food and Feed: Comparison of Different Approaches for Extraction, Purification, and Calibration. *J. AOAC Int.* **2018**, *101*, 647–657. [CrossRef] [PubMed]
28. European Union. Summary Report of the Standing Committee on Plants, Animals, Food and Feed Held in Brussels on 17 September 2018. Available online: https://ec.europa.eu/food/sites/food/files/safety/docs/reg-com_toxic_20180917_sum.pdf?_cldee=bXZhbndlZWRlQGZubGkubmw%3d&recipientid=contact-2800eeaa6e2be8118127e0071b65ce91-45527089537b4f53a6f83c9233e6de1d&esid=3b6db41a-2660-e911-a96d-000d3ab490f3 (accessed on 30 September 2019).

29. SANTE/12089/2016. *Guidance Document on Identification of Mycotoxins in Food and Feed*; European Commission Directorate General for Health and Food Safety: Brussels, Belgium, 2016.
30. SANTE/11813/2017. *Guidance Document on Analytical Quality Control and Method Validation Procedures for Pesticide Residues and Analysis in Food and Feed*; European Commission Directorate General for Health and Food Safety: Brussels, Belgium, 2017.

MDPI
St. Alban-Anlage 66
4052 Basel
Switzerland
Tel. +41 61 683 77 34
Fax +41 61 302 89 18
www.mdpi.com

Toxins Editorial Office
E-mail: toxins@mdpi.com
www.mdpi.com/journal/toxins

www.ingramcontent.com/pod-product-compliance
Lightning Source LLC
Chambersburg PA
CBHW061810210926
51850CB00006B/661
9783039362066